T0138637

Neural Networks for Applied Sciences and Engineering

OTHER AUERBACH PUBLICATIONS

Agent-Based Manufacturing and Control Systems: New Agile Manufacturing Solutions for Achieving Peak Performance
Massimo Paolucci and Roberto Sacile
ISBN: 1-5744-4336-4

Curing the Patch Management Headache
Felicia M. Nicastro
ISBN: 0-8493-2854-3

Cyber Crime Investigator's Field Guide, Second Edition
Bruce Middleton
ISBN: 0-8493-2768-7

Disassembly Modeling for Assembly, Maintenance, Reuse and Recycling
A. J. D. Lambert and Surendra M. Gupta
ISBN: 1-5744-4334-8

The Ethical Hack: A Framework for Business Value Penetration Testing
James S. Tiller
ISBN: 0-8493-1609-X

Fundamentals of DSL Technology
Philip Golden, Herve Dedieu,
and Krista Jacobsen
ISBN: 0-8493-1913-7

The HIPAA Program Reference Handbook
Ross Leo
ISBN: 0-8493-2211-1

Implementing the IT Balanced Scorecard: Aligning IT with Corporate Strategy
Jessica Keyes
ISBN: 0-8493-2621-4

Information Security Fundamentals
Thomas R. Peltier, Justin Peltier,
and John A. Blackley
ISBN: 0-8493-1957-9

Information Security Management Handbook, Fifth Edition, Volume 2
Harold F. Tipton and Micki Krause
ISBN: 0-8493-3210-9

Introduction to Management of Reverse Logistics and Closed Loop Supply Chain Processes
Donald F. Blumberg
ISBN: 1-5744-4360-7

Maximizing ROI on Software Development
Vijay Sikka
ISBN: 0-8493-2312-6

Mobile Computing Handbook
Imad Mahgoub and Mohammad Ilyas
ISBN: 0-8493-1971-4

MPLS for Metropolitan Area Networks
Nam-Kee Tan
ISBN: 0-8493-2212-X

Multimedia Security Handbook
Borko Furht and Darko Kirovski
ISBN: 0-8493-2773-3

Network Design: Management and Technical Perspectives, Second Edition
Teresa C. Piliouras
ISBN: 0-8493-1608-1

Network Security Technologies, Second Edition
Kwok T. Fung
ISBN: 0-8493-3027-0

Outsourcing Software Development Offshore: Making It Work
Tandy Gold
ISBN: 0-8493-1943-9

Quality Management Systems: A Handbook for Product Development Organizations
Vivek Nanda
ISBN: 1-5744-4352-6

A Practical Guide to Security Assessments
Sudhanshu Kairab
ISBN: 0-8493-1706-1

The Real-Time Enterprise
Dimitris N. Chorafas
ISBN: 0-8493-2777-6

Software Testing and Continuous Quality Improvement, Second Edition
William E. Lewis
ISBN: 0-8493-2524-2

Supply Chain Architecture: A Blueprint for Networking the Flow of Material, Information, and Cash
William T. Walker
ISBN: 1-5744-4357-7

The Windows Serial Port Programming Handbook
Ying Bai
ISBN: 0-8493-2213-8

AUERBACH PUBLICATIONS

www.auerbach-publications.com
To Order Call: 1-800-272-7737 • Fax: 1-800-374-3401
E-mail: orders@crcpress.com

Neural Networks for Applied Sciences and Engineering

From Fundamentals to Complex Pattern Recognition

Sandhya Samarasinghe

Auerbach Publications
Taylor & Francis Group
Boca Raton   New York

Auerbach Publications is an imprint of the
Taylor & Francis Group, an informa business

Auerbach Publications
Taylor & Francis Group
6000 Broken Sound Parkway NW, Suite 300
Boca Raton, FL 33487-2742

© 2007 by Taylor & Francis Group, LLC
Auerbach is an imprint of Taylor & Francis Group, an Informa business

No claim to original U.S. Government works
Printed in the United States of America on acid-free paper
10 9 8 7 6 5 4 3 2 1

International Standard Book Number-10: 0-8493-3375-X (Hardcover)
International Standard Book Number-13: 978-0-8493-3375-0 (Hardcover)

Library of Congress Cataloging-in-Publication Data

Samarasinghe, Sandhya.
 Neural networks for applied sciences and engineering : from fundamentals to complex pattern recognition / Sandhya Samarasinghe.
     p. cm.
 Includes bibliographical references and index.
 ISBN-13: 978-0-8493-3375-0 (alk. paper)
 ISBN-10: 0-8493-3375-X (alk. paper)
 1. Neural networks (Computer science) 2. Pattern recognition systems. I. Title.

QA76.87.S255 2006
006.3'2--dc22
                                                                2006007265

Visit the Taylor & Francis Web site at
http://www.taylorandfrancis.com

and the Auerbach Web site at
http://www.auerbach-publications.com

Dedication

To Don

My husband

For your constant love, support, and encouragement

To do the best I can do in all

My endeavors as a

Woman and a

Scholar!

Contents

6   Data Exploration, Dimensionality Reduction, and Feature Extraction 245

Preface

This book is an exploration of neural networks for pattern recognition in scientific data. An important highlight is the extensive visual presentation of neural networks concepts throughout. This book is motivated by the necessity for a text that caters to both researchers and students from a wide range of backgrounds, one that puts neural networks into a multi-disciplinary scientific context. For the last seven years, I have taught neural networks to graduate students from diverse backgrounds, including biology, ecology, applied sciences, engineering, computing, and commerce at Lincoln University in New Zealand. My interactions with these students evolved my presentation of the material in such a way that it makes networks and their internal details transparent, thereby building confidence in the methods. Visual presentation became an invaluable tool in making difficult mathematical concepts easier to grasp. This book is a reflection of these efforts and of my own interest in exploring neural networks.

My intent is to provide a sound theoretical background within an applied context. My experience has shown that learning combined with hands-on applications using neural networks software provides the best outcome. Additionally, practical tutorial sessions to complement the theoretical treatments have been very successful in presenting this material.

I have designed this book to introduce neural networks to senior under-graduate and graduate students from applied fields of research with some mathematical and basic calculus background. Simple presentations in conjunction with visual aids make it possible to unravel a network to understand the mathematical concepts and derivations, and to appreciate the internal workings of neural networks that are considered to be a 'black box' by many.

Chapter 1 begins with an introductory discussion of the role neural networks play in scientific data analysis and a detailed layout of the book is presented. Many scientists are interested in determining the advantage of neural networks over classical statistical methods. In this book, statistical methods are addressed in detail in relation to neural networks; Chapter 2 illustrates that the two approaches are equivalent in linear data analysis and then begins to build a solid foundation of basic neural network concepts, instilling a deep understanding to continue forth with confidence.

Chapter 3 through Chapter 5 address nonlinear data analysis with neural networks using multilayer networks that are the most popular networks for nonlinear pattern recognition. Multilayer networks are a class of networks that have layers of neurons with nonlinear processing capabilities. The book provides extensive coverage of these networks because their potential and usefulness in systems modeling are increased if their limitations in relation to robustness and extensive trial-and-error requirements are addressed. The advantages of neural networks over statistical methods in nonlinear modeling are illustrated in these chapters. Specifically, these chapters address in detail nonlinear processing in networks, network training, and optimization of network configurations. Examples and case studies are presented so that these chapters can be easily understood. The material in these chapters is intended for both regular lectures and independent study.

Chapter 6 is a discussion of data exploration and preprocessing; it involves a significant number of statistical methods, some of which are available on commonly known statistical programs. The objective of the chapter is to extract relevant and independent inputs for effective model development and it can be used in conjunction with hands-on problem solving on statistical software.

Chapter 7 discusses uncertainty assessment in neural networks and relies heavily on statistical methods; neural networks are examined on a rigorous statistical foundation. Although neural networks are powerful nonlinear processes, tools to assess their robustness have been limited. In this chapter, neural networks are put into the context of Bayesian statistics for a rigorous assessment of their uncertainty in relation to network parameters, errors, and sensitivities. The material presented in this chapter requires a basic understanding of the concepts of simple, joint, and conditional probabilities, as well as the neural networks concepts developed in Chapter 3 through Chapter 5. Uncertainty assessment presented in the chapter can be invaluable for gaining confidence in the neural network models and then using them in decision making.

In my experience, students are particularly interested in self-organizing maps—unsupervised networks for discovering unknown clusters and relationships in multidimensional data. Chapter 6 presents this material in a step-by-step manner that highlights the important concepts. These can be used as both lecture material and for independent study. The essential concepts are presented incrementally and many features of unsupervised data clustering and its relation to some statistical clustering methods are illustrated using examples. Specifically, the chapter covers topics including competitive learning and topology preservation, one- and two-dimensional maps, map training and validation, map quality assessment, cluster formation on maps, and evolving self-organizing maps using extensive graphical illustrations.

The last chapter treats linear and nonlinear time-series forecasting with neural networks. It extensively covers concepts of recurrent and feedforward networks for short-term and long-term time-series forecasting, and the majority of the material can be used as both lecture material and for independent study. A variety of practical example case studies highlight all new concepts introduced. The similarity of linear neural networks and the relevant classical statistical methods are illustrated and the advantages of nonlinear neural networks are demonstrated.

The examples presented in the book have been developed mainly on Neural Networks for *Mathematica*® and Machine Learning Framework for *Mathematica*—two *Mathematica* add-on programs—and NeuroShell2™, a commercial software. There are many commercial and free neural networks software programs available on the World Wide Web to complement the material in the book.

It was my intention to present the material in this book in such a way that the fundamentals gained from it will help the reader apply this knowledge and understanding to a variety of other networks that are not covered in the book, as well as any other new developments in this fast-growing field. My experience has shown that the approach used in this book has helped many diverse researchers learn and apply neural networks in their individual fields of research. I hope that the readers will find this to be true for themselves as well.

How to Use the Book

This book covers a number of important issues in model development with neural networks and is suitable as a research-focused textbook or as a reference for researchers interested in independent study. The book has been written for applied scientists and engineers, and as a textbook for students in these fields. The material may be presented over two semesters: Chapter 2 through Chapter 5 may be covered in the first semester, and Chapter 6 through Chapter 9 in the second. Although there is a seamless and logical progression of material from Chapter 2 through Chapter 7, Chapter 8 relies on the concepts developed in Chapter 2. The final chapter has a strong relationship to Chapter 3 through Chapter 7. I have used NeuroShell2 (Ward Systems, Inc., USA), Neural Connnection™ (SPSS, Inc., USA), and NeuroSolutions™ (NeuroDimension, Inc.) in the past and each is suitable for class tutorials as well as independent research.

In a multidisciplinary audience, expected outcomes of the participants are quite broad. Some wish to learn how to use neural networks as a tool in their research, and for them it is essential to have user-friendly software such as those mentioned above. Others prefer to experiment with neural networks concepts; for these, *Mathematica* (Wolfram Research, Inc.), MATLAB® (The MathWorks Inc., USA), and C++ programming environments have been useful.

Acknowledgments

First and foremost, I must express my heartfelt gratitude to my husband, Don Kulasiri, Professor of Computer Modeling and Simulations at Lincoln University in New Zealand. It was he who encouraged me to pursue the field of neural networks and who inspired me to write this book. His moral, professional, and personal support has been invaluable, not only in writing this book, but also in developing my career as a scholar.

I wrote the book proposal and some early chapters while on study leave at the Princeton Environmental Institute at Princeton University in New Jersey, USA. I am thankful to Professor Simon Levin, Moffette Professor of Biology and the Director of the Center for BioComplexity at Princeton University, for inviting me to be a Visiting Scholar in 2004.

My graduate student, Erica Wang, prepared all of the figures used in this book, and I sincerely thank her for her contribution. I would also like to thank the neural networks researchers who assisted me in various ways. My special thanks go to Dr. Filipe Aires at Université Pierre et Marie Curie in France and Professor Jonas Sjoberg (developer of Neural Networks for *Mathematica*), Chalmers University of Technology in Sweden.

I am grateful to my parents for their encouragement and support, especially to my father, whose constant dedication has been invaluable in my education.

About the Author

Sandhya Samarasinghe received a MSc (Hons) in mechanical engineering from Lumumba University in Russia before attending Virginia Tech in the United States, where she earned her MS and PhD in engineering. She is currently a senior lecturer at the Department of Natural Resources Engineering and a founding member of the Centre for Advanced Computational Solutions (C-fACS), both at Lincoln University, New Zealand. Her research interests include neural networks, statistics, soft computing, and utilizing artificial intelligence, statistical methods, computer vision, and complex systems modelling to solve real-world problems in engineering, biology, ecology, environmental and natural systems, and applied sciences. Dr. Samarasinghe has been involved in diverse scientific and industrial projects in those fields and has published extensively in international journals and conference proceedings. The focus of her neural networks research is on theoretical understanding and advancements, as well as practical implementations. She is involved in reviewing grant proposals for various national and international funding organizations, reviewing journal papers, organizing conferences, and acting as a member of editorial boards. She has spent sabbaticals at Stanford University and Princeton University in the United States and Commonwealth Scientific and Industrial Research Organization (CSIRO) in Australia.

Chapter 1

From Data to Models: Complexity and Challenges in Understanding Biological, Ecological, and Natural Systems

1.1 Introduction

Nature is complex. It sustains many interacting and interdependent systems to maintain biological and ecological diversity as well as natural and environmental processes. Many problems researchers currently face are related to one or more of these interdependent systems. In solving biological, ecological, and environmental problems, scientists attempt to develop models to predict an outcome, understand or explain a process, or classify a process's outcome. A major hindrance in modeling real problems is the lack of understanding of their underlying mechanisms because of complex and nonlinear interactions among various aspects of the problem. For example, individuals in an ecosystem make up species that coexist with other species to form a community that depends on their habitat for sustenance and regeneration. Ecosystems are maintained through complex interactions between various players in the system that make it impossible to gain a complete understanding of this system to develop predictive models. Even the simplest mathematical models of population dynamics, for example, can exhibit oscillatory and even chaotic behavior, making it impossible to predict precise dynamics of populations governed by such equations [1]. In many cases, the best solution is to learn system behavior through observations or data. Specifically, researchers collect data that characterizes a system, and they attempt to extract complex nonlinear and multidimensional patterns and relationships embedded in the data.

There are many reasons for complex system behavior. Many natural systems display randomness, heterogeneity, multiple causes and effects, and noise [2]. For example, there are many plant species, and their growth depends on genetic, environmental, and soil conditions that are also variables with an element of randomness. Thus, plant growth is a dynamic process that makes all aspects of growth and properties vary in space and time. Therefore, understanding the underlying mechanism of plant growth and its interaction with the environment is a complex problem.

Many situations exist where researchers rely on data to study system behavior. In ecosystem management, researchers want to know which plant or animal species is invasive, their habitat's characteristics, and the risks they pose to health, crops, commerce, and the management of the ecosystem. Through biological study, researchers now know the complete human genome that contains the blueprint for life. However, how genes express in response to various harmful agents by resisting or giving in to diseases, how enzymes build proteins, how complex protein structures fold into compact forms, and many related issues are complex and the only way to understand these processes is through data. Complex interactions are also found in environmental management, including water quality, air pollution, and contaminant transport in porous media, as well as in the management of natural resources, rivers, lakes, forests, fisheries, and wildlife [2–4].

The nature of the aforementioned problems requires a systems approach where the most essential features of a complex problem with multiple interactions are modeled so that the system behavior can be predicted reliably even under random and noisy conditions. Neural networks are flexible, adaptive learning systems that follow the observed data freely to find patterns in the data and develop nonlinear system models to make reliable predictions; they provide a promising approach for solving many real-world problems [5,6].

Information processing in the brain inspired neural networks. The brain processes information incrementally and learns concepts over time. In this process, the brain attains a remarkable ability to make decisions and draw conclusions when presented with complex, noisy, irrelevant, or partial information [7,8]. Neural networks are popular because of their ability to imitate some of the brain's creative processes, albeit in a simplistic way, that cannot be imitated by existing mathematical or logical methods. Such capabilities are essential for solving many complex problems.

The remarkable capabilities in the brain arise from its massive networks of interconnected neurons that incrementally process information transmitted from the external or its internal environment to develop robust internal representations of the external phenomena. This is called learning, and the brain can be trained or left to learn on its own. Artificial neural networks, the subject of this book, are a system of interconnected neurons organized into a network where each neuron processes data locally using the concepts of learning in the brain [5,8]. Thus, the networks can be either specifically trained or left to self-organize and learn on their own. This is accomplished by repeated exposure to data representing the studied system, so that the network learns system behavior from data. Once trained, networks can be used to make pragmatic decisions with regard to the nature, behavior, use, or management of the system. When they are trained with samples of input–output data in supervisory mode, they can make predictions, classifications, or forecasts of future events. In self-organization, networks learn in an unsupervisory mode and can learn to discover unknown clusters. For example, they may cluster similar species, groups, protein structures, etc., and they can provide insight into the internal structure and relations in the data.

Neural networks have been successfully developed to solve problems in a variety of applied fields, and a list of examples to demonstrate the diversity of applications includes

- Plant ecosystems (growth, health, and interaction with the environment) [9]
- Plant disease identification and prediction of disease spread [10]
- Study of the dynamics of plant and animal communities and their habitat characteristics [11]
- Study of the effects of deforestation and habitat change on ecosystems
- Classification of plant and animal species
- Micro-array data analysis and protein structure prediction [12]
- Meat quality and tenderness characterization [13–15]
- Animal disease diagnosis and stages of severity to produce animal health indicators [16,26]
- Prediction of properties and behavior of biological materials [17,23,24,27]
- Modeling land use change
- Properties and behavior of natural systems such as ground water systems, time and space variation of properties, contamination of aquifers and atmospheric systems [3,4]
- Chemicals in the environment and their local and global consequences [18]

- Forecasting inflows into rivers and lakes [19]
- Reservoir management
- Integration of many management parameters to provide an effective solution for the management of a system
- Understanding waste generation factors and long-term forecasting of waste production [20,21]
- Electricity load forecasting [22,25]
- Economic predictions [28]

1.2 Layout of the Book

This book is an exploration of neural networks for pattern recognition in scientific data, and a major component of the book is the extensive visual presentations illustrating neural network concepts. Starting with the basics, the book provides instruction on a variety of neural networks' internal workings, and it shows how to apply them to solve real problems. A thorough explanation of the fundamentals provides a solid foundation for understanding many types of neural networks. Once this is achieved, we will explore, in detail, Multilayer Perceptron for predictions and classification, Self-Organizing Feature Maps for unsupervised clustering and Recurrent Networks for time-series understanding and forecasting. Other selected networks such as Generalized Neuron models and Generalized Regression Networks are also presented. The importance of, and approaches to, data preprocessing, model validation, and uncertainty assessment that are crucial to successful model development are also addressed. Relevant statistical concepts are presented alongside the neural network concepts throughout the book. All new concepts are explained using hands-on examples, and the use and behavior of all network types are demonstrated through practical application case studies. Following is a summary of the rest of the book's chapters.

Chapter 2 introduces neural networks and relevant concepts from biological neural networks. It demonstrates the operation of single neurons and several models developed to capture information processing in a neuron. Neurons presented in this chapter are linear models and their performance is compared with linear statistical models. Concepts of neuron activation functions and connection strength (weights) between neurons are introduced. Weights are the free parameters, and they are the most important feature of networks because they hold internal representations (memory) of the model. Learning involves optimizing the weights. This chapter offers an exploration of learning in these neurons using the learning theories that have been proposed over time to adjust weights incrementally. Specifically, a study of the classification capabilities of threshold limited neurons as well as classification and predictive capabilities of linear neurons are conducted. The performance of networks made up of multiple numbers of these neurons organized in a layer is discussed. Their relationship to discriminant function analysis and linear and multiple linear regression will be demonstrated. The chapter also introduces supervised learning, including Hebbian learning and delta rule, as well as unsupervised learning.

Neural networks draw their nonlinear modeling capabilities from the flexible processing in nonlinear neurons that are organized into layers in networks. There are several variants of such networks, and Chapter 3 deals with the operation of a multiple layer network, popularly known as a Multilayer Perceptron, where more complex neurons are organized in several layers that make it possible for them to do complex nonlinear mapping tasks. In-depth discussion within the chapter will provide a solid foundation for understanding the operation of other similar networks as well as more complex networks. Here, all the details of data processing by a network are illustrated through examples. These examples include how a network processes information and how learning organizes the internal aspects of a network through activities such as neuron activation functions, connection strengths, and hidden neuron layers to produce the desired outcome.

Actual learning mechanisms are covered in Chapter 4, which explores the internal workings of multilayer networks, and it pays particular attention to how a network can be trained to learn using learning methods. Learning involves optimizing the free parameters (i.e., weights) of a network, and the most widely used approach is minimizing mean square error. The delta rule is one such approach and this chapter illustrates its use for simultaneously adapting all the weights in a network. Every detail of the delta rule's application is explained through graphs and hands-on examples. Extensive coverage of several other variants of delta rule—back propagation with momentum, adaptive learning rate or delta-bar-delta, steepest descent, and second-order learning methods including QuickProp, Gauss–Newton and Levenberg–Marquardt methods—is given in this chapter. Each learning method is explained with a hand calculation and a computer experiment, and learning methods are compared to assess their efficiency. This investigation is complemented with case studies comparing different learning methods and assessing their performance on complex data.

Neural networks' power comes at a cost. There are many possible ways to configure networks and train them. In other words, with a large enough number of free parameters, neural networks can be trained to rigidly fit data that may also include noise. Therefore, it is important to understand how to optimize the structure and learning of the networks to develop models that generalize well to unseen data and that are reliable for decision making. Chapter 5 treats these aspects extensively for multilayer feedforward networks. Specifically, methods for improving the generalization ability of networks and the effect of data, noise, and initial network weights on the generalization ability of networks are illustrated. The effect of the internal structure, such as the number of neurons and connections, on the performance of the networks is explored graphically, and several approaches to pruning networks to reduce their complexity are illustrated with examples. Reasons for the instability of weights (i.e., multiple solutions) in general in a network are explored, and the robustness and uncertainty of the networks is assessed by analyzing the resistance of network weights to perturbations.

An important aspect of model development is finding the essential features that must be incorporated into the model and omitting unnecessary information or noise. This process is called feature extraction. Using only the relevant inputs helps reduce the complexity of models and makes the model parameters robust. Chapter 6 is devoted to data exploration and preprocessing. It starts with a presentation on approaches to data visualization and proceeds to discuss correlation and covariance between variables to identify correlated data. Several approaches to data normalization are offered to improve the representation of the variables in the model. Various statistical tools, including partial correlation, best subsets regression, and principal component analysis, are presented for selecting inputs into a neural network. The correlated inputs give rise to multicollinearity, which can severely affect the model's accuracy and robustness. Therefore, removing multicollinearity to reduce input dimensionality can greatly improve model accuracy. In this chapter, several approaches to addressing multicollinearity are illustrated, including principal component analysis and partial least-square regression. Outlier detection and noise removal in multivariate data are addressed for cleaning the data. The input selection is illustrated through a case study that highlights the positive effects of dimensionality reduction on model accuracy of feedforward networks.

Chapter 7 is devoted to uncertainty assessment of feedforward networks using Bayesian statistics. First, it puts network learning in the context of maximum likelihood parameter estimation in statistics. It then puts the optimum parameters (weights) obtained from training of networks in a probabilistic framework so the uncertainty of weights can be properly assessed. Specifically, for a trained network, weight probability distribution is attained and is used to assess the uncertainty of other parameters, such as model output, error due to intrinsic noise, and network sensitivity to inputs. A case study is presented where the uncertainty of networks' sensitivities are explored to assess the relevance of inputs, and the uncertainty of output errors are assessed to ascertain the robustness of the model's output. This chapter systematically illustrates the significance of the principal component-based dimensionality reduction on the robustness of models.

Chapter 8 presents self-organizing map (SOM) networks, also called unsupervised networks, that discover cluster structures and relationships in multidimensional data that are not initially known. These networks have an input layer and an output layer (the map), which has a predefined structure (i.e., number of neurons). Input layer is connected to output layer neurons with weights that reorganize themselves in a way so that inputs that are similar are clustered together. The structure and training of these networks using competitive and self-organization learning are discussed, and illustrated using hand calculations, computer experiments, and real case studies. One- and two-dimensional maps and relevant learning issues are discussed in greater detail. A trained map is a compact preservation of the input probability distribution. Ways to assess the quality of the map as well as defining specific numbers of clusters on a trained map are also presented. Evolving SOMs that allow a flexible map structure to grow as dictated by the data are presented and illustrated through examples.

Time-series forecasting with neural networks is the focus of Chapter 9. Time-series are auto-correlated, and an outcome at an instance of time has a strong correlation to past observations (lags) of the same series. First, a detailed analysis of linear models is presented with examples, then, nonlinear neural networks for time-series forecasting are discussed. Specifically, a modified back propagation and several variants of recurrent networks are analyzed extensively to demonstrate their ability to capture temporal dynamics in data, and each network is illustrated with an example case study. These networks are presented as an extension of the classical linear Autoregressive (AR) and Moving Average (ARMA) models. Network development for extended long-term forecasting is presented and illustrated using a case study. Networks' bias and variance components for time-series forecasting are analyzed with respect to input lags and network structure. Finally, approaches for input selection in time-series forecasting are presented and illustrated using a practical example case study.

References

1. Levin, S.A. Population dynamics in models in heterogeneous environments, *Annual Review of Ecology and Systematics*, 7, 287, 1976.
2. Kulasiri, D. and Verwoerd, V. *Stochastic Dynamics: Modeling Solute Transport in Porous Media*, North Holland Series in Applied Mathematics and Mechanics, Vol. 44, Elsevier, Amsterdam, 2002.
3. Rajanayaka, C., Kulasiri, D., and Samarasinghe, S. A comparative study of parameter estimation in hydrology modelling: Artificial neural networks and curve fitting approaches, *Proceedings of International Congress on Modelling and Simulation (MODSIM'03)*, D.A. Post, ed., Vol. 2, Modelling and Simulation Society of Australia and New Zealand, Townsville, Australia, p. 843, 2003.

4. Rajanayake, C., Samarasinghe, S., and Kulasiri, D. Solving the inverse problem in stochastic groundwater modelling with artificial neural networks, *Proceedings of the 1st Biennial Congress of the International Environmental Modelling and Software Society,* A.E. Rizzoli and A.J. Jakeman, eds., Vol. 2, Servizi Editorial Association, Lugano, Switzerland p. 154, 2002.
5. Haykin, S. *Neural Networks: A Comprehensive Foundation,* 2nd Ed., Prentice Hall, Upper Saddle River, NJ, 1999.
6. Aires, F., Prigent, C., and Rossow, W.B. Neural network uncertainty assessment using Bayesian statistics with application to remote sensing: 2. Output error, *Journal of Geophysical Research,* 109, D10304, 2004.
7. Mind and brain, *Readings from "Scientific American" Magazine,* W.H. Freeman, New York, 1993.
8. Rumelhart, D.E. and McClelland, J.L. Foundations, *Parallel Distributed Processing—Explorations in the Microstructure of Cognitio,* Vol. 1, MIT Press, Cambridge, MA, 1986.
9. Spencer, M., McCullagh, J., Whitfort, T., and Reynard, K. An application into using artificial intelligence for estimating organic carbon, *Proceedings of International Congress on Modelling and Simulation (MODSIM'05),* A. Zerger and R.M. Argent, eds., Modelling and Simulation Society of Australia and New Zealand, Melbourne, Australia, 84, 2005.
10. De Wolf, E.D. and Francl, L.J. Neural networks that distinguish infection periods of wheat tan spot in an outdoor environment, *Phytopathalogy,* 87, 83, 1997.
11. Lek, S. and Jean-Francois, G. eds. *Artifical Neuronal Networks: Application to Ecology and Evolution,* Springer Environmental Science Series, Springer, New York, 2000.
12. Wu, C.H. and McLarty, J.W. *Neural Networks and Genome Informatics,* Series on Methods in Computational Biology and Biochemistry, Elsevier Science, Oxford, 2002.
13. Chandraratne, M.R., Samarasinghe, S., Kulasiri, D., and Bickerstaffe, R. Prediction of lamb tenderness using image surface texture features, *Journal of Food Engineering,* 2005 (in press).
14. Chandraratne, M.R., Kulasiri, D., Frampton, C., Samarasinghe, S., and Bickerstaffe, R. Prediction of lamb carcass grades using features extracted from lamb chop images, *Journal of Food Engineering,* 74, 116, 2005.
15. Chandraratne, M., Kulasiri, D., Samarasinghe, S., Frampton, C., and Bickerstaffe, R. Computer vision for meat grading: Neural networks and statistical approaches, *Proceedings of the 48th International Meat Congress,* Italy, p. 756, 2002.
16. Lopez-Benavides, M., Samarasinghe, S., and Hickford, J.G.H. The use of artificial neural networks to diagnose mastitis in dairy cattle, *Proceedings of the International Joint Conference on Neural Networks,* IEEE, Los Alamitos, CA, 5, p. 100. 2003.
17. Samarasinghe, S., Kulasiri, D., Rajanayake, C., and Chandraratne, M. Three neural network application case studies in biology and natural

resource management, *Proceedings of the 9th International Conference on Neural Information Processing,* W. Lipo, ed., IEEE, Los Alamitos, CA, 5, 2279, 2002.

18. Jiang, D., Zhang, Y., Hu, X., Zeng, Y., Tan, J., and Shao, D. Progress in developing an ANN model for air pollution index forecast, *Atmospheric Environment,* 38, 7055, 2004.

19. Chiang, Y.M., Chang, L.C., and Chang, F.J. Comparison of static feedforward and dynamic feedback neural networks for rainfall-runoff modeling, *Journal of Hydrology,* 290, 297, 2004.

20. Ordonez, E., Samarasinghe, S., and Torgeson, L. Neural networks for assessing waste generation factors and forecasting future waste generation in Chile, *Proceedings of the Waste and Recycle 2004 Conference,* Waste Management and Environment (WME), Western Australia, Australia, p. 205, 2004.

21. Ordonez, E., Samarasinghe, S., and Torgeson, L., Relations and recovery of domiciliary solid waste using artificial neural networks: a case study of Chile, *Proceedings of the 19th International Conference on Solid Waste Technology and Management,* R.L. Merseky, ed., Widener University School of Engineering, Philadelphia, PA, p. 1273, 2004.

22. Chaturvedi, D.K., Mohan, M., Singh, R.K., and Karla, P.K. Improved generalized neuron model for short-term load forecasting, *Soft Computing,* 8, 370, 2004.

23. Samarasinghe, S. and Kulasiri, D. *High Speed Video Imaging and Neural Networks for Determination of Fracture Toughness of Wood,* Applied Computing Mathematics and Statistics Publication Series, Division of Applied Computing and Management, Lincoln University, New Zealand, ISSN 1174, Serial QA75.5 Res no. 2002/03, 2002.

24. Samarasinghe, S. and Kulasiri, D. *Stress Intensity Factor of Wood from Crack-tip Displacements and Predicting Fracture Toughness Using Neural Networks,* Applied Computing Mathematics and Statistics Publication Series, Division of Applied Computing and Management, Lincoln University, New Zealand, ISSN 1174, Serial QA75.5 Res no. 2002/05, 2002.

25. Rayudu, R. and Samarasinghe S. A network of neural nets to model power system fault diagnosis, *Proceedings of the Fourth International Conference on Neural Information Processing,* IEEE, Los Alamitos, CA, 164, 1997.

26. Wang, E. and Samarasinghe, S. On-line detection of mastitis in dairy herd using neural networks, *Proceedings of International Congress on Modelling and Simulation (MODSIM'05),* A. Zerger and R.M. Argent, eds., Modelling and Simulation Society of Australia and New Zealand, Melbourne, Australia, 273, 2005.

27. Tian, X., Samarasinghe, S., and Murphy, G. An integrated algorithm for detecting position and size of knots on logs using texture analysis, *Proceedings of the International Conference on Image and Visions Computing,* P. Bones, ed., University of Canterbury, New Zealand, 60, 1999.

28. Limsombunchai, V. and Samarasinghe, S. House price prediction: hedonic price model vs. artificial neural networks, *Kasetsar University Journal of Economics,* Kasetsart University, Thailand, ISSN 0858-9291, 61, 2005.

Chapter 2

Fundamentals of Neural Networks and Models for Linear Data Analysis

2.1 Introduction and Overview

Neural networks are an evolving field with origins in neurobiology. Neural networks are models that attempt to mimic some of the basic information processing methods found in the brain. Because our brains perform complex tasks, neural networks modeled after the brain have also been found useful in solving complex problems. The field of neural networks has grown from the modeling of simple processing elements or neurons to massively parallel neural networks. This chapter demonstrates the basic concepts of neural networks by following the evolution of some neural networks concepts. Specifically, the chapter will look broadly at what comprises a neural network and will present a detailed study of what neurons are, how they have been modeled, and how to interpret the model outcomes. It will also give an introduction to the foundation of learning methods, and to important major developments that laid the groundwork for the development of powerful neural network models. At the end of the chapter, the reader will have a solid understanding of information processing in single-neuron models and linear neural network models, which will aid the study of nonlinear neural network models in subsequent chapters. Along the way, the chapter will also relate several statistical concepts of linear analysis to neural network concepts that will be developed incrementally.

Section 2.2 introduces the concepts of neurons and neural networks; Section 2.3 presents the fundamental concepts from neurobiology that inspired the development of neural networks. The modeling of neurons is introduced in Section 2.4, and learning strategies are discussed in Section 2.5, along with two single-neuron models for linear data analysis—perceptron and linear neuron—as well as linear neural networks. Corresponding statistical methods are also highlighted in Section 2.5. A chapter summary is presented in Section 2.6.

2.2   Neural Networks and Their Capabilities

A broader definition of a practical neural network is that it is a collection of interconnected neurons that incrementally learn from their environment (data) to capture essential linear and nonlinear trends in complex data, so that it provides reliable predictions for new situations containing even noisy and partial information. Neurons are the basic computing units that perform local data processing inside a network. These neurons form massively parallel networks, whose function is determined by the network structure (i.e., how neurons are organized and linked to each other), the connection strengths between neurons, and the processing performed at neurons.

Haykin [1] states that "A neural network is a massively parallel distributed processor that has a natural propensity for storing experiential knowledge and making it available for use. It resembles the brain in two respects: 1. Knowledge is acquired by the network through a learning process; 2. Interconnection strengths between neurons, known as synaptic weights or weights, are used to store knowledge."

Neural networks perform a variety of tasks, including prediction or function approximation, pattern classification, clustering, and forecasting, as shown in Figure 2.1 [2]. Neural networks are very powerful when fitting models to data (Figure 2.1a). They can fit arbitrarily complex nonlinear models to multidimensional data to any desired accuracy; consequently, neural network predictors are called universal approximators [3]. From a functionality point of view, they can be thought of as extensions to some multivariate techniques, such as multiple linear regression and nonlinear regression.

Neural networks are also capable of complex data and signal (time-series) classification tasks involving arbitrarily complex nonlinear classification boundaries (Figure 2.1b). In situations in which the naturally formed clusters in the data are unknown *a priori*, neural networks are useful in unsupervised

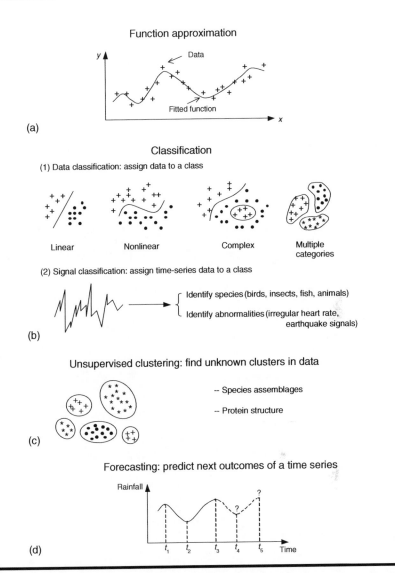

Figure 2.1 Some functions of neural networks suitable for scientific data modeling: (a) fitting models to data, (b) complex classification tasks, (c) discovering clusters in data, and (d) time-series forecasting.

clustering, in which they use the internal properties of the data to discover unknown cluster structures (Figure 2.1c). A powerful feature of the unsupervised neural clustering method called self-organization is that it can also simultaneously reveal spatial relations between clusters of data while finding the clusters. Neural networks are also capable of time-series forecasting, in which the next outcome or outcomes for the next several time steps are predicted (Figure 2.1d). This is accomplished by capturing temporal patterns in the data in the form of past memory, which is embedded in the model. In forecasting, this knowledge about the past defines future behavior.

There are a variety of neural networks with special features that have been developed to accomplish the above tasks; some of the most relevant for scientific data modeling are shown in Figure 2.2, which illustrates the organization of the individual processing elements or neurons (denoted by circles), the links between them, and the manner in which this structure

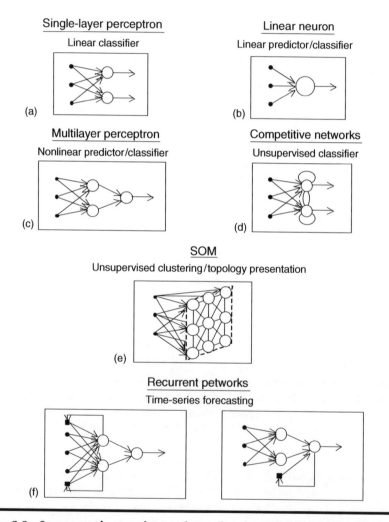

Figure 2.2 Some neural network types for performing tasks depicted in Figure 2.1: (a) single-layer perceptron, (b) linear neuron, (c) multilayer perceptron, (d) competitive networks, (e) self-organizing feature map, (f) recurrent networks.

links inputs to outputs. The perceptron network in Figure 2.2a, which will be explored in this chapter, is a linear classifier and is functionally similar to simple- and multiple-discriminant function analysis in statistics.

The linear neuron shown in Figure 2.2b is a linear classifier and predictor whose predictive capabilities are equivalent to simple and multiple linear regression models; as a classifier, it resembles simple- and multiple-discriminant function analysis in statistics. These aspects will also be explored in this chapter.

The multilayer perceptron (MLP) model shown in Figure 2.2c is the most well-known neural network for the nonlinear prediction and classification tasks shown in Figure 2.1a and Figure 2.1b. This, in fact, is an extension of the perceptron network. Chapter 3 and Chapter 4 are devoted entirely to these networks. The competitive networks (Figure 2.2d) are unsupervised networks that can find clusters in the data. The self-organizing feature map (SOFM) competitive network shown in Figure 2.2e not only finds unknown clusters in the data but also preserves the topological structure (spatial relations) of the data and clusters [4]. Two popular neural networks for time-series forecasting are the Jordan and Elman Networks [5] presented in Figure 2.2f. These networks contain feedback links that help to capture temporal effects. All the networks illustrated in Figure 2.2 will be discussed in various chapters throughout this book.

The common element between all these networks is that they each contain many links connecting inputs to neurons and neurons to outputs. These links are called weights, and they facilitate a structure for flexible learning that allows a network to freely follow the patterns in the data. The weights are called free parameters, and the neural networks are therefore parametric models involving the estimation of optimum parameters. The flexible structure of these neural networks is what makes them capable of solving such a variety of complex problems.

To illustrate this point, Figure 2.3 shows the structure of an MLP network (also shown in Figure 2.2c) that is capable of complex input–output mapping, shown in Figure 2.1a and Figure 2.1b. It has an input layer, a hidden layer, and an output layer of neurons, denoted by I, H, and O, respectively. These three layers are linked by connections whose strength is

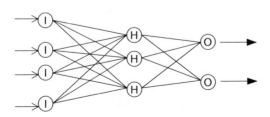

Figure 2.3   An example multilayer neural network.

called weight. Thus there are two sets of weights, the input-hidden layer weights and the hidden-output layer weights. These weights provide the network with tremendous flexibility to freely adapt to the data; they are the free parameters, and their number is equivalent to the degrees of freedom of a network.

The input layer transmits input data to the hidden neurons through input-hidden layer weights. Inputs are weighted by the corresponding weights before they are received by the hidden neurons. The neurons in the hidden layer accumulate and process the weighted inputs before sending their output to the output neurons via the hidden-output layer weights, where the hidden-neuron output is weighted by the corresponding weights and processed to produce the final output. This structure is trained to learn by repeated exposure to examples (input–output data) until the network produces the correct output. Learning involves incrementally changing the connection strengths (weights) until the network learns to produce the correct output. The final weights are the optimized parameters of the network.

Some of the key features of neural networks can therefore be summarized as follows: they process information locally in neurons; neurons operate in parallel and are connected into a network through weights depicting the connection strength; networks acquire knowledge from the data in a process called learning, which is stored or reflected in the weights; a network that has undergone learning captures the essential features of a problem and can therefore make reliable predictions. These are essentially the functions of the brain, and they illustrate the manner in which the functioning of the brain has inspired neural networks. To understand the internal workings of neural networks, the next section will briefly examine how the brain processes information.

2.3 Inspirations from Biology

The human brain can be thought of as an information-processing entity. It receives information from the external environment via the senses and processes them to form internal models of external phenomena. The brain is particularly capable of adjusting these models, as well as interpolating or extrapolating them to suit new situations with such agility that it can make reliable decisions, including recognizing patterns, understanding concepts, and making predictions even with partial information that may be random or noisy.

The local information processing in brain cells or neurons, which form a large number of parallel networks in the cortex of the brain, is central to these activities. The cortex is the thin outer layer of the brain that contains a large number of neurons, in the order of 100 to 500 billion [6]. Neurons are organized into about 1000 main clusters, each with about 500 networks [7]. A single network may have in the order of 10 000 neurons, and it is known that some networks are organized in a hierarchical or layered fashion. The brain has a variety of specialized neurons, and depending on the type, each neuron can send signals to anywhere from a hundred to several thousand other neurons. It is now known that the repeated excitation of neurons leads to the growth of new connections between them, thus creating and expanding a massively interconnected network that holds memory. The memory, or the acquired knowledge, is known to be stored as the connection strengths between neurons.

A biological neuron consists of three main components, as shown in Figure 2.4a: (i) dendrites that channel input signals, which are weighted by connection strengths, to a cell body; (ii) a cell body, which accumulates the weighted input signals and further processes these signals; and (iii) an axon, which transmits the output signal to other neurons that are connected to it. The computing process in a neuron is idealized in a model neuron shown in Figure 2.4b, in which signals are received, accumulated, or summed (Σ) in the cell body and processed further [$f(\Sigma)$] to produce an output. These aspects of neural operation will be examined in detail later in this chapter. The other neurons that receive this output signal (and the output signals from other neurons) in turn process the information locally and pass the output signal to other neurons until the process is completed and a concept is generated or reviewed, or an action is taken. This process is shown in detail in Figure 2.4c, in which each input is first weighted appropriately and the weighted inputs are summed and processed through an input–output

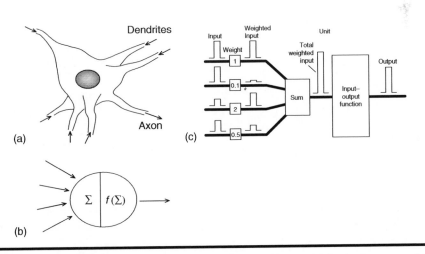

Figure 2.4 A biological neuron and its representation: (a) biological neuron, (b) neuron model, (c) detailed workings of a single neuron.

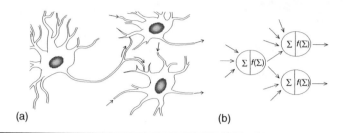

Figure 2.5 Communication between neurons: (a) network of three biological neurons, (b) neural network model.

function inside the neuron to produce an output [6]. This figure also highlights the effect of the weights, which must undergo adaptation if the neuron were to learn patterns in the information it receives.

The signals (inputs) from one neuron are passed to another neuron through connections between them; Figure 2.5a shows the communication between three neurons comprising a small network [6]. The first neuron sends signals to the latter two, which receive additional signals from the other neurons connected to them, as indicated by the extra arrows. Figure 2.5b shows a model of this network [6]. By organizing neurons in this fashion, massively parallel networks are formed in the brain. In biological neurons, signals are electrical in nature and are generated as a result of the concentration differential in potassium (K^+) and sodium (Na^+) ions within and outside of cells. The signal passes from one neuron to the next through the release of neurotransmitters, which leads to the generation of an electric potential in the receiving neuron.

2.4 Modeling Information Processing in Neurons

An interest in modeling biological neural networks emerged in the 1940s. Initial modeling efforts were in biology, the cognitive sciences, and related fields, and were stimulated by the possibility that the models might explain brain function based on the observations made by neurobiologists and cognitive scientists. To mimic biological networks, it is important to model the information processing in individual neurons. This involves understanding how signals are synthesized by a neuron to produce an output signal, and how neurons work collectively to produce an outcome. It also necessitates an understanding of the mechanisms of learning in neural networks from data, which is crucial to the development of memory or knowledge in the network, which in turn enables a model to perform some brain-like functions.

A lack of detailed knowledge of the mechanisms of neural information processing provided the researchers with an opportunity to experiment with new ideas for these networks, resulting in a rich array of neural networks, some of which are simple approximations of biological neural networks, and others are highly useful for problem solving but bear little resemblance to the actual operation of the brain. These efforts led to the development of artificial neural networks, which are widely used for solving a variety of problems in many fields remotely related to neurobiology, such as ecology, biology, engineering, agriculture, environmental and resource studies, and commerce and marketing. In this book, the term "neural networks" represents artificial neural networks, and "neurons" denotes artificial neurons. In this chapter, biological and artificial neural networks are intertwined, but the aim is to demonstrate the development of artificial neural networks that are useful for practical problem solving.

The next section presents an incremental introduction to neural network concepts, neuron models, mechanisms of learning from data, and other fundamental issues of neural networks, so that the reader can better appreciate and understand the neural networks in the rest of the book. These discussions will also facilitate the exploration of deeper aspects of the nature of data modeling as relevant to many applied fields of study.

2.5 Neuron Models and Learning Strategies

Neural computing has undergone several distinct stages. Early attempts occurred from the beginning of the 20th century to about 1969; 1969 to 1982 were quieter years, and 1982 marks the resurgence of activities that propelled a growth of neural networks that continues to this day [3]. This section highlights some of the important conceptual developments that are important for understanding and applying neural networks.

During the early 20th century, William James, an eminent American psychologist, provided two important clues to neural modeling: (1) If two neurons are active together, or in immediate succession, then on reoccurrence they repeatedly excite each other and the intensity between them grows; (2) the amount of activity of a neuron is the sum of the signals it receives, with signals being proportional to the strength of the connection through which a signal is received [7]. This basically suggests that a signal from one neuron going to the cell body of another is weighted in proportion to how strongly one neuron excites the other. The more intensely one neuron excites another, the larger the weight between them. These fundamental ideas have been implemented and incrementally advanced throughout the development of artificial neural networks.

2.5.1 Threshold Neuron as a Simple Classifier

A threshold neuron is a simple model, developed as a simple approximation to biological neurons by McCulloch-Pitts in 1940 [8]. It provided the stepping stone for the development of neural networks and learning methods that followed later [7]. It uses the threshold function in the neuron to transform inputs to an output, producing an output of either 0 or 1. This is simplistic from a biological point of view, because real neurons seem to have continuous signal outputs. The model neuron also has fixed weights, so it does not learn. This is because, at the time, it was not known how to adapt the connection strengths (weights) between the neurons. In the original design, threshold neurons used binary inputs (1 or 0) and McCulloch and Pitts [8] demonstrated that even these simple neurons could be organized into parallel networks that can perform some complex classifications tasks [3].

This chapter will study the general properties of a single-threshold neuron that takes real values as inputs, as it highlights some basic aspects of neural processing and it has some interesting basic features of a classifier. To illustrate the operation of this neuron, a simple classification problem, given in Table 2.1 and plotted in Figure 2.6a, will be solved. This problem includes two inputs (x_1 and x_2), one target output (t) that belongs to one of two categories (0 or 1), and four sets of input–output pairs. One row or one set of inputs is called an input vector or input pattern. The task is to correctly classify the input patterns into two groups (1 or 0). This task will be solved using a threshold neuron, shown in Figure 2.6b, to model the data. Here the neuron receives the two inputs through weights w_1 and w_2, but both weights will be fixed at 1.0 for this exercise, which means that there is no learning.

There are two aspects to the computations in this neuron. It first calculates the net input u and then decides the output y using a threshold function. These two calculations are performed as follows: The net input u to neuron is the sum of the weighted inputs, calculated as

$$\Sigma = u = w_1 x_1 + w_2 x_2, \tag{2.1}$$

where x_1 and x_2 are inputs. Because w_1 and w_2 are both equal to 1,

Table 2.1 Classification Data

x_1	x_2	t
0.2	0.3	0
0.2	0.8	0
0.8	0.2	0
1.0	0.8	1

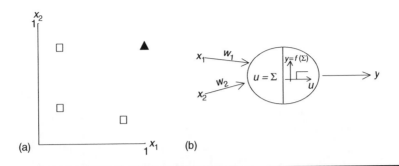

Figure 2.6 Classification by a threshold neuron: (a) classification data and (b) configuration of a linear threshold neuron for this task.

$$u = x_1 + x_2. \tag{2.2}$$

The value of u decides the activation threshold. Because the sum of the two inputs for one category is 2 and for the other category is 0, 1, and 1, respectively, for the inputs in Table 2.1, the threshold should be placed anywhere between 1.0 and 2.0. A threshold of 1.3 will be arbitrarily chosen for this case. Then the threshold function computes the activation or output (y) of the neuron as a function of u, such that

$$f(\Sigma) = y = \begin{cases} 0 & u < 1.3 \\ 1 & u \geq 1.3. \end{cases} \tag{2.3}$$

Using this simple classifier, it is now possible to check its performance. For the four inputs, the neuron output is computed following the same procedure described above; the results are presented in Table 2.2.

The neuron correctly classifies the data immediately due to the prior decision to set the threshold function at $u = 1.3$. As stated earlier, for this case u can be fixed anywhere between 1.0 and 2.0 to obtain a correct classifier for the data. Because the location of the threshold function defines the two categories, $u = 1.3$ decides a classification boundary that can be formulated as

Table 2.2 Performance of the Threshold Classifier

Input (x_1, x_2)	u	y
(0.2, 0.3)	0.5	0
(0.2, 0.8)	1.0	0
(0.8, 0.2)	1.0	0
(1.0, 0.8)	1.8	1

$$u = x_1 + x_2 = 1.3$$
$$x_2 = 1.3 - x_1. \tag{2.4}$$

This boundary line is superimposed on the data in Figure 2.7. The data on one side of the classification boundary belong to one category, and those on the other side of the boundary are classified into the other category. This is a simple classifier neuron that accumulates inputs and produces a bounded output (0 or 1) using a threshold function.

Key aspects of the above threshold neuron classifier can be summarized as follows: It does not learn from the environment (weights are equal to 1), but it can be designed to perform a classification task if the designer carefully positions the threshold function at a particular location (ideally, the neuron would decide this position by itself). The threshold neuron also classifies the data regions that are linearly separable. This means that a straight line can separate the two classes, and the threshold fixes this line as the classification boundary. Any input to the left of the boundary produces an output of 0, and those to the right of and on the boundary line yield an output of 1.

Inputs and weights as vectors. For simplicity, vector notation will be used. In this notation, an input vector is represented by upper case **x** as

$$\mathbf{x} = \{x_1, x_2\}.$$

Thus, the four input vectors can be represented as

$$\mathbf{x}_1 = \{0.2, 0.3\}, \ \mathbf{x}_2 = \{0.2, 0.8\}, \ \mathbf{x}_3 = \{0.8, 0.2\}, \ \mathbf{x}_4 = \{1.0, 0.8\}.$$

Similarly, the weight vector can be denoted in vector form as

$$\mathbf{w} = \{w_1, w_2\}.$$

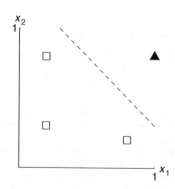

Figure 2.7   Classification boundary of the threshold neuron superimposed on the data.

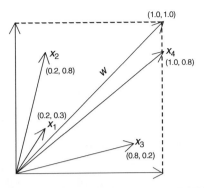

Figure 2.8   Representation of the input data and the weights as vectors.

The four input vectors and the weight vector are graphically presented in Figure 2.8.

The weighted sum of the inputs, $u = x_1.w_1 + x_2.w_2$, can be represented as multiplication of the input and weight vectors or dot product as

$$u = \mathbf{w} \cdot \mathbf{x} = \{w_1, w_2\} \cdot \{x_1, x_2\}$$
$$= w_1 x_1 + w_2 x_2$$

For $\mathbf{w} = \{1, 1\}, \quad u = x_1 + x_2.$

Refer to the Appendix for a brief introduction to vectors and vector processing.

2.5.2   Learning Models for Neurons and Neural Assemblies

2.5.2.1   Hebbian Learning

A major drawback of the threshold neuron considered in the previous section is that it does not learn. In 1949, Donald Hebb, a psychologist, proposed a mechanism whereby learning can take place in neurons in a learning environment. In his book *The Organization of Behavior*, Hebb [9] defined a method to update weights between neurons that came to be known as Hebbian learning. Key points of his contribution are: (1) He stated that the information in a network is stored in weights or connections between the neurons. (2) He postulated that the weight change between two neurons is proportional to the product of their activation values (neuron outputs), thereby enabling a mathematical formulation of the concept that stronger excitation between neurons leads to the growth in weights between them. (3) He proposed a neuron assembly theory, suggesting that as learning takes place by repeatedly and simultaneously activating a group of weakly connected neurons, the strength and patterns of the weights between them undergo incremental changes, leading to the formation of assemblies of strongly connected neurons.

The above ideas of learning were motivated by well-known concepts of classical, or Pavlovian, conditioning, established through animal experiments that supported the fact that through repeated exposure to a stimulus, learning takes place in the brain. These findings strongly supported the hypothesis that learning involves the formation of new connections between neurons that grow in strength through repeated exposure to the stimulus. Hebb [9] developed this into a learning method that allows neurons to learn by adjusting their weights in a learning environment.

Formulation of Hebbian learning. Hebbian learning can be expressed as follows: If two neurons have activations, or outputs, of x and y, and if x excites y (or moves in the same direction), the connection strength between them increases. Therefore, the change in weight between two neurons, Δw, is proportional to the product of x and y, as given in Equation 2.5:

$$\Delta w \propto x \cdot y. \tag{2.5}$$

The symbol $\propto$ denotes proportionality, which can be removed by using a coefficient, β, so that

$$\Delta w = \beta\, x \cdot y. \tag{2.6}$$

The new value of the weight, w_{new}, is

$$w_{new} = w_{old} + \Delta w = w_{old} + \beta\, x \cdot y, \tag{2.7}$$

where w_{old} is the initial value of the weight prior to learning. This concept was later logically extended to inhibitory connections, in which one neuron inhibits another and the connection strength decreases (i.e., $w_{new} = w_{old} - \Delta w$) [7]. The constant of proportionality, β, is termed the "learning rate," and determines the speed at which learning takes place. The larger the β, the faster the weights change, and vice versa. With repeated exposure to stimuli (learning environment or example data), this mechanism allows for incremental learning from the environment. However, it took several years before this learning was incorporated into the next model neuron—perceptron—which not only became very popular and caused a great stir in the research community, but was also the platform on which many later developments were made in artificial neural networks. The next section will present an example of Hebbian learning as it is applied to perceptron.

Implementation of learning in a neural assembly (perceptron). Research on modeling of the learning process in neural networks dates back to Frank

Rosenblatt, who, during the 1950s, laid the foundation for the field of adaptive neural computing. He recognized that the threshold neuron is not suitable for modeling brain functions such as cognition because it is not flexible enough to learn from and adapt to the environment. He focused on how the brain learns from experience, responds in similar ways to similar experiences, recognizes patterns, groups similar experiences together, and differentiates them from dissimilar experiences, despite the imprecision in initial wiring in the brain [3]. In a landmark paper, Rosenblatt [10] proposed the first neural model, called perceptron, which was capable of learning to classify certain pattern sets as similar or dissimilar by modifying its connections. Essentially, he made threshold neurons learn using Hebbian learning.

Rosenblatt [10,11] used biological vision for his network model. He demonstrated many possible network structures for this task, but this section will look at one that has interesting features (structure and learning) relevant to all the networks that will be studied later in this book. In this configuration, the perceptron network has three layers, as shown in Figure 2.9. The input layer consists of a set of sensory cells in the retina, randomly and partially connected to neurons in the next higher association layer. The association layer neurons are connected bidirectionally in a partial and random manner to neurons in the next higher response layer. With bidirectional connections, association and response neurons can excite or inhibit each other. Moreover, all response neurons are interconnected with inhibitory connections, causing them to competitively inhibit each other by sending inhibitory signals. The neurons in the association and response layers were threshold neurons, with a threshold function set at the origin. The goal of the perceptron network was to activate the correct response neurons for each input pattern class. Learning only happened between the association layer and the response layer. A popular version of this network (single-layer perceptron) is shown in Figure 2.2a, which is presented in Chapter 3 and Chapter 4.

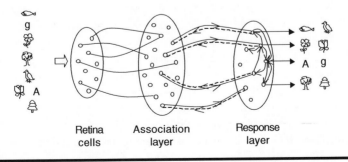

Figure 2.9   A schematic diagram capturing some essential features of a perceptron network that learns through competition.

Rosenblatt [10,11] introduced two important learning concepts—unsupervised and supervised learning—to train perceptron networks. These concepts are still the basis for neural learning today. Their fundamental aspects will be reviewed in the next section in relation to his perceptron model.

2.5.2.2   Unsupervised or Competitive Learning

Competitive learning, also called unsupervised learning, involves a network learning to respond correctly on its own without the involvement of an external agent (actual output). When an input is presented to the perceptron shown in Figure 2.9, various association neurons become active, and they in turn activate some response neurons. The response neuron that receives the largest input inhibits the other cells in the same layer and becomes the winner. The winner produces the network output, which may be an implicit action of perception such as pattern classification. The association neurons that activate the winner response neuron grow in connection strength, and those that do not send signals to it get inhibited and decrease in strength. In this manner, the response neurons become increasingly sensitive to the type of input they initially respond to; over time, various response neurons learn to specialize by responding to specific inputs. This basic idea is currently used in competitive networks and SOFMs, which were briefly discussed earlier and are shown in Figure 2.2d and Figure 2.2e. These networks will be covered in detail in Chapter 8.

2.5.2.3   Supervised Learning

Supervised learning does not involve competition, but uses an external agent (actual output) for each input pattern that guides the learning process. There are several forms of supervised learning. Its simplest form, forced learning, works as follows: At the same time as an input pattern (one input vector) is presented, an appropriate response neuron is forced into action from outside. The active association neurons feeding this neuron will grow in connection strength and, over time, this response neuron becomes more sensitive to that input pattern and learns to classify it correctly without any outside force. Another form of forced learning is reinforcement learning, in which the network receives feedback whether the output is positive or negative and uses this information to improve its response over time. These two methods imply Hebbian learning.

A third, more advanced, form of supervised learning that grew out of the above methods is error correction learning, now generally known as supervised learning, in which the actual value of the correct output is shown to the network and the weights are adjusted until the actual difference between the output of response neurons and correct output becomes acceptable. This idea is more complex than Hebbian learning, and has been developed into more powerful learning methods based on error gradient; it is currently used in the widely popular MLP discussed in detail in Chapter 3 and Chapter 4.

2.5.3 Perceptron with Supervised Learning as a Classifier

The neurons in perceptron are threshold neurons working together. However, the difference is that a perceptron network learns from example data and the weights change during learning. This section examines a simple version of a response neuron without feedback or competition, as shown in Figure 2.10. This neuron receives multiple inputs and processes them to produce an output. One of the inputs can be made equal to 1 and called bias; it can be considered as incorporating the effects that are not accounted for by the input variables feeding the neuron. This is similar to the intercept in regression analysis. Learning requires a set of inputs and the corresponding output(s), which together are called a training set. Thus, the true output class must be known for each input vector prior to learning, and supervised learning involves repeated exposure to training data and iterative modification of the weights that are set to random values, until the model learns to perform the task properly. This process is called training a network.

Processing in this simple perceptron takes place as follows: Weights are initialized with random values. The first input pattern is presented and the inputs are weighted by the corresponding weights, summed, and transformed through a threshold function to produce an output. The threshold function is a unit step function positioned at the origin, such that the perceptron output is either 1 or 0 (on or off). If the weighted sum is greater than or equal to 0, the output is 1; otherwise, it is 0. Consequently, a one-output perceptron can indicate only two output classes (1 or 0). The perceptron output, y, is compared with the target output, t. If the

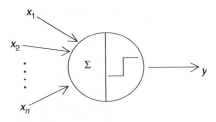

Figure 2.10 A single-neuron perceptron model.

classification result is wrong, the learning algorithm adjusts the weights to map the input data correctly to one of the two classes. The next input pattern is presented, and the process is repeated for all input patterns until learning is complete. This iterative learning process is demonstrated below.

2.5.3.1 Perceptron Learning Algorithm

Assume n input connections representing n input variables $x_1, x_2, ..., x_n$. The input vector, $\mathbf{x}$ is $\{x_1, x_2, ..., x_n\}$. The corresponding weight vector, $\mathbf{w}$, is $\{w_1, w_2, ..., w_n\}$. The net input u for an input vector x is:

$$u = w_1 x_1 + w_2 x_2 + \cdots + w_n x_n. \tag{2.8}$$

The threshold function set at the origin produces an output y, such that

$$y = \begin{cases} 0 & u < 0 \\ 1 & u \geq 0. \end{cases} \tag{2.9}$$

If the classification is correct, the perceptron has classified correctly and the weights are not adjusted. Otherwise, the individual weights are adjusted using a perceptron learning algorithm, which is a modified form of Hebbian learning that incorporates the error as follows:

$$\text{Error} = E = t - y \tag{2.10}$$

where t is the target output. The new value for any single weight is

$$w_{\text{new}} = w_{\text{old}} + \beta \mathbf{x} E \tag{2.11}$$

where $\mathbf{x}$ is the input. This results in the following for the three possible conditions of error, E:

$$w_{\text{new}} = \begin{cases} w_{\text{old}} & E = 0 & (\text{i.e., } t = y) \\ w_{\text{old}} + \beta \mathbf{x} & E = 1 & (\text{i.e., } t = 1, y = 0) \quad \text{Rule 1,} \\ w_{\text{old}} - \beta \mathbf{x} & E = -1 & (\text{i.e., } t = 0, y = 1) \quad \text{Rule 2} \end{cases} \tag{2.12}$$

where w_{new} is the new value of any weight, w_{old} is the old or initial value of the weight, and $\mathbf{x}$ is the input vector. β is the learning rate, a constant between 0 and 1, that adjusts how fast learning should take place. Smaller values indicate a slower weight adjustment, requiring a longer period of time to complete training; larger values accelerate the rate of weight increments. Accelerated weight adjustment is not necessarily better because it may cause the solution (i.e. weights) to oscillate around the optimum, leading to instability, as will be explored in later chapters.

Next, this perceptron learning algorithm will be used in a modified form to train a simple perceptron classifier with two inputs and one output, as shown in Figure 2.11b, in which x_1 and x_2 are input variables and y is the perceptron output. The task is to classify the two-dimensional patterns plotted in Figure 2.11a, which belong to two output classes (Δ denotes class A and ■ denotes class B).

For this example, assume that $\beta = 0.5$. Random initial values will be assigned for the weights; $w_1^0 = 0.8$ and $w_2^0 = -0.5$ (the superscript 0 denotes initial.) Thus the initial weight vector is $W^0 = \{0.8, -0.5\}$, and is superimposed on the data in Figure 2.12. The length L^0 of the initial weight vector, which is denoted by $\|w^0\|$ in vector notation, is

$$L^0 = \sqrt{(w_1^0)^2 + (w_2^0)^2} = \sqrt{(0.8)^2 + (-0.5)^2} = 0.94.$$

Present input pattern 1—A1: $x_1 = 0.3$, $x_2 = 0.7$, $t = 1$; $w_1^0 = 0.8$, $w_2^0 = -0.5$. The weighted input is

$$u = (0.8)(0.3) + (-0.5)(0.7) = -0.11,$$
$$u < 0 \Rightarrow y = 0$$
$$\Rightarrow \text{classification INCORRECT;}$$
$$\Rightarrow \text{weights should be adjusted using Rule 1 in Equation 2.12.}$$

Suppose that the increment of the two weights is denoted by Δw_1^1 and Δw_2^1, the superscript 1 denoting the first weight increment. Then

$$\Delta w_1^1 = \beta x_1 = (0.5)(0.3) = 0.15$$
$$\Delta w_2^1 = \beta x_2 = (0.5)(0.7) = 0.35.$$

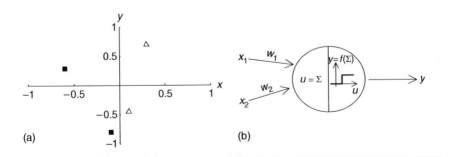

(a) (b)

Figure 2.11 **Perceptron learning: (a) plot of the input data belonging to two classes for perceptron learning and (b) the perceptron configuration for this task.**

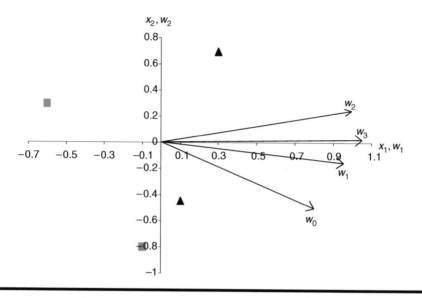

Figure 2.12 Progression of weight adaptation in a two-input perceptron.

New weights after the first increment will be denoted by w_1^1 and w_2^1. Then

$$w_1^1 = w_1^0 + \Delta w_1^1 = 0.8 + 0.15 = 0.95$$
$$w_2^1 = \Delta w_2^0 + w_2^1 = -0.5 + 0.35 = -0.15.$$

In vector form:

$$\mathbf{w}^1 = [w_1^1, w_2^1] = [0.8, -0.5] + [0.15, 0.35] = [0.95, -0.15].$$

Thus the new weight vector is {0.95, −0.15} and the length of the weight vector is

$$\mathbf{L}^1 = \sqrt{(w_1^1)^2 + (w_2^1)^2} = \sqrt{0.95^2 + (-0.15)^2} = 0.96.$$

In Figure 2.12, the initial and modified weights are superimposed on the input data. The wrong classification causes the weights to change. It can be seen that if the perceptron underpredicts (i.e., $t=1$, $y=0$), the weights move closer to the input vector; in this case, they move closer to one that is at the upper right corner.

Reapplying the input vector {0.3, 0.7}:

$$u = (0.3)(0.95) + (0.7)(-0.15) = 0.18 > 0$$
$$\Rightarrow y = 1 \Rightarrow \text{classification CORRECT.}$$

Now, the next input pattern is presented to the modified perceptron:

Present input pattern 2—B1: $x_1 = -0.6$, $x_2 = 0.3$, $t = 0$;
$$w_1^1 = 0.95, \; w_2^1 = -0.15,$$

$$u = (-0.6)(0.95) + (0.3)(-0.15) = -0.615$$
$$u < 0 \Rightarrow y = 0$$
$\Rightarrow$ classification CORRECT $\Rightarrow$ weights are not modified.

Thus the neuron correctly classifies the two input patterns.

Present input pattern 3—B2: $x_1 = -0.1$, $x_2 = -0.8$,
$$t = 0; \; w_1^1 = 0.095, \; w_2^1 = -0.15.$$

$$u = (-0.1)(0.95) + (-0.8)(-0.15) = 0.025$$
$$u > 0 \Rightarrow y = 1$$
$\Rightarrow$ classification INCORRECT

$\Rightarrow$ perceptron over-predicts (i.e., $t = 0$, $y = 1$); weights need adjustment using Rule 2 in Equation 2.12.

Suppose that the increment of the two weights is denoted by Δw_1^2 and Δw_2^2, with the superscript 2 denoting the second weight increment. Then

$$\Delta w_1^2 = -\beta x_1 = -(0.5)(-0.1) = 0.05$$
$$\Delta w_2^2 = -\beta x_2 = -(0.5)(-0.8) = 0.4.$$

Denoting new weights after second increment by w_1^2, and w_2^2:

$$w_1^2 = w_1^1 + \Delta w_1^2 = 0.95 + 0.05 = 1.0$$
$$w_2^2 = w_2^1 + \Delta w_2^2 = -0.15 + 0.4 = 0.25.$$

In vector form

$$\mathbf{w}^2 = \left[w_1^2, w_2^2 \right] = [0.95, -0.15] + [0.05, 0.4] = [1.0, 0.25].$$

The new weight vector $\mathbf{w}^2 = (w_1^2, w_2^2) = \{1.0, 0.25\}$. The magnitude or the length of the new weight vector denoted by $\|\mathbf{w}^2\|$ is

$$\mathbf{L}^2 = \|\mathbf{w}^2\| = \sqrt{1.0^2 + 0.25^2} = 1.03.$$

The new weight vector is superimposed on the data in Figure 2.12, which shows that overprediction (i.e., $t = 0$, $y = 1$) pushes the weight away from that input.

Present input pattern 4—A2: $x_1 = 0.1$, $x_2 = -0.45$, $t = 1$; $w_1^2 = 1.0$, $w_2^2 = 0.25$.

$$u = (0.1)(1.0) + (-0.45)(0.25) = -0.0125$$
$$u < 0 \Rightarrow y = 0 \text{ underprediction, classification INCORRECT}$$
$$\Rightarrow \text{weights need adjustment using Rule 1 in Equation 2.12.}$$

Suppose that the increment of the two weights is denoted by Δw_1^3 and Δw_2^3, with the superscript 3 denoting the third weight increment. Then

$$\Delta w_1^3 = \beta x_1 = (0.5)(0.1) = 0.05$$
$$\Delta w_2^3 = \beta x_2 = (0.5)(-0.45) = -0.225.$$

Denoting new weights after the third increment by w_1^3 and w_2^3:

$$w_1^3 = w_1^2 + \Delta w_1^3 = 1.0 + 0.05 = 1.05$$
$$w_2^3 = w_2^2 + \Delta w_2^3 = 0.25 - 0.225 = 0.025$$

In vector form:

$$\mathbf{W}^3 = \left[w_1^3, w_2^3 \right] = [1.0, 0.25] + [0.05, -0.225] = [1.05, 0.025].$$

The new weight vector $\mathbf{W}^3 = (w_1^3, w_2^3) = \{1.05, 0.025\}$. The magnitude or the length of the weight vector $\mathbf{L}^3$ denoted by $\|\mathbf{W}^3\|$ is

$$\mathbf{L}^3 = \sqrt{1.05^2 + 0.025^2} = 1.05.$$

The new weight vector is superimposed on the data in Figure 2.12, which again shows that underprediction pulls the vector towards the input vector.

Thus the final weight vector is $\mathbf{W}^3 = (w_1^3, w_2^3) = (1.05, 0.025)$. The final weight vector and the all the preceding weights are superimposed on the data in Figure 2.12.

It is now possible to see if all the input patterns are correctly classified using the final weights. The results are presented in Table 2.3, which shows that the perceptron classified all the data correctly.

This example demonstrates how the perceptron learns by adjusting the weights. In this case, learning is complete in one iteration of the dataset due to the linear nature of the problem. One presentation or iteration of all input data is called an epoch. Weights move until the perceptron classifies all the data correctly. Therefore, learning involves iteratively finding the appropriate weights so that the perceptron classifies the data perfectly. For any input vector that the neuron underpredicts, the new weight vector is pulled towards that input, and for any input vector that it overpredicts, the weight vector is pushed away from that input. Also, incremental learning via the perceptron learning algorithm employed here results in an increasing length

Table 2.3 Classification Accuracy of the Trained Perceptron

Input Pattern	x_1, x_2	t	u	y	Classification Accuracy
A1	0.3, 0.7	1	$0.3 \times 1.05 + 0.7 \times 0.025 = 0.3325 > 0$	1	Correct
B1	$-0.6, 0.3$	0	$-0.6 \times 1.05 + 0.3 \times 0.025 = -0.6225 < 0$	0	Correct
B2	$-0.1, -0.8$	0	$-0.1 \times 1.05 + (-0.8) \times 0.025 = -0.125 < 0$	0	Correct
A2	$0.1, -0.45$	1	$0.1 \times 1.05 + (-0.45) \times 0.025 = 0.09375 > 0$	1	Correct

of the weight vector with each weight adjustment. This is not a desirable quality when a large number of input patterns are used in training, and also is not correct from a biological point of view in that the connection strength between biological neurons does not grow infinitely large. In biological neurons, there appear to be corrective mechanisms [12], but for artificial neurons, this can lead to computational problems. This issue is addressed in more advanced networks later.

What are the features of the trained perceptron? Examine the classification boundary that divides the two groups. Recall that the boundary is defined by the position of the threshold function, which in this case is at $u = 0$. (Note that u, being the weighted sum of the inputs, compactly encapsulates all inputs in one single quantity.) Thus, the boundary line can be expressed as

$$u = w_1 x_1 + w_2 x_2 = 0$$

$$x_2 = -\left(\frac{w_1}{w_2}\right) x_1. \tag{2.13}$$

The above equation defines a straight line. Each weight adjustment results in a different classification boundary; therefore, learning is a process that searches a set of weights (w_1 and w_2) that produce the correct classification boundary. The final classification boundary for this problem is found by inserting final weight values into Equation 2.13:

$$x_2 = -(1.05/0.025)x_1 = -42x_1,$$

$$x_2 = -42x_1. \tag{2.14}$$

The slope of this line is -42, which is equal to the tangent of the angle that the slope makes with respect to the horizontal. The inclination of the boundary line to the negative horizontal axis is therefore tan(42), and is equal to 88.6°.

The boundary line is superimposed on the data in Figure 2.13. It can be seen that the boundary line separates the data into two categories with a straight line, and that the linear classification boundary means that the perceptron can be trained to solve any two-class classification problem in which the classes can be separated by a straight-line boundary (i.e., linearly separable). In higher-dimensional problems in which more than two input variables are involved, the classes are still separable by a hyperplane that the reader could intuitively understand; the learning algorithm and the concepts discussed here still apply. However, they cannot be visually demonstrated beyond three dimensions.

Note from Figure 2.13 that the final weight vector is perpendicular to the classification boundary. Therefore, learning finds the optimum weight vector to fix a boundary line perpendicular to itself and that correctly divides the input domain into two regions to yield an accurate classification.

For the benefit of mathematically inclined readers, it can be shown that the weights always define a boundary that is perpendicular to the weights. This can be explained using vector algebra (refer to the Appendix for an introduction to vectors). For example, for an input vector $\mathbf{x} = \{x_1, x_2\}$ and a weight vector $\mathbf{w} = \{w_1, w_2\}$,

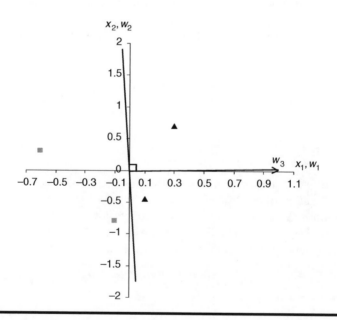

Figure 2.13 **The perceptron classification boundary and the final weight vector superimposed on the data.**

$$u = x_1 w_1 + x_2 w_2,$$

which can be expressed using vectors as

$$u = \|\mathbf{x}\|\|\mathbf{w}\|\cos\boldsymbol{\theta},$$

where $\|\mathbf{x}\|$ and $\|\mathbf{w}\|$ are the magnitude of an input vector and weight vector, respectively, and θ is the angle between the two. Because $u = 0$ at the boundary

$$u = \|\mathbf{x}\|\|\mathbf{w}\|\cos\boldsymbol{\theta} = 0.$$

Because $\|\mathbf{x}\|$ and $\|\mathbf{w}\|$ are not equal to zero, for the above expression to be valid, $\cos\theta$ must be zero, which means that $\theta = \pm 90°$. Because $\cos\theta$ is positive in the range between $-90°$ and $+90°$, any input that makes an angle from $-90°$ to $+90°$ with the weight vector (i.e., lying on one side of the boundary line) will result in $u \geq 0$ and thus $y = 1$. Therefore, the maximum angle between the weight vector and an input vector that belongs to the class with $t = 1$ is $\pm 90°$. Those input vectors yielding an angle greater than $\pm 90°$ (i.e., lying on the other side of the boundary line) will result in $u < 0$ and produce $y = 0$. Thus, the weights fix a boundary line that is perpendicular to themselves.

2.5.3.2 A Practical Example of Perceptron on a Larger Realistic Data Set: Identifying the Origin of Fish from the Growth-Ring Diameter of Scales

A simple but realistic problem will be solved using perceptron. Environmental authorities concerned with the depletion of salmon stocks decided to regulate the catches. To do this, it is necessary to identify whether a fish is of Alaskan or Canadian origin. Fifty fish from each place of origin were caught, and the growth-ring diameter of the scales was measured for the time they lived in freshwater and for the subsequent time they lived in saltwater [13]. The aim is to identify the origin of a fish from its growth-ring diameter in freshwater and saltwater. This section will study how well perceptron classifies the fish as being of Alaskan or Canadian origin.

Figure 2.14 shows the relationship between the two variables, which are separated into two distinct groups. Because there are two inputs, a two-input classifier must be trained. There are 50 input patterns each of Class 1 (Canadian) and Class 2 (Alaskan). Only one output node represents the two classes—with an output of 0 representing one class, and 1 the other. Due to the large number of patterns, only the outcomes are shown, but the process is exactly the same as that studied in the previous section.

The evolution of the classification boundary for the same data at a few selected points in the training process is shown in Figure 2.15, in which more continuous lines show the progressive improvement of the classifier's

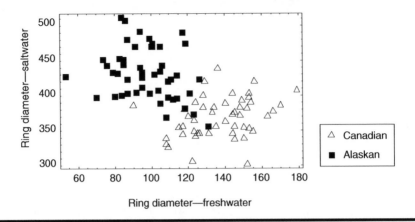

Figure 2.14   The growth-ring diameter of salmon in freshwater and saltwater for Canadian and Alaskan fish.

performance. It can be seen that the initial classification boundary is far from where it should be (at the lower right corner in Figure 2.15) and that learning incrementally repositions the boundary until the correct classification boundary is obtained. The correct classification has put one class of data on one side and the other class of data on the other side.

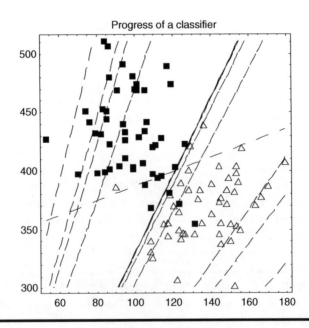

Figure 2.15   The refinement of the classification boundary of a perceptron with training.

Figure 2.16 depicts the training results, showing how classification improves with each epoch. Recall that an epoch is one pass of all input patterns through the perceptron. By now, the reader must know that an input pattern is the same as an input vector, or one record of input data. In many real problems, the input dataset must be repeatedly presented over many epochs, and the weights are adjusted until the correct classification is achieved. This is because each wrongly classified input repositions the boundary line and can offset the changes caused by some previous input patterns. Only after the perceptron has seen the whole dataset numerous times is it able to correctly position the final boundary line. Figure 2.16a shows the number of input patterns classified as Class 1, and Figure 2.16b shows the number classified as belonging to Class 2 after each epoch. The solid line represents the number of patterns actually belonging to the specified class, and the dashed lines show the input patterns being wrongly classified as belonging to that same class. For example, the top image in Figure 2.16 shows that the perceptron learns to classify Class 1 data immediately, but that it wrongly classifies some Class 2 data as belonging to Class 1.

However, the perceptron incrementally learns to classify Class 2 data correctly, and after 80 epochs, it classifies all data correctly, and no input is wrongly classified as belonging to Class 1. The bottom figure of Figure 2.16 shows the classification accuracy for Class 2. In the initial iterations, the perceptron has difficulties differentiating and classifying patterns, but it later

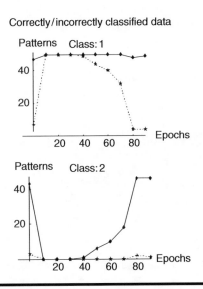

Figure 2.16   The improvement of classification accuracy during training for Class 1 and Class 2.

learns to correctly classify Class 2 data and eliminates the Class 1 data wrongly classified as belonging to Class 2.

Figure 2.17 shows the final classification boundary superimposed on the data, which shows that the perceptron has found the best linear classification boundary for this problem. After training is finished, the accuracy of classification can be assessed. Accuracy can be defined either for each class as the percentage of patterns correctly classified for that class, or for the whole dataset as the percentage of patterns correctly classified across the set. In this example, there were 50 patterns in each class, and the perceptron correctly classified 97 samples, with only three misclassifications. Thus, the classification accuracy for Canadian salmon is 100 percent, and for Alaskan salmon is 94 percent; the overall classification accuracy is 97 percent.

The learning rate, β, specifies how fast the learning takes place. A value of zero indicates no learning, and as the value gets larger, the perceptron learns at a faster rate. Larger values for the learning rate can accelerate the training process; however, they also may induce oscillations that could slow down the convergence to an acceptable solution. It has been common to use a value between 0 and 1.

2.5.3.3 Comparison of Perceptron with Linear Discriminant Function Analysis in Statistics

Discriminant analysis is a multivariate statistical method used to simultaneously analyze the difference between categories in terms of several independent numerical variables. It can also be used as a classifier in which a class is assigned to the values of a set of input variables [14]. Therefore, discriminant function analysis can be used to classify input data.

In discriminant analysis, the centroid or the mean of the categories is calculated and an input is classified as belonging to the class whose center is

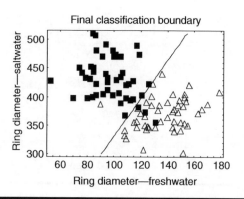

Figure 2.17   The final classification boundary produced by the perceptron.

the closest to the input. This analysis can be done on any commercial statistical software and a linear discriminant function that separates the classes can be obtained from the analysis.

As a classifier, perceptron and linear discriminant analysis are equivalent. To demonstrate this, discriminant analysis was performed on the salmon dataset used for classification using perceptron in the previous section. The resulting classification boundary is shown in Equation 2.18, superimposed on the data in Figure 2.18, along with the perceptron classification boundary given in Equation 2.15a and already shown in Figure 2.17. The dashed line represents the discriminant classifier boundary and the solid line depicts the perceptron classifier.

The perceptron classification boundary is

$$x_2 = 0.026 + 3.31x_1. \tag{2.15a}$$

The discriminant function classification boundary is

$$x_2 = 106.9 + 2.5x_1 \tag{2.15b}$$

Figure 2.18 shows the similarity between the two classifiers. However, the perceptron classification accuracy is slightly better than that of the discriminant function classifier. For the linear discriminant classifier (dashed line in Figure 2.18), the classification accuracy for Alaskan salmon is 90 percent, with five misclassifications, the accuracy for Canadian salmon is 100 percent, and the overall classification accuracy for the entire dataset is 95 percent. These values for the perceptron classifier (solid line in Figure 2.18) were 94 percent, 100 percent, and 97 percent, respectively. This was intended to be a brief comparison of the linear discriminant function classifier

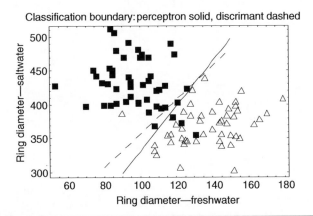

Figure 2.18 Perceptron and linear discriminant classifiers superimposed on the data.

with perceptron. A detailed treatment of other aspects of linear discriminant analysis can be found in books on multivariate statistical analysis [15,16].

2.5.3.4 Multi-Output Perceptron for Multicategory Classification

Classification problems involving more than two output classes can be solved by a multi-output perceptron that has one output neuron for each class in the output layer. The training process for a multi-output perceptron involves mapping each input pattern to the correct output class by iteratively adjusting the weights to produce an output of 1 at the corresponding output neuron and 0 at all the remaining ones. In some problems, it is quite possible that a number of input patterns map to several classes, indicating that they belong to more than one class. This may be due to the nature of data, which could require nonlinear classification boundaries, or to the insufficient complexity of the network structure itself.

This section will now examine the two-dimensional multiclass mapping performance of a perceptron network. Figure 2.19 shows data belonging to three classes [17]; this requires three output perceptron neurons, one for each class, as shown in Figure 2.20. There are 20 data points in each class.

The perceptron network was trained using the perceptron learning algorithm. For a particular input, the trained network activates output neurons with a response of either 0 or 1. The process of training is similar to that used for single perceptron, except for the fact that the required output from the three perceptrons for Class 1 would be $1, 0, 0$, for Class 2 they would be $0, 1, 0$, and for Class 3 they would be $0, 0, 1$, respectively. During training, the actual output of the three perceptrons is compared with the target outputs, and any perceptron yielding incorrect classifications has its weights adjusted using the learning rule over several epochs until all the perceptrons correctly classify the data that they represent and do not misclassify other data as belonging to them.

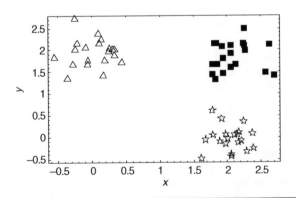

Figure 2.19   Data belonging to three categories.

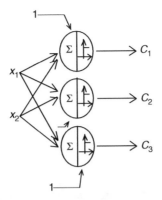

Figure 2.20 Multiple-perceptron network configuration for multiclass classification in which each perceptron represents a class.

The final classifications boundaries obtained from training are superimposed on the data in Figure 2.21. Because each perceptron defines a linear classification boundary, there are three linear boundaries, as shown in this figure. Each classification boundary separates one class from the other two classes. Because this problem is suited to linear classification, the perceptron network correctly classifies all the data.

In Figure 2.21, there are overlapping regions in which two neurons will produce an output of 1. This is because there is no data in those regions with which to train the network. To demonstrate that the network has learned to classify correctly, one input pattern from each class was randomly chosen to be presented to the network. The input and the corresponding response are shown in Table 2.4.

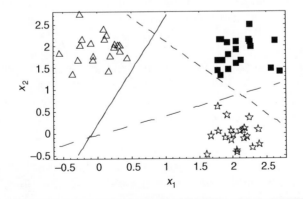

Figure 2.21 The classification boundaries for the three neurons in the multiple-perceptron network shown in Figure 2.20 superimposed on the data.

Table 2.4   Inputs and the Corresponding Output from the Multi-Output Perceptron

Input	Output	Target
$(-0.56, 1.84)$	$(1, 0, 0)$	$(1, 0, 0)$
$(2.23, 2.15)$	$(0, 1, 0)$	$(0, 1, 0)$
$(2.00, -0.04)$	$(0, 0, 1)$	$(0, 0, 1)$

Thus, the trained network classifies all three input patterns. The classification boundary shown in Figure 2.21 for each neuron can be easily constructed from the corresponding weights from the trained network. Recall that each neuron is a threshold neuron and the process that was used to extract the boundary line for one perceptron neuron applies here for each of the three neurons. The extracted equation for each classification line is given as

$$-3.8 - 42.8x_1 + 17.5x_2 = 0$$
$$-78.2 + 26.9x_1 + 24.3x_2 = 0 \qquad (2.16)$$
$$-1.35 + 16.6x_1 - 38.0x_2 = 0.$$

The coefficients in the equations denote the weights w_1 and w_2 for each neuron, and the intercept denotes the weight associated with the bias input of $+1$.

Figure 2.22 shows how classification improves with each epoch for each of the three neurons representing Class 1, 2, and 3, respectively. It shows, for example, that neurons one and three initially misclassify but learn correct classification reasonably quickly, whereas neuron two learns Class 2 data with some difficulty. Neuron two misclassifies a large number of patterns belonging to the other two classes and sheds them slowly over the epochs. After 25 epochs, all three neurons produce the correct classification for the 20 data points represented by each neuron, and produce the individual boundary lines already shown in Figure 2.21 and given in Equation 2.16.

Comparison with multiple discriminant classifier. When there are more than two categories of the dependent variable, multiple linear discriminant analysis can be used. Multiple discriminant analysis results on the same data shown in Figure 2.19 were obtained from SPSS statistical software [18], and are presented in Figure 2.23. In the figure, three Fisher's discriminant functions corresponding to three classes, the canonical discriminant function classifier, and the cluster centroids are superimposed on the data in Figure 2.23a through Figure 2.23c, respectively. As can be seen from Figure 2.23a and Figure 2.22, the multiple Fisher's discriminant functions are functionally similar to the multi-output perceptron classifier, in that each

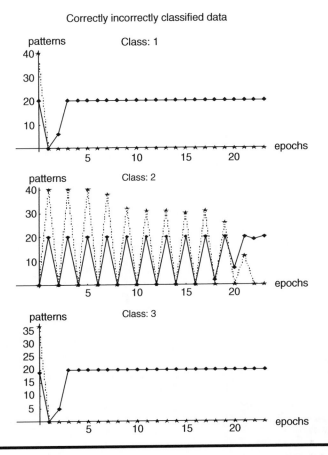

Figure 2.22 The performance of a multi-output perceptron with training epochs.

function attempts to separate one class from the other two classes. Generally, two discriminant functions can classify three classes, and these functions can be obtained from the Fisher's discriminant functions. Canonical discriminant functions yield classifier such as shown in Figure 2.23b. Overall, perceptron performance is perfect in this example and discriminant analysis performance is slightly inferior. The perceptron does not find the cluster centroids as given by canonical function coefficients (Figure 2.23c) in statistical analysis.

The equations for the three Fisher's discriminant functions are

$$-31.4 + 29.7x_1 - 0.953x_2 = 0$$
$$-49.5 + 30.0x_1 + 18.5x_2 = 0 \qquad (2.17)$$
$$-20.6 - 0.231x_1 + 20.1x_2 = 0$$

Each function in Equation 2.17 corresponds to a class.

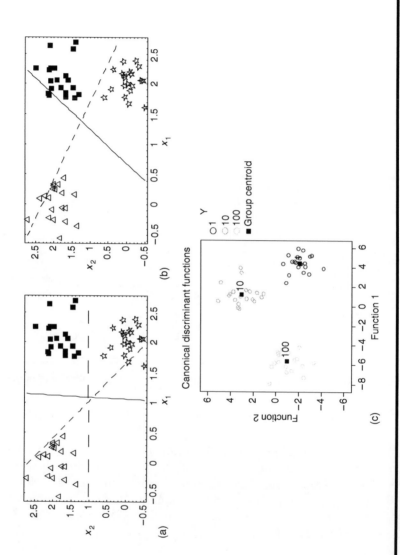

Figure 2.23 Discriminant analysis on data: (a) Fisher's multiple discriminant functions superimposed on the data, (b) canonical discriminant function classifier, (c) cluster centroids given by canonical discriminant functions.

2.5.3.5   *Higher-Dimensional Classification Using Perceptron*

The perceptron network can be extended to higher-dimensional classification when the dimensions of the input pattern are larger than two. This is done by adding the appropriate number of inputs and the corresponding connection weights. However, it is no longer possible to illustrate the results for dimensions higher than three, although various two-dimensional projections of data can still be viewed. Classification plots similar to Figure 2.22 can still be obtained, and the results can be assessed as for two-dimensional classification.

As can be deduced from the previous two comparisons between the perceptron and the discriminant function analysis, the higher-dimensional perceptron classifier is functionally equivalent to the multiple discriminant function analysis for multidimensional data.

If the classification problem is not linearly separable, then it is impossible for the perceptron to classify all patterns correctly. If some misclassifications are acceptable, then the perceptron could still be a good linear classifier. However, the perceptron linear classifier is often inadequate as a model for many nonlinear problems. Linearity comes from the classification boundary $u = 0$ associated with the threshold function. This simplicity of the perceptron came under attack by Minsky and Papport [19], but later advances in MLP networks led to the development of models that perform complex nonlinear classification and prediction tasks, shown earlier in Figure 2.1; more details of this will be seen later in this book. Nevertheless, perceptron could still apply to many classification problems that have simple solutions, or even to nonlinear problems that can be transformed to linear problems. More importantly, the simple learning mechanism in perceptron can provide important insights and shed some light onto the issues involved in developing more complex neural network models.

2.5.3.6   *Perceptron Summary*

The perceptron learning presented in the above sections can be summarized as follows: Inputs are weighted and processed by a threshold function and, therefore, the output is bounded between 0 and 1. The perceptron learns if its classification is wrong. If it underpredicts, it compensates by moving the weight vector towards the input and vice versa, using a modified form of Hebbian learning called perceptron training algorithm, which incorporates a learning rate that controls the rate of weight adjustment. The perceptron learns incrementally until the input patterns are classified correctly. The classification boundary defined by the perceptron is perpendicular to the weight vector. More importantly, the classification boundary is linear and, therefore, the perceptron can classify only linearly separable patterns, i.e., patterns that can be separated by a straight line, as shown in Figure 2.17.

Perceptron can be used for two or multicategory classification involving two or more inputs. In all these aspects, the perceptron network classifier is functionally equivalent to the multiple discriminant function analysis in multivariate statistics, and can even outperform it in some cases.

2.5.4 Linear Neuron for Linear Classification and Prediction

To this point, the neurons studied in this book have had only two outputs (0 or 1). In a linear neuron, output is continuous, i.e., it can take many values. Linear means that output is a linear sum of weighted inputs, and thus the activation function in the neuron is linear, as opposed to a threshold in perceptron. This is depicted in a neuron model with several inputs and a bias input that is equal to one, as shown in Figure 2.24. The linear output makes this neuron capable of both linear classification and, more importantly, linear function approximation. Widrow and Hoff [20] developed the first adaptive linear neuron model (ADALINE) and, for the first time, implemented supervised error correction learning, known as gradient descent or the delta rule, in neural learning. It is implemented such that the square error between the target and the network output is minimized. This concept is called least square error minimization, and is also a criterion used in many statistical methods.

The linear neuron shown in Figure 2.24 has two key attributes: (1) It uses supervised learning as in perceptron, which means the neuron must have the desired (target) output for the inputs on which it is trained, and (2) learning is based on the delta rule or gradient descent, which adjusts the weights in the direction in which error goes down most steeply. It was developed as an engineering model, and in its original design, inputs were either 1 or -1 and outputs were also 1 or -1, so it was used as a classifier. This section, however, looks at both the classification and predictive capabilities of a general linear neuron to explore some of its broader capabilities and to examine how it is trained using delta rule.

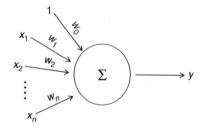

Figure 2.24 A linear neuron model.

A linear neuron processes data as follows. For a given input pattern, x_1, $x_2, ..., x_n$, with corresponding weights $w_1, w_2, ..., w_n$, and a bias input of $+1$ with corresponding weight w_0, it first calculates the net input u as before:

$$u = \sum_{1}^{n} w_i x_i + w_0 = w_1 x_1 + w_2 x_2 + \cdots + w_n x_n + w_0. \qquad (2.18)$$

Linear transformation results in

$$y = u;$$

therefore, mathematically, the linear neuron gives rise to a linear model that has the following simple equation for the output:

$$y = w_1 x_1 + w_2 x_2 + \cdots + w_n x_n + w_0. \qquad (2.19)$$

Thus the linear neuron model is analogous to multiple linear regression models in statistics. In statistical terminology, w_0 is the intercept, which accounts for factors that are not accounted for by the input variables; in neural networks terminology, it is called a bias and has the same meaning. The output is said to regress on the inputs $x_1, x_2, ..., x_n$. It is possible to make this neuron a classifier by restricting the output to 1 if u is greater than or equal to 0, and 0 otherwise, as was done in perceptron. Both of these capabilities will be examined after an exploration of how supervised learning with the delta rule works.

2.5.4.1 Learning with the Delta Rule

This section will examine how to train a linear neuron using the delta rule. For simplicity, the section will begin with the derivation of the delta rule for a linear neuron with one input, x, and without a bias input, as shown in Figure 2.25. The target output is t and the predicted output is y.

Suppose the neuron is presented with an input pattern. It first calculates the net input u and output y as before:

$$\begin{aligned} u &= w_1 x \\ y &= u = w_1 x. \end{aligned} \qquad (2.20)$$

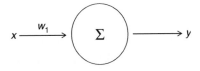

Figure 2.25 A one-input linear neuron.

The error E is

$$E = t - y = t - w_1 x \tag{2.21}$$

and the square error ε for an input pattern is

$$\varepsilon = \frac{1}{2}E^2 = \frac{1}{2}(t - w_1 x)^2. \tag{2.22}$$

The indicator of error used in the delta rule is the square of the error, which is preferred to simple error because squaring eliminates the sign of the error, which could cause a problem when the error is summed over all input patterns. The fraction ½ is arbitrary, and is used for mathematical convenience. It is used in the statistics community, but not in some engineering and computing fields. The square error function has a parabolic shape in relation to the weight, as shown in Figure 2.26.

To draw this curve using the above equation, the square error for a range of values of w has been calculated for a fixed pair of values for input x and target output t of 2.0 and 1.53, respectively, whose values are not important in this demonstration of the relationship between the square error and the weight. It can be seen that the optimum weight is at the bottom of the bowl, and that the delta rule proposes to reach it from the initial weight values by descending down the bowl in the opposite direction to the gradient during training. The solid arrow in the figure represents the gradient at an arbitrary initial weight of 1.28. To find the gradient at a particular weight after each presentation of an input, it is necessary to differentiate the square error function with respect to w, yielding

$$\frac{d\varepsilon}{dw_1} = \frac{2}{2}(t - y)(-x) = -Ex \tag{2.23}$$

which gives the magnitude of the error gradient and the direction in which the error increases most rapidly at the current weight value, as indicated by

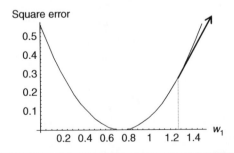

Figure 2.26 Parabolic square error curve for a one-input linear neuron.

the arrow in Figure 2.26. According to the delta rule, weight change is proportional to the negative of the error gradient because it is necessary to go down the error curve to minimize the error; therefore, the weight change is

$$\Delta w_1 \propto Ex \tag{2.24}$$

where $\propto$ denotes proportionality, which can be replaced with learning rate, β, as

$$\Delta w_1 = \beta\, Ex. \tag{2.25}$$

The learning rate determines how far along the gradient it is necessary to move to fix the new weights. Larger values of β accelerate the weight change and smaller values slow it down. The new weight after the ith iteration can be expressed as

$$w_1^{i+1} = w_1^i + \Delta w_1 = w_1^i + \beta\, Ex. \tag{2.26}$$

Using the same logic, it is possible to extend this idea for a multiple-input neuron with a bias input; the new weights for this situation would be

$$
\begin{aligned}
w_0^{i+1} &= w_0^i + \beta\, E \\
w_j^{i+1} &= w_j^i + \beta\, x_j E,
\end{aligned}
\tag{2.27}
$$

where w_0 is the bias weight and w_j is the weight corresponding to input x_j. In this case, the square error function is a multidimensional surface with respect to weights. However, by slicing it with respect to each weight axis, the square error and individual weight relationship can still be viewed as one-dimensional curves or two-dimensional surfaces.

Example-by-example learning versus batch learning. Recall that one pass of the whole training dataset is called one epoch, and that training can take many epochs to complete learning. Learning can be performed after each input pattern (iteration), or after an epoch. Adjusting the weights after each presentation of an input pattern is called example-by-example learning. For some problems, this can cause weights to oscillate due to the fact that the adjustment required by one input vector may be canceled by that of another input; however, this method works well for some other problems. Epoch or batch training is more popular because it generally provides stable solutions.

Batch learning. In many situations, it is preferable to wait until all the input patterns (or some portion of them) have been processed and then adjust weights in an average sense. This is called batch learning. A batch can be the entire training dataset (epoch-based learning) or some portion of the dataset. Generally, the goal is to reduce the average error over all the patterns, which is called the mean square error (MSE) and can be expressed as

$$\text{MSE} = \frac{1}{2n} \sum_{i=1}^{n} E_i^2 \tag{2.28}$$

where E_i is error for ith input pattern, and n is the total number of input patterns in the batch. If the weights are adjusted after all the patterns have been processed, this approach is called epoch-based learning, and there is only one weight adjustment after each epoch. The basic idea with this method is to obtain the error gradient for each input pattern as it is processed, average them at the end of the epoch, and use this average value to adjust the weights using the delta rule. For the linear neuron this results in a weight increment of

$$\Delta w_1 = \beta \left(\frac{1}{n} \sum_{i=1}^{n} E_i x_i \right) \tag{2.29}$$

after an epoch. This can facilitate a smoother climb down the error surface for many problems.

In summary, supervised learning using the delta rule is implemented as follows: An input pattern $(x_1, x_2, ..., x_n)$ is transmitted through connections whose weights are initially set to random values. The weighted inputs are summed, the output y is produced, and y is compared with the given target output (t) to determine error (E) for this pattern. Inputs and target outputs are presented repeatedly, and the weights are adjusted using the delta rule at each iteration or after an epoch until the minimum possible square error is achieved. This may involve the iterative presentation of the entire training dataset many times.

The delta rule greatly simplifies for a linear neuron and, in example-by-example learning, it is similar in form to the perceptron learning rule, except that the method of computing the error is different for the two methods. However, the two methods have different origins and the gradient descent concept in the delta rule has a far greater implication for modeling more complex networks than the simple perceptron rule, as will be discussed later. This is due to the delta rule being more meaningful and powerful because it follows the error curve in the direction in which error decreases most. More importantly, it has made it possible to train a neuron to learn continuous linear or nonlinear mapping of inputs to output(s). In the above derivation, only linear mapping of x to y was examined, but the delta rule is especially useful in the nonlinear mapping of a set of inputs to output(s), as discussed in Chapter 3 and Chapter 4.

2.5.4.2 Linear Neuron as a Classifier

This section will examine how a linear neuron can be trained as a classifier using the delta rule. It is possible to convert the linear neuron into a classifier by passing the output through a threshold function, as shown in Figure 2.27b. The error, however, is calculated based on the linear output, y. Because the output, y, of the linear neuron is continuous, the output of the classifier is specified as

$$y' = \begin{cases} 1 & y \geq 0 \\ 0 & y < 0. \end{cases} \tag{2.30}$$

This section will demonstrate how to train a linear neuron to classify the data shown in Figure 2.27a. The same data was used earlier in Section 2.5.3 to train a perceptron. Because there are two input variables, it is necessary to use a linear neuron with two inputs and one output, as shown in Figure 2.27b.

The learning process for example-by-example learning is

$$u = w_1 x_1 + w_2 x_2$$
$$y = u$$
$$E = t - y \tag{2.31}$$
$$\Delta w = \beta \, Ex$$
$$w_{\text{new}} = w_{\text{old}} + \Delta w.$$

Assume that the initial weights are $w_1 = 0.8$, $w_2 = -0.5$, and $\beta = 0.5$, as were used for the perceptron.

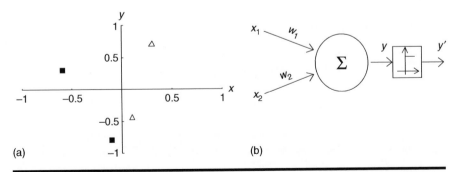

(a)

(b)

Figure 2.27 Classification with a linear neuron: (a) two-dimensional data and classes and (b) two-input linear neuron classifier.

Present input pattern 1—$A1$: $x_1 = 0.3$, $x_2 = 0.7$, $t = 1$; $w_1^0 = 0.8$,
$$w_2^0 = -0.5.$$
$$u = (0.3)(0.8) + (0.7)(-0.5) = -0.11$$
$$y = u = -0.11 < 0 \Rightarrow \text{Category 0}$$
$$\Rightarrow \text{WRONG CLASSIFICATION}$$
$$E = 1 - (-0.11) = 1.11.$$

The first weight increments are

$$\Delta w_1^1 = (0.5)(1.11)(0.3) = 0.1665$$

$$\Delta w_2^1 = (0.5)(1.11)(0.7) = 0.3885.$$

The new weights after the first increment are

$$w_1^1 = 0.8 + 0.1665 = 0.9665$$

$$w_2^1 = -0.5 + 0.3885 = -0.1115.$$

Thus, the new weight vector $\mathbf{w}^1 = [0.9665, -0.1115]$.

Present input pattern 2—$B1$: $x_1 = -0.6$, $x_2 = 0.3$, $t = 0$; $w_1^1 = 0.9665$,
$$w_2^1 = -0.1115.$$

$$u = (-0.6)(0.9665) + (0.3)(-0.1115) = -0.61335$$
$$y = u = -0.61335 < 0 \Rightarrow \text{Category 0}$$
$$\Rightarrow \text{CORRECT CLASSIFICATION} \Rightarrow \text{WEIGHTS DO NOT CHANGE.}$$

Present input pattern 3—$B2$: $x_1 = -0.1$, $x_2 = -0.8$, $t = 0$; $w_1^1 = 0.9665$,
$$w_2^1 = -0.1115.$$

$$u = (-0.1)(0.9665) + (-0.8)(-0.1115) = -0.00745$$
$$y = u = -0.00745 < 0 \Rightarrow \text{Category 0}$$
$$\Rightarrow \text{CORRECT CLASSIFICATION} \Rightarrow \text{WEIGHTS DO NOT CHANGE.}$$

Present input pattern 4—$A2$: $x_1 = 0.1$, $x_2 = -0.45$, $t = 1$; $w_1^1 = 0.9665$,
$$w_2^1 = -0.1115.$$

$$u = (0.1)(0.9665) + (-0.45)(-0.1115) = 0.1468$$
$$y = u = 0.1468 > 0 \Rightarrow \text{Category 1}$$
$$\Rightarrow \text{CORRECT CLASSIFICATION} \Rightarrow \text{WEIGHTS DO NOT CHANGE.}$$

The linear neuron classifies all four patterns correctly. The results demonstrate that for this simple classification task, the linear neuron finds the best weights using the delta rule more quickly than the perceptron. This is because the error is calculated based on the linear neuron output, y, not on the threshold function output, y'.

The classification boundary for the trained linear neuron is given by $y = 0$. Because $y = u$

$$y = u = w_1 x_1 + w_2 x_2 = 0$$
$$x_2 = -(w_1 x_1)/w_2$$

(2.32)

For the neuron trained above, the final weights are $w_1 = 0.9665$ and $w_2 = -0.115$. Thus, the boundary is

$$x_2 = -(0.9665 x_1)/(-0.1115)$$
$$x_2 = 8.67 x_1.$$

(2.33)

This boundary line (dashed line) is superimposed on the data in Figure 2.28, along with the classification boundary for the perceptron (solid line).

The boundary lines found for the linear neuron and the perceptron have different slopes. This is because the two methods are fundamentally different and the data is sparse. With more data they should produce identical results.

The linear neuron can be trained to solve any two-class classification problem in which the classes are linearly separable. The linear neuron finds the best weights using the delta rule more quickly than the perceptron. In higher-dimensional problems in which more than two input variables are involved, the classes are separable by a hyperplane that the reader could intuitively understand; the learning concept applies without change.

2.5.4.3 Classification Properties of a Linear Neuron as a Subset of Predictive Capabilities

Classification is only half the capability of a linear neuron. In the above section, a classifier was trained, but a linear neuron is more capable and can also be a predictor. To highlight this, it is useful to look at the general form of

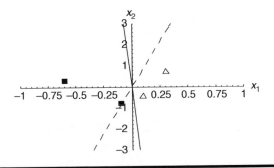

Figure 2.28 Linear neuron and perceptron classification boundaries super-imposed on the data (linear neuron: dashed line; perceptron: solid line).

the input–output relationship established by the linear neuron to put classification into context and to understand its predictive capability, as well as the effect of learning, i.e., weight change, on the output. The output of the one-input one-output neuron trained in Section 2.5.4.2 is

$$y = w_1 x_1 + w_2 x_2, \tag{2.34}$$

where y is a plane called the solution plane and shown in Figure 2.29 for the linear neuron trained in the previous section. Here, w_1 is the slope with respect to the x_1 axis, and w_2 is the slope with respect to the x_2 axis. The solution plane has a unique output for each input vector $\{x_1, x_2\}$. In fact, this solution plane is used to classify patterns in classification problems. The classification boundary previously established is drawn in the figure as the dashed line that the solution plane cuts through the x_1, x_2 plane. This is the same as the dashed line representing the classification boundary for the linear neuron shown in the x_1, x_2 plane in Figure 2.28. Thus, unlike in perceptron, the classification boundary now lies in a continuous solution plane. What is seen here is that weight adjustment alters the slope of the solution plane until the error between the target and the predicted output is minimized. For classification, positive outputs ($y > 0$) classify the patterns into one category, and negative outputs ($y < 0$) classify the patterns into the other categories; this again produces results comparable to the linear discriminant analysis in statistics.

It can be seen that the linear neuron is capable of the continuous mapping of inputs to outputs, which is required in prediction or function approximation. Classification is a simplified form of prediction. The next section examines in detail the predictive capabilities of a linear neuron.

2.5.4.4   Example: Linear Neuron as a Predictor

The linear neuron can approximate a linear function, as was shown in Equation 2.34, and is functionally similar to linear regression in statistics. In a

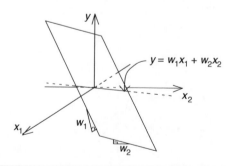

Figure 2.29   The classification boundary superimposed on the linear neuron solution plane.

Table 2.5 Linear Function Data

x	t
0	0
1.0	0.75
2.0	1.53
3.0	2.34
4.0	3.2

function approximation situation, the actual magnitude of the output (y) for a given input pattern is of concern, not the category or classification boundary. Specifically, during training, it is the linear function that minimizes the prediction error is sought. This section presents a simple example in which a linear neuron is used to fit a linear function to the data shown in Table 2.5 and plotted in Figure 2.30, in which there is one input and one output. The data was generated from the function $y = 0.8x$, and a small noise was added to the data to introduce variability. However, the linear neuron does not know that the data came from this function, and its task is to find it iteratively from the data. Because each input has a unique output, the linear neuron learns continuously until the error decreases to zero or an acceptable limit.

For simplicity, the bias will be disregarded because the desired result is a function that goes through the origin. Thus, there is one weight and the linear neuron output takes the form of

$$y = wx. \tag{2.35}$$

In this case, the weight defines the slope of the straight line and by beginning with any initial random value for the weight, it should eventually settle down to 0.8. Thus, learning requires finding the accurate value of the

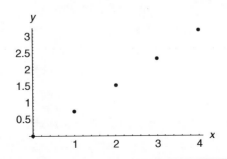

Figure 2.30   Plot of the linear data.

weight (0.8) that fixes the slope of the line. The computation process for example-by-example learning is now

$$y = wx$$
$$E = t - y = t - wx$$
$$\Delta w = \beta Ex$$
$$w_{new} = w_{old} + \Delta w.$$

(2.36)

If batch learning is employed, Δw is defined as $\beta \left(\dfrac{1}{n} \sum_{i=1}^{n} E_i x_i \right)$. A random value of 0.5 is assigned to the initial weight and it is assumed that the learning rate is 0.1.

Example-by-example learning (recursive or on-line learning). The response to the first input, $x = 1$, is

$$y = 0.5 \times 1 = 0.5.$$

The target output is 0.75. Thus, the error, E, is

$$E = t - y = 0.75 - 0.5 = 0.25$$
$$w^1 = w + \beta xE = 0.5 + 0.1 \times 0.25 \times 1 = 0.5 + 0.025 = 0.525.$$

The weight has now been adjusted to 0.525 and the next input, $x = 2$, can be presented to the neuron. The target for this input is 1.53. The predicted output is

$$y = 0.525 \times 2 = 1.05$$
$$E = t - y = 1.53 - 1.05 = 0.48$$
$$w^2 = 0.525 + 0.1 \times 0.48 \times 2 = 0.525 + 0.096 = 0.621.$$

The presentation of the next input, $x = 3$, for which $t = 2.34$, results in

$$y = 0.621 \times 3 = 1.863$$
$$E = 2.34 - 1.863 = 0.477$$
$$w^3 = 0.621 + 0.1 \times 0.477 \times 3 = 0.621 + 0.1431 = 0.7641.$$

The last input, $x = 4$ with $t = 3.2$ yields

$$y = 0.7641 \times 4 = 3.0564$$
$$E = 3.2 - 3.0564 = 0.1436$$
$$w^4 = 0.7641 + 0.1 \times 0.1436 \times 4 = 0.7641 + 0.05744 = 0.8215.$$

The final weight after processing the four inputs is very close to the required 0.8. During learning, the weight changes steadily in increments, from 0.525, 0.621, 0.7641, to the final value of 0.8215.

The section will now train the neuron on the same data using batch learning to demonstrate how batch learning is implemented.

Batch learning. The response to the first input, $x_1 = 1$, is

$$y_1 = 0.5 \times 1 = 0.5.$$

The error for the first input pattern is

$$E_1 = t - y_1 = 0.75 - 0.5 = 0.25.$$

The response to the second input, $x_2 = 2$, is

$$y_2 = 0.5 \times 2 = 1.0.$$

The error for the second input pattern is

$$E_2 = t - y_2 = 1.53 - 1.0 = 0.53.$$

The response to the third input, $x_3 = 3$, is

$$y_3 = 0.5 \times 3 = 1.5.$$

The error for the third input pattern is

$$E_3 = t - y_3 = 2.34 - 1.5 = 0.84.$$

The response to the last input, $x_4 = 4$, is

$$y_4 = 0.5 \times 4 = 2.0.$$

The error for the last input pattern is

$$E_4 = t - y_4 = 3.2 - 2.0 = 1.2.$$

The weight increment Δw^1 after first epoch is

$$\Delta w^1 = \beta \left(\frac{1}{n} \sum_{i=1}^{n} E_i x_i \right) = 0.1 \times \left[\frac{1}{4} (E_1 x_1 + E_2 x_2 + E_3 x_3 + E_4 x_4) \right]$$

$$= 0.1 \times \left[\frac{1}{4} (0.25 \times 1 + 0.53 \times 2 + 0.84 \times 3 + 1.2 \times 4) \right]$$

$$= 0.2158.$$

The new weight, w^1, after first increment is

$$w^1 = w + \Delta w = 0.5 + 0.2158 = 0.7158.$$

The same process is repeated for next epoch, with the weight adjusted to 0.7158; the results are

$$y_1 = 0.7158 \qquad E_1 = 0.0342$$

$$y_2 = 1.4316 \qquad E_2 = 0.0984$$

$$y_3 = 2.1474 \qquad E_3 = 0.1926$$

$$y_4 = 2.8632 \qquad E_4 = 0.3368$$

$$\Delta w^2 = 0.0539$$

$$w^2 = 0.7697.$$

The weight has now been adjusted to 0.7697; the results for the third epoch are

$$y_1 = 0.7697 \qquad E_1 = -\,0.0197$$

$$y_2 = 1.5394 \qquad E_2 = -\,0.0094$$

$$y_3 = 2.3091 \qquad E_3 = 0.0309$$

$$y_4 = 3.0788 \qquad E_4 = 0.1212$$

$$\Delta w^3 = 0.013475$$

$$w^3 = 0.783175.$$

The neuron has settled to a final weight of 0.821 after four iterations in example-by-example learning, and 0.7832 after three epochs in batch learning with the four inputs. The final weight is very close to the target of around 0.8. Note that with batch learning, the first weight increment itself brings the weights up close to the target in one epoch (0.7158) and in the next epochs the weight is fine tuned and adjusted much more slowly. This is because the average error over an epoch contains information about the entire dataset. On the other hand, in example-by-example learning the weight adjustment is gradual from the beginning to the end because the neuron is incrementally exposed to the entire dataset. Further training with the same data will not substantially alter the weight, but will fine tune it around this value. You may try training with another epoch and check the results yourself.

The equation of the line thus derived from example-by-example learning is

$$y = 0.821x \qquad\qquad (2.37)$$

and is superimposed on the data in Figure 2.31.

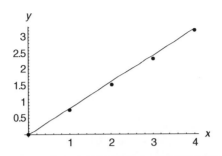

Figure 2.31 The linear neuron output superimposed on the data.

It is now possible to examine the overall prediction error in the form of MSE over all the input patterns by averaging the error

$$MSE = \frac{1}{2n} \sum_{i=1}^{n} E_i^2 \qquad (2.38)$$

where i indicates the pattern number and n is the total number of input patterns.

First compute the square error for each input pattern and then find the average. The input, predicted output, target output, and square error over each pattern are

$$x_1 = 1, \; y_1 = 0.821, \; t_1 = 0.75, \; (t_1 - y_1)^2 = (0.75 - 0.821)^2 = 0.00504$$

$$x_2 = 2, \; y_2 = 0.821 \times 2 = 1.642, \; t_2 = 1.53, \; (t_2 - y_2)^2 = (1.53 - 1.642)^2 = 0.0125$$

$$x_3 = 3, \; y_3 = 0.821 \times 3 = 2.463, \; t_3 = 2.34, \; (t_3 - y_3)^2 = (2.34 - 2.463)^2 = 0.0151$$

$$x_4 = 4, \; y_4 = 0.821 \times 4 = 3.284, \; t_4 = 3.2, \; (t_4 - y_4)^2 = (3.2 - 3.284)^2 = 0.0071,$$

$$MSE = (0.00504 + 0.0125 + 0.0151 + 0.0071)/(2 \times 4) = 0.00497.$$

Thus, the average predicted error of the trained neuron over all the input patterns is 0.00495, which is acceptably small.

It is possible to visualize the square error for an input pattern in graphical format:

$$E = t - y = t - wx$$
$$\varepsilon = E^2 = (t - wx)^2. \qquad (2.39)$$

For convenience, $\frac{1}{2}$ has been removed from the square error. For a particular input, for instance $x = 1$, $t = 0.75$, the error function becomes

$$\varepsilon = (0.75 - w)^2 \qquad (2.40)$$

and Figure 2.32 illustrates this graphically, where the abscissa represents the weight and the ordinate represents the square error. The error function takes

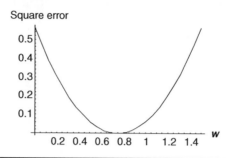

Figure 2.32 **Error surface for a one-input linear neuron without bias.**

the form of a parabola, and the error is at a minimum for the weight that indicates the lowest point on this parabola. In the example, it is seen that the delta learning rule involves descending the gradient of this error curve, incrementally searching the weight that minimizes the prediction error. It has found the target weight value in one epoch.

In the case where there is a bias (or intercept), two weights, w_0 and w_1, must be found to minimize the square error. The w_0 is the weight associated with the bias input which is typically $+1$, and w_1 is that associated with the input x. The output of a linear neuron for this case is

$$y = w_0 + w_1 x. \tag{2.41}$$

The square error function is

$$\varepsilon = E^2 = [t - (w_0 + w_1 x)]^2. \tag{2.42}$$

To illustrate this function graphically, an intercept of approximately 1 will be introduced to the linear function by shifting all the data points vertically by one unit. Now, for $x = 4$, the target is 4.2 (i.e., $3.2 + 1$). It is possible to plot the error function for this input–output pair for a range of w_0 and w_1 values as

$$\varepsilon = E^2 = [4.2 - (w_0 + 4w_1)]^2, \tag{2.43}$$

as shown in Figure 2.33.

The error function is a two-dimensional surface with the minimum error at approximately $w_0 = 1$ and $w_1 = 0.8$. These correspond to the intercept and the slope, respectively, of the linear function in Equation 2.41.

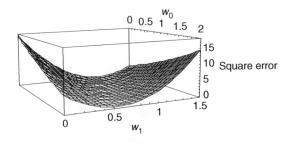

Figure 2.33   Error surface for a one-input neuron with bias.

2.5.4.5   A Practical Example of Linear Prediction: Predicting the Heat Influx in a Home

This section will solve a practical example using a linear neuron model to estimate the heat influx into a home. It is known that many factors, such as insulation, northern, southern, and eastern aspect, etc., affect heat influx; however, of all these factors, northern exposure has been found to be the most important. We are going to model the relationship between heat influx and northern exposure on data collected to study this behavior [13]. Due to the large number of observations involved, only the final results are shown here. A linear neuron with northern exposure and a bias input was trained using the delta rule, starting with random initial weights, until the error reached the minimum possible level. The resulting model is superimposed on the data in Figure 2.34.

The simpler linear relationship made it possible to find the optimum weights in one epoch; they were found to be 607 (bias) and −21.4. Therefore, the linear neuron model fits the data well, with a slope of −21.4 and an intercept of 607. Note that for visual clarity, the origin of the axes in the plot is not set at $(0, 0)$, hence the smaller intercept. The relationship between the heat influx and the northern exposure is

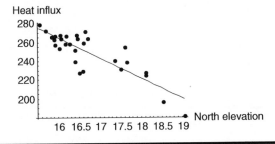

Figure 2.34   Linear neuron model predictions of heat influx superimposed on the data.

therefore modeled by

$$\text{Heat Influx} = 607 - 21.4 \text{ Northern Elevation} \qquad (2.44)$$

with a coefficient of determination (R^2) of 0.72 and MSE of 152.

2.5.4.6 Comparison of Linear Neuron Model with Linear Regression

As a predictor, the linear neuron is functionally equivalent to linear regression in statistics. To demonstrate this, the same heat influx problem was solved using linear regression. In linear regression, the coefficients (intercept and slope) that minimize the sum of square error for the whole sample are sought using one pass of the dataset. The problem was solved using Minitab statistical software [13], and the analysis of variance (ANOVA) results are shown in Table 2.6.

The slope and intercept from this analysis are -21.4 and 607, respectively, which are identical to the results obtained from the linear neuron model. The R^2, which is a measure of the amount of variation of the dependent variable accounted for by the model (i.e., the ratio of the sum of the squares for the model to the total sum of the squares) is 0.72, and MSE is 152; these results are again identical to those obtained from the neuron model. The statistical methods used to ascertain the significance of the coefficients can be used to test the significance of those given by the linear neuron. The linear neuron does not make any assumptions about the distribution of the data, whereas linear regression assumes that the variables are normally distributed and that the variance of the dependent variable is uniform across the range of the independent variables (homoskedasticity).

Table 2.6 ANOVA Table for Heat Influx–Northern Aspect Relationship

		Estimate	*SE*	*T Stat*	*P Value*
Parameter Table®	1	607.103	42.9061	14.1496	5.24025×10^{-2}
	x	-21.4025	2.56525	-8.34323	5.93501×10^{-9}

RSquared® 0.720524, Adjusted RSquared® 0.710173, Estimated Variance® 151.972

		DF	*Sum of Sq*	*Mean Sq*	*Fratio*	*P Value*
ANOVA Table®	Model	1	10 578.7	10 578.7	69.6094	5.93501×10^{-9}
	Error	27	4103.24	151.972		
	Total	28	14 681.9			

When statistical methods are used to test the significance of the coefficients of neuron models, however, the concept of sampling distribution should apply.

2.5.4.7   Example: Multiple Input Linear Neuron Model—Improving the Prediction Accuracy of Heat Influx in a Home

To improve the accuracy of the heat flux prediction in the problem addressed in the previous section, the southern aspect was included as a second variable, as it is thought to be the second most important predictor variable. This necessitates the use of a two-input (and bias) neuron, which produces a model in the form

$$y = b_0 + w_1 x_1 + w_2 x_2. \tag{2.45}$$

Training was completed using a linear neuron with random initial weights; the final bias weight and the weights associated with the inputs were found to be 483.7, -24.21, and 4.79, respectively, resulting in the following model:

Heat Influx $= 483.7 - 24.21$ North Elevation $+ 4.79$ South Elevation.

$$\tag{2.46}$$

This model has an R^2 of 0.86, which is an improvement over the one-input model; the MSE for this model is reduced to 79.78. The optimum weights were found in one epoch.

2.5.4.8   Comparison of a Multiple-Input Linear Neuron with Multiple Linear Regression

The multiple-input neuron model is functionally equivalent to multiple linear regression in statistics. To demonstrate this, the same heat influx prediction problem was solved using both the northern and the southern exposures. In multiple linear regression, the coefficients (intercept and slopes) of a relationship between a dependent variable and several independent variables are sought such that the sum of the square error for the whole dataset is minimized. Table 2.7 shows the ANOVA table for this analysis, which demonstrates that the intercept is 483.67 and that the slopes with respect to the northern southern aspects are -24.215 and 4.79, respectively. These are identical to those obtained from the linear neuron.

The R^2 for the model is 0.86, which is also identical to that given by the linear model.

Table 2.7 ANOVA Table for Relationship between Heat Influx and the Northern and Southern Aspects

		Estimate	SE	T Stat	P Value
Parameter Table®	1	483.67	39.5671	12.2241	2.77867×10^{-12}
	x1	−24.215	1.94054	−12.4785	1.75336×10^{-12}
	x2	4.79629	0.951099	5.04289	0.0000300103

RSquared® 0.858715, Adjusted RSquared® 0.847847,
Estimated Variance® 79.7819

		DF	Sum of Sq	Mean Sq	F Ratio	P Value
ANOVA Table®	Model	2	12 607.6	6303.8	79.0129	8.93785×10^{-12}
	Error	26	2074.33	79.7819		
	Total	28	14 681.9			

2.5.4.9 Multiple Linear Neuron Models

Many linear neuron units can be combined to form a multi-output linear classifier or a predictor network consisting of many linear neurons, as illustrated in Figure 2.35. In classification, each neuron represents a class; the reader should now understand that the multi-output linear classifier is equivalent to a linear multiple discriminant function classifier.

As a predictor, multiple linear neural networks can be used to simultaneously model the linear relationship between one or more dependent variables and several predictor variables. This allows for the study of the simultaneous effects of inputs on several dependent variables. Given the input variables and the output variables, the delta rule adjusts the weights incrementally, simultaneously reducing the prediction error over all

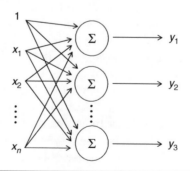

Figure 2.35 A multi-output linear neuron model.

the dependent variables. The resulting model is actually a set of models that are a linear combination of inputs whose relative importance to each output is captured by the weights; the model preserves the correlation of the inputs to the predicted variables. The *i*th model is

$$y_i = b_0 + w_{i1}x_1 + w_{i2}x_2 + \cdots + w_{in}x_n \qquad (2.47)$$

where y_i is the predicted value of the *i*th dependent variable and w_{ij} is the weight associated with the *j*th input and the *i*th dependent variable.

2.5.4.10 Comparison of a Multiple Linear Neuron Network with Canonical Correlation Analysis

As a predictor, a multiple linear neural network has a close correspondence to a canonical correlation in statistics. Canonical correlation is a multivariate statistical method designed to study the simultaneous effect of a set of independent variables on a set of dependent variables. Canonical correlation is similar to multiple linear regression in that they both develop linear combinations of inputs for predicting an outcome. The difference is that canonical correlation analysis does this simultaneously for several dependent variables [15]. Canonical correlation involves obtaining the set of weights for the dependent and independent variables that provides the maximum simple correlation for the two sets of variables.

2.5.4.11 Linear Neuron and Linear Network Summary

This chapter has presented the concepts of the delta rule as applied to a linear neuron and has examined its classification and predictive capabilities. Learning amounts to the alteration of weights, which in essence fixes the slope of the solution plane. The delta rule was used to minimize error; through an example it was found that this method can approximate a linear function well. It was also demonstrated that, as a classifier, a linear neuron is functionally equivalent to simple linear discriminant function classifier, and as a predictor is analogous to a simple linear and multiple linear regression for single- and multiple-input cases, respectively. In a multiple-input case, the weights adjust the slopes of a hyperplane. Many linear neurons can be combined to form a network that is suitable for multiclass classification and that is similar to multiple discriminant function analysis. When many linear neurons are used for simultaneous function approximation in which the effect of a set of inputs on several outputs is modeled, this model resembles canonical correlation analysis in statistics.

2.6 Summary

This chapter presented the fundamentals of neural networks through examples. Specifically, it examined threshold neuron models, linear neuron models, and learning strategies, including Hebbian, perceptron, supervised, and unsupervised learning methods. It highlighted, through this coverage of neural network fundamentals, the linear analysis capabilities of single and multiple neuron models of perceptron and linear neurons, and compared these with equivalent statistical methods for linear analysis. Specifically, it was shown that a single perceptron and linear neuron classifiers are equivalent to a linear simple discriminant function classifier in statistics, and that linear analysis of a single linear neuron is equivalent to simple and multiple linear regression in statistics. Multiple-perceptron and multiple-linear neuron classifiers are equivalent to multiple-discriminant function analysis, and the predictive capabilities of the multiple-linear neuron models are equivalent to canonical correlation analysis.

The next chapter will examine neural networks for nonlinear data analysis. One of the early criticisms of neural networks focused on their inability to solve nonlinear problems involving the complex, nonlinear mapping of inputs to outputs [19]. This is essential for solving many real problems, which are generally complex in nature. As a response to these criticisms, neural networks for nonlinear analysis emerged and the next chapter begins a discussion on MLP networks, which are among the most important neural networks developed for the nonlinear analysis of data. The essential conceptual developments that took place during the growth period of neural networks are presented in a three-volume series by Rumelhart and McClelland [21–23]; these are considered to be the most comprehensive reference on the foundation of neural networks.

Problems

1. In linear analysis of data with neural networks, what aspects of neurons make them linear models? What does the term "linear" refer to?
2. What are the advantages and limitations of threshold neurons?
3. What is a classification boundary, and how is it obtained for a threshold and a linear neuron?
4. What is a "linearly separable" classification problem?
5. Use the data in Table 2.5 to train the linear classifier in Section 2.5.4.2 in continuous learning mode, in which it adjusts weights even if the classification is correct, to move the weights closer to the inputs. Compare the difference between learning only when a

mistake is made and continuous learning in terms of the final weights and the classification boundary.

6. On a dataset of choice, perform a linear classification using perceptron, linear neuron, and discriminant analysis, and compare the results in terms of classification accuracy.

7. Describe the delta rule and show how it is applied on a two-input linear model.

8. On a dataset of choice, perform a linear prediction using a linear neuron and a simple or multiple linear regression, as appropriate. Compare the results from the two methods. Extract the model parameters from the linear neuron and explain what they mean in relation to the inputs and the output.

9. What is the difference between the perceptron learning algorithm and the delta rule?

10. What are the advantages of the delta rule in learning?

References

1. Haykin, S. *Neural Networks: A Comprehensive Foundation*, MacMillan College Publishing, New York, 1994.

2. Jain, A.K., Mao, J., and Mohiuddin, K.M. *Artificial Neural Networks: A Tutorial*, IEEE Computer, 29(3), 56, March 1996.

3. Smith, M. *Neural Networks for Statistical Modeling*, International Thompson Computer Press, London, UK; Boston, MA, 1996.

4. Kohonen, T. *Self-Organizing Feature Maps*, 3rd Ed., Springer-Verlag, Heidelberg, 2001.

5. Gaupe, D. *Principles of Artificial Neural Networks*, Vol. 3, Chap. 2, Fundamentals of Biological Networks, Advance Series in Circuits and Systems, World Scientific, Singapore, pp. 4–7, 1997.

6. Mind and Brain, *Readings from "Scientific American" Magazine*, W.H. Freeman, New York, 1993.

7. Eberhart, R.C. and Dobbins, R.W. *Neural Networks PC Tools: A Practical Guide*, Chap. 1, Background and History, R.C. Eberhart and R.W. Dobbins, eds., Academic Press, San Diego, CA, 1990.

8. McCulloch, W.S. and Pitts, W. A logical calculus of ideas immanent in nervous activity, *Bulletin of Mathematical Bio-Physica*, 5, 115, 1943.

9. Hebb, D. *The Organization of Behavior*, Wiley, New York, 1949.

10. Rosenblatt, F. The perceptron: A probabilistic model for information storage and organization in the brain, *Psychological Review*, 65, 386, 1958.

11. Rosenblatt, F. *Principles of Neurodynamics: Perceptron and the Theory of Brain Mechanisms*, Sparton Books, Washington, DC, 1961.

12. Gurney, K. *An Introduction to Neural Networks*, UCL Press, London, 2003.

13. *Minitab 14*, Minitab Inc., State College, PA, www.minitab.com, 2004.

14. Diamantopoulos, A. and Schlegelmilch, B. *Taking the Fear Out of Data Analysis: A Step-by-Step Approach*, Business Press, London, 2000.
15. Hair, Jr., J.F., Anderson, R.E., Tatham, R.L., and Black, W.C. *Multivariate Data Analysis*, 5th Ed., Prentice Hall, Upper Saddle River, NJ, 1998.
16. Johnson, R. and Wichern, D. *Applied Multivariate Statistical Methods*, 3rd Ed., Prentice Hall, Englewood Cliffs, NJ, 1992.
17. *Mathematica-Neural Networks*, Wolfram Research, Inc., Champaign, IL, 2002.
18. *SPSS 13.0*, SPSS Inc., Chicago, USA., www.spss.com, 2004.
19. Minski, M. and Papert, S. *Perceptrons*, MIT Press, Cambridge, MA, 1969.
20. Widrow, B. and Hoff, M.E. Adaptive switching circuits, *1960 IRE WESCON Convention Record: Part 4, Computers: Man-Machine Systems*, New York, p. 96, 1960.
21. Rumelhart, D.E. and McClelland, J.L. Foundations, *Parallel Distributed Processing—Explorations in the Microstructure of Cognition*, Vol. 1, MIT Press, Cambridge, MA, 1986.
22. Rumelhart, D.E. and McClelland, J.L. Psychological and biological models, *Parallel Distributed Processing—Explorations in the Microstructure of Cognition*, Vol. 2, MIT Press, Cambridge, MA, 1986.
23. Rumelhart, D.E. and McClelland, J.L. Explorations in parallel distributed processing, *A Handbook of Models Programs and Exercises*, MIT Press, Cambridge, MA, 1986.

Chapter 3

Neural Networks for Nonlinear Pattern Recognition

3.1 Overview and Introduction

This chapter will extend the discussion about linear analysis to nonlinear analysis using neural networks. There are several methods suitable for nonlinear analysis, including multilayer perceptron (MLP) networks, radial basis function (RBF) networks, support vector machines (SVMs), generalized model for data handling (GMDH), also called polynomial nets, generalized regression neural network (GRNN) and generalized neural network (GNN). Most of these networks have several processing layers that give them nonlinear modeling capability.

The topic of this chapter, the MLP, is the most popular and widely used nonlinear network for solving many practical problems in applied sciences, including ecology, biology, and engineering [1–3]. It is conceptually similar to RBF networks in that the intermediate processing is done by one or more layers of hidden neurons with nonlinear activation functions. RBFs use Gaussian functions as activation functions and MLPs use a range of activation functions. The GNN, presented in Chapter 9 (Section 9.6), is a network with only two hidden neurons, one performing linear analysis and the other performing nonlinear analysis. The GRNN is used for nonparametric estimation of the probability density of data; it does not require iterative training and is useful for relatively nonlinear data processing. The GRNN is presented in Chapter 9 (Section 9.10). The GMDH is another nonlinear network that builds polynomials from input variables in successive stages to obtain a more complex polynomial that contains the most influential input variables, much like in regression. The SVM is a statistical method that transforms a nonlinear, multidimensional problem into a linear problem in a higher dimensional space.

The reason for the popularity of the MLP network is that it is very flexible and can be trained to assume the shape of the patterns in the data, regardless of the complexity of these patterns. In this chapter, MLP networks are presented in detail in order to highlight nonlinear processing in neural networks. It is presented in such a way that the processing in other networks can be understood with relative ease. MLP is a powerful extension of the perceptron; these networks are called universal approximators due to their ability to approximate any nonlinear relationship between inputs and outputs to any degree of accuracy [4]. The power of these networks comes from the hidden layer of neurons located between the input layer and output layer of neurons. The hidden layer may consist of one or many nonlinear neurons and, more importantly, it performs continuous, nonlinear transformations of the weighted inputs, in contrast with the linear mapping in the linear neuron and the step function mapping used in the perceptron.

Recall from Chapter 2 that the perceptron and linear neuron are only capable of classifying linearly separable patterns and therefore cannot form the arbitrary nonlinear classification boundaries required by complex data. Furthermore, the linear neuron is capable of performing only a linear mapping of input data to output data in prediction. In contrast, nonlinear mapping done locally in each neuron gives nonlinear networks the flexibility and power to approximate many complex relationships inherent in the data. Founding concepts and studies on these ideas were developed in the 1980s and are presented in McClelland and Rumelhart [5,6] and Werbos [7], who laid the foundation for implementing nonlinear mapping in neural networks. Kohonen [8] contributed greatly in implementing non-linear projections of multidimensional data onto one- and two-dimensional maps in a self-organizing manner so that unknown data clusters and classification boundaries can be discovered. Some later developments of these concepts are elaborated in detail in Haykin [9], Principe [10], Fausett [11], Kohonen [12], and Eberhart et al. [13,14].

Both this chapter and the next concentrate on MLP networks. This chapter examines in detail how data is processed by individual neurons and how the whole network assembles individual neurons and synthesizes their outputs to produce a final output. It begins with a brief overview of the operation of MLPs, and then, in Section 3.2, moves on to a detailed study of processing in a nonlinear neuron, with examples. Also, the nonlinear neuron model is compared with nonlinear regression. In Section 3.3 and Section 3.4, networks for single- and two-dimensional input processing are presented in detail and illustrated with examples. These sections explore fundamental processing in neurons and the ways in which these neurons work together in a network to perform nonlinear mapping of inputs to outputs where highly nonlinear mapping functions are formed and complex classification boundaries are created. The next chapter explains how networks learn to perform these tasks.

3.1.1 Multilayer Perceptron

The layout of an MLP network with one hidden layer is shown in Figure 3.1. In Figure 3.1, $x_1, ..., x_n$ are input variables comprising the input layer. The first set of arrows represent the weights (or input-hidden neuron connections) that link this layer to the hidden middle layer, consisting of one or many hidden neurons—so called because they are not exposed to the external environment (data), as are the input and output neurons. Hidden neurons sum the corresponding weighted inputs as denoted by Σ in Figure 3.1; this is similar to the initial processing in the linear neuron and perceptron. However, unlike those systems, each hidden neurons passes its weighted sum through a nonlinear transfer function, denoted by σ. The outputs of the hidden neurons are fed through the second set of weights (hidden-output neuron connections) into the output neuron(s), which assemble the outputs by computing the weighted sum and passing it through a linear or nonlinear function. The output of these neurons makes up the network output, which is usually a single output in prediction (or function approximation), and one or many outputs in classification, indicating the class to which each input belongs. There are many choices for the neuron activation (transfer) function, and this chapter will spend a great deal of time studying these choices, their characteristics, and the ways they turn multilayer networks into powerful classifiers and predictors.

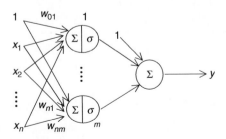

Figure 3.1   Configuration of a multilayer neural network.

Learning or training in MLPs is supervised by showing the network the desired output for a particular input. Learning involves presenting the input vectors, one at a time, at the input layer and passing them through the hidden layer and output layer, where the final network output is generated. At this point, the network output is compared with the desired output and the difference, which is the error, is calculated. If the absolute error is larger than an acceptable threshold, the error is backpropagated through the network; this process adjusts the weights between the input and hidden layers and those between the hidden and output layers using an appropriate learning method to minimize the error in the repeated processing of the inputs by the network. The error threshold, or acceptable limit for error, defines the accuracy of the model and depends on practical considerations.

One popular learning method for error correction is the delta rule, also called steepest descent, presented in Chapter 2 in relation to learning in linear neurons. Error correction methods are continually being improved; some of these methods will be studied in Chapter 4. The training process is repeated, adjusting weights until the optimum network performance (resulting in minimum error) is achieved, at which point training stops. For simple linear problems, hidden neurons are not required, as can be seen from the analysis of the linear neuron in Chapter 2. For simple nonlinear problems, one or few hidden neurons may be sufficient. However, for highly nonlinear problems involving many input variables, a larger number of neurons may be necessary to correctly approximate the desired input–output relationship.

The following sections will systematically examine data processing in MLPs by looking in detail at the internal workings of these networks. Specifically, they will look at characteristics of activation functions and hidden neuron processing, and process data through networks of increasing complexity to highlight how individual neurons process information locally and how an ensemble of neurons collectively produce complex mappings of data. In essence, this chapter is a study of the ways in which networks process data, called forward passing of input–output data. Once data processing has been studied thoroughly, the entirety of Chapter 4 is devoted to a discussion of how to train MLP networks to perform these complex mappings. Through these discussions, it is expected that the reader will not only understand the fundamentals of nonlinear neural processing but also be able to study other networks with ease.

3.2   Nonlinear Neurons

MLP derives its power from nonlinear processing in the hidden neurons. Crucial to this task are the nonlinear activation functions that transform the weighted input of a neuron nonlinearly to an output. This section will first

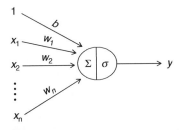

Figure 3.2 Nonlinear neuron.

look at some of these nonlinear activation functions in detail and then study how they themselves are transformed during training to map inputs to output(s) nonlinearly. Figure 3.2 shows a hidden neuron receiving n inputs, $x_1, \ldots, x_n$.

The output of the neuron is given in Equation 3.1, where the weighted sum of the inputs (as shown within brackets) is passed through a nonlinear function σ as

$$\sigma\left(\sum_{j=1}^{n} w_j x_j + b\right) \tag{3.1}$$

where b denotes the weight associated with the bias input and w_j represents the weight associated with the jth input. The most widely used function is sigmoid—a family of curves that includes logistic and hyperbolic tangent functions—and is used for modeling in population dynamics, economics, and so on. Other functions that are used are Gaussian, sine, arc tangent, and their variants. Some of these are presented in Figure 3.3 and explored in the next section.

3.2.1 Neuron Activation Functions

The neuron activation functions shown in Figure 3.3 have some important characteristics that make them vital to neural information processing.

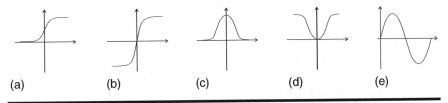

(a)   (b)   (c)   (d)   (e)

Figure 3.3 Some nonlinear neuron activation functions: (a) logistic, (b) hyperbolic-tangent, (c) Gaussian, (d) Gaussian complement, (e) sine function.

They are nonlinear, continuous functions that remain within some upper and lower bounds. Nonlinear means that the output of the function varies nonlinearly with the input; this aspect makes it possible for neural networks to do nonlinear mapping between inputs and outputs. Continuity of the functions implies that there are no sharp peaks or gaps in the function, so that they can be differentiated throughout, making it possible to implement the delta rule to adjust both input-hidden and hidden-output layer weights in backpropagation of errors, as Chapter 4 will study in detail. These two important properties are instrumental in shifting the neural networks application domains from simple linear to complex nonlinear domains. The term "bounded" means that the output never reaches very large values, regardless of the input. This means that the output activation remains bounded even if the net input to a neuron is large. These developments were a direct result of the attempts to develop models that mimic the biological neurons, where the outputs are nonlinear, continuous, bounded signals. Due to their popularity, sigmoid activation functions will be studied next in more detail. The concepts apply equally well to other functions.

3.2.1.1 Sigmoid Functions

The sigmoid functions are a family of S-shaped functions whose characteristics are described in the previous section; two of them are shown in Figure 3.3a and Figure 3.3b. The most widely used sigmoid function is the logistic function, shown in Figure 3.4 for the range of input values u from -10 to 10. $L(u)$ denotes the output for an input u.

The logistic function has a lower bound of zero and upper bound of 1. This means that the function value (or the output) range is [0,1]. At the input $u = 0$, the output is the midpoint (0.5), and the slope of the function, which indicates how fast the function is changing, is the greatest at this point. The slope at $u = 0$ is 0.25 (14°). The output increases relatively quickly in the

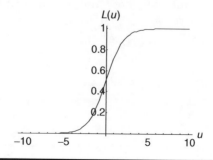

Figure 3.4 Logistic activation function.

vicinity of $u = 0$, as input increases and approaches the upper bound much more slowly. For inputs below zero, the output initially decreases more rapidly, then more slowly as the lower bound is approached. Smith [4] gives a more detailed graphical analysis of the components of the logistic function. The logistic function has the following mathematical formulation:

$$y = L(u) = \frac{1}{1 + e^{-u}} \tag{3.2}$$

where e is the base of natural logarithm, which is a constant with a value of 2.71828.

Another commonly used sigmoid function is the hyperbolic tangent function shown in Figure 3.3b and given below:

$$\tanh(u) = \frac{1 + e^{-u}}{1 - e^{-u}} \tag{3.3}$$

As shown in Figure 3.5, the hyperbolic tangent function has a lower bound of -1 and an upper bound of 1, making its output range $[-1,1]$ in contrast to the $[0,1]$ range for the logistic function. Another difference is that the output at $u = 0$ is zero. The slope of the hyperbolic tangent is also higher at $u = 0$, meaning that it reaches the bounds more quickly than the logistic function. The slope at $u = 0$ here is 1.0 (i.e., 45°).

Another related function is the inverse tan ($\tan^{-1}$ or arctan) function, shown in Figure 3.6. It has a more gradual variation than the above two functions, with a slope at boundary in between those of the logistic and hyperbolic tangent functions.

Many of these functions have been used for neural information processing. When they are used in output neurons, the actual output must be scaled to fit the output range of the transfer function because the

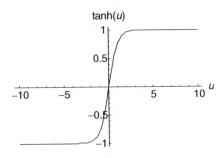

Figure 3.5 Hyperbolic tangent function.

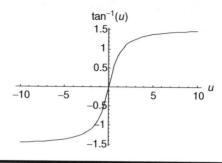

Figure 3.6   Arctan function.

network output is compared to the actual output. For example, if a neuron with a logistic activation function is used for the output neuron, the output will be in the range of [0,1]. Therefore, the actual outputs have to be scaled to fit this range. The inputs, however, can take any value regardless of the bounds of the sigmoid function; however, in some situations it may be necessary and advantageous to scale the input data to the range of the transfer function receiving the inputs. These aspects will be explored later.

3.2.1.2   Gaussian Functions

Standard normal curve. The standard normal curve, shown in Figure 3.7, has a symmetric bell shape and is the commonly known standard normal distribution (Equation 3.4). It represents input data with mean zero and standard deviation one. Its range is [0,1], it peaks at input $u = 0$, is highly sensitive to u values around zero, and is almost insensitive to those

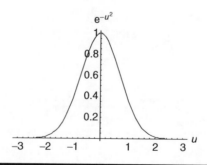

Figure 3.7   Gaussian function.

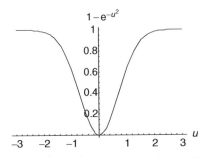

Figure 3.8   Gaussian complement function.

at the tails. Thus, it amplifies the mid-range of the input distribution. Therefore, when this function is used in a neuron as a component of a neural network, it is more sensitive to the weighted inputs that are close to zero:

$$y = e^{-u^2}. \tag{3.4}$$

Gaussian complement. This is the inverted Gaussian function so it peaks at the tails and assumes a value of zero when $u = 0$, as shown in Figure 3.8 and Equation 3.5. Thus, it has a larger output for the inputs at the upper and lower ends. When it is used in a neuron as part of a neural network, the network is more sensitive to the weighted inputs that are at the two extreme ends:

$$y = 1 - e^{-u^2}. \tag{3.5}$$

The working of these functions in a neural network can now be explored, beginning with a one-input network and examining thoroughly all aspects of the network in order to completely understand what enables MLPs to approximate any nonlinear function or classify data with arbitrarily nonlinear classification boundaries.

3.2.2   Example: Population Growth Modeling Using a Nonlinear Neuron

Exponential functions are popularly used for modeling population growth. Presented below is the way in which a single neuron with a logistic function learns to model population growth through training using the data shown in Figure 3.9.

A single neuron model was trained, as shown in Figure 3.10, using the delta rule starting with a small random initial weight value; the progress of

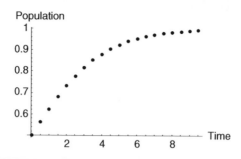

Figure 3.9 Population growth over time.

the model at four epochs of training is shown in Figure 3.11, where the solid line represents the model output [15].

It can be seen that the initial model output shown for iteration zero is far from what it should be, and the training quickly corrects most of the error in the first epoch itself. (In the figure, iteration is an epoch.) The final optimum values obtained for the bias and input-neuron weights are -0.00002 and 0.5, respectively, resulting in the following model output:

$$y = \frac{1}{1 + e^{-0.5x}} \tag{3.6}$$

Thus, a single, nonlinear neuron with logistic transfer function is capable of modeling simple, nonlinear functions such as growth models. It also highlights the fact that a single neuron can model any region of output where the output is monotonically (continuously) increasing or decreasing. A single neuron with multiple inputs will produce a multidimensional, nonlinear model with an output in the form of

$$y = \frac{1}{1 + e^{-(w_1 x_1 + w_2 x_2 + \cdots + w_n x_n)}} . \tag{3.7}$$

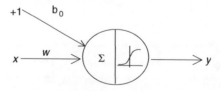

Figure 3.10 Nonlinear neuron with a logistic transfer function for modeling population growth.

Function estimate after

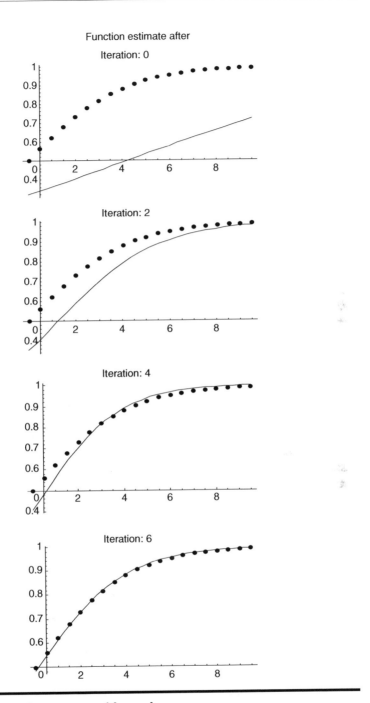

Figure 3.11   Learning progress with epochs.

Table 3.1   Results from Nonlinear Regression on Population Growth Data

Parameter	Estimate	Asymptotic SE	CI
Theta 1	1.39×10^{-16}	1.59×10^{-16}	$\{1.95 \times 10^{-16}, 4.73 \times 10^{-16}\}$
Theta 2	-0.5	7.35×10^{-17}	$\{-0.5, -0.5\}$

3.2.3   Comparison of Nonlinear Neuron with Nonlinear Regression Analysis

Functionally, the performance of the single, nonlinear neuron is similar to the nonlinear regression in statistics. The same population growth data used in the previous section were used to fit a nonlinear regression [16] model with the results presented in Table 3.1 where theta 1 and theta 2 refer to b_0 and w, respectively, of the nonlinear neuron. It can be seen that the parameter estimates from the nonlinear neuron and nonlinear regression are identical.

3.3   One-Input Multilayer Nonlinear Networks

3.3.1   Processing with a Single Nonlinear Hidden Neuron

In this section we study a simple MLP network, shown in Figure 3.12, with one input in the input layer, one hidden neuron in the hidden neuron layer, and one output in the output layer. Therefore, it has one input-hidden layer weight denoted by a_1 and one hidden-output layer weight denoted by b_1. The input is x and the network output is z. The hidden neuron has a bias input of $+1$ with an associated weight of a_0 and the output neuron has a bias weight of b_0. This notation for the weights is also used by Smith [4]. This section examines the most common activation function, logistic function, in the hidden neuron and for simplicity assumes a linear function in the output neuron, which in theory can be any activation function previously described. However, it will be shown that a linear activation function is more appropriate for the output neuron in prediction, and logistic or a bounded function is more appropriate for classification.

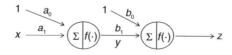

Figure 3.12   One-input, one-hidden neuron, one-output neural network.

The input is fed into the network and the hidden neuron calculates the weighted sum of inputs (including bias) and passes it through the logistic function to produce the hidden-neuron output, y. This output is fed as input into the output neuron through the associated connection link, where it is weighted. The weighted input is passed through the neuron's activation function and the output of this neuron becomes the network output. The most important part of processing takes place in the hidden neuron, whose details are shown in Figure 3.13.

The weighted input is

$$u = a_0 + a_1 x. \tag{3.8}$$

The result of this operation is to map x linearly to u, as shown in Figure 3.14, with a slope a_1 and intercept a_0.

As the weight a_1 changes, the slope of the line changes, and as a_0 changes, the vertical position of the line changes. Thus, the weights (a_0, a_1) fix a line in two-dimensional space of $(u-x)$. The weight a_0 can be thought of as incorporating the effects of all inputs other than x that are not explicitly involved in the model.

The second task of the hidden neuron is to pass the weighted sum u through the logistic function. The logistic function has u as its argument and is always a standard function with $y = 0.5$ at $u = 0$. It will be more useful to express the output y in terms of input x to illustrate how x is mapped to y through u. Substituting u into the logistic function, the hidden-neuron output y is

$$y = \frac{1}{1 + e^{-(a_0 + a_1 x)}}. \tag{3.9}$$

Figure 3.13 **Hidden neuron details.**

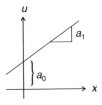

Figure 3.14 **Hidden neuron weighted input *u* as a function of *x*.**

Because learning is about adjusting weights, it is essential to see the effect of learning on the hidden-neuron output. Several cases will be presented to illustrate this concept, starting with the simplest case.

1. $a_0 = 0$, $a_1 = 1$

The hidden neuron output becomes

$$y = \frac{1}{1 + e^{-x}}.$$

This is the familiar standard logistic function, as shown in Figure 3.4, with the weighted sum $u = x$. Thus, when there is no bias input and the input-hidden weight is one, the hidden-neuron output with respect to the network input is represented by the standard logistic function. The function is such that when $x = 0$, $y = 0.5$. When x is greater than zero, y is greater than 0.5, and when x is less than zero, y is less than 0.5. Thus the position of $x = 0$ can be thought of as a boundary that divides the input space into two equal regions: one in which y is closer to one and the other in which y is closer to zero. This point of x is called the boundary point; this concept will be used later in classification. The slope of the curve at the boundary point is 0.25.

2. $a_0 = 0$ *and* $a_1 = -1$

The hidden-neuron output now becomes

$$y = \frac{1}{1 + e^{x}}. \tag{3.10}$$

The resulting function is plotted in Figure 3.15, which shows that the slope of the curve is now reversed and the curve is the mirror image of the standard logistic function.

3. $a_0 = 0$, a_1 *varies from* -1 *to* 2

Changing a_1 from -1 to 2 yields curves of varying slope at the boundary point, as shown in Figure 3.16.

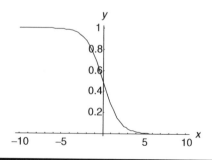

Figure 3.15 Logistic function for $a_0 = 0$, $a_1 = -1$.

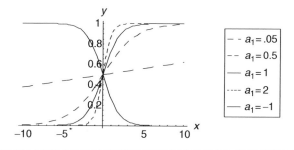

Figure 3.16   Logistic functions for $a_0 = 0$, a_1 varies from -1 to 2.

The larger the a_1, the steeper the slope of the curve at the boundary point. Thus, the effect of adjustment of a_1 during learning is to fix the slope of the hidden-neuron output function around the boundary point. The overall effect of a_1 is that its magnitude adjusts the slope of the curve and its sign determines whether the direction of the slope is positive or negative.

4. $a_1 = 1$, $a_0 = -3, 0, 4$

Here the effect of the bias weight alone is illustrated. Figure 3.17 shows the hidden-neuron output function for three values of a_0: $-3, 0, 4$. As can be seen, negative values of a_0 push the curve forward and positive values pull it backwards. A value of zero produces the standard logistic function. Thus the effect of adjusting a_0 during learning is to control the horizontal position of the curve. Because the boundary point is where the function value is 0.5, the change in a_0 essentially serves to move the boundary point horizontally. The position of the boundary point is not a function of a_0 alone, but varies with a_1 as well, as demonstrated below.

5. $a_1 = 0.3$, $a_0 = -3, 0, 4$

Figure 3.18 shows a hidden-neuron output for the same values of a_0 as in the previous figure, but with $a_1 = 0.3$. This highlights a dramatic effect of a_1

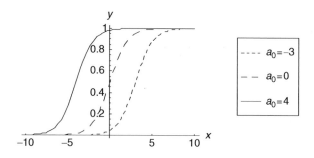

Figure 3.17   Logistic functions for $a_1 = 1.0$, $a_0 = -3, 0, 4$.

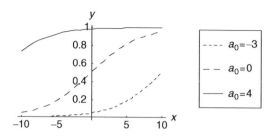

Figure 3.18 Logistic functions for $a_1 = 0.3$, $a_0 = -3, 0, 4$.

on the position of the curves, demonstrating that the most active regions (boundary regions) of the curves are placed at various locations along the x-axis. This feature is used for modeling various local nonlinear features of an input–output function, as will be seen later.

What are the boundary points for the three curves shown in Figure 3.18? The boundary point will be denoted by x'. The boundary point is where $u = 0$, which always gives a function value of 0.5, but $u = a_0 + a_1 x$, therefore, at the boundary point, $a_0 + a_1 x' = 0$, resulting in

$$x' = -a_0/a_1. \tag{3.11}$$

Thus, the boundary point for $a_0 = -3$ is $-(-3)/0.3 = 10$, for $a_0 = 0$ is zero, and for $a_0 = 4$ is $(-4)/0.3 = -13.3$. The above equation shows that the smaller the magnitude of a_1, the larger the shift that curve makes horizontally for a given value of a_0, as can be seen in Figure 3.17 and Figure 3.18 for $a_1 = 1$ and $a_1 = 0.3$, respectively.

Now it is possible to summarize the combined effect on the hidden-neuron output of adjusting a_0 and a_1 during learning. By plotting the hidden-neuron output as a function of input x, not only can the relationship between the inputs and hidden-neuron output be seen, but also the ways in which this relationship is transformed during learning as a_0 and a_1 change. Basically, a_0 and a_1 alter the position and shape, respectively, of the logistic function with respect to the inputs, and learning involves finding the appropriate a_0 and a_1 incrementally. Because there is a lot of scope for changing the slope and the position of the curve, it can be tailored to take any desired form by adjusting its slope, direction, and horizontal position.

When the desired function is more complex for approximation by a single hidden neuron, the capability of the neural network can be greatly enhanced by adding more neurons that act in parallel. Each neuron processes information in a similar fashion, but due to different initial weights, they begin using logistic functions with different slopes and positions. During training, each of these undergoes transformations in shape and position through changes in the corresponding weights to model various aspects of the desired function, so that collectively they

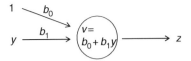

Figure 3.19 Output neuron and network output.

approximate the desired function. This gives MLP networks the power of nonlinear processing to approximate any function to any desired degree of accuracy.

Output of the network. Now the way in which the network produces the final output will be explained (Figure 3.19).

The output neuron, like the hidden neuron, first computes the weighted sum of the inputs it receives, denoted here by v, and then produces the final output z, which is equal to v for the case where a linear activation function is used. This yields

$$v = b_0 + b_1 y$$
$$z = v,$$

(3.12)

where b_0 is the bias weight and b_1 is the hidden-output neuron weight. The output is linear with respect to v but is still nonlinear with respect to the original input due to the nonlinear processing in the hidden neuron.

For the general case in which there is more than one (say, n) hidden neurons, the output of each of the neurons, $y_1, y_2, ..., y_n$, is fed into the output neuron through the corresponding weights, $b_1, b_2, ..., b_n$, along with the bias input $+1$ through bias weight b_0, as shown in Figure 3.20.

The weighted sum of inputs (v) received by the output neuron from many hidden neurons and the output z is

$$v = b_0 + \sum_{j=1}^{n} b_j y_j$$
$$z = v,$$

(3.13)

where b_j is the weight linking hidden neuron j and the output neuron.

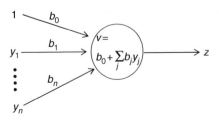

Figure 3.20 Network output for multiple hidden neurons.

For classification problems, it is more appropriate to use a logistic activation function. The output then is

$$z = \frac{1}{1 + e^{-v}}. \tag{3.14}$$

Using a logistic function, the output is bounded between zero and one, and output is nonlinear with respect to both v and the original input.

This concludes the forward processing in a one-input and one-output network with more than one hidden neuron. This network with one output neuron receiving inputs from many hidden neurons, which in turn receive a single input, can approximate arbitrarily complex single-input single-output functions.

3.3.2   Examples: Modeling Cyclical Phenomena with Multiple Nonlinear Neurons

Now two examples involving cyclical phenomena will be explored. Many natural processes are cyclical in nature. The first example is a single cycle square wave that can be thought of as an idealization of phenomena such as seasonal plant growth. The second example involves modeling two-cycle phenomena, such as spring and autumn species migration, which are complex, nonlinear functions to model. A deeper exploration into the networks modeling these phenomena sheds light on the internal transformations that finally produce the desired nonlinear function.

3.3.2.1   Example 1: Approximating a Square Wave

First the square wave function, as presented in Smith [4] and shown in Figure 3.21, will be explored, in which t is the desired output and x is the input. The function is constant at 0.25 for x is less than 0.3 or greater than 0.7. Within the range of 0.3 to 0.7, the value of the target function is 0.75.

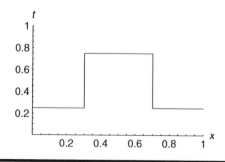

Figure 3.21   Square wave function *t* in relation to *x*.

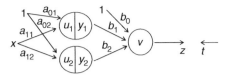

Figure 3.22 Network with two nonlinear hidden neurons for square wave approximation.

The first stage of the development of a neural network model is to generate the data. For a function approximation or classification, this involves generating input–output pairs that describe the problem. For the square wave problem, any number of input–output pairs can easily be generated within the range of x between zero and one, as it is known from the function in Figure 3.21 what the target value should be. As training has not yet been discussed in detail, it will be assumed that this function will be modeled using two hidden neurons with a logistic activation function and one output neuron with a linear function, as shown in Figure 3.22, where x is the input, a_{01} and a_{02} are bias weights associated with the two hidden neurons, and a_{11} and a_{12} are input-hidden neuron weights for the two hidden neurons. For the output neuron, b_0 represents the bias weight and b_1 and b_2 are the weights associated with links to the output neuron from the two hidden neurons. Thus, this problem has seven unknown weights that require incremental adjustment during training. The hidden-neuron weighted sum is u and the output is y; the weighted sum for the output neuron is v and the network output is z.

Random initial values between ± 0.3 were assigned to the weights, and the network predictions before training (solid line) were superimposed on the data in Figure 3.23.

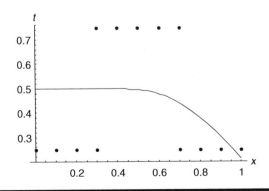

Figure 3.23 Model output with initial random weights.

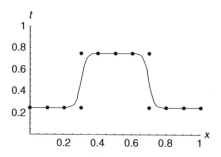

Figure 3.24 Trained network predictions superimposed on data.

As can be seen from Figure 3.23, the network with initial weights poorly models the data. The network was trained using the delta rule until error did not change appreciably; this occurred in 15 epochs, and the predictions from the trained network superimposed on data are shown in Figure 3.24.

Figure 3.24 shows that the trained network follows the data very well except for the two steep areas with infinite slope. These steep areas are generally very difficult to model, but this network has done it rather well. The final values obtained for the weights are shown in Table 3.2.

Now a forward pass of two input–output pairs selected from Figure 3.21 will be performed, and the network output for these values will be examined.

1. $x = 0$, $t = 0.25$

For this pair, the first hidden-neuron output is

$$u_1 = a_{01} + a_{11}x = -20.8 + 0 = 20.8$$

$$y_1 = \frac{1}{1 + e^{-u_1}} = \frac{1}{1 + e^{-20.8}} = 0.999.$$

The second hidden-neuron output is

$$u_2 = a_{02} + a_{12}x = 47.6 + 0 = 47.6$$

$$y_2 = \frac{1}{1 + e^{-u_2}} = \frac{1}{1 + e^{-47.6}} = 1.$$

Table 3.2 Weights for the Two-Neuron Nonlinear Model Approximating Square Wave

a_{01}	a_{02}	a_{11}	a_{12}	b_0	b_1	b_2
20.8	47.6	−69	−68	0.25	−0.5	0.5

The net input v and output z of output neuron are

$$v = b_0 + b_1 y_1 + b_2 y_2 = 0.25 - (0.5 \times 0.999) + (0.5 \times 1) = 0.25$$

$$z = v = 0.25.$$

The desired target $t = 0.25$, so the predicted and desired results are identical. Now the second input–output pair will be examined.

2. $x = 0.5$, $t = 0.75$ (This point belongs to the stepped portion of the function.)

For this pair, the first hidden-neuron output is

$$u_1 = a_{01} + a_{11} x = 20.8 - (69 \times 0.5) = -13.7$$

$$y_1 = \frac{1}{1 + e^{-u_1}} = \frac{1}{1 + e^{-(-13.7)}} = 1.122 \times 10^{-6}.$$

The second hidden-neuron output is

$$u_2 = a_{02} + a_{12} x = 47.6 - 68 \times 0.5 = 13.6$$

$$y_2 = \frac{1}{1 + e^{-u_2}} = \frac{1}{1 + e^{-13.6}} = 0.999.$$

The net input and output of output neuron is

$$y = b_0 + b_1 y_1 + b_2 y_2 = 0.25 - 0.5 \times 1.122 \times 10^{-6} + 0.5 \times 0.999 = 0.75$$

$$z = v = 0.75.$$

The desired target is $t = 0.75$, so the predicted and target values are identical.

The reader may wish to try another input–output pair from the latter portion of the function to see how the predicted and desired values compare.

Hidden neuron outputs in relation to the input. Now that the model behaves satisfactorily, it is time to examine the way that the training has positioned and adjusted the slope of the hidden-neuron activation functions in the input space of x. The predicted output of the network will also be considered.

Because $u_1 = a_{01} + a_{11} x$ and $y_1 = 1/(1 + e^{-(u_1)})$, substituting for u_1 into y_1 yields

$$y_1 = \frac{1}{1 + e^{-(a_{01} + a_{11} x)}}. \tag{3.15}$$

Similarly, for the second hidden neuron, $u_2 = a_{02} + a_{12} x$ and $y_2 = 1/(1 + e^{-(u_2)})$, resulting in

$$y_2 = \frac{1}{1 + e^{-(a_{02} + a_{12} x)}}. \tag{3.16}$$

Network output in relation to the input. The predicted model output z is

$$z = \frac{1}{1+e^{-v}} = \frac{1}{1+e^{-\left(b_0 + \sum_{j=1}^{2} b_j y_j\right)}}. \tag{3.17}$$

By substituting u_j into y_j in the above equation, the output z as a function of x is obtained, together with all the weights, a_{01}, a_{02}, a_{11}, and a_{12}, as

$$z = \frac{1}{1+e^{-\left(b_0 + \sum_{j=1}^{2} b_j(1/1+e^{-(a_{0j}+a_{1j}x)})\right)}}$$

$$= \frac{1}{1+e^{-(b_0 + [b_1(1/1+e^{-(a_{01}+a_{11}x)}) + b_2(1/1+e^{-(a_{02}+a_{12}x)})])}}, \tag{3.18}$$

which becomes the model that maps inputs nonlinearly to the output through weights that are free parameters of the network.

By substituting for values of weights from Table 3.2, the final output of the trained network as a function of input x becomes

$$z = \frac{1}{1+e^{-(0.25 + [(-0.5)(1/1+e^{-(20.8+(-69)x)}) + 0.5(1/1+e^{-(47.6+(-68)x)})])}}. \tag{3.19}$$

Graphical illustration of steps in network processing. Figure 3.25 and Figure 3.26 show plots of all intermediate steps in the calculation to show graphically how the original data is mapped to u_1, u_2, y_1, y_2, v, and finally z. This was done by plotting the above functions of u_j, v_j, and z in relation to input x. Exploring these graphs one by one illuminates the neural information processing in MLP in general, and in this network in particular. Figure 3.25 shows u_1 and u_2 as a function of x.

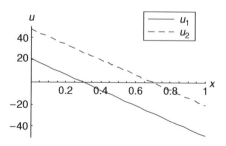

Figure 3.25 **Mapping of inputs to weighted sum u of inputs in hidden neurons.**

As previously discussed, u_1 is the weighted sum of the inputs to the first hidden neuron and must be a linear function of x with intercept equal to bias weight a_{01}, and slope equal to input-first hidden-neuron weight a_{11}. Learning in the network has configured the final values of these weights to 20.8 and -69, respectively, showing a positive intercept and negatively inclined line, as shown in Figure 3.25. Similarly, u_2 is the weighted sum of the inputs to the second hidden neuron, and this function has an intercept equal to bias weight, a_{02}, and slope equal to weight, a_{12}. Learning in the network has configured the final values of these weights to 47.6 and -68, respectively, showing a positive intercept and negatively inclined line (Figure 3.25). These are the values of weights in Table 3.2.

When u_1 and u_2 are mapped to corresponding hidden-neuron outputs y_1 and y_2, a standard logistic function results, but in the input space we know that the effect of a_{01} and a_{11}, for example, is to control the position and slope, respectively, of y_1 to follow the desired target function. Similarly, the effect of a_{02} and a_{12} is to control the corresponding aspects of y_2 to represent the desired target function jointly with y_1. These two outputs are plotted against x in Figure 3.26.

Figure 3.26 shows how the active regions of the two hidden neurons have been positioned in the input space to perform the required mapping. In Figure 3.26 the two curves have been shifted in the horizontal direction such that the boundary point that divides y_1 in half is $x' = -a_{01}/a_{11} = (-20.8)/(-69) = 0.3$. Note that this is the point where the original function steps up. The slope of y_1 at the boundary point is -17.29 (87°), which is an approximation of the infinite slope in the original function. This can be seen in Figure 3.27, in which hidden-neuron outputs are superimposed on the output of the network, z. Because an activation function must be smooth and differentiable, a step function will never be modeled perfectly. The output of the second hidden neuron y_2 in relation to x, as shown in Figure 3.26, shows that its boundary point is $-a_{02}/a_{12} = -47.6/-(68) = 0.7$. This is exactly the point at which the target function steps down

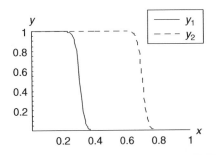

Figure 3.26 **Mapping of input to two hidden-neuron outputs.**

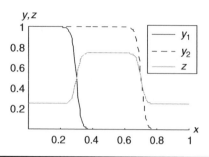

Figure 3.27 Mapping of input to output by the joint activity of the two hidden neurons.

(Figure 3.27). This logistic function also has a slope of 17.0 (87°), fixed by a_{12}, and this is an approximation of the infinite slope of the original function at this location. Note that the slope of the logistic is not equal to the weight but is controlled by it. Comparison of the output z and the target t in Figure 3.24 shows that the network has found the correct positioning of the curves and models the data well.

Next the activation of the hidden neurons will be closely studied. The first neuron is initially fully active ($y_1 = 1$) up to about $x = 0.25$ (Figure 3.27) and then decreases its output up to about $x = 0.4$; from that point on, the first neuron remains inactive with an output of zero. The second neuron also starts with full activation ($y_2 = 1$), but remain fully active until x reaches 0.7, where the activity slows down and ceases at $x = 0.8$. The overall effect of neuronal activation is that the neurons cooperate by taking care of separate features of the target function and crafting their own logistic function to mimic the target function.

How do the neurons work together to produce the final outcome? To answer this question, the network outcome will be analyzed. The predicted output z of the network is

$$z = b_0 + b_1 y_1 + b_2 y_2 = 0.25 - 0.5y_1 + 0.5y_2. \qquad (3.20)$$

The activation of both hidden neurons, y_1 and y_2, is initially 1. Because hidden-output weights are 0.5 and -0.5, their weighted sums are canceled and thus the target value of 0.25 in the input range of 0 to 0.3 comes entirely from the bias output weight of 0.25. Closer to $t = 0.3$, y_1 begins decreasing to accommodate the step function at $x = 0.3$. This results in an overall increase in z according to Equation 3.20, which continues until y_1 becomes zero at $x = 0.4$. At this point $z = 0.75$, which is equal to the required target and the first hidden neuron is responsible for the step function at $x = 0.3$. In the input range of around 0.3 to 0.7, only hidden neuron 2 is active at the full capacity that keeps z constant. At x closer to 0.7, neuron 2 decreases its activity to accommodate the second step and its effect is to reduce the z

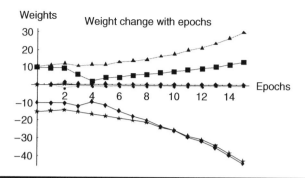

Figure 3.28 Change in network weights during training with number of epochs.

value; this continues until $x = 0.8$, when both neurons are inactive. The required output of 0.25 in the last input range is solely provided by the bias weight. Note that because both neurons are active simultaneously to produce the required alterations of the shape of the output function, removal of one neuron affects the output of the network for the entire input range.

The above example also highlights the number of neurons required to model the output. Because there are two distinct regions in which the direction of output changes, two logistic functions are required to model these two regions. This is because a single neuron can model only monotonic changes, not reversals in direction.

Figure 3.28 shows how all the network weights change during training until the desired weights are achieved. Each line represents adaptation in one weight. With further training, these weights do not change, but reach a plateau. Values of the seven weights (Table 3.2) of the trained network are those for the 15th epoch in Figure 3.28. Note that in this figure all of the output neuron weights are very small compared to hidden-neuron weights, and therefore cannot be distinguished from one another.

In summary, all of the stages of neural information processing were examined in detail in this example. It was shown graphically how learning crafts hidden neuron activation functions such that the final output follows the target function. The slopes of the original function are modeled by placing logistic functions at appropriate locations and adjusting the slopes of these functions that are primarily responsible for specific regions of the input space. The final output is produced by the joint activation of the hidden neurons, which are combined and processed by the output neuron. Thus, the whole network operates as an integrated whole, but the internal computation is distributed among the hidden neurons that do most of the work and, therefore, removal of a neuron affects the entire output.

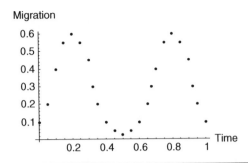

Figure 3.29   A pattern of bimodal seasonal species migration.

3.3.2.2   Example 2: Modeling Seasonal Species Migration

In this example, seasonal species migration is modeled. Many species show bimodal migratory patterns depicting spring and autumn migrations. Figure 3.29 shows data from such a pattern.

This pattern requires more than two hidden neurons. Since there are four distinct regions in which monotonicity is broken, it was approximated using four hidden neurons, as shown in Figure 3.30. The network comprises one input, four hidden neurons, and one output neuron. Consequently, there are 13 unknown weights, including the bias weights, to be estimated in the training. The data shown in Figure 3.29 was used to train the network using the delta rule until the error was minimized. The model output with random weights before training and the output with final optimum weights are shown in Figure 3.31a and Figure 3.31b, respectively. In Figure 3.31b, target data (black dots) is superimposed on the predictions (solid line) and there is a high degree of accuracy in the prediction, which shows that the network has mapped input data to the output perfectly. The prediction error will be studied in detail later. The final weights are shown in Table 3.3.

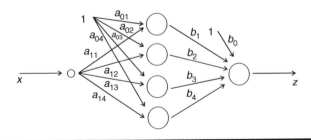

Figure 3.30   Multilayer perceptron network for modeling bimodal pattern of species migration.

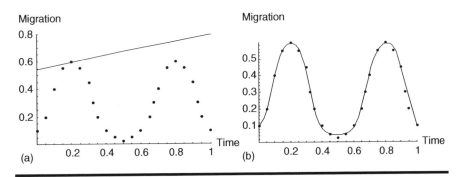

Figure 3.31 **Network prediction superimposed on data: (a) for initial weights and (b) for final optimum weights.**

First, the forward pass of data will be studied to explore the hidden workings of this more complex network. The data is processed similarly to that in the previous network containing two hidden neurons, except that this network contains four neurons. Here, each neuron sees the same input, but due to their different initial weights, the position and the shape of the activation functions in the input space will be different. This can be thought of as four activation functions with different slopes being placed at random locations in the input space. With random starting positions, weights are adjusted incrementally during learning until they make the activation functions assume final position and shape. This is done in such a way that each function takes care of a different region of the input space appropriately, so that their joint activity produces the network output that attempts to mimic the target function, which in this case is the bimodal pattern of species migration. A few input–output pairs will be selected and the inputs will be passed through the network. Table 3.4 shows the intermediate results of the processing of three inputs by the network.

Table 3.4 and Figure 3.31b indicate that the network error is very small. Now that the network models the data very well, the way in which the network produces the final outcome will be explored. First of all, the hidden

Table 3.3 Input-Hidden and Hidden-Output Neuron Weights; Bias $b_0 = 0.65$

Neuron, i	Bias Weight, a_0	Weight, a_{1i}	Weight, b_{1i}
1	2.7	−31.6	−0.55
2	−10.4	32.4	−0.58
3	35.0	−38.1	0.53
4	20.0	−29.5	−0.57

Table 3.4 Intermediate Results of Processing Three Inputs by the Network

Time	u_1	y_1	u_2	y_2	u_3	y_3	u_4	y_4	$z = v$	t	E
0.2	−3.59	0.026	−3.96	0.019	27.4	1.0	14.2	0.999	0.592	0.6	0.008
0.6	−16.2	0.00	9.0	0.999	12.1	0.999	2.39	0.996	0.087	0.1	0.013
0.9	−25.7	0.000	18.7	1.0	0.71	0.699	−6.45	0.001	0.433	0.45	0.017

neuron outputs will be examined. The plotting of u will be skipped because now it is known know that each u represents a line with an intercept equal to the bias weight and a slope equal to the input-hidden neuron weight. Figure 3.32 shows the final shape and the position of the four logistic functions after training. Neuron 2 has a positive slope and neurons 1, 3, and 4 have negative slopes.

According to Figure 3.32, neurons 3 and 4 are fully active from the beginning, and are joined by neuron 2, which becomes fully active when input is around 0.4. Neuron 2's initial activity is zero. Neuron 1 has a high initial activity that deceases quickly to zero at input around 0.2, where neuron 2 begins to show an increase. To show how these functions work together to produce the bimodal pattern, the output neuron activity is superimposed on the four hidden-neuron outputs in Figure 3.33. Note that each neuron is the most active near its boundary region, where it contributes mostly to the final output z. These boundary points for neurons 1, 2, 3, and 4 are 0.08, 0.32, 0.92, and 0.68, respectively.

The network output is

$$z = v = b_0 + b_1 y_1 + b_2 y_2 + b_3 y_3 + b_4 y_4$$
$$z = 0.65 - 0.55 y_1 - 0.58 y_2 + 0.53 y_3 - 0.57 y_4. \tag{3.21}$$

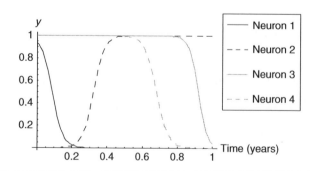

Figure 3.32 Activation functions of the four hidden neurons in the nonlinear network.

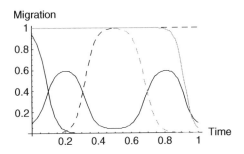

Figure 3.33   Activation of the four hidden neurons superimposed on the network output.

As both y_3 and y_4 are initially 1 and they have similar weights with opposite signs (0.53, −0.57), they do not contribute to the initial rise of the model output shown in Equation 3.21. Inactive neuron 2 does not contribute either. Therefore, the initial rise is due mainly to the bias and the rapidly decreasing activation of neuron 1, i.e., $z = 0.65 − 0.55y_1$. The fall of the output after the initial rise is provided by neuron 2 alone because neuron 1 becomes inactive at that time and because active neurons 3 and 4 cancel each other. In this region, $z = 0.65 − 0.58y_2$. Because y_2 is increasing, z continues to fall. When it comes to the third rise, neurons 2 and 3 are fully active, but activation of neuron 4 is decreasing. Activation of neurons 2 and 3 cancel each other, so only neuron 4 contributes in this region to produce the increasing output by decreasing y_4, i.e., $z = 0.65 − 0.57y_4$. When it comes to the last fall, neurons 1 and 4 are inactive, 2 is fully active ($y_2 = 1$), and neuron 3 decreases its activation. Thus, in this region neurons 2 and 3 jointly contribute to produce the output, i.e., $z = 0.65 − 0.58y_2 + 0.53y_3 = 0.07 + 0.53y_3$. The decreasing activity of neuron 3 gradually brings the output to a minimum.

In summary, neurons process data locally but interact globally to produce the output. Furthermore, they become active at various locations in the input space to produce the required intensity or to offset a continuing trend. The pattern that forms out of this complex and smoothly propagating activity is the desired target function, which models the bimodal migratory data.

The MSE for all the data points for this network was calculated by

$$\text{MSE} = \frac{1}{2N} \left[\sum_{i=1}^{N} (t_i − z_i)^2 \right], \qquad (3.22)$$

where N is the number of data points, t is the target, and z is the network output. The resulting MSE is 0.085. The square root of MSE (RMSE) is 0.291. Chapter 4 will use this concept of network error to train a network.

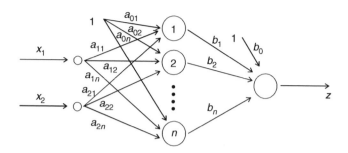

Figure 3.34 Two-input multilayer network structure.

3.4 Two-Input Multilayer Perceptron Network

The understanding gained with one input can be extended to two or more inputs. By looking at two inputs, it is possible to build a solid foundation for the understanding of networks with many inputs, because the knowledge we gained from one and two inputs (one- and two-dimensional problems) generalize to many inputs (multidimensional problems). This is possible because the basic principles underlying neural information in MLP can be extracted from these examples. A network with two inputs can approximate any function or predict any output that depends on two independent variables. Hence, it can solve any two-dimensional prediction or classification problem.

The structure of a two-input network is shown in Figure 3.34, in which there are two inputs, one or more hidden neurons, and one output. For classification problems involving more than two classes, it is necessary to use one output neuron for each class. However, for most prediction problems, only one neuron is needed. With two inputs, the network has many extra weights, thus making learning more complex in terms of number of weights to be optimized.

3.4.1 Processing of Two-Dimensional Inputs by Nonlinear Neurons

The discussion of this network is treated in the same way as the one-dimensional network, beginning with the processing of one isolated, hidden neuron, as shown in Figure 3.35. Here, x_1 and x_2 are inputs, and a_0, a_1, and a_2 are the bias and the input-hidden neuron weights. The u represents the weighted sum of inputs, and y is the hidden neuron output. All activation functions are assumed to be logistic.

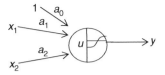

Figure 3.35 One hidden neuron of the two-dimensional nonlinear network.

Each hidden neuron receives two inputs and the bias input, which are weighted by the corresponding weights and summed during the first stage of computing. The weighted sum u for the above neuron is

$$u = a_0 + a_1 x_1 + a_2 x_2. \tag{3.23}$$

This relationship is a plane in two-dimensional space of x_1 and x_2, as shown in Figure 3.36. The weight a_1 controls the slope of the plane with respect to the x_1 axis, and a_2 controls its slope with respect to the x_2 axis. Therefore, the effect of learning is to map inputs x_1 and x_2 to a two-dimensional plane and to completely control the position and orientation of the plane in two-dimensional space through a_0, a_1, and a_2.

The weighted sum u is passed through the logistic function to obtain the hidden node output, y, as follows:

$$y = \frac{1}{1 + e^{-u}}, \tag{3.24}$$

where y is a standard logistic function with respect to u. Once again, the boundary point is where $u = 0$, at which point $y = 0.5$. By substituting for u, it is possible to find how the weights map inputs x_1 and x_2 to the hidden-node output as follows:

$$y = \frac{1}{1 + e^{-(a_0 + a_1 x_1 + a_2 x_2)}}. \tag{3.25}$$

The output y is now a two-dimensional logistic function, in contrast with the one-dimensional function seen in the one-input case. The

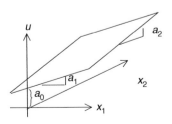

Figure 3.36 Mapping of two-dimensional inputs to weighted sum.

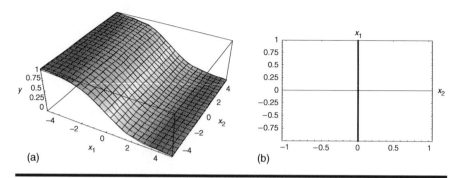

Figure 3.37 Two-dimensional logistic function characteristics for $a_0 = 0$, $a_1 = 1$, $a_2 = 0$: (a) logistic function and (b) boundary line that symmetrically divides the input space.

boundary is still defined by $u = 0$; however, it is no longer a point, but a line defined by

$$u = a_0 + a_1 x_1 + a_2 x_2 = 0. \tag{3.26}$$

Now y and u will be explored for several cases of a_0, a_1, and a_2 values to link the one-dimensional logistic with the two-dimensional case.

1. $a_0 = 0$, $a_1 = 1$, $a_2 = 0$

The plot of y as a function of x_1 and x_2 for this case is shown in Figure 3.37a, which depicts a standard logistic function in two-dimensional space. The value of the weight a_1 controls the slope of the function with respect to the x_1 axis. Because $a_2 = 0$, the slope with respect to x_2 is zero. Because $a_0 = 0$, the function is centered at $x_1 = 0$ and $x_2 = 0$. The equation for the boundary line that results from passing a horizontal plane through the middle of the logistic function can be obtained by solving $u = 0$, and is plotted in Figure 3.37b as a straight vertical line. Neuron activation increases below the line and decreases above it.

The points where the boundary line crosses the x_1 and x_2 axes can be found using simple algebra. Denoting the boundary point on x_1 axis by x_1' and that on x_2 axis by x_2', x_1' can obtained by substituting $x_2 = 0$ into Equation 3.26 as follows:

$$x_1' = \frac{-a_0}{a_1}.$$

Similarly, by substituting $x_1 = 0$ into Equation 3.26:

$$x_2' = \frac{-a_0}{a_2}.$$

2. $a_0 = 0$, $a_1 = 0$, $a_2 = 1$

This case is plotted in Figure 3.38a, which demonstrates that a_2 controls the slope of the function with respect x_2 and the slope with

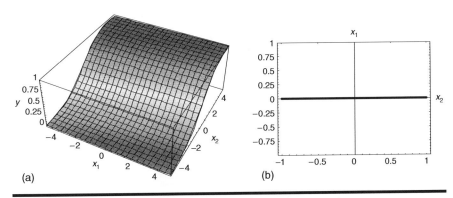

(a)

(b)

Figure 3.38   Two-dimensional logistic function characteristics for $a_0 = 0$, $a_1 = 0$, $a_2 = 1$: (a) logistic function and (b) boundary line that symmetrically divides the input space.

respect to x_1 is zero due to a_1 being zero and the function being centered at $x_1 = 0$ and $x_2 = 0$. The boundary line in this case is a straight horizontal line, as plotted in Figure 3.38b. The neuron activation increases above the line and decreases below it.

3. $a_0 = 0$, $a_1 = 1$, $a_2 = 2$

In this case, where a_1 and a_2, which control the slopes, are nonzero, a more complex logistic function is produced, as shown in Figure 3.39a. The boundary line is, predictably, along the diagonal, symmetrically dividing the input space as shown in Figure 3.39b. Neuron activity is greater above the line and lesser below it.

4. $a_0 = -0.5$, $a_1 = 1$, $a_2 = -1$

In this case, the slope with respect x_1 is positive, and that with respect to x_2 is negative, as shown in Figure 3.40a. The effect of a_0 is to offset the boundary line, which essentially shifts the region of the highest activity from the center, as shown in Figure 3.40b.

The above graphical illustrations show that two inputs are mapped to a two-dimensional logistic function of y whose slopes are controlled by the weights a_1 and a_2. The weight a_0 shifts the region of the highest activation of the logistic function, depicted by the boundary line, in the two-dimensional input space. Comparing this with the one-input case, it is possible to visualize how several neurons can act together to approximate a two-dimensional function or a model that predicts or classifies an outcome from two independent variables. Basically, each node crafts its own two-dimensional sigmoid function, whose shape and position are controlled by its weights a_0, a_1, and a_2, depending on the nonlinear nature of the function being approximated. In this way, several neurons assuming their form in a flexible manner adds the power and tremendous flexibility to the network

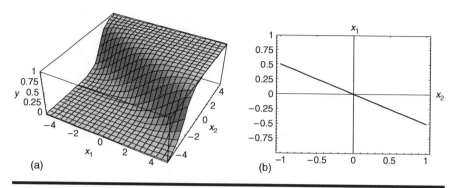

(a)

(b)

Figure 3.39 Two-dimensional logistic function characteristics for $a_0 = 0$, $a_1 = 1$, $a_2 = 2$: (a) logistic function and (b) boundary line that symmetrically divides the input space.

that allows it to approximate any two-dimensional function for prediction or classification.

3.4.2 Network Output

The last stage of the processing is to synthesize the hidden-neuron outputs by computing their weighted sum, v, and then processing v through the output neuron activation function. The weighted sum produces the desired form of the target function in a similar fashion to the one-input case, except that the hidden neuron output is now produced by two-dimensional logistic functions that work together. The output activation, z, adjusts the weighted sum to approximate the target function, t. The z in the case of prediction can be an arbitrarily complex, nonlinear surface. In classification, z values

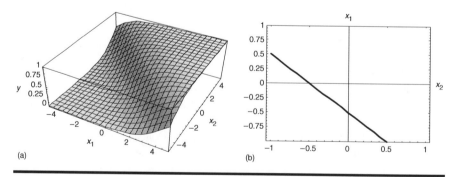

(a)

(b)

Figure 3.40 Two-dimensional logistic function characteristics for $a_0 = -0.5$, $a_1 = 1$, $a_2 = -1$: (a) logistic function and (b) boundary line that symmetrically divides the input space.

above 0.5 (or any user-defined threshold) are adjusted to one and classified as one class, and those below it are adjusted to 0 and classified as another. Prediction and classification are basically the same problem except for this final adjustment. Thus, classification is a subset of prediction problems. The final classification boundary is obtained by passing a plane across the model output surface horizontally at the boundary value of the output activation, and this boundary can be arbitrarily complex and nonlinear, dividing the input space into classes in a complex manner. The next section will visually explore nonlinear model surfaces and complex classification boundaries of a two-input network. It will also aid in understanding the power of the trained networks, as well as how proficiently they perform the prediction or classification task. First, a prediction problem will be discussed.

3.4.3 Examples: Two-Dimensional Prediction and Classification

3.4.3.1 Example 1: Two-Dimensional Nonlinear Function Approximation

In this example, a complex, two-dimensional nonlinear function is approximated to illustrate the powerful feature of approximating highly nonlinear functions by multilayer networks. The target data was generated from the function shown in Figure 3.41, which shows that the target outcome has a nonlinear relationship with x_1 and x_2. The data was modeled

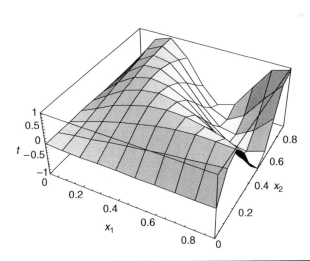

Figure 3.41 Two-dimensional nonlinear surface to be approximated by the neural network. (Adapted from *Mathematica—Neural Networks*, Wolfram Research, Champaign, IL, 2002.)

by a neural network with one hidden layer of four neurons using logistic activation functions. The output activation function is linear [15].

Figure 3.42a shows the network output with random initial weights before training, and Figure 3.42b shows that of the fully-trained network. Training took 20 epochs with mean square error minimization using the delta rule.

Function approximation accuracy of the network. According to Figure 3.42, the approximated surface is quite close to the target surface, which illustrates how neural networks are capable of approximating highly nonlinear problems. The prediction error for each input–output pair can be obtained in the usual way by subtracting predicted outcome from the target. The histogram of the error (or the error distribution) for the above problem is shown in Figure 3.43. It shows that the error distribution is approximately

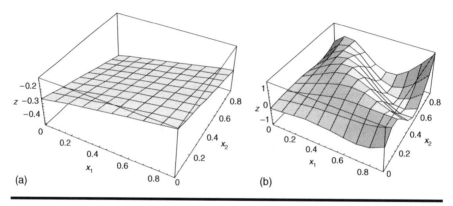

(a)   (b)

Figure 3.42 Neural network approximation of the two-dimensional nonlinear surface: (a) surface created by initial random weights, (b) predicted surface from a trained neural network.

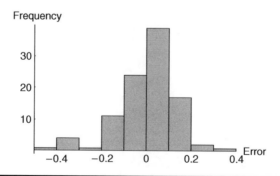

Figure 3.43 Error histogram for the network output.

normal with a mean around zero, indicating that the network has captured the trend, leaving the unaccounted-for variance in output as noise.

The nonlinear surface in Figure 3.42 is generated by the combined response of the two-dimensional logistic activation functions of the four hidden neurons. Due to the adaptation of the free parameters (weights) associated with each function, a unique configuration is assumed by these functions in the input space in such a way that arbitrarily nonlinear surfaces can be created by their combined response. For the same reason, multilayer networks can form complex classification boundaries and in the next example, the outcome of an MLP classification problem will be visualized.

3.4.3.2 Example 2: Two-Dimensional Nonlinear Classification Model

This example solves a two-dimensional classification of data requiring complex nonlinear classification boundaries. Data shown in Figure 3.44 comprises three classes, each containing two clusters. Each class has 20 observations, producing a total of 60 observations. The data was sourced from *Neural Networks for Mathematica* [15]. This problem was solved with an MLP network with six hidden neurons using a logistic activation function. Because there are three classes, three output nodes are required to represent these classes. The output activation function is logistic; this is more suitable than a linear function for classification, because a class is represented in the output as either 0 or 1, with one indicating class membership and zero indicating nonmembership. The task of the classifier is to sort the data into three classes by creating appropriate boundaries. Learning involves

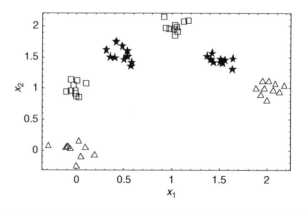

Figure 3.44 Three classes of data with two clusters in each class. (From *Mathematica—Neural Networks,* **Wolfram Research, Champaign, IL, 2002.)**

controlling the form of the hidden neuron activation functions through weight adjustment until the desired boundaries are created.

Figure 3.45 and Figure 3.46 show the performance of the developed classifier superimposed on the data. In Figure 3.45, hidden neuron boundary lines are superimposed on the data to show how learning has evolved network weights in such a way that the boundary line of hidden neurons has separated classes and clusters. Careful examination of the boundary lines reveals that clusters belonging to classes have been properly identified.

A clearer view of classification can be seen in Figure 3.46, which illustrates the classification boundaries produced by the output neurons whose function is to combine the output of hidden neurons. The figure shows more clearly the nonlinear classification boundaries crafted by the hidden

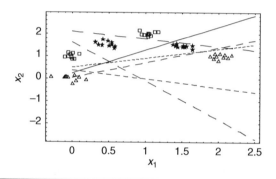

Figure 3.45 Boundary lines of the six hidden neurons superimposed on data.

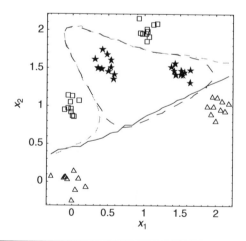

Figure 3.46 Final nonlinear classification boundaries created by the network.

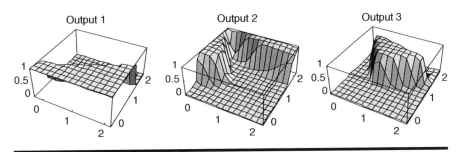

Figure 3.47 Output function for each of the output neurons shown in the input space.

neurons to accommodate clusters within classes. For example, the network has correctly grouped clusters into appropriate classes, and the three classes are distinctly separated from each other.

The surfaces generated by the network can explain how such complex boundaries are formed. Recall that MLPs construct nonlinear surfaces or functions of input data to predict the target outcome. With three output neurons, there are three output surfaces, one for each neuron; they are shown in Figure 3.47. The classification boundaries in Figure 3.46 are obtained by passing a plane through the output of each neuron shown in Figure 3.47 in such a way that it divides the output in half, and projecting the resulting boundaries of the three neurons onto the input space.

Observing the output functions and the boundary lines for each neuron, it can be noted in Figure 3.46 that neuron 1 represents the class indicated by Δ, neuron 2 represents the class indicated by □, and neuron 3 classifies data belonging to the class denoted by ★. There is a very small overlap between the classes denoted by ★ and Δ. However, for the given data, this is not an issue because there is no data in the overlapping region and classification of the data will always be correct. However, if the overlap must be removed, more data from this region must be generated and the network retrained, in which case shape of at least the two affected output functions will change to accommodate the new data and eliminate the overlap.

Network classification accuracy. The classification error is defined by the number of patterns that are wrongly classified, and can be visualized in a three-dimensional bar graph with the number of patterns classified plotted against the corresponding actual class and the predicted class for each pattern. Figure 3.48 illustrates the progress at three instances during training of the classifier. (Iteration is the same as epoch.) For correct classification, actual and predicted classes must be the same. At the initial stages of training (first plot), classification is not accurate; however, the fully trained MLP classifier (last plot) works perfectly with zero

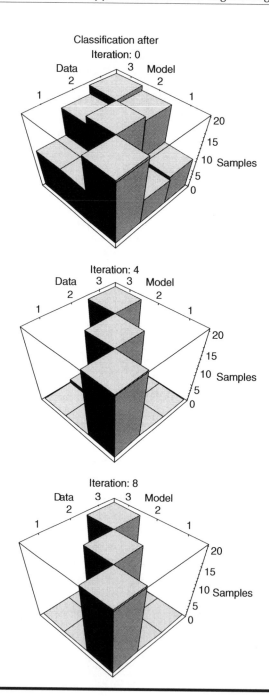

Figure 3.48   Progress of the MLP classifier for data belonging to three classes with two clusters in each.

classification error, indicated by the fact that there are no instances in which the actual class and the predicted class differ, as depicted by off-diagonal entries.

To this point, information processing in one-input and two-input MLPs capable of one-dimensional and two-dimensional prediction or classification have been studied. The concepts learned from these two cases can be extended to networks with multiple inputs. The problem is that the logistic functions become multidimensional and therefore cannot be visualized graphically. However, it is possible to intuitively understand the whole process of information processing even in these networks, based on an understanding of data processing in one- and two-dimensional networks. As the pattern of the formulation is now clear, it is possible to write the equations with relative ease, as shown in the next section.

3.5   Multidimensional Data Modeling with Nonlinear Multilayer Perceptron Networks

An MLP, in its most general form, can have many inputs and many outputs. In the case of prediction, there is usually one output neuron; multiclass classification requires more than one. There can be one or several hidden layers and any number of hidden neurons in each layer. In the general case where there are n inputs, m hidden neurons, and k output neurons, the intermediate stages of processing within an MLP can be constructed as follows.

The hidden neuron input u_j and output y_j of the jth neuron are

$$u_j = a_{0j} + \sum_{i=1}^{n} a_{ij} x_i$$

$$y_j = f(u_j),$$

(3.27)

where x_i is the ith input, a_{ij} is the weight associated with the input i and neuron j, a_{0j} is the bias weight of hidden neuron j and $f(u_j)$ can be any activation function that transforms u_j into a hidden neuron output y_j.

The weighted sum of inputs v_k and the output z_k of the kth output neuron can be written as

$$v_k = b_{0k} + \sum_{j=1}^{m} b_{jk} y_j$$

$$z_k = f(v_k),$$

(3.28)

where m and k are the number of hidden neurons and output neurons, respectively, b_{0k} is the bias weight of output node k, b_{jk} is the weight of the connection between the jth hidden neuron and the kth output neuron, and

$f(v_k)$ is the activation function of the kth output neuron, which transforms v_k into its final output.

3.6 Summary

This chapter covered an in-depth study of information processing in multiple-layer perceptron networks. Specifically, nonlinear processing in a single-input neuron with logistic activation function was examined, and it was shown that the bias and the input-hidden weights alter the slope and horizontal position, respectively, of the logistic function so that it assumes the shape of the data. A single neuron with a logistic function can approximate any monotonically changing trend in data. When there are reversals in the trend of the data, more than one neuron is needed to take care of each of these changes in trend. The training alters the slope and position of the corresponding activation functions through changes in weights such that they are located in the critical regions of the input space. Only the performance of networks with logistic functions was specifically demonstrated, but the concepts apply exactly to other functions as well.

The next chapter covers the ways in which networks learn to produce the outcomes presented in this chapter. Learning involves adjusting weights that control the configuration of activation functions in input space. Weights are the free parameters or the degrees of freedom of a network, and these free parameters are optimized through learning.

Problems

1. What does "nonlinear" refer to in nonlinear data analysis?
2. Discuss how nonlinear processing is incorporated into neurons. For a single-input neuron, plot the input–output relationship mapped by different activation functions.
3. What criteria are used in selecting an activation function? Comment on the similarities or differences between activation functions.
4. What is an active region of an activation function and what is its significance?
5. How does learning alter activation functions and what drives this process?
6. What is a boundary point (or line) for an activation function and what is its significance?
7. Explain in detail how nonlinear mapping is performed by the network in Figure 3.22.
8. What is the difference between one-dimensional and multi-dimensional input mapping in relation to internal workings of a network?

9. Select an input–output pair from the latter part of the square wave depicted in Figure 3.21 and compute the output of the nonlinear network in Figure 3.22 that approximates this function.

10. Draw the following three sigmoid functions for u ranging from -10 to 10 on the same plot and answer the questions that follow regarding the nature of these functions:

   (i) logistic function: $g(u) = \dfrac{1}{1 + e^{-u}}$

   (ii) $b(u) = 2g(u) - 1 = \dfrac{1 - e^{-u}}{1 + e^{-u}}$

   (iii) hyperbolic tangent: $\tanh(u) = \dfrac{e^u - e^{-u}}{e^u + e^{-u}}$

   (a) Comment on the differences between these functions in terms of the upper and lower bounds and slope.
   (b) Compute the function values for $u = -0.5$ and $u = 7.0$.

11. For the logistic function $y = g(a_o + a_1 x)$, analyze the following cases for x ranging from -10 to 10:

   (a) $a_o = 0 \rightarrow$ Draw the function for $a_1 = 0.1, 1, 3, -1$ and calculate the slope at the boundary point.
   (b) $a_1 = 1 \rightarrow$ Draw the function $a_o = 0, -3, 6$.
   (c) $a_1 = 0.2 \rightarrow$ Plot the function and calculate the boundary point for $a_o = 0, -3, 6$.

12. Use a neural network software to train a multilayer network to approximate the sine function (sin x) using one input neuron, five hidden-layer neurons, and one output neuron. This should predict $\sin(x)$ for a given value of x

   (a) Extract 100 or more input–output pairs from the sine function. Call this the target function (t).
   (b) Train a network that predicts $\sin(x)$ from x.
   (c) Extract weights (i.e., input-hidden neuron weights and hidden-output neuron weights) from the software.
   (d) Plot the hidden neuron output y and output neuron output z as a function of x. Find the boundary point for each hidden neuron output function and comment on the participation of the hidden neurons in producing the output.
   (e) Calculate the MSE for the network using the following formula:

$$MSE = \frac{1}{2N} \left[\sum_{i=1}^{N} (t_i - z_i)^2 \right]$$

where t_i is the target output, z_i is the network output for the ith data point, and N is the number of data points.

References

1. Lek, S. and Jean-Francois, G. ed. *Artificial Neuronal Networks: Application to Ecology and Evolution*, Springer Environmental Science Series, Springer, New York, 2000.
2. Lek, S. and Jean-Francois, G. Artificial neural networks as a tool in ecological modeling—An introduction, *Ecological Modelling*, 120, 65, 1999.
3. Wu, C.H. and McLarty, J.W. *Neural Networks and Genome Informatic, Methods in Computational Biology and Biochemistry*, Elsevier, Oxford, 2000.
4. Smith, M. *Neural Networks for Statistical Modeling*, International Thompson Computer Press, London, UK; Boston, MA, 1996.
5. Rumelhart, D.E and McClelland, J.L. *Foundations. Vol. 1 of Parallel Distributed Processing—Explorations in the Microstructure of Cognition*, MIT Press, Cambridge, MA, 1986.
6. Rumelhart, D.E. and McClelland, J.L. *Psychological and Biological Models. Vol. 2 of Parallel Distributed Processing—Explorations in the Microstructure of Cognition*, MIT Press, Cambridge, MA, 1986.
7. Werbos, P.J. *The Roots of Backpropagation. From Ordered Derivatives to Neural Networks and Political Forecasting*, Wiley Series on Adaptive and Learning Systems for Signal Processing, Communication, and Control, Wiley, New York, 1994.
8. Kohonen, T. *Self Organization and Associative Memory*, Springer-Verlag, Berlin, 1984.
9. Haykin, S. *Neural Networks: A Comprehensive Foundation*, 2nd Ed., Prentice Hall, Upper Saddle River, NJ, 1999.
10. Principe, J.C., Euliano, N.R., and Lefebvre, W.C. *Neural and Adaptive Systems—Fundamentals Through Simulations*, Wiley, New York, 2000.
11. Fausett, L. *Fundamentals of Neural Networks—Architectures, Algorithms and Applications*, Prentice Hall, Upper Saddle River, NJ, 1994.
12. Kohonen, T. *Self-Organizing Feature Maps*, 3rd Ed., Springer-Verlag, Berlin, 2001.
13. Eberhart, R.C. and Dobbins, R.W. *Neural Networks PC Tools—A Practical Guide*, Academic Press, London, 1990.
14. Eberhart, R., Simpson, P., and Dobbins, R. *Computational Intelligence PC Tools—An Indispensable Resource for the Latest in Fuzzy Logic, Neural Networks, Evolutionary Computing*, Academic Press, London, 1996.
15. *Mathematica—Neural Networks Version 1.0.2*, Wolfram Research Inc., Champaign, IL, 2002.
16. *Mathematica Version 5.0*, Wolfram Research Inc., Champaign, IL, 2002.

Chapter 4

Learning of Nonlinear Patterns by Neural Networks

4.1 Introduction and Overview

Multilayer networks can perform complex prediction and classification tasks. Chapter 3 detailed examples of one-dimensional and two-dimensional predictions involving highly nonlinear relationships as well as nonlinear classification boundaries. Such complex approximations are facilitated by nonlinear activation functions in hidden neurons whose features are controlled by the weights of the networks. Learning involves the simultaneous and incremental adjustment of these weights in such a way that the activation functions gradually assume features that help collectively approximate the desired response. In the process, the network prediction error goes down incrementally until it falls below a specified error threshold. This process is called training a network.

This chapter treats the concepts of learning in depth and illustrates them in detail. Specifically, beginning with mean square error (MSE), it graphically portrays the error surface in relation to the weights in order to define the error minimization problem in Section 4.2. It looks deeply into the error gradient with respect to each of the weights that are to be optimized, and, in Section 4.2 and Section 4.3, examines how learning methods operate on these gradients. Specifically, it will examine several variants of gradient descent, namely backpropagation, delta-bar-delta (or adaptive learning rate), steepest descent, QuickProp, Gauss–Newton, and the Levenberg–Marquardt (LM) learning methods. Backpropagation, delta-bar-delta, and steepest descent are first-order error minimization methods based solely on the gradient of the error surface. Gauss–Newton and LM learning methods are second-order error minimization methods in which the gradient descent concept is extended to include the curvature (second derivative) of the error surface.

Each learning method is treated in detail with a hand calculation and a computer experiment, and the methods are compared to one another to ascertain their relative effectiveness. Specifically, backpropagation is presented in Section 4.4, delta-bar-delta in Section 4.5, and steepest descent in Section 4.6. The concept of second-order methods of error minimization is presented in Section 4.7, in which QuickProp (Section 4.7.1), the Gauss–Newton method (Section 4.7.3), and the LM method (Section 4.7.4) are treated in detail. First-order and second-order error minimization methods are compared in relation to the efficiency of error minimization in Section 4.7.5, and convergence characteristics are discussed in Section 4.7.6.

4.2 Supervised Training of Networks for Nonlinear Pattern Recognition

The training of feedforward networks such as the multilayer perceptron (MLP) is supervised in that, for each input, the corresponding output is also presented to the network. The initial weights are set at random. The network processes each input vector and the network output is compared with the desired or target output. Initially, the error would be large, due to the random assignment of values to weights. The MSE, the most commonly used error indicator, of the prediction over all the training patterns for a network with one output neuron can be written as

$$E = \frac{1}{2N} \sum_i^N (t_i - z_i)^2 \tag{4.1}$$

where E denotes MSE, t_i and z_i are the target and the predicted output for the ith training pattern, and N is the total number of training patterns. The division by 2 is a mathematical convenience and is conventionally used in statistics, although in some disciplines it is not used. As illustrated in Chapter 3, z depends on the output neuron activation function, as well all of the outputs of the hidden neurons, which in turn depend on the hidden-neuron activation functions and inputs.

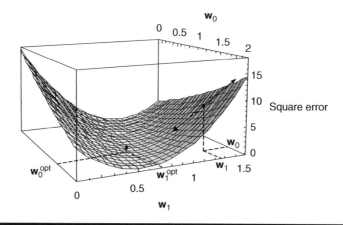

Figure 4.1   A two-dimensional error surface with respect to weights w_0 and w_1 and slope of error surface.

The best way to start probing into learning is to look at the features of the error surface. Figure 4.1 shows the MSE surface generated in Chapter 2 for a single neuron with bias weight $\mathbf{w}_0$ and weight $\mathbf{w}_1$. In this problem, there are two parameters to be adjusted, $\mathbf{w}_0$ and $\mathbf{w}_1$, and the error surface shows the amount of error that would result for each combination of $\mathbf{w}_0$ and $\mathbf{w}_1$. The lowest point on the surface gives the optimum set of weights, and the learning challenge is to find the weights that produce the minimum error for the whole training set. In a practical network, there are more than two weights, so it is not possible to simultaneously visualize the error function with respect to all weights; however, it is possible to visualize two weights at a time, and the above concept would hold equally true for all of them. In linear regression, the least square error method is used to directly obtain the coefficients of a linear equation that minimize the error on a set of data. Unfortunately, for highly nonlinear problems, there is no such direct method to find the weights, and they must be established iteratively. Trying all of the possible combinations of weights randomly would be prohibitively costly in time and effort. The gradient descent approach is an efficient method to find the bottom of the error surface more quickly during network training.

4.3   Gradient Descent and Error Minimization

Gradient descent, as the name implies, uses the error gradient to descend the error surface. Here, the gradient is the slope of the error surface, which indicates how sensitive the error is to changes in the weights; this sensitivity can be exploited to incrementally guide the changes in the weights towards the optimum. Starting with random values for the weights, this method finds

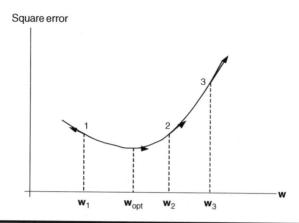

Figure 4.2 A slice through the error surface at three values of a weight.

the slope of the error surface at these weights. For example, the up-arrow in Figure 4.1 shows the slope at a point on the error surface. The gradient descent method proposes to change the weights in the direction in which the error decreases most rapidly, i.e., in the opposite direction to the gradient, as shown by the down arrow in Figure 4.1. By continuing to do this iteratively, the bottom of the error surface will eventually be reached, and the optimum set of weights will be found. This is illustrated in Figure 4.2, in which a slice through the error function along one weight is shown for clarity. This shows how the error changes with one weight. If the starting point on the error surface is at 1, the slope is negative, as shown, and the value of the weight needs to be increased to reach the optimum. If the current point is at 2, the slope is positive and the weight has to be decreased. If the current point is at 3, the slope is positive, but the weight is still far too large and has to be decreased.

It is first necessary to find the slope of the error surface. Then it is necessary to know how much to go down the error surface and exactly how the weights should be adjusted after each iteration. The methods that are used to adjust the weights are called learning rules, and this chapter will spend a great deal of time exploring these issues. Other issues to be addressed in this chapter include when to stop training and what governs this decision; whether the solution (i.e., weights) are suboptimal; and how good the final model is and how to assess its goodness of fit.

4.4 Backpropagation Learning

The error derivative, or the slope of the error surface in relation to weights, is crucial to the adjustment of the weights. During the training of a network, such as the one shown in Figure 4.3, all of the output and hidden-neuron

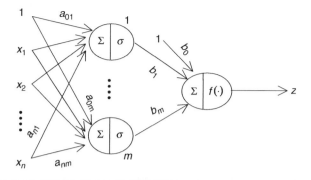

Figure 4.3 Multilayer perceptron and weights (free parameters) to be adapted.

weights must be adjusted simultaneously. Therefore, it is necessary to find the error derivative with respect to all these weights. Denote the derivative with respect to output node weights by $\partial E/\partial b$, and the derivatives with respect to the hidden node weights by $\partial E/\partial a$. Since E is not directly linked to b and a, the concept of chain rule is used that finds the derivatives when the link from the error E to a weight is not direct by following through the associations one by one. Chapter 2 illustrated that E is linked to z (the network output) and z is related to v (the weighted sum of hidden-neuron outputs y), which in turn depends on the weights b. So it is possible to use the chain rule of differentiation to obtain the $\partial E/\partial b$. Then, v is also related to y (the hidden-node output), which links the inputs and the input-hidden neuron weights, a. Thus it is possible to follow this chain of association from E, z, v, y to the inputs to obtain $\partial E/\partial a$. This concept is called backpropagation, and was first proposed by Werbos [1] and later by Rumelhart [2,3].

4.4.1 Example: Backpropagation Training—A Hand Computation

This section includes an example to familiarize the reader with the training and derive error gradients for a simple one-input, one-output network with one hidden neuron using a logistic activation function; the network is shown in Figure 4.4a. The function to be approximated is the first quarter of the sine wave, shown in Figure 4.4b. Two input–output pairs {(0.7853, 0.707) and (1.57, 1.0)} were chosen from Figure 4.4b. (Note that this example is continued from here to the end of the chapter in various sections, as appropriate, to highlight relevant aspects of the discussion.)

Once the network configuration has been decided, the first step in training is to initialize the weights to random values. The random initial values chosen for this problem are given in Table 4.1, along with the input x

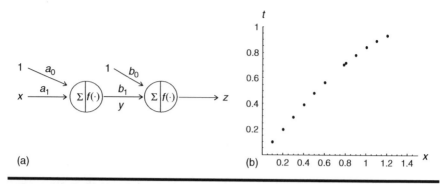

Figure 4.4 A network training example: (a) simple one-input, one-output, and one-hidden-neuron networks, (b) first quarter of sine wave to be approximated.

and target t. Then an input value is presented to the network and a forward pass of the input is made. The order of this process is to first determine u (the weighted sum of the inputs to the hidden neurons) and to transform this to y, the hidden-neuron output. Next, the weighted sum of inputs, v, to output neuron is calculated and transformed to z, which completes the forward pass. At this point, the target output is presented to the network and the MSE is calculated.

Following the forward pass to determine the network output for $x = 0.7853$, $t = 0.707$:

$$u = a_0 + a_1 x = 0.3 + 0.2(0.7853) = 0.457$$

$$y = \frac{1}{1 + e^{-u}} = \frac{1}{1 + e^{-(0.457)}} = 0.612$$

$$v = b_0 + b_1 y = -0.1 + 0.4(0.612) = 0.143$$

$$z = \frac{1}{1 + e^{-v}} = \frac{1}{1 + e^{-(0.143)}} = 0.536.$$

The predicted output z and target t are not equal, so there is a prediction error, which can be represented by the square error. For the

Table 4.1 Initial Weights and Two Example Input–Output Patterns (x, t)

a_0	a_1	b_0	b_1	x	t
0.3	0.2	−0.1	0.4	0.7853	0.707
				1.571	1.00

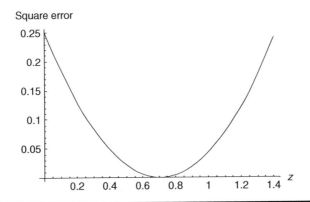

Figure 4.5 **The square error as a function of network output z.**

single input–output pair of {*x*, *t*} this is

$$E = \frac{1}{2}(z - t)^2, \tag{4.2}$$

where *z* is the network output and *t* is the target, or desired output. Note that (*z*−*t*) is used instead of (*t*−*z*). There is no real effect in this change other than the convenience of not having to carry a minus sign later in the calculation. For the given value of *t* = 0.707, the error function *E* with respect to *z* is quadratic, with a minimum at the point where *z* = *t* = 0.707, as shown in Figure. 4.5.

For the example problem, the square error is

$$E = \frac{1}{2}(0.536 - 0.707)^2 = 0.0146.$$

In fact, Equation 4.2 can be expanded by substituting for *z*, which requires *v*, which in turn involves, *y*, *u*, and *x* (see the expressions used above for the hand calculations) to express the square error for the input–output pair {*x*, *t*} as a function of the weights. The output *z* and *E* expressed this way are

$$z = \frac{1}{1 + e^{-(b_0 + b_1)\left\{\frac{1}{1 + e^{-(a_0 + a_1 x)}}\right\}}}$$

$$E = \frac{1}{2}\left\{\frac{1}{1 + e^{-(b_0 + b_1)\left\{\frac{1}{1 + e^{-(a_0 + a_1 x)}}\right\}}} - t\right\}^2. \tag{4.3}$$

The first part of Equation 4.3 is the output z as a function of the weights (the free parameters) to be estimated and the input. This is called the neural network model, and it defines the form of the relationship between the output and the input. From this, the input–output relationship can be further investigated. This will be done later in the chapter. In the second part of Equation 4.3, z is substituted into Equation 4.2 to express the error as a function of the weights and the input. From this, it is possible to ascertain how sensitive the error is to each of the free parameters. It is this sensitivity (error derivative or gradient) that is used in gradient descent learning which will now be explored in detail.

4.4.1.1 Error Gradient with Respect to Output Neuron Weights

According to the chain rule, the error derivative for any hidden-output weight b is

$$\frac{\partial E}{\partial b} = \frac{\partial E}{\partial z} \cdot \frac{\partial z}{\partial v} \cdot \frac{\partial v}{\partial b}, \tag{4.4}$$

which consists of three parts. First is the partial derivative of the error with respect to the network output, second is the partial derivative of z with respect to the weighted sum, v, of the inputs to the output neuron, and third is the partial derivative of v with respect to the hidden-output weight b. This section will now examine each of these terms, following the approach used by Smith [4].

Continuing with the original format of the error and differentiating E with respect to z yields

$$\frac{\partial E}{\partial z} = z - t, \tag{4.5}$$

which highlights the reason for using 0.5 in the error formulation and $(z-t)$ for error, which avoids the necessity of carrying a minus sign.

For the example problem

$$\frac{\partial E}{\partial z} = z - t = 0.536 - 0.707 = -0.171.$$

The $\partial E/\partial z$ is basically the slope of this error surface with respect to the network output, i.e., the sensitivity of the error to the network output, as can be depicted by a tangent drawn at a point on the error surface, shown in Figure 4.6. As illustrated, the slope is negative at this point of z.

The second derivative $\partial z/\partial v$ is the slope (derivative) of the activation function used in the output neuron, and is illustrated for the logistic function in Figure 4.7. This indicates the sensitivity of z to any changes in the weighted sum, v, of inputs to the output neuron. As stated in Chapter 3, the

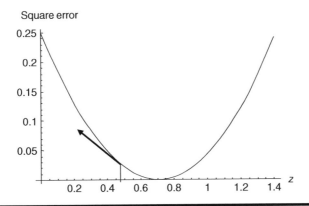

Figure 4.6 ∂*E*/∂*z* with respect to *z*.

steepest slope of the logistic function is equal to 0.25 at the boundary point, where $z = 0.5$. The slope increases continuously up to the boundary point and then slowly decreases, reaching zero again at the upper bound.

The derivative of the sigmoid can be obtained in the standard way as follows:

$$z = \frac{1}{1 + e^{-v}}$$

$$z'(v) = \frac{\partial z}{\partial v} = -\frac{e^{-v}(-1)}{(1 + e^{-v})^{-2}} = \frac{e^{-v}}{(1 + e^{-v})^{-2}}.$$

(4.6)

A graph of the derivative $z'(v)$ is presented in Figure 4.8, which depicts a Gaussian curve. The largest derivative is at $v = 0$, which is the boundary point.

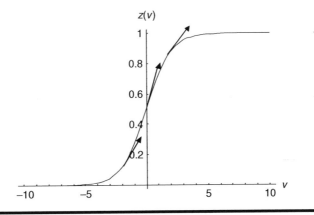

Figure 4.7 **Illustration of the network output gradient ∂*z*/∂*v*.**

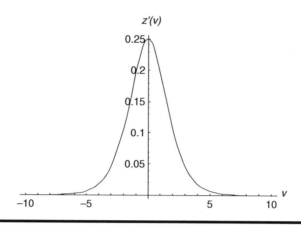

Figure 4.8   Derivative $\partial z/\partial v$.

However, $1 + e^{-v} = 1/z$ and $e^{-v} = \dfrac{1}{z} - 1 = \dfrac{1-z}{z}$. Substituting these into the above equation:

$$\frac{\partial z}{\partial v} = \frac{(1-z)/z}{1/z^2} = z(1-z). \tag{4.7}$$

The third derivative is $\partial v/\partial b$, which indicates the sensitivity of the weighted sum of the inputs v to any changes in the output neuron weights. There are two weights, b_0 and b_1. Because

$$v = b_0 + b_1 y, \tag{4.8}$$

$$\frac{\partial v}{\partial b_1} = y$$

$$\frac{\partial v}{\partial b_0} = 1. \tag{4.9}$$

Thus, the derivative depends on the weight, whether it is the bias or the hidden-output neuron weight.

Now that all three derivative components have been obtained, the required error derivative with respect to the two weights can be presented as

$$\frac{\partial E}{\partial b_0} = (z-t)\,z\,(1-z) = p$$

$$\frac{\partial E}{\partial b_1} = (z-t)\,z\,(1-z)\,y = py. \tag{4.10}$$

The symbol p is used in the equation in subsequent calculations for simplicity, as it was used by Smith [4].

Example problem continued (see Section 4.4.1). For the example problem ($z = 0.536$, $t = 0.707$, $y = 0.612$)

$$\frac{\partial E}{\partial b_0} = (z - t)\,z\,(1 - z) = p = (0.536 - 0.707)(0.536)(1 - 0.536) = -0.042$$

$$\frac{\partial E}{\partial b_1} = py = (-0.042)(0.6120) = -0.026.$$

A graphical illustration of the square error surface (Equation 4.3) with respect to b_0 and b_1 is presented in Figure 4.9 with the current weights ($b_0 = -0.1$ and $b_1 = 0.4$) and resultant error gradient denoted on the surface. For this graph, the current input $x = 0.7853$, the target output $t = 0.707$, and the input-hidden weights of $a_0 = 0.3$ and $a_1 = 0.2$. The optimum weights seem to be further down from this point. By taking a slice of the error surface at these fixed values of b_0 and b_1 separately, as shown in Figure 4.10 and Figure 4.11, it can be seen that the gradients just calculated are indeed negative at this location on the surface. The error curve is more nonlinear with respect to b_0 than b_1.

4.4.1.2 The Error Gradient with Respect to the Hidden-Neuron Weights

The required derivatives here are $\partial E/\partial a_0$ and $\partial E/\partial a_1$. Again, there is no direct link, so the chain rule must be used. Starting from the error, this derivative can be formulated as

$$\frac{\partial E}{\partial a} = \left(\frac{\partial E}{\partial z} \cdot \frac{\partial z}{\partial v} \cdot \frac{\partial v}{\partial y}\right) \cdot \frac{\partial y}{\partial u} \cdot \frac{\partial u}{\partial a} \tag{4.11}$$

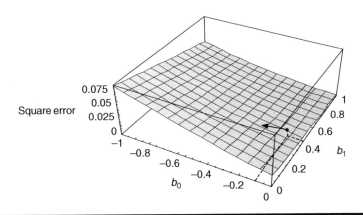

Figure 4.9 The square error surface with respect to b_0 and b_1.

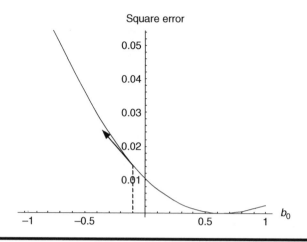

Figure 4.10 The square error with respect to b_0 depicted on a slice through the error surface at $b_1 = 0.4$.

where a is either a_0 or a_1. It can be seen that in this formulation, the chain rule is extended far back to the input. This section will again examine these components separately.

The set of derivatives within brackets in Equation 4.11 is $\partial E/\partial y$. The first two derivatives of this set are already known as p, and only the last component, which indicate the sensitivity of the weighted sum of the inputs, v, to the changes in the hidden-neuron output, is required. This can be easily calculated because $v = b_0 + b_1 y$, resulting in

$$\frac{\partial v}{\partial y} = b_1. \tag{4.12}$$

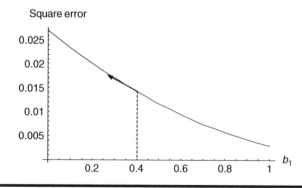

Figure 4.11 The square error with respect to b_1 depicted on a slice through the error surface at $b_0 = -0.1$.

Substituting the three derivatives for the bracketed expression in Equation 4.11

$$\frac{\partial E}{\partial y} = \frac{\partial E}{\partial z} \cdot \frac{\partial z}{\partial v} \cdot \frac{\partial v}{\partial y} = pb_1 \tag{4.13}$$

The second component of $\partial E/\partial a$ is $\partial E/\partial u$, which is simply the derivative of the hidden-neuron logistic function with respect to the weighted sum of the inputs to the hidden neuron. This component is analogous to the previous derivation of the derivative of the output neuron logistic function, and can be written as

$$\frac{\partial y}{\partial u} = y(1 - y), \tag{4.14}$$

where y is the hidden-neuron output. The last component is $\partial u/\partial a$, and it depends on the type of weight considered. Because $u = a_0 + a_1 x$:

$$\frac{\partial u}{\partial a_1} = x$$

$$\frac{\partial u}{\partial a_0} = 1 \tag{4.15}$$

Putting the three components together yields the error derivatives for the bias and the input-hidden-neuron weights as

$$\frac{\partial E}{\partial a_1} = pb_1 y(1 - y) x = qx$$

$$\frac{\partial E}{\partial a_0} = pb_1 y(1 - y) = q. \tag{4.16}$$

Again for simplicity, the symbol q has been used to represent $pb_1 y(1-y)$.

Example problem continued (see Section 4.4.1 and Section 4.4.1.1). For this example problem ($p = -0.042$, $b_1 = 0.4$, $y = 0.612$, $x = 0.7853$)

$$\frac{\partial E}{\partial a_0} = pb_1 y(1 - y) = q = (-0.042)(0.4)(0.612)(1 - 0.612) = -0.004$$

$$\frac{\partial E}{\partial a_1} = pb_1 y(1 - y)x = qx = (-0.004)(0.7853) = -0.00314.$$

The square error surface with respect a_0 and a_1 and the gradient at current weights for the given x and t and the given hidden-output weights of $b_0 = -0.1$ and $b_1 = 0.4$ is shown in Figure 4.12. Slices of the error surface taken at the current weights ($a_0 = 0.3$ and $a_1 = 0.2$) are presented in Figure 4.13 and Figure 4.14. The figures show that the gradients are indeed

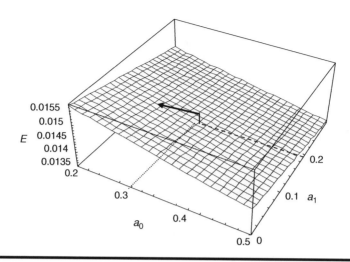

Figure 4.12 The square error surface with respect to a_0 and a_1.

negative, and that there is quite a distance to traverse down the error surface to reach the optimum values for a_0 and a_1, for which the error is at a minimum.

Repeating the procedure for the second input–output pair $\{x, t\} = (1.571, 1.00)$ yields the following values for the gradients:

$$\frac{\partial E}{\partial a_0} = -0.0104$$

$$\frac{\partial E}{\partial a_1} = -0.0163$$

$$\frac{\partial E}{\partial b_0} = -0.1143$$

$$\frac{\partial E}{\partial b_1} = -0.0742.$$

The output and square error are

$$z = 0.5398$$

$$\text{Square error } E = 0.1059. \qquad \text{(from Equation 4.2)}$$

The MSE for the two patterns are $(0.0146 + 0.1059)/2 = 0.0602$.

The error derivatives with respect to all the weights that must undergo transformation during training have now been obtained for two examples. These are a_0, a_1, b_0, and b_1, and their respective error derivatives

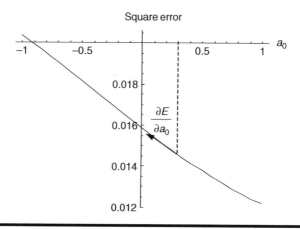

Figure 4.13 **The square error with respect to weight a_0 for $a_1 = 0.2$.**

are $\partial E/\partial a_0$, $\partial E/\partial a_1$, $\partial E/\partial b_0$, and $\partial E/\partial b_1$. With these, it is now possible to apply the learning rules to modify the weights in the next section.

4.4.1.3 Application of Gradient Descent in Backpropagation Learning

The error gradients derived in the previous section are the components of the total derivative at a point on the error surface, with respect to each of the individual weights that define the error surface. Gradient descent dictates that the error should be minimized in the direction of the steepest descent, indicated by the opposite direction to the previously calculated gradients; to accomplish an incremental adjustment of all of the weights, this must be done simultaneously with each gradient. This can be done in two ways: example-by-example or on-line learning, in which the weights

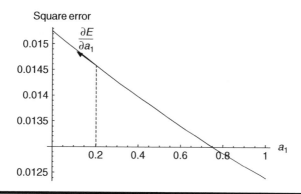

Figure 4.14 **The square error with respect to weight a_1 for $a_0 = 0.3$.**

are adjusted after every training pattern; and batch or off-line learning, in which learning (i.e., weight adjustment) occurs after all of the training examples have been presented to the network once.

The most widely used method is batch learning, in which learning happens in such a way that the overall error with respect to the whole training set decreases incrementally in an average sense, whereas in example-by-example learning, the network learns to minimize error for every example. For complex problems, this may lead to oscillations or instability, and could take longer than batch learning to arrive at the optimum weight values. Both approaches will be treated here.

4.4.1.4   Batch Learning

Because batch learning involves learning after the whole training set has been presented to the network, it is necessary to store the gradients for all of the examples, and to find the average or the resultant gradient after the whole set has been processed. The error is minimized in the direction of the descent indicated by this resultant gradient. To highlight how this method works, imagine a slice of the error surface along one weight axis. Then imagine presenting one training example and calculating the square error E and the gradient $\partial E/\partial w$ for this weight. The w can be either a hidden or an output neuron weight. Now present the next training example, calculate E, find the gradient, and repeat this process for all examples.

Because the square error varies with the input, as shown in Equation 4.3, there is a new error surface, corresponding new values of error, and gradients for each training pattern for the weights at their current locations. This is illustrated in Figure 4.15a for a few hypothetical input patterns, in which the complete error curves for two input patterns are shown (for clarity, those for the other patterns are not shown). The gradients $\partial E/\partial w$ are denoted by short arrows. All the gradients thus obtained for an epoch are shown in Figure 4.15b, in which the length of the arrow indicates the magnitude and the arrow head indicates the direction ($+$ or $-$) of the error gradient with respect to the weight at the current location. The total of all the gradients is shown as a dashed arrow in Figure 4.15b, and represents the average sensitivity of the error to changes in this one weight for the whole training set. This total gradient, d_m, for epoch m (note that an epoch is one pass of the whole training set), can be presented mathematically as

$$d_m = \sum_{n=1}^{N} \left[\frac{\partial E}{\partial w_m} \right]_n, \qquad (4.17)$$

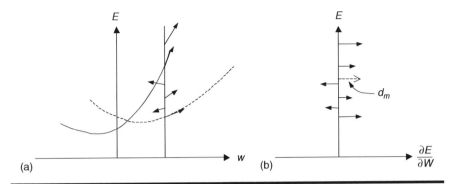

Figure 4.15 **An error gradient with respect to one weight: (a) gradient for several input patterns with error curves for only two input patterns illustrated, and (b) gradient for the whole training set and total derivative d_m (dashed arrow) for epoch *m*.**

which is basically the sum of the gradient for each of the n examples of the mth epoch and the result is the gradient for the mth epoch.

Now imagine that the same has been done for another weight by taking a slice of the error surface with respect to this second weight and determining the resultant gradient. Recall that the magnitude of the error for each input pattern is the same for all weights at the current location because this process is slicing a multidimensional error surface. Denote the two weights by w_0 and w_1. Placing these two average gradients, $\partial E/\partial w_0$ and $\partial E/\partial w_1$, at the appropriate point on the error surface shows these mutually perpendicular gradients, as seen in Figure 4.16. Because the gradient is being used for the whole batch, the error curve in the figure represents the MSE, which is the average error across all of the training patterns, represented as

$$E = \frac{1}{2N} \sum_{i}^{N} (t_i - z_i)^2. \tag{4.18}$$

Therefore, when the error is minimized using the batch gradient, this method minimizes the MSE over the whole training set. The overall or the resultant gradient, $\partial E/\partial W$, of the two batch gradients is shown by the diagonal arrow in Figure 4.16. Its length is the magnitude of the resultant, and the head indicates the direction of the slope. This is the gradient used in the gradient descent method for batch learning. It is necessary to move in the opposite direction from this arrow, as indicated in Figure 4.16, which will be the direction of the steepest descent. It can be seen that during descent of the slope, the two weights decrease in magnitude in search of the

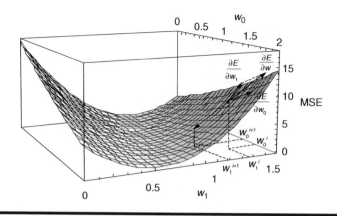

Figure 4.16 An error gradient with respect to two weights, resultant gradient and direction of steepest descent.

bottom of the error surface. The question is how far to descend in this direction; this is controlled by the learning rate, ε.

4.4.1.5 Learning Rate and Weight Update

The learning rate indicates how far in the direction of steepest descent the weights must be shifted per epoch. For example, if ε is 1.0, the distance of the descent will be equal to the total arrow length of the resultant gradient; after this descent, the new weights will be calculated before the next batch is presented. For practical problems in which the error surface is generally more nonlinear than that in Figure 4.16, a smaller learning rate must be used to slowly and smoothly guide the descent towards the optimum weights; therefore, ε is normally between 0 and 1, and indicates the proportion of the arrow length that will be traversed in the direction of the deepest descent. If the new increment of a weight after the epoch m is denoted by Δw_m, the new weight for the epoch $m+1$, w_{m+1}, can be presented as

$$w_{m+1} = w_m + \Delta w_m$$
$$\Delta w_m = -\varepsilon d_m,$$

(4.19)

where a $(-)$ sign indicates the descent, and εd_m represents the distance of the descent as a portion of the gradient, d_m, for the epoch.

Now that the method has been explained, this section will address the question of what the optimum learning rate should be. This depends on the problem and there is unfortunately no direct way to set it other than trial and error. Basically, the optimum learning rate should decrease the error as quickly as possible while guiding the process smoothly down the

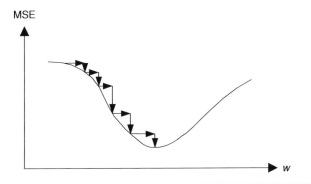

Figure 4.17   The optimal learning rate for efficient error minimization.

error surface, as shown in Figure 4.17 on a slice of the error surface taken through a weight axis. In the figure, the vertical step is the amount of error reduction during each epoch, and the horizontal step is the corresponding adjustment of the weight. Because the actual error surface is multidimensional in weights and because there is an error surface for each input pattern, it is not possible to construct the average error surface, although it can be thought of conceptually. Thus, in order to find the optimum learning rate, it is necessary to understand the effect of sub-optimal learning rates, i.e., the effect of a too-small or too-large learning rate on an average error surface.

Figure 4.18 shows the effect of a large learning rate. It shows that the error may initially decrease, but due to the large learning rate, the weight increments may also be so large that the global minimum of the error surface may either be reached after a long time or the solution may oscillate around the minimum and never reach it. Too-large a learning rate could never reach

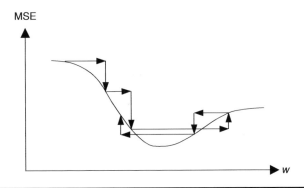

Figure 4.18   The effect of a high learning rate on learning.

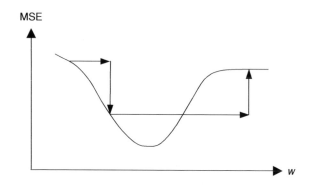

Figure 4.19 The effect of too high a learning rate on learning.

the global minimum if the weight increment is large enough to throw the new weights to the wing of the error surface, as demonstrated in Figure 4.19. Because the gradient is so small or closer to zero in this region, there may not be a large enough weight increment to push it to the steeper regions, and as a result, learning may halt [4].

Example problem continued (see Section 4.4.1 and Section 4.4.1.2). Returning to the two-example problem to adjust the weights after the first epoch, this section will continue with the gradients calculated in Section 4.4.1.3 and the assumed learning rate $\varepsilon = 0.1$.

Epoch 1. The sum of error gradients is

$$d_1^{a_0} = \sum \frac{\partial E}{\partial a_0} = -0.004 - 0.0104 = -0.0144$$

$$d_1^{a_1} = \sum \frac{\partial E}{\partial a_1} = -0.00314 - 0.0163 = -0.01944$$

$$d_1^{b_0} = \sum \frac{\partial E}{\partial b_0} = -0.042 - 0.1143 = -0.1563$$

$$d_1^{b_1} = \sum \frac{\partial E}{\partial b_1} = -0.026 - 0.0742 = -0.1002.$$

The weight change after the first epoch is

$$\Delta a_0 = -\varepsilon d_1^{a_0} = -0.1(-0.0144) = 0.00144.$$

The new value of the bias weight carried forward to epoch 2 is

$$a_0^2 = a_0 + \Delta a_0 = 0.3 + 0.00144 = 0.30144.$$

Similarly, the new values for a_1, b_0, and b_1 can be determined as

$$a_1^2 = a_1 + \Delta a_1 = a_1 - \varepsilon d_1^{a_1} = 0.2 - (0.1)(-0.01944) = 0.2019$$

$$b_0^2 = b_0 + \Delta b_0 = b_0 - \varepsilon d_1^{b_0} = -0.1 - (0.1)(-0.1563) = -0.0844$$

$$b_1^2 = b_1 + \Delta b_1 = b_1 - \varepsilon d_1^{b_1} = 0.4 - (0.1)(-0.1002) = 0.410.$$

Thus, the new adjusted weights after epoch 1 are

$$a_0^2 = 0.3014$$

$$a_1^2 = 0.2019$$

$$b_0^2 = -0.0844$$

$$b_1^2 = 0.410.$$

These new weights are used to calculate the network output for the next epoch and the process is repeated until the error decreases to an acceptable level, or until the network output reaches the target value. The square error from Equation 4.2 for the two patterns and the MSE after epoch 1 with the new weights are

$$\text{Pattern 1: } (0.7583, 0.707) \Rightarrow 0.01367$$

$$\text{Pattern 2: } (1.571, 1) \Rightarrow 0.1033$$

$$\text{MSE} = (0.01367 + 0.1033)/2 \Rightarrow 0.0585,$$

which is smaller than the initial MSE of 0.0602.

Epoch 2. Using the above weights, the average error gradient was calculated using the same procedure; it was found to be

$$d_2^{a_0} = \sum \frac{\partial E}{\partial a_0} = -0.00399 - 0.0105 = -0.01449$$

$$d_2^{a_1} \sum \frac{\partial E}{\partial a_1} = -0.00312 - 0.0165 = -0.0196$$

$$d_2^{b_0} = \sum \frac{\partial E}{\partial b_0} = -0.04106 - 0.1127 = -0.1538$$

$$d_2^{b_1} = \sum \frac{\partial E}{\partial b_1} = -0.0251 - 0.0732 = -0.0983,$$

and the weight changes and new weights after epoch 2 are

$$\Delta a_0^2 = 0.00145 \quad a_0^3 = 0.3029$$

$$\Delta a_1^2 = 0.00196 \quad a_1^3 = 0.2039$$

$$\Delta b_0^2 = 0.01538 \quad b_0^3 = -0.069$$

$$\Delta b_1^2 = 0.00983 \quad b_1^3 = 0.4198.$$

The total error for the two patterns is

$$\text{MSE} = (0.0128 + 0.10085)/2 = 0.0568,$$

which is smaller than that after the first epoch. The above results may be verified by repeating the process followed in epoch 1.

4.4.1.6 Example-by-Example (Online) Learning

The previous example illustrated the weight change after two epochs with a sample of two training patterns. It is also possible to change the weights after each presentation of a training pattern. By changing the weights after each pattern, they could bounce back and forth with each iteration, possibly resulting in a substantial amount of wasted time. This happens because the training examples are randomized and therefore the next example could be from anywhere in the input–output space, requiring random changes to the weights, which might offset the changes already made to the weights. For some problems, this method may yield effective results [5] and it may be the most suitable method for online learning, in which learning or updating is required as and when data arrives in real time. However, for complex mapping problems, the random movement of weights may cause instability problems [4].

The random oscillation of the weights can be minimized by batch training, in which weight adjustment is based on the total error derivative over the whole training set as described in the previous two sections. This allows for weight changes that are correct on average, and the error over the whole training set decreases in an average sense. This section does not include a hand calculation of example-by-example learning, but the reader is encouraged to try the method on the two-input patterns and to compare the difference between batch and example-by-example learning. Another averaging method, called momentum, will be discussed next.

4.4.1.7 Momentum

As illustrated in Section 4.4.1.5 and Section 4.4.1.6, batch learning using MSE helps to improve the stability in the gradient descent approach in its search for optimum weights. Momentum is another averaging approach that provides stability when reaching the optimum weights during learning [4,14], and can be very useful for some problems, especially in online learning. This method basically tags the average of the past weight changes onto the new weight increment at every weight change, thereby smoothing out the net weight change. Recall that batch learning used the sum of current error derivatives to compute the current weight increment. The idea behind momentum is to use the exponential average of all of the previous weight changes to guide the current change. This is presented mathematically in Equation 4.20:

$$\Delta w_m = \mu \Delta w_{m-1} - (1 - \mu)\varepsilon d_m^w, \qquad (4.20)$$

where μ is a momentum parameter that should be between 0 and 1, and Δw_{m-1} is the previous weight change during the preceding epoch. Therefore, μ indicates the relative importance of the past weight change on the new weight increment, Δw_m. The second term on the right contains the usual amount of weight change for epoch m, considering the current total derivative alone for the weight w, d_m^w. However, it is now weighted by $(1 - \mu)$, indicating the amount of influence it has on the proposed weight change (Δw_m) relative to that of the past change. Thus, the current gradient and the past weight change together decide how much the new weight increment will be. For example, if μ is equal to 0, momentum does not apply at all, and the past history has no place. If it is equal to 1, the current change is totally based on the past change. Values of μ between 0 and 1 result in a combined response to weight change. However, note that the influence of the past weight change incorporates that of all previous weight changes as well, because Equation 4.20 is recursive in that each previous weight change would depend on the change prior to that, all the way back to the first change, as shown below:

$$\Delta w_m = \mu \Delta w_{m-1} - (1 - \mu)\varepsilon d_m^w$$
$$\Delta w_{m-1} = \mu \Delta w_{m-2} - (1 - \mu)\varepsilon d_{m-1}^w$$
$$\Delta w_{m-2} = \mu \Delta w_{m-3} - (1 - \mu)\varepsilon d_{m-2}^w, \qquad (4.21)$$
$$\cdot$$
$$\cdot$$
$$\Delta w_1 = (1 - \mu)\varepsilon d_1^w.$$

By recursively substituting the relevant Δw components in each of the above equations up to the first one, the past weight change Δw_{m-1} can be expressed as

$$\Delta w_{m-1} = \mu^{m-1}(1-\mu)\varepsilon d_1^w - \mu^{m-2}(1-\mu)\varepsilon d_2^w - \mu^{m-3}(1-\mu)\varepsilon d_3^w$$
$$- \cdots - \mu(1-\mu)\varepsilon d_{m-1}^w. \tag{4.22}$$

Thus, with momentum, all of the past weight changes are exponentially averaged with the current required change. The recent weight changes influence the current weight update much more than the very distant changes. For example, the proportion of the influence of the weight change in the epoch $m-1$ to that of epoch 1 is $\mu^{m-1}{:}\mu$. Momentum can be used with both batch and online learning. In batch learning, it can provide further stability to the gradient descent. Momentum can be especially useful in online learning to minimize oscillations in error after the presentation of each pattern.

Example problem continued (see Section 4.4.1 and Section 4.4.1.5). This section will now return to the practical example to apply momentum. Using the previously calculated weight changes after the first epoch {0.00144, 0.00194, 0.01563, 0.01002} and the gradients for epoch 2 presented in Section 4.4.1.5, the new weights for epoch 2 obtained for $\varepsilon = 0.1$ and $\mu = 0.5$, for example, would be

$$\Delta a_0^2 = \mu \Delta a_0^1 - (1-\mu)\varepsilon d_2^{a_0}$$
$$\Delta a_0^2 = 0.5(0.00144) - (1-0.5)(0.1)(-0.01449) = 0.00144.$$

Similarly

$$\Delta a_1^2 = 0.5(0.00194) - (1-0.5)(0.1)(-0.0196) = 0.00195$$
$$\Delta b_0^2 = 0.5(0.01563) - (1-0.5)(0.1)(-0.1538) = 0.0155$$
$$\Delta b_1^2 = 0.5(0.01002) - (1-0.5)(0.1)(-0.0983) = 0.00992.$$

The new weights after the momentum-based adjustment will be

$$a_0^3 = a_0^2 + \Delta a_0^2 = 0.30144 + 0.00144 = 0.3029$$
$$a_0^3 = a_1^2 + \Delta a_1^2 = 0.2019 + 0.00195 = 0.2038$$
$$b_0^3 = b_0^2 + \Delta b_0^2 = -0.0844 + 0.0155 = -0.0689$$
$$b_1^3 = b_1^2 + \Delta b_1^2 = 0.410 + 0.00992 = 0.4199.$$

The MSE from Equation 4.18 for the two patterns is now

$$MSE = (0.0128 + 0.1008)/2 = 0.0564,$$

which is slightly smaller than the MSE after the second weight update (0.0568) without momentum.

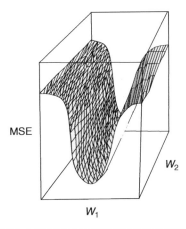

Figure 4.20 A steep error surface in the form of a ravine requiring stabilized descent.

In practical terms, momentum stabilizes the learning process. Basically, if the previous cumulative change has been in the same direction as suggested by the current direction, momentum accelerates the current weight change (hence the name momentum) and if the previous cumulative change has been in the opposite direction, it dampens the current change. This is particularly useful for error surfaces on which the optimum weights are at the bottom of a ravine that has steep sides and a flat floor, as shown in Figure 4.20 [4].

When the search path is partway down a side of the ravine, the direction of steepest descent points across the ravine, whereas the optimum weights lie almost perpendicular to it, in the direction of the ravine. Even batch averaging may not solve this, because the search path could still oscillate, jumping back and forth across the ravine until it hits the floor and then slowly find its way towards the optimum. Momentum can help quickly subdue the oscillations and slowly guide the search down the steep slopes until the optimum weights are reached with minimum oscillation. This is possible because, as the search jumps back and forth across the ravine, the required weight changes have alternating signs, thus averaging out to a small actual weight change during an epoch, which allows the network to settle down to the bottom of the ravine where it can accelerate, picking up momentum in the flatter region and periodically cruising up the side of the ravine. However, it will be squashed again due to the alternating signs of the required weight change, producing small actual changes.

Smith [4] states that such error surfaces, in which the optimum weights lie in a ravine whose floor becomes gradually flatter and flatter, are not uncommon in practice. Thus, momentum can be useful in some cases. However, it can be detrimental in cases in which the error surface takes the shape of a bowl. In this case, by the time the search approaches the bottom of the bowl, the process has accumulated enough momentum to push the search path off to the other side of the bowl, where it would turn back and return to the other side; it would oscillate in this fashion indefinitely.

So far, the mechanism behind backpropagation learning has been presented. Specifically, this chapter has examined the learning rate that controls the step size along the steepest descent. Weights are adjusted by adding this portion of the weight increment to the previous weight. Training can use either example-by-example or batch learning. Example-by-example learning may lead to oscillations and wasted time in some situations, which can be prevented by averaging. Batch learning and momentum are two such methods.

In batch learning, the error derivative of a weight during an epoch is summed over all of the training patterns to produce a weight change that is correct on average. The momentum, in contrast, exponentially averages all of the previous weight increments and tugs it to the current required weight change to produce a final weight change that is stable enough to facilitate even a descent down an error surface resembling a ravine to reach its minimum. Batch learning and momentum can be combined in such a way that the learning rate and momentum apply to epoch-based gradients and weight changes, respectively. Batch learning with or without momentum is the preferred method when error backpropagation learning is used. Momentum could provide stability in online learning; however, it should be used with caution, as it could lead to large oscillations when the error surface is a smoothly shaped bowl.

4.4.2 Example: Backpropagation Learning Computer Experiment

Illustration of the effect of the learning rate and momentum. In the previous example started in Section 4.4.1, a small neural network with one input, one output, and one hidden neuron was trained by hand, using two training patterns from the first quarter of the sine function. Here, the same network and the same function will be used, but more input–output examples will be added to the existing two to allow for a computer experiment to find out the effect of the learning rate and the momentum. Figure 4.21 shows the data; 12 input–output patterns are now being used.

The weights will be initialized with the same random values used earlier for this example: $a_0 = 0.3$, $a_1 = 0.2$, $b_0 = -0.1$, $b_1 = 0.4$. The network

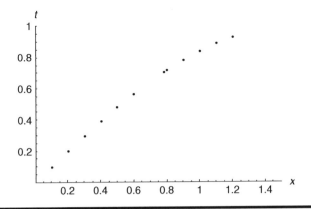

Figure 4.21   Training data for backpropagation.

approximation of the target output with the initial weights is shown as a solid line in Figure 4.22, and it is extremely poor.

The network was first trained by varying the momentum while the learning rate was held constant at 0.1. The error reduction as the training progresses in each of these cases is shown in Figure 4.23. One example is also given in which the learning rate has increased to 0.2, with a momentum of 0.9. Recall that momentum exponentially averages all of the past weight changes, and that if the direction of the average past changes and the current suggested direction are the same, momentum allows for rapid changes, but that if they are different (i.e., when oscillating around a minimum in the error surface), it dampens the rate of weight change.

Figure 4.23 shows that the momentum has a significant effect on the network performance, indicated by the behavior of the mean error and

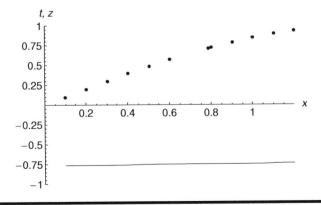

Figure 4.22   Initial network performance superimposed on training data.

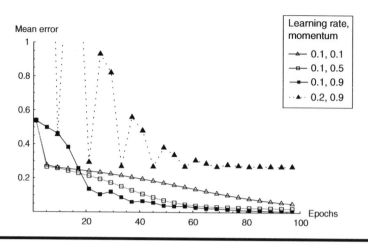

Figure 4.23   The effect of momentum on learning.

the minimum mean error reached. For the learning rate of 0.1, a momentum of 0.9 gives the best performance, and 0.1 gives the worst performance. This is because this problem is simple, and with a smaller learning rate high momentum can be used. The dashed line shows that increasing the learning rate further to 0.2 with momentum held at 0.9 has a drastic effect, with very large oscillations that take the error far beyond the scale of the y-axis. For this example, a learning rate and momentum of 0.1 and 0.9 provide the best solution. In Figure 4.24, further trials around this best parameter set are made to study the effect of changing the learning rate for a constant momentum of 0.9. From this figure it is evident that the original learning

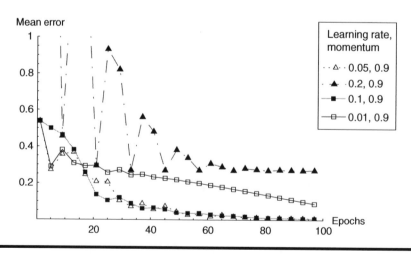

Figure 4.24   The effect of learning rate on error.

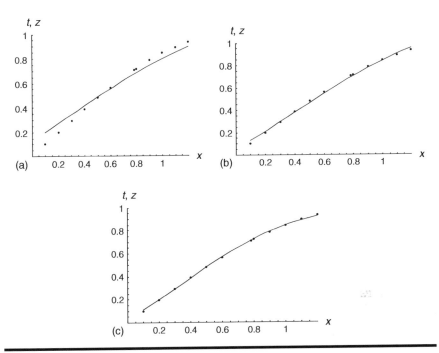

Figure 4.25 Network output superimposed on data for learning rate of 0.1: (a) momentum = 0.1, (b) momentum = 0.5, and (c) momentum = 0.9.

parameter set of 0.1 and 0.9 for the learning rate and momentum, respectively, is still the best.

Figure 4.25 shows the trained network performance for the three momentum values of 0.1, 0.5, and 0.9 for a learning rate of 0.1. It shows that a momentum of 0.9 is the best, but that between 0.5 and 0.9, there is only a slight change in the model.

So far, a simple one-input, one-hidden neuron, and one-output network has been used to derive the formulae for the gradients and to illustrate the concepts of learning. A similar approach applies to any network configuration; in the following section these formulae are derived for networks with many inputs, hidden neurons, and output neurons.

4.4.3 Single-Input Single-Output Network with Multiple Hidden Neurons

A network with multiple hidden neurons can approximate any nonlinear function of one input variable. Two such examples are presented in Chapter 3, Section 3.3.2. These involve modeling a square wave and seasonal species migration requiring two or more hidden neurons. The formulae

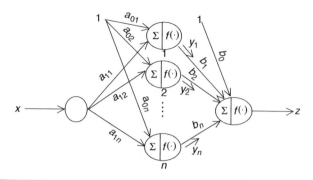

Figure 4.26 One-input, many hidden-neuron, one-output network.

derived in the previous section can be easily extended to this and to more general cases. For this particular case, there are as many hidden-output weights as there are hidden neurons, as shown in Figure 4.26, and derivatives must be computed for each of these. Analogous to the one-hidden neuron case, these can be derived as follows:

$$\frac{\partial E}{\partial b_0} = p = (z - t)z(1 - z)$$

$$\frac{\partial E}{\partial b_1} = py_1; \frac{\partial E}{\partial b_2} = py_2; \dots; \frac{\partial E}{\partial b_n} = py_n,$$ (4.23)

where $y_1, \dots, y_n$ are hidden-neuron outputs. Similarly, the error gradient with respect to the hidden-neuron weights can be derived as

$$\frac{\partial E}{\partial a_{01}} = q_1; \frac{\partial E}{\partial a_{02}} = q_2; \dots; \frac{\partial E}{\partial a_{0n}} = q_n$$

$$\frac{\partial E}{\partial a_{11}} = q_1 x; \frac{\partial E}{\partial a_{12}} = q_2 x; \dots; \frac{\partial E}{\partial a_{1n}} = q_n x,$$ (4.24)

where

$$q_1 = pb_1 y_1(1 - y_1); q_2 = pb_2 y_2(1 - y_2); \dots; q_n = pb_n y_n(1 - y_n).$$ (4.25)

4.4.4 Multiple-Input, Multiple-Hidden Neuron, and Single-Output Network

When more than one input is used in a multilayer network, a multidimensional function can be approximated. The details of processing in such networks are given in Chapter 3, Section 3.4 for a two-dimensional

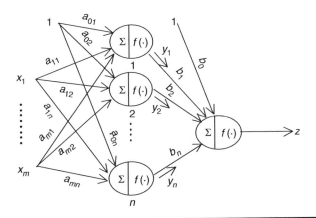

Figure 4.27 **Weights for *m*-input, *n*-hidden neuron, and one-output network.**

network using two-dimensional logistic functions. Examples of two-dimensional prediction and classification are presented in Chapter 3, Section 3.4.3. For *m* inputs and *n* hidden neurons, there are $m \times n$ input-hidden-neuron weights, *n* hidden-neuron bias weights, $n \times 1$ hidden-output neuron weights, and one bias weight for the output neuron (Figure 4.27), with respect to which an error derivative must be calculated. However, because many of these have already been dealt with in the previous cases, only the connection weights from the extra inputs to the hidden neurons need be discussed. This section will demonstrate that these can be added easily:

$$\frac{\partial E}{\partial a_{11}} = q_1 x_1; \frac{\partial E}{\partial a_{12}} = q_2 x_1; \dots; \frac{\partial E}{\partial a_{1n}} = q_n x_1$$

.

. (4.26)

.

$$\frac{\partial E}{\partial a_{m1}} = q_1 x_m; \frac{\partial E}{\partial a_{m2}} = q_2 x_m; \dots; \frac{\partial E}{\partial a_{mn}} = q_n x_m.$$

4.4.5 *Multiple-Input, Multiple-Hidden Neuron, Multiple-Output Network*

This situation represents the general case of a network with many output classes, used for classification as shown in Figure 4.28. An example of a three-class classification is presented in Chapter 3, Section 3.4.3.2, in which each class is represented by two clusters. The network error for a multioutput network is the combined error at each of the *k* output neurons, represented by $E_1, E_2, \dots, E_k$.

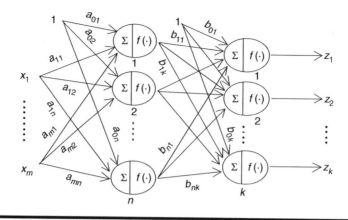

Figure 4.28   A general case of a network with several outputs.

The total network error, E, is sum of the MSE at each output neuron and is expressed as

$$E = E_1^2 + E_2^2 + \cdots + E_k^2 = \frac{1}{2NK} \sum_{i=1}^{K} \left[\sum_{i=1}^{N} (z_{ik} - t_{ik})^2 \right], \qquad (4.27)$$

where N is the total number of training patterns, K is the number of output neurons, and z_{ik} and t_{ik} are the predicted output and target output of the kth output neuron for the input pattern i. With this modification, the process remains similar to the previous derivations. The main difference is that each hidden neuron contributes to the error at each output node. The derivatives with respect to output neuron bias weights now take the form of

$$\frac{\partial E}{\partial b_{01}} = p_1 = (z_1 - t_1)z_1(1 - z_1), \ldots, \frac{\partial E}{\partial b_{0k}} = p_k$$

$$= (z_k - t_k)z_k(1 - z_k), \qquad (4.28)$$

where z_1, t_1; z_k, and t_k are the network output and target output values for nodes $1, \ldots, k$, respectively. It can easily be seen that the hidden-output weights produce following error derivatives:

$$\frac{\partial E}{\partial b_{11}} = p_1 y_1; \ldots; \frac{\partial E}{\partial b_{1k}} = p_k y_1$$

.

.

. (4.29)

$$\frac{\partial E}{\partial b_{n1}} = p_1 y_n; \ldots; \frac{\partial E}{\partial b_{nk}} = p_k y_n.$$

The hidden-neuron weight derivatives are now affected by the error at all output neurons. This is because the hidden-neuron outputs are sent to all of the output neurons, and thus contribute to the error at each output neuron. The error contributed by each hidden neuron to the output neurons is accumulated at each hidden neuron by backpropagation. The resulting error gradients for the bias weights are expressed as

$$\frac{\partial E}{\partial a_{01}} = q_1 = [(p_1 b_{11} + \cdots + p_k b_{1k})] y_1 (1 - y_1) = \left[\sum_{i=1}^{k} p_i b_{1i}\right] y_1 (1 - y_1)$$

.

.

$$\frac{\partial E}{\partial a_{0n}} = q_n = [(p_1 b_{n1} + \cdots + p_k b_{nk})] y_n (1 - y_n) = \left[\sum_{i=1}^{k} p_i b_{ni}\right] y_n (1 - y_n),$$

$$(4.30)$$

where the sum of the contributed error by the hidden neuron i is shown in square brackets. The gradients with respect to the other input-hidden weights $\partial E/\partial a_{ij}$ are derived in exactly the same manner as given in Equation 4.26, using the new values of $q_1, \ldots, q_n$ calculated from Equation 4.30. This concludes the determination of the error derivative for a general case, and the backpropagation learning that was illustrated for a simple network applies to any network.

4.4.6 Example: Backpropagation Learning Case Study—Solving a Complex Classification Problem

This section will present an example to explain how backpropagation works on a complex problem. It will look at a classification problem provided by Haykin [5]. The task is to classify the data represented by two normal distributions, depicted in Figure 4.29, into two classes. There are two dimensions, x_1 and x_2, so the problem involves two-dimensional Gaussian distributions. For the first distribution, representing Class 1, the mean values of both x_1 and x_2 are zero; thus the mean vector is [0, 0]. The variance of the distribution is 1.0. For the second distribution, the two mean values are 2.0 and 0.0, and variance is equal to 4.0. Both classes are equiprobable (i.e., the class probabilities $p_1 = p_2 = 0.5$). The training data shown in the bottom image of Figure 4.30 consists of uniformly distributed random points generated from each of the two distributions. The data for individual classes is shown separately in the top and middle images of Figure 4.30. There is a significant overlap between the two classes, and inevitably there will be a significant probability of misclassification.

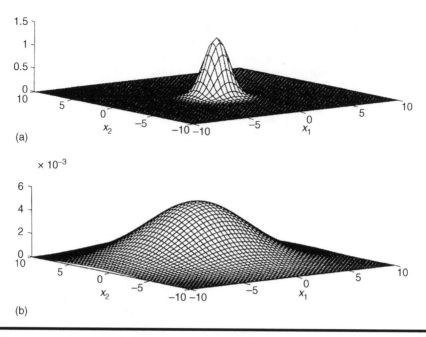

(a)

(b)

Figure 4.29 Input probability distributions for two overlapping classes. (From Haykin, S., *Neural Networks: A Comprehensive Foundation*, 2nd Ed., Prentice Hall, Upper Saddle River, NJ, 1999.)

The solution is not simple as it requires nonlinear classification boundaries to separate the two classes.

Because this is a two-dimensional problem, there are two inputs, x_1 and x_2. The first simulation results are for two hidden neurons, and the results from three runs for three different training dataset sizes are summarized in Table 4.2. The learning rate is a small nominal value set at 0.1, and momentum is 0. The number of epochs for each run is set so that the total number of training iterations (total number of individual input patterns over all epochs) is constant. This way, any irregularity arising from the use of different training dataset sizes is averaged out. The classification rate in Table 4.2 is for an independent test set containing a total of 32 000 data points equally representing the two distributions. The training results and the classification rate, based on the independent data set for a network containing four hidden neurons, were found to be similar to the two-hidden neuron case, indicating that there was no additional gain from the use of four neurons.

In the next stage of the simulation, the effect of different combinations of learning rate and momentum on the two-hidden neuron network performance will be investigated. Each combination was trained with the same set of initial weights and the same training datasets. Figure 4.31a

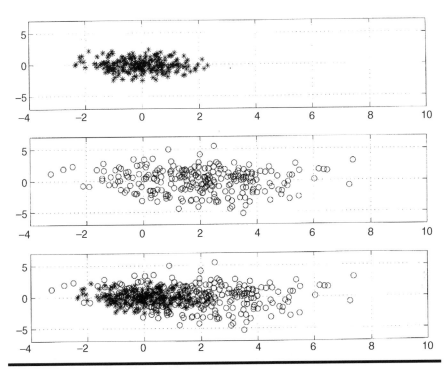

Figure 4.30 Data for classification: Class 1 data (top), class 2 data (middle), and joint Class 1 and Class 2 data (bottom) (From Haykin, S., *Neural Networks: A Comprehensive Foundation*, 2nd Ed., Prentice Hall, Upper Saddle River, NJ, 1999.)

through Figure 4.31d show the decrease in the ensemble-averaged MSE as learning progresses for various learning rates and momentum values.

Figure 4.31a shows the four ensemble-averaged learning curves corresponding to four momentum values (0, 0.1, 0.5, and 0.9) for a fixed learning rate of 0.01. It shows that for all cases, the MSE drops as the number of training epochs increases. However, the drop is fastest for $\mu = 0.9$. The drop becomes progressively slower as momentum decreases,

Table 4.2 Classification Results for Two Hidden Neurons

Run Number	Training Get Size	Number of Epochs	MSE	Probability of Correct Classification (percent)
1	500	320	0.2199	80.36
2	2000	80	0.2108	80.33
3	8000	20	0.2142	80.47

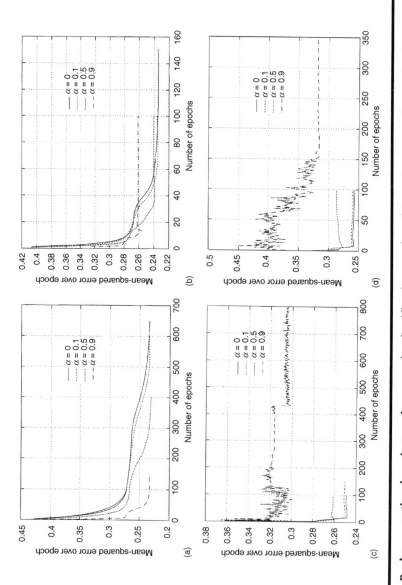

Figure 4.31 Backpropagation learning performance in classification: (a) learning rate = 0.01 and momentum varies (b) learning rate = 0.1 and momentum varies, (c) learning rate = 0.5 and momentum varies, and (d) learning rate = 0.9 and momentum varies. η denotes learning rate, α denotes momentum. (From Haykin, S., *Neural Networks: A Comprehensive Foundation*, 2nd Ed., Prentice Hall, Upper Saddle River, NJ, 1999.)

and the top-most curve is for zero momentum, meaning that only the learning rate applies. What these results reveal is that, for this problem, when a very small learning rate is used, a very high momentum helps to converge at optimum solution much faster than lower momentum values. The best results are achieved in 100 epochs, whereas with no momentum term, more than 600 epochs are required to produce a similar output.

Figure 4.31b through Figure 4.31d graphically present the effect of increasing the learning rate on the outcome of the same problem. In Figure 4.31b, learning performance is illustrated for a fixed learning rate of 0.1, which is ten times larger than in the previous case. Here, the first noticeable difference is that the high momentum ($\mu = 0.9$) has failed to provide the minimum error, indicating that a high momentum is not appropriate as the learning rate increases. The best learning performance is given by $\mu = 0.5$, for which case the minimum possible MSE is achieved in 20 epochs. As the learning rate increases to 0.5 (five times larger than in the last case), the high momentum has a devastating effect on the learning performance, as shown in Figure 4.31c. Not only does $\mu = 0.9$ fail to reach even its prior limits of MSE, but the high momentum has also led to oscillations, indicating that the acceleration is large enough to make the search path oscillate around the minimum indefinitely, never converging to the optimum. In this case, a small momentum ($\mu = 0.1$) produces the best result and reaches a constant MSE level in about ten epochs.

The last panel of Figure 4.31d shows the damaging influence of a high momentum (0.9) when accompanied by a larger learning rate (0.9). Here, the oscillations are far too great to reach even the values attained in the previous stage. However, the best solution is achieved in about ten epochs without momentum. Now, what is the best combination of learning rate and momentum for this problem? In Figure 4.32, the best learning curves from the four cases are plotted together to determine the overall best learning curve. The figure shows that the best curve is produced by the learning rate and momentum combination of 0.5 and 0.1. The fact that the range of the minimum MSE achieved for the various combinations of these two parameters is not large indicates that the error surface for this problem is relatively smooth [5]. This conclusion is also supported by the fact that the high momentum is counterproductive for this problem when accompanied by a high rate of learning.

Twenty-fold cross validation of model performance. The above presentation provided an in-depth look at the influence of the learning parameters on the backpropagation learning performance. The actual performance of the optimal network must also be measured in practical terms, which in this case would be an indication of classification accuracy.

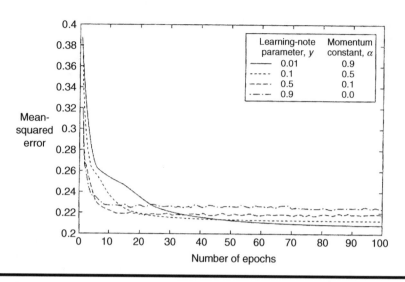

Figure 4.32 Best learning curves from different training trials. (From Haykin, S., *Neural Networks: A Comprehensive Foundation*, 2nd Ed., Prentice Hall, Upper Saddle River, NJ, 1999.)

However, a network output trained with a finite training set is stochastic in nature. This is typical of complex data drawn from natural or biological systems. Therefore, in the next stage of simulation, 20 independent networks with two hidden neurons and different initial random weights were trained with the optimum learning parameters found in the previous stage for a new training dataset size increased to 1000 from each class. These networks were tested on the same set of test data that had previously been used and which comprised 32 000 observations, equally representing the two classes.

The average classification accuracy on the test data was determined using the ensemble average, as presented in Table 4.3, which shows that the network average accuracy is approximately 80 percent, with a minimum average MSE of 0.2277. The table also shows the standard deviation of the classification accuracy and the MSE for the 20 networks. The extremely small standard deviation for both parameters indicates that all of the networks have produced very similar results. Figure 4.33

Table 4.3 Ensemble Performance Measured from a 20-Fold Cross Validation

Performance Measure	Mean	Standard Deviation
Probability of correct classification	79.7 percent	0.44 percent
Final MSE	0.2277	0.0118

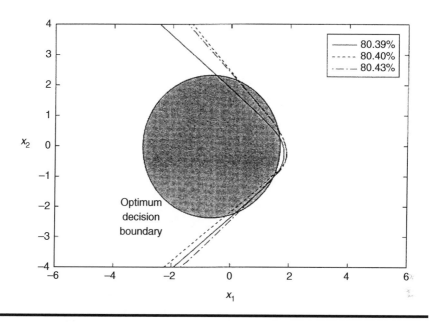

Figure 4.33 Classification boundary for three best networks superimposed on Bayesian classification boundary (solid circle). (From Haykin, S., *Neural Networks: A Comprehensive Foundation*, 2nd Ed., Prentice Hall, Upper Saddle River, NJ, 1999.)

shows the classification boundaries (in input space) and classification accuracy for the three best networks from the set of 20 networks. For visual clarity, only the classification boundaries are shown without input data. Figure 4.33 also shows the classification boundary for the Bayesian statistical classifier (circular) on the same data. The Bayesian model has provided a classification accuracy of 81.51 percent. Thus, the network performance in this case is comparable to the performance of the Bayesian Classifier.

A Bayesian classifier is a statistical classifier that involves the class probability distributions (likelihood) to obtain a posteriori probabilities, i.e., the classification of an input [5,14]. The Bayesian classification boundary is the intersection between the two Gaussian data distributions. Because in this case the two class distributions are symmetric Gaussian distributions, the classification boundary is circular, as illustrated in Figure 4.33. Figure 4.34 shows the classification boundaries from the three poorest networks from the 20-network ensemble, along with their classification rates and Bayesian classification boundaries. As stated earlier, because there is a considerable class overlap, a significant probability of misclassification is inevitable.

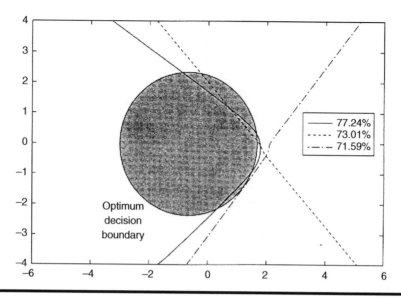

Figure 4.34 **Classification boundary for the three poorest networks superimposed on Bayesian classification boundary (solid circle). (From Haykin, S., *Neural Networks: A Comprehensive Foundation*, 2nd Ed., Prentice Hall, Upper Saddle River, NJ, 1999.)**

4.5   Delta-Bar-Delta Learning (Adaptive Learning Rate) Method

In backpropagation learning, discussed above, the same learning rate applies to all of the weights. Therefore, all of the weights change at the same rate. However, in reality, some weights may be closer to the optimum or have a stronger influence on the error than the others and, therefore, more flexibility and a higher speed of convergence could be achieved if each weight were to be adjusted independently in an adaptive manner. The adaptive learning rate method, popularly known as delta-bar-delta and also as TurboProp, developed by Jacob [6], proposes such variable learning rates for different weights. In this method, each weight has its own learning rate and is adjusted during each iteration as follows: if the direction in which the error decreases at the current point, as indicated by the error gradient, is the same as the direction in which the error has been decreasing recently, then the learning rate is increased. However, if the current direction in which the error decreases is opposite to the recent direction in which the error has been decreasing, the learning rate is decreased [4].

Basically, this method requires that the sign of the current error gradient for a weight be compared with its recent history. In batch learning, which is the most preferred method, the current error gradient for an epoch m is the resultant derivative over the whole training set, d_m. The recent history of the direction in which the error has been decreasing up to epoch m is expressed by f_m as

$$f_m = \theta f_{m-1} + (1 - \theta)d_{m-1}, \tag{4.31}$$

which is basically the exponential average of all of the past error derivatives, a concept which is similar to that used in momentum. The θ is the weighting on the exponential average of the past derivatives and $1 - \theta$ is the weighting on the last derivative; these weightings determine whether the most recent gradients or the distant ones have a stronger influence on the f_m, i.e., the direction in which the error has been decreasing recently. The duration of "recent" is determined by the value of θ, which is a constant between 0 and 1.

If θ is equal to zero, only the gradient in the previous epoch defines what 'recent' means, and there is no effect of the earlier gradients. If it is equal to one, the recent direction is totally defined by all of the gradients up to and excluding the very last gradient, and is equal to the exponential average of those derivatives. Intermediate values of θ put intermediate weightings on the most recent and past derivatives. For example, if θ is 0.3, $f_m = 0.3f_{m-1} + 0.7d_{m-1}$; but because $f_{m-1} = 0.3f_{m-2} + 0.7d_{m-2}$ and so on, by back substituting, the current $f_m = 0.7d_{m-1} + 0.21d_{m-2} + 0.063d_{m-3} + \ldots$ back to epoch 1. Thus, the more recent derivatives are weighted more heavily than the early ones. In this way, f_m can be calculated for each weight in the network. To see whether or not the recent direction is similar to that indicated by the current derivative, f_m is multiplied by d_m. If $f_m d_m$ is positive, the recent and the current directions are the same and the weights can be adjusted at an increased rate. If it is negative, the current derivative is opposite to the recent average direction, indicating that some minimum on the error surface has been passed, and in consequence, the weights should be adjusted more slowly. The delta-bar-delta method reflects this concept by allowing larger changes in the learning rate (ε) in the former case, in which $f_m d_m$ is positive, and allowing only smaller adjustments in the latter case, for which $f_m d_m$ is negative, as shown in Equation 4.32:

$$\varepsilon_m = \begin{cases} \varepsilon_{m-1} + \kappa & \text{for} \quad d_m f_m > 0 \\ \varepsilon_{m-1} \times \phi & \text{for} \quad d_m f_m \leq 0, \end{cases} \tag{4.32}$$

where κ and φ are parameters whose values are between 0 and 1. Once the new learning rate is decided for epoch m, the backpropagation algorithm is used to determine the new weight change as:

$$\Delta w = -\varepsilon_m d_m \tag{4.33}$$

or with the momentum term:

$$\Delta w = \mu \Delta w_{m-1} - (1 - \mu)\varepsilon_m d_m. \tag{4.34}$$

Smith [4] states that the learning process is not highly sensitive to the choice of the values for κ, ϕ, and θ, and suggests a set of values that work well for a wide range of problems as

$$\kappa = 0.1, \quad \phi = 0.5, \quad \theta = 0.7. \tag{4.35}$$

Because the method uses a separate adaptive learning rate for each weight, learning can occur rapidly. Smith [4] states that it is not uncommon to achieve the target error in one tenth of the time used by back-propagation learning with an optimum learning rate. A notable advantage of the delta-bar-delta method is that training generally does not require searching for the optimum parameters through trial and error, and therefore only one training session is required. For some problems, however, this may not be the case, and some trial and error may be required to find the parameters.

4.5.1 Example: Network Training with Delta-Bar-Delta— A Hand Computation

In this section, the delta-bar-delta method will be applied to the problem that was started in Section 4.4.1 and continued throughout this chapter to Section 4.4.1.7. Recall that it involves a one-input, one-output network with one hidden neuron, shown in Figure 4.4a, and is used to model the data from the quarter of the sine wave plotted in Figure 4.4b. This computation will begin with the same initial weights and the two input–output patterns extracted from the data, as given in Table 4.1 in Section 4.4.1. These two input–output patterns are {0.7853, 0.707} and {1.571, 1.0}, and the initial weights are $a_0 = 0.3$, $a_1 = 0.2$, $b_0 = -0.1$, and $b_1 = 0.4$.

Here, the delta-bar-delta method will be applied after the first weight update using backpropagation (Section 4.4.1.5), which provides a history of weight update. Usually when this method is applied, the initial weight update is either picked up as random values or set to zero to initiate the process, and after several epochs it begins to perform as it should, using the correct history of the weight updates.

Two input–output pairs are given, and assume that the current weights are the weights after the first epoch using standard backpropagation

(Section 4.4.1.5). These weights are

$$a_0^2 = 0.3014$$
$$a_1^2 = 0.2019$$
$$b_0^2 = -0.0844$$
$$b_1^2 = 0.410.$$

(4.36)

This calculation will use an initial learning rate of 0.1 for all of the weights, as was used to adjust the above weights. In practice, they can be set to random initial values. To apply this method it is necessary to calculate the total error gradient for the second epoch, which has already been done during the second epoch in Section 4.4.1.5, in the case in which backpropagation was used:

$$d_2^{a_0} = -0.01449$$
$$d_2^{a_1} = -0.0196$$
$$d_2^{b_0} = -0.1538$$
$$d_2^{b_1} = -0.0983.$$

(4.37)

Having obtained the current gradients d_2, all that remains to do is to calculate f_2 from Equation 4.31. Assume that f_1 is zero. From Equation 4.31:

$$f_2 = \theta f_1 + (1 - \theta)d_1$$

This needs to be applied to each weight; d_1 is the overall gradient for each weight in the first epoch, which for a_0, a_1, b_0, and b_1 were found to be -0.0144, -0.01944, -0.1563, and -0.1002, respectively (see Section 4.4.1.5). Calculate f_2 values for each weight with $\theta = 0.7$:

$$a_0 \rightarrow f_2^{a_0} = (1 - 0.7)(-0.0144) = -0.00432$$
$$a_1 \rightarrow f_2^{a_1} = (1 - 0.7)(-0.01944) = -0.00583$$
$$b_0 \rightarrow f_2^{b_0} = (1 - 0.7)(-0.1563) = -0.0469$$
$$b_1 \rightarrow f_2^{b_1} = (1 - 0.7)(-0.1002) = -0.03.$$

Next, it is necessary to find whether the direction in which the error has been decreasing is the same as the direction given by the current batch error gradient. For this, f_2 is multiplied by the corresponding d_2. For the four weights, $f_2 d_2$ will be

$$f_2^{a_0} . d_2^{a_0} = (-0.00432)(-0.01449) = 0.0000625$$

$$f_2^{a_1} . d_2^{a_1} = (-0.00583)(-0.0196) = 0.000114$$

$$f_2^{b_0} . d_2^{b_0} = (-0.0469)(-0.1538) = 0.00721$$

$$f_2^{b_1} . d_2^{b_1} = (-0.03)(-0.0983) = 0.00295.$$

Because all of the above values are greater than zero, the previous and the current directions of error decrease are the same, and the weights can be adjusted by a large amount, using the learning rates given by the first condition in Equation 4.32. The new learning rates for the four weights are

$$e^{a_0} = 0.1 + 0.1 = 0.2$$

$$e^{a_1} = 0.1 + 0.1 = 0.2$$

$$e^{b_0} = 0.1 + 0.1 = 0.2$$

$$e^{b_1} = 0.1 + 0.1 = 0.2.$$

The new learning rate is the same for all of the weights for the next epoch, because $f.d$ is positive for all of the weights. However, in repeated training, $f.d$ for some weights can become negative as they adapt to the target function; in that situation, the learning rate for these weights must be cut down by half (i.e., by φ, which is taken to be 0.5). Returning to the problem, the new weight changes after the second epoch are

$$\Delta a_0^2 = -\varepsilon^{a_0} d_2^{a_0} = -(0.2)(-0.0145) = 0.0029$$

$$\Delta a_1^2 = -(0.2)(-0.0196) = 0.00392$$

$$\Delta b_0^2 = -(0.2)(-0.1538) = 0.0308$$

$$\Delta b_1^2 = -(0.2)(-0.0983) = 0.0197.$$

The new weights are

$$a_0^3 = a_0^2 + \Delta a_0^2 = 0.30144 + 0.0029 = 0.3043$$

$$a_1^2 = 0.2019 + 0.00392 = 0.2058$$

$$b_0^2 = -0.0844 + 0.0308 = -0.0536$$

$$b_1^2 = 0.410 + 0.0197 = 0.4297.$$

The new MSE for the two input–output patterns with the new weights is

$$E = (0.0119 + 0.0984)/2 = 0.0551,$$

which is smaller than that obtained after the second epoch from back-propagation with the learning rate only (0.0568) and the combined learning rate and momentum (0.0564).

Although for the two examples used here there was no visible change in the MSE between the backpropagation and delta-bar-delta methods, it is common for the delta-bar-delta method to achieve a target error level in one-tenth of the number of epochs required by backpropagation with an optimal learning rate [4]. This is because in the delta-bar-delta method, the learning rate is adjusted for each weight depending on how it contributes to the reduction in error, whereas in backpropagation, the same learning rate applies regardless of the relevance of the individual weights.

4.5.2 Example: Delta-Bar-Delta with Momentum— A Hand Computation

Momentum can also be applied to the weight change, as was shown in Equation 4.34 and repeated here:

$$\Delta w = \mu \Delta w_{m-1} - (1 - \mu)\varepsilon_m d_m.$$

The past weight changes for epoch 1 from the backpropagation example were 0.00144, 0.001944, 0.01563, and 0.01002 for a_0, a_1, b_0, and b_1, respectively (see Section 4.4.1.5). Using a momentum, $\mu = 0.5$, and substituting the past weight changes, the current learning rate calculated for each weight, and the average error gradients into the above equation, new weight increments would be

$$\Delta a_0^2 = \mu \Delta a_0^1 - (1-\mu)\varepsilon^{a_0} d_2^{a_0} = (0.5)(0.00144) - (1-0.5)(0.2)(-0.01449) = 0.00217$$

$$\Delta a_1^2 = \mu \Delta a_1^1 - (1-\mu)\varepsilon^{a_1} d_2^{a_1} = (0.5)(0.001944) - (1-0.5)(0.2)(-0.0196) = 0.00293$$

$$\Delta b_0^2 = \mu \Delta b_0^1 - (1-\mu)\varepsilon^{b_0} d_2^{b_0} = (0.5)(0.01563) - (1-0.5)(0.2)(-0.1538) = 0.0232$$

$$\Delta b_1^2 = \mu \Delta b_1^1 - (1-\mu)\varepsilon^{b_1} d_2^{b_1} = (0.5)(0.01002) - (1-0.5)(0.2)(-0.0983) = 0.0148.$$

The new weights are

$$a_0^3 = a_0^2 + \Delta a_0^2 = 0.30144 + 0.00217 = 0.3036$$

$$a_1^2 = 0.2019 + 0.00293 = 0.2048$$

$$b_0^2 = -0.0844 + 0.0232 = -0.0612$$

$$b_1^2 = 0.410 + 0.0148 = 0.4248.$$

The new MSE for the two input–output patterns is

$$E = (0.0123 + 0.0996)/2 = 0.0559,$$

which is slightly higher than the MSE obtained without momentum (0.0551).

Thus, the momentum is too high for this case; however, there are only two input patterns (see Table 4.1 in Section 4.4.1) in this hand calculation.

4.5.3 Network Training with Delta-Bar-Delta— A Computer Experiment

This section will apply the delta-bar-delta method to the problem that was started in Section 4.4.1 and continued throughout this chapter up to Section 4.5.1. Recall that it involves a one-input, one-output network with one hidden neuron, as shown in Figure 4.4a, and is used to model data in Figure 4.4b, which was extracted from the first quarter of the sine wave. This section will use the same initial weights given in Table 4.1, Section 4.4.1; these are $a_0 = 0.3$, $a_1 = 0.2$, $b_0 = -0.1$, and $b_1 = 0.4$. The initial learning rate for each weight is set to 0. Figure 4.35 shows the decrease of the square root of mean square error (RMSE) over 100 epochs.

As illustrated in Figure 4.35, the error initially shows a sharp decrease, then decreases more moderately, slowing towards the end. It also shows that for the current learning parameter settings, the MSE oscillates initially, but is smoothed out as training progresses. The final RMSE reached was 0.00527, and the final learning rates were 0.644, 3.59, 1.488, and 1.403 for a_0, a_1, b_0, and b_1, respectively. The final weights for these connections were -0.85, 2.345, -0.413, and 1.528, respectively.

Figure 4.36 shows how the learning rate adapts from random starting points during the course of learning for input-hidden and hidden-output weights. It can be seen that all four weights have distinct learning rate values; moreover, the regular weights (a_1 and b_1) have higher learning rates than do the corresponding bias weights, indicating that the regular weights influence the output more significantly than do the bias weights.

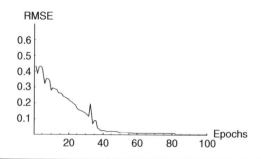

Figure 4.35 An error (RMSE) decrease with training for the delta-bar-delta (adaptive learning rate) method for 100 epochs.

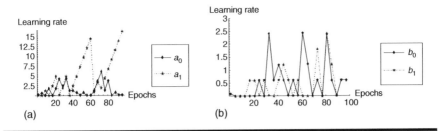

Figure 4.36 The learning rate adaptation for individual weights in the delta-bar-delta method: (a) input-hidden weights and (b) hidden-output weights.

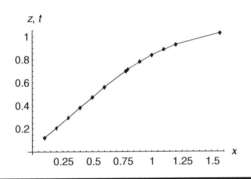

Figure 4.37 A trained network output from the delta-bar-delta method (solid line) superimposed on data.

The network output superimposed on the data is shown in Figure 4.37, which demonstrates a perfect match.

4.5.4 Comparison of Delta-Bar-Delta Method with Backpropagation

By adjusting κ, φ, and θ, the smoothness of the curves and the convergence rate can be altered; however, this exercise illustrates how the delta-bar-delta method iteratively adapts the learning rate to control an efficient descent down the error surface to reach a global minimum. Figure 4.38 compares the performance of backpropagation with delta-bar-delta, showing that the delta-bar-delta method can reduce the error quickly, but that for this example, the backpropagation method is faster initially. However, at the end of 100 epochs, the delta-bar-delta error is 12 percent smaller than that yielded by backpropagation. This can be significant for more complex problems.

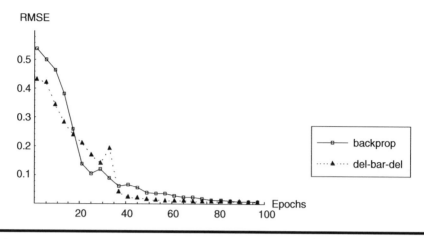

Figure 4.38 A comparison of the performance of backpropagation and delta-bar-delta methods.

4.5.5   Example: Network Training with Delta-Bar-Delta— A Case Study

Demonstration of the performance of delta-bar-delta on complex data. To demonstrate the performance of the delta-bar-delta method and to compare it with standard backpropagation, this section provides the results of a study conducted by Haykin [5] to classify two-dimensional data distributed in two classes, as shown in Figure 4.39a. Class C_1 consists of points inside the area marked C_1 and Class C_2 contains data from the area marked C_2. The task for the neural network is to decide whether an input pattern belongs to Class C_1 or C_2. The following network with two hidden-neuron layers has been used for this problem:

Number of inputs = 2

Number of neurons in the first hidden layer = 12

Number of neurons in the second hidden layer = 4

Number of neurons in the output layer = 2.

Figure 4.39b shows the randomly selected training data and Figure 4.39c shows the randomly selected test data used to evaluate the performance of the network after training. There are 100 points from each class in the training set and the test dataset contains 482 points from Class C_1 and 518 from Class C_2, for a total of 1000 data points in the test set.

The network has been trained with both backpropagation and delta-bar-delta methods using the same initial weights. For the backpropagation, both the learning rate and the momentum are 0.75, and for the delta-bar-delta

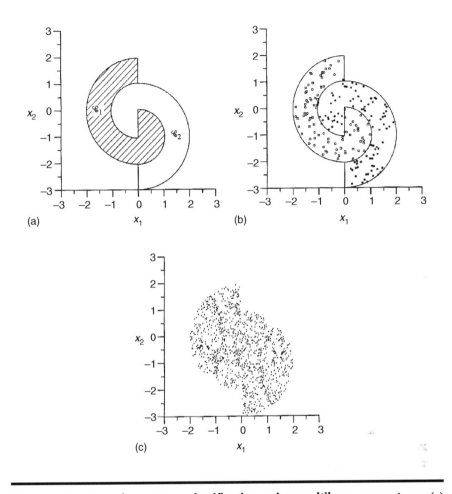

Figure 4.39 Data for pattern classification using multilayer perceptron: (a) distribution of classes C_1 and C_2; (b) 100 training data points each from classes C_1 and C_2; (c) test data for evaluating trained networks (482 from C_1 and 518 from C_2).

method, the momentum is 0.75, the initial learning rate for all of the weights is 0.75, $\kappa = 0.01$, $\Phi = 0.2$, and $\theta = 0.7$. A comparison of the training performance with respect to the mean square error reduction for the two methods for 3000 epochs is shown in Figure 4.40, which shows that for the same initial conditions, the delta-bar-delta outperforms the backpropagation by an order of magnitude.

The performance of the two methods has been compared at an instance of training after 150 epochs using the previously mentioned test dataset containing 1000 randomly selected data from the two classes. Figure 4.41a and Figure 4.41b illustrate the correctly classified data for backpropagation and the delta-bar-delta method, respectively.

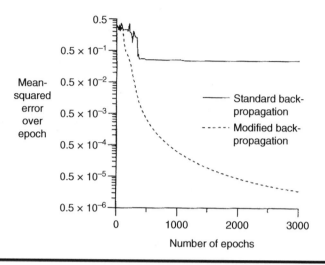

Mean-squared error over epoch

Standard back-propagation

Modified back-propagation

Number of epochs

Figure 4.40 A comparison of batch update learning curves from backpropagation and delta-bar-delta methods.

These demonstrate that the delta-bar-delta method learns the classification more quickly. For example, Figure 4.41a indicates that, with back-propagation, the shapes of the classes have not been fully developed in 150 epochs and thus the network misclassifies a total of 474 of the 1000 data points in the test dataset (i.e., 52.6 percent classification accuracy). In contrast, the well-developed shapes of the two classes can be seen in

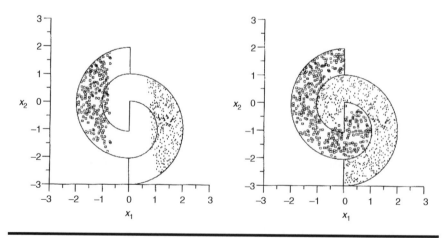

Figure 4.41 A comparison of classifiers trained with delta-bar-delta and backpropagation: (a) classification success for backpropagation; (b) classification success for delta-bar-delta after 150 epochs.

Figure 4.41b, indicating that with the delta-bar-delta method, the network has learned to separate the overlapping regions and to classify the data in these regions rather early with much greater success, with only 96 misclassified data points, resulting in a classification accuracy of 90.4 percent.

4.6 Steepest Descent Method

In the steepest descent method, the error is reduced along the negative gradient of the error surface, similar to the backpropagation and delta-bar-delta methods. However, the learning rate ε, which is the same for all of the weights, is adapted internally during training [7]. Recall that in the delta-bar-delta method, the learning rate is adapted for each weight, and in backpropagation, it is fixed throughout learning. The method of updating the learning rate in steepest descent is different from that used in delta-bar-delta. Specifically, starting with an initial value, ε is doubled in each step (trial epoch). This yields a preliminary update for the weights. The MSE is calculated as usual for the updated weights corresponding to the current learning rate. If the MSE does not decrease with this learning rate, the weights return to their original values, the learning rate is halved, and training is continued. If the MSE still does not decrease, ε is halved repeatedly until a learning rate is reached at which the MSE decreases. The final weight adjustment is made only after a learning rate that reduces the MSE is obtained. At this point, ε is doubled again and a new step is started; the whole process is repeated over and over. The search continues in this fashion and terminates within the predefined number of training epochs if the decrease in error with respect to the previous step is smaller than a specified level, E_{min}, or if the value of ε falls below a specified limit, ε_{min}, as shown in Equation 4.38:

$$\frac{E(w_m) - E(w_{m+1})}{E(w_m)} \leq E_{min} \tag{4.38}$$

$$\varepsilon \leq \varepsilon_{min}.$$

where $E(w_m)$ and $E(w_{m+1})$ are the errors for the previous and current epochs, respectively. E_{min} and ε_{min} are thresholds specified by the user.

4.6.1 Example: Network Training with Steepest Descent—Hand Computation

In this section, this method will be applied to the problem started in Section 4.4.1 and continued throughout this chapter. Recall that it involves a one-input, one-output network with one hidden neuron, shown in Figure 4.4a, and is used to model the data from a quarter sine wave, plotted in Figure 4.4b. The calculations begin using the same initial weights and the two input–output patterns extracted from the data given in Table 4.1 in Section 4.4.1. These two input–output patterns are {0.7853, 0.707} and {1.571, 1.0}; the initial weights are $a_0 = 0.3$, $a_1 = 0.2$, $b_0 = -0.1$, and $b_1 = 0.4$.

Because the reader already knows how to compute the sum of the error gradients, only a summary of the training results will be presented here in Table 4.4 to show how and when the learning rate changes. Starting with an initial rate of 20, it is doubled at the start of the first epoch. Then, within an epoch, several trial epochs (steps) are used to obtain the weight increments for that epoch.

4.6.2 Example: Network Training with Steepest Descent—A Computer Experiment

Comparison with delta-bar-delta and backpropagation. This section will apply the steepest descent method to the problem started in Section 4.4.1 and continued throughout this chapter up to Section 4.6.1. Recall that it involves a one-input, one-output network with one hidden neuron, shown in Figure 4.4a, and is used to model the data in Figure 4.4b, which was extracted from the first quarter of the sine wave. This section will use the same initial weights given in Table 4.1, Section 4.4.1; these are $a_0 = 0.3$, $a_1 = 0.2$, $b_0 = -0.1$, and $b_1 = 0.4$. All of the data in Figure 4.4b will be used to iteratively train the network.

In Figure 4.42, the RMSE for steepest descent training is presented, along with those from the delta-bar-delta and backpropagation methods. It shows that the error in steepest descent decreases dramatically in five epochs, demonstrating the effectiveness of the method. Note, however, that for this

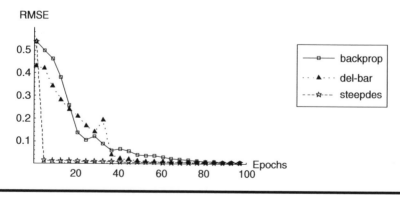

Figure 4.42 Comparison of steepest descent learning with the backpropagation and delta-bar-delta methods.

Table 4.4 Demonstration of Training Results from Steepest Descent Learning

Epoch/Step	ε	Initial Weights	MSE Initial	Average Gradient, d_m	Weight Change	New Weight	MSE After
Epoch 1/ step 1	40	{0.3, 0.2}, {−0.1, 0.4}	0.511	{−0.13, −0.16}, {−1.4, −0.89}	{5.2, 6.5}, {56.1, 5.6}	{5.5, 6.7}, {56.0, 36.0}	8304—no update
1/2	20	{0.3, 0.2}, {−0.1, 0.4}	8304	{−0.13, −0.16}, {−1.4, −0.89}	{2.6, 3.2}, {28.0, 17.8}	{2.9, 3.4}, {27.9, 18.2}	2047—no update
1/3	10	{0.3, 0.2}, {−0.1, 0.4}	2047	{−0.13, −0.16}, {−1.4, −0.89}	{1.3, 1.6}, {14.0, 8.9}	{1.6, 1.8}, {13.9, 9.3}	488—no update
.	.					.	.
1/7	0.625	{0.3, 0.2}, {−0.1, 0.4}	3.559	{−0.13, −0.16}, {−1.4, −0.89}	{0.08, 0.1}, {0.88, 0.55}	{0.38, 0.30}, {0.78, 0.95}	0.338 update
Epoch 2/ step 1	40	{0.38, 0.3}, {0.78, 0.95}	0.338	{0.24, 0.26}, {1.14, 0.76}	{−9.6, −10.3}, {−45.5, −30.5}	{−9.3, −10.14}, {−45.6, −30.1}	2159

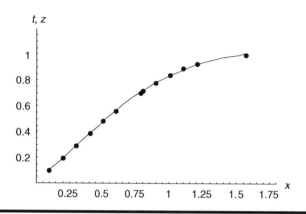

Figure 4.43 Output of network trained with steepest descent superimposed on data.

method, one epoch consists of many trail epochs, within which the learning rate is incrementally adjusted until the error decreases below that of the previous step. Therefore, a direct comparison with the other two methods on the basis of the number of epochs is not wholly appropriate. At the end of 100 epochs, the RMSE has gone down to 0.0051 compared with 0.00527 and 0.0067 from the delta-bar-delta and backpropagation methods, respectively. The final error reached in steepest descent is 28 percent less than that reached by backpropagation.

The final weights obtained from the steepest descent are $\{a_0, a_1\}$ $=\{-0.78, 2.5\}$ and $\{b_0, b_1\}=\{-0.465, 1.54\}$. For comparison, these weights from delta-bar-delta are $\{a_0, a_1\}=\{-0.85, 2.345\}$ and $\{b_0, b_1\}$ $=\{-0.413, 1.528\}$. The weights produced by backpropagation are $\{a_0, a_1\}=\{-1.087, 2.82\}$ and $\{b_0, b_1\}=\{-0.303, 1.347\}$. The network performance is superimposed on the data in Figure 4.43, showing that this set of weights also produces a network that fits the data perfectly.

4.7 Second-Order Methods of Error Minimization and Weight Optimization

In second-order methods, the curvature of the error surface, denoted by the second derivative of the error surface, is used to more efficiently guide the error down the error surface. This section will turn to a simpler second-order method, called "QuickProp," which was proposed to speed up the convergence process of backpropagation learning. It illustrates the concepts of second-order error minimization in a simple way, using only the first derivative. The subject of second-order methods will be dealt with properly in Section 4.7.2.

4.7.1 QuickProp

QuickProp was developed by Scott Fahlman in 1988 [8]; the method implicitly uses the curvature and directly involves the slope of the error surface at a point defined by the current weights. Because the curvature is implicit, the second derivative is not used in the calculation, as it will be in more advanced methods discussed later. The implementation of QuickProp is illustrated on a slice through the error surface along one weight axis in Figure 4.44.

The method follows this argument: the objective of learning is to quickly find the optimum weights at which the error derivative is zero. Suppose the derivative after the last epoch, $m-1$, was d_{m-1}, and that it led to a weight change of Δw_{m-1}, as shown in Figure 4.44. If the derivative for the current epoch m is d_m, the required weight change Δw_m that leads to a zero derivative is calculated using basic algebraic concepts as follows:

$$\Delta w_m = \frac{d_m}{d_{m-1} - d_m} \Delta w_{m-1}. \tag{4.39}$$

In Equation 4.39, the term $(d_{m-1}-d_m/\Delta w_{m-1})$ is an approximation of the curvature, which is the derivative of the gradient of the error surface at weight w. Thus, the higher the curvature, the lower the weight update, and vice versa. Because the error surface is defined by the actual data, the weight change after an epoch normally does not lead to the optimum weights where the gradient of the error is zero. Therefore, the method is repeated

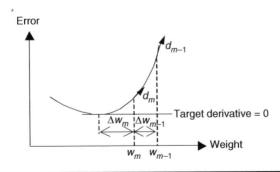

Figure 4.44 Illustration of QuickProp learning method that implicitly involves curvature of error surface.

over many epochs until the optimum weights producing the target level of minimum error are reached. Thus, the weight changes are obtained without the use of a learning rate and momentum, or any of the learning parameters of the adaptive learning rate approach. According to Smith [4], the rate of convergence with QuickProp is similar to that for adaptive learning rate (delta-bar-delta).

4.7.1.1 Example: Network Training with QuickProp— A Hand Computation

In this section, QuickProp will be applied to the problem started in Section 4.4.1 and continued throughout this chapter. Recall that it involves a one-input one-output network with one hidden neuron, shown in Figure 4.4a, and is used to model the data plotted in Figure 4.4b. The calculations will begin with the same initial weights and the two input–output patterns extracted from the data, given in Table 4.1, Section 4.4.1. There are two bias weights and two regular weights in the network, which must be optimized. The two input–output patterns are {0.7853, 0.707} and {1.571, 1.0}, and the initial weights are $a_0 = 0.3$, $a_1 = 0.2$, $b_0 = -0.1$, and $b_1 = 0.4$.

Because the method requires a previous weight change, it can be either assumed to be zero or some small random initial value; in this calculation, weight updates after the first epoch using backpropagation involving only the learning with a value of 0.1 will be used. Previously, the derivatives, weight changes, and actual modified weights were obtained after epoch 1 for this example (see Section 4.4.1.5 and Section 4.4.1.2); these are repeated here.

For epoch 1:

$$d_1^{a_0} = -0.0144 \quad d_1^{a_1} = -0.01944 \quad d_1^{b_0} = -0.1563 \quad d_1^{b_1} = -0.1002$$

$$\Delta a_0^1 = 0.00144; \quad \Delta a_1^1 = 0.001944; \quad \Delta b_0^1 = 0.01563; \quad \Delta b_1^1 = 0.01002.$$

The adjusted weights after epoch 1:

$$a_0^2 = 0.30144; \quad a_1^2 = 0.2019; \quad b_0^2 = -0.0844; \quad b_1^2 = 0.410.$$

The MSE after first epoch 1 is 0.0585.

For epoch 2:

$$d_2^{a_0} = \sum \frac{\partial E}{\partial a_0} = -0.01449; \quad d_2^{a_1} = -0.0196; \quad d_2^{b_0} = -0.1538;$$

$$d_2^{b_1} = -0.0983.$$

Now that all the information about the error derivatives and weight updates after epoch 1 and the error derivates for epoch 2 have been obtained, QuickProp can be used to calculate the update for one weight after epoch 2:

$$\Delta a_0^2 = \frac{d_2^{a_0}}{d_1^{a_0} - d_2^{a_0}} \Delta a_0^1 = -0.01449 \times 0.00144/(-0.0144 - (-0.01449))$$

$$= -0.2318.$$

Similarly

$$\Delta a_1^2 = \frac{d_2^{a_1}}{d_1^{a_1} - d_2^{a_1}} \Delta a_1^1 = -0.0196 \times 0.001944/(-0.01944 - (-0.0196))$$

$$= -0.2381$$

$$\Delta b_0^2 = \frac{d_2^{b_0}}{d_1^{b_0} - d_2^{b_0}} \Delta b_0^1 = -0.1538 \times 0.01563/(-0.1563 - (-0.1538)) = 0.9615$$

$$\Delta b_1^2 = \frac{d_2^{b1}}{d_1^{b_1} - d_2^{b_1}} \Delta b_1^1 = -0.0983 \times 0.01002/(-0.1002 - (-0.0983)) = 0.518.$$

Thus, the new weights are

$$a_0^3 = 0.30144 - 0.2318 = 0.0694$$

$$a_1^3 = 0.2019 - 0.2381 = -0.0362$$

$$b_0^3 = -0.0844 + 0.9615 = 0.8771$$

$$b_1^3 = 0.410 + 0.5184 = 0.9284$$

$$\text{MSE} = 0.02519.$$

The new MSE is much smaller than that achieved by the other methods, which shows that QuickProp reduces the error more efficiently. The process is repeated over and over through the epochs until the desired error level is achieved or the error does not change any further.

4.7.1.2 Example: Network Training with QuickProp— A Computer Experiment

The QuickProp method will be applied to the problem previously started in Section 4.4.1 and continued throughout this chapter up to Section 4.7.1.1. Recall that it involves a one-input, one-output network with one hidden neuron, as shown in Figure 4.4b and used to model the data in Figure 4.4b. This is extracted from the first quarter of the sine wave. The same initial weights will be used as given in Table 4.1 in Section 4.4.1. These are $a_0 = 0.3$, $a_1 = 0.2$, $b_0 = -0.1$, and $b_1 = 0.4$. All data in Figure 4.4b will be used to train the network iteratively. There are two bias weights and two regular weights in the network to be optimized.

Starting with the same initial weights, the network was trained using QuickProp for 100 epochs. Figure 4.45 shows how error decreases with an increasing number of epochs. There are large oscillations in error that are stabilized towards the end of training.

4.7.1.3 Comparison of QuickProp with Steepest Descent, Delta-Bar-Delta, and Backpropagation

The performance of QuickProp is compared with the other three methods—backpropagation, delta-bar-delta, and steepest descent—in Figure 4.46. It is evident from this example that QuickProp is very efficient because there are no parameters to adjust at all and the error goes down much faster initially than for backpropagation and delta-bar-delta. However, this momentum is not continued because the oscillations of the MSE and the final error after 100 epochs is only 0.015. This error is larger than what is obtained from backpropagation (0.0067), delta-bar-delta (0.00527), or

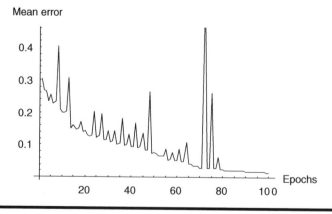

Figure 4.45 Error reduction during training with QuickProp.

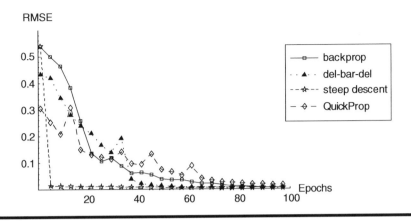

Figure 4.46 Comparison of QuickProp with backpropagation, delta-bar-delta, and steepest descent.

steepest descent (0.0051). These methods outperform QuickProp at the end because of their ability to fine tune their parameters, something that QuickProp cannot do.

The predicted output from the network trained with QuickProp is shown in Figure 4.47 and is superimposed on the actual data showing some misfit.

In QuickProp, the curvature of the error surface is implicitly involved. The more advanced methods that explicitly use the curvature information are known to produce greater acceleration and accuracy. The following section will address second-order methods and will put backpropagation in context by showing that it is a simplified version of a general problem of optimization. In order to put error minimization in context, the concepts involved in second-order methods of error minimization will be addressed.

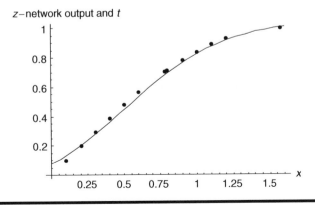

Figure 4.47 Network output from QuickProp superimposed on data.

4.7.2 General Concept of Second-Order Methods of Error Minimization

In second-order methods, the slope and the curvature at the current point in weight space is determined [11,12,14]. The slope, which is the first derivative of error, indicates the rate of change of error (i.e., how fast the error is decreasing at the current point). The curvature indicates the rate at which the slope itself changes (i.e., curves) at the current weights. Therefore, the curvature indicates the deceleration of error. For example, consider the two error curves in Figure 4.48. The slopes are equal at the point where the two curves touch as indicated by the arrow. However, their curvatures are different. If the current weight is at the point where the two curves touch, it is closer to the minimum of the dashed curve with a larger curvature than it is to the solid curve with a smaller (flatter) curvature. This idea is used in the second-order error minimization methods.

The curvature of the error surface at a point is expressed by the second derivative of error with respect to the weights i.e., $\partial^2 E/\partial w^2$, which is obtained by differentiating the error derivative $\partial E/\partial w$, with respect to a weight. In general, the distance to the optimum weights can be estimated by dividing the derivative by the second derivative. This gives the distance required for the deceleration (curvature) of the error to bring the speed of error change (slope) to zero. Thus, the change in a weight can be expressed as

$$\Delta w = -\frac{\partial E/\partial w}{\partial^2 E/\partial w^2}. \tag{4.40}$$

Before proceeding with the second-order methods, all error minimization methods should be put in perspective first so that the first-order methods can be seen in light of the second-order methods.

In training feedforward networks, the structure of the network is chosen first. This includes the number of inputs, number of hidden layers, and

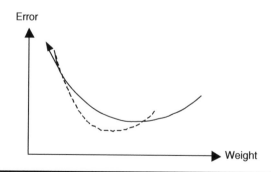

Figure 4.48 Two error surfaces of different curvature.

hidden neurons in each layer, the number of output neurons, and the activation functions of all neurons. Weights are then initialized to random values, the training data is presented to the network repeatedly, and the weights are adjusted incrementally until the MSE decreases gradually. The MSE depicted as a function of weights w, $E(w)$, is

$$E(w) = \frac{1}{N}\sum_{i=1}^{N}[t_i - z_i]^2 = \frac{1}{N}\sum_{i=1}^{N}[t_i - f(w, x_i)]^2. \qquad (4.41)$$

Where the network output z_i for the ith input pattern x_i is expressed as a function f of weights and the input, t_i, is the target output, and N is the number of training patterns. Note that 2 has not been used in the denominator of the formula in order to be compatible with the neural network program used to illustrate the application of the second-order methods [7]. Training or learning involves finding the set of weights w that minimize the MSE. All error minimization methods are alike in that they are iterative. Starting with the initial values for the weights, they incrementally update the weights in the negative direction of the gradient as

$$w_m = w_{m-1} - \varepsilon R d_m, \qquad (4.42)$$

where m is the current epoch, ε is the learning rate, and d_m is the sum of error derivatives over an epoch or batch. The only new parameter in second-order methods is **R**, which has several variants of second derivative of error. This is very useful in changing the search direction from a negative gradient to a more favorable direction. Recall that backpropagation, delta-bar-delta and steepest descent all use negative gradient [7]. Thus, in Equation 4.42 two parameters, ε and **R**, can change, and it is in choosing ε and **R** that various training methods differ.

When **R** = 1 a gradient descent method is obtained that solely follows the direction of the negative gradient of the error surface without using any curvature information. In the steepest descent method, one learning rate, ε, applies to all weights. During learning, it is adapted by starting with a larger value, and ε is halved in each epoch until a value that reduces the error is reached. Backpropagation is another variant of gradient descent and also uses a single learning rate for all weights; however, it stays constant during learning. Delta-bar-delta, or the adaptive learning rate method, is another variant of gradient descent in which the learning rate for each weight is unique and is adapted during learning, thereby efficiently altering individual weights according to how significantly each affects the output. If **R** is not equal to 1 in Equation 4.42 but contains curvature information, more advanced second-order learning methods result. Examples include the Gauss–Newton method, the Levenberg–Marquardt method, and conjugate gradient methods. The following section will discuss the Gauss–Newton and

Levenberg–Marquardt methods in further detail. Backpropagation, delta-bar-delta, and steepest descent are all special cases (first-order) of a general optimization problem of weight adaptation [9,10].

4.7.3 Gauss–Newton Method

The Gauss–Newton method is a fast and reliable method for a wide range of problems [5,7,14] and it explicitly involves the curvature of the error surface. The curvature is specified by the second derivative of the error function with respect to a weight. If the second derivative is denoted by d^s, then $d^s = \partial^2 E/\partial w^2$. The **R** in Equation 4.42 for the Gauss–Newton method is the inverse of the second derivative, i.e., $1/d^s$. Therfore, the weight change Δw_m for a particular epoch m is given by

$$\Delta w_m = -\varepsilon \frac{d_m}{d_m^s}. \tag{4.43}$$

Closer examination shows that this formula is similar to that used for weight adaptation in QuickProp; however, the actual, not an approximation of the curvature is used. At each epoch, the learning rate ε is set to 1, and is only accepted if the MSE decreases for this value. Otherwise, it is halved over and over again until a value for which the MSE decreases is attained. Then the weights are adjusted and a new epoch begins. Because this whole process is done automatically as part of the algorithm, it avoids having to search for an optimal learning rate parameter through trial and error by the user. The termination criteria are similar to those in the steepest descent method and are repeated below:

$$\frac{E(w_m) - E(w_{m+1})}{E(w_m)} \le E_{\min}$$

$$\varepsilon \le \varepsilon_{\min}, \tag{4.44}$$

where $E_{\min}$ and $\varepsilon_{\min}$ are minimum acceptable levels for the MSE and learning rate, respectively.

The next section shows in detail how this method is implemented. As already discussed, the idea of training when there is an arbitrary number of weights becomes what is shown in Equation 4.42 and is repeated below:

$$w_m = w_{m-1} - \varepsilon R d_m,$$

where **R** is the inverse of the second derivative of error, and d_m is the sum of the error gradient for each weight across all patterns. It is already known how to obtain d_m and the next step is to obtain **R**.

If there is only one weight, **R** is simply $1/d_s$. When there are many weights, it is easier to represent the whole set of second derivatives for all weights in a matrix form. This particular matrix of second derivatives is called the Hessian matrix and is denoted by $\mathbf{H}$. $\mathbf{R}$ in Equation 4.42 is the inverse of $\mathbf{H}$, or $1/\mathbf{H}$, also denoted as $\mathbf{H}^{-1}$. This matrix formulation can be used efficiently to obtain the necessary weight change for all weights simultaneously. This idea is presented in Equation 4.45 for an arbitrary network. The MSE E is

$$E = \frac{1}{N} \sum_{i=1}^{N} (t_i - z_i(x_i, w_i))^2,$$ (4.45)

where t_i and z_i are the target and network output, respectively, for the ith input pattern, and N is the number of input patterns. The network output is a function of weights, w_i, that are adjusted during training and inputs, x_i. By differentiating the error twice with respect to the weights, the second derivative or curvature of error surface at the current location of each weight is obtained. The first derivative of error is

$$\frac{\partial E}{\partial w_i} = \frac{2}{N} \sum_{i=1}^{N} \left\{ (t_i - z_i)\left(\frac{-\partial z_i}{\partial w_i}\right) \right\},$$ (4.46)

which simplifies to

$$\frac{\partial E}{\partial w_i} = \frac{2}{N} \sum_{i=1}^{N} \left\{ e_i \left(\frac{-\partial z_i}{\partial w_i}\right) \right\},$$ (4.47)

where e_i is the error $(t_i - z_i)$ for an individual input pattern i. Differentiating Equation 4.46 with respect to the weights again will have

$$\mathbf{H} = \frac{\partial^2 E}{\partial w_i \partial w_j} = \frac{2}{N} \sum_{i=1}^{N} \left\{ (t_i - z_i)\left(\frac{-\partial^2 z_i}{\partial w_i \partial w_j}\right) + \left(-\frac{\partial z_i}{\partial w_i}\right)\left(-\frac{\partial z_i}{\partial w_j}\right) \right\}.$$ (4.48)

There are two parts to Equation 4.48 that describe the curvature of the error function. The first part can cause computational instability problems when $\mathbf{H}$ is inverted to obtain $\mathbf{R}$., To avoid these problems, the Hessian matrix is approximated by the second part of the equation as

$$\mathbf{H} = \frac{\partial^2 E}{\partial w_i \partial w_j} \approx \frac{2}{N} \sum_{i=1}^{N} \left\{ \left(-\frac{\partial z_i}{\partial w_i}\right)\left(-\frac{\partial z_i}{\partial w_j}\right) \right\},$$ (4.49)

which for a network with a single weight becomes

$$\mathbf{H} = \frac{2}{N} \sum_{i=1}^{N} \left(\frac{\partial z_i}{\partial w_i}\right)^2.$$ (4.50)

For a single weight, the computation is simple because the first derivative is squared for each training pattern and the curvature is obtained by calculating the mean, or $\mathbf{H}$, $\mathbf{R}$ is simply the inverse of $\mathbf{H}$ (i.e., $\mathbf{R} = (1/\mathbf{H}) = \mathbf{H}^{-1}$). For an arbitrary number of weights, the full form of the approximated Hessian is a matrix containing the product of the derivative of output, with respect to each pair of weights, w_i and w_j, as shown in Equation 4.49.

4.7.3.1 Network Training with the Gauss–Newton Method—A Hand Computation

The Gauss–Newton method will now be applied to the problem previously started in Section 4.4.1 and continued throughout this chapter for illustrating learning concepts. Recall that it involves a one-input, one-output network with one hidden neuron, as shown in Figure 4.4a, that is used for modeling the data plotted in Figure 4.4b. Start with the same initial weights and the two input–output patterns extracted from the data as given in Table 4.1 in Section 4.4.1. The two input–output patterns are: {0.7853, 0.707} and {1.571, 1.0} and the initial weights are $a_0 = 0.3$, $a_1 = 0.2$, $b_0 = -0.1$, and $b_1 = 0.4$. There are two bias weights and two regular weights in the network to be optimized.

The sum of the first derivative d_m for the first epoch of the two input patterns as found in previous learning methods are (see Section 4.4.1.5)

$$d_m = \{-0.0144, -0.01944, -0.1563, -0.1002\}. \qquad (4.51)$$

With respect to the four weights, the first derivative of the error is (from Equation 4.3)

$$\frac{\partial E}{\partial w_i} = \begin{bmatrix} \dfrac{2.718^{-(a_0 + a_1 x)} b_1}{(1 + 2.718^{-(a_0 + a_1 x)})^2} \\[2ex] \dfrac{2.718^{-(a_0 + a_1 x)} b_1 x}{(1 + 2.718^{-(a_0 + a_1 x)})^2} \\[2ex] 1 \\[2ex] \dfrac{1}{1 + 2.718^{-(a_0 + a_1 x)}} \end{bmatrix} \qquad (4.52)$$

where e is replaced with 2.718.

To obtain $\mathbf{H}$, multiply the derivative with respect to each weight by itself and other derivatives to obtain the two bracketed derivatives in

Equation 4.49, $(\partial z_i/\partial w_i)(\partial z_i/\partial w_j)$. Thus

$$\frac{\partial z}{\partial w_i}\frac{\partial z}{\partial w_j} = \begin{bmatrix} \frac{(0.16)2.718^{-(0.6+0.4x)}}{1+2.718^{-(0.3+0.2x)4}} & \frac{(0.16)2.718^{-(0.6+0.4x)}x}{1+2.718^{-(0.3+0.2x)4}} & \frac{(0.4)2.718^{-(0.3+0.2x)}}{1+2.718^{-(0.3+0.2x)2}} & \frac{(0.4)2.718^{-(0.3+0.2x)}}{1+2.718^{-(0.3+0.2x)3}} \\ & \frac{(0.16)2.718^{-(0.6+0.4x)}x^2}{1+2.718^{-(0.3+0.2x)4}} & \frac{(0.4)2.718^{-(0.3+0.2x)}x}{1+2.718^{-(0.3+0.2x)2}} & \frac{(0.4)2.718^{-(0.3+0.2x)}x}{1+2.718^{-(0.3+0.2x)3}} \\ & & 1. & \frac{1}{1+2.718^{-(0.3+0.2x)}} \\ & & & \frac{1}{(1+2.718^{-(0.3+0.2x)})^2} \end{bmatrix}$$

$$(4.53)$$

By substituting the two inputs to each component in the matrix, summing the two outcomes, and multiplying by $(2/N)$ as in Equation 4.49, the Hessian matrix is obtained as

$$\mathbf{H} = \begin{bmatrix} 0.0011 & 0.0012 & 0.0115 & 0.0072 \\ 0.0012 & 0.0016 & 0.0134 & 0.0085 \\ 0.0115 & 0.0134 & 0.1235 & 0.0779 \\ 0.0072 & 0.0085 & 0.0779 & 0.0491 \end{bmatrix}, \qquad (4.54)$$

where the first entry is the $(2/N)\sum_{i=1}^{N}(\partial z/\partial a_0)_i^2$ where N is the number of input patterns which is 2 in this example . The second entry in the first row is $(2/N)\sum_{i=1}^{N}(\partial z/\partial a_0)_i^2(\partial z/\partial a_1)_i$, and so on. It is difficult to invert this matrix by hand; therefore, *Mathematica*® [13] is used to obtain it. **R** becomes

$$\mathbf{R} = \mathbf{H}^{-1} = 10^{17} \begin{bmatrix} -109 & 57 & 79 & -119 \\ -16 & 5.5 & 8.5 & -12 \\ 0 & -3.2 & -3.4 & 6 \\ 19 & -4.3 & -7.6 & 10 \end{bmatrix}. \qquad (4.55)$$

By substituting for **R** and d_m from Equation 4.51 into Equation 4.42 with a learning rate of 1, the preliminary weight increments for the four weights after the first trial epoch are

$$\Delta w_m = -\mathbf{R}d_m = -10^{17} \begin{bmatrix} -109 & 57 & 79 & -119 \\ -16 & 5.5 & 8.5 & -12 \\ 0 & -3.2 & -3.4 & 6 \\ 19 & -4.3 & -7.6 & 10 \end{bmatrix} \begin{bmatrix} -0.0144 \\ -0.01944 \\ -0.1563 \\ -0.1002 \end{bmatrix}$$

$$= \begin{bmatrix} -256 \\ -48 \\ 8 \\ 16 \end{bmatrix}. \tag{4.56}$$

The new weights obtained by adding the increments to the original weights are {−255.7, −47.8, 7.9, 16.4}. The resulting MSE (from Equation 4.41) is 0.021. Thus, the preliminary update has decreased the error from 0.0602; therefore, a new epoch is started with a learning rate of $\varepsilon = 1.0$ and the process is repeated. Although the error is low, weights are too high due to large values in $\mathbf{R}$ that are caused by numerical instabilities because only two input patterns were used. However, this is corrected when more inputs are used in the computer experiment conducted in the next section.

4.7.3.2 Example: Network Training with Gauss–Newton Method—A Computer Experiment

In this section, the Gauss–Newton method is applied to the problem previously started in Section 4.4.1 and continued throughout this chapter up to Section 4.7.3.1. Recall that it involves a one-input, one-output network with one hidden neuron, as shown in Figure 4.4a, and used to model data in Figure 4.4b extracted from the first quarter of the sine wave. We use the same initial weights as given in Table 4.1 in Section 4.4.1, which are $a_0 = 0.3$, $a_1 = 0.2$, $b_0 = -0.1$, and $b_1 = 0.4$. All of the data in Figure 4.4b will be used to train the network iteratively. There are two bias weights and two regular weights in the network to be optimized.

The training performance is shown in Figure 4.49 that illustrates how the error decreases with the epochs. The figure shows that the termination criteria are achieved in seven epochs that produce a minimum RMSE of 0.005, similar to the error reached in the steepest descent method. The Gauss–Newton method, however, reaches the error quicker.

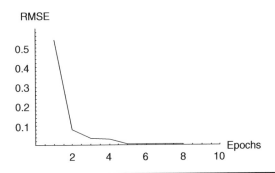

Figure 4.49   Training performance with the Gauss–Newton method.

The final weights reached by the Gauss–Newton method are $a_0 = -0.84$, $a_1 = 2.56$, $b_0 = -0.43$, and $b_1 = 1.5$.

The network output and the target data are superimposed in Figure 4.50 and shows a perfect agreement.

One possible drawback of this method is that there might be a situation where it is difficult to always achieve learning rates that decrease error, causing the weight updates to sometimes be in the positive gradient direction. This is due to the nature of the second derivative of the error surface. For example, if the first derivative is positive at a point but the second derivative is negative, indicating that the curve is getting concave approaching maxima, then the weight update would be positive and would lead to an increased error. This is illustrated in Figure 4.51.

The slope d_m at the point indicated in the figure is positive but the curvature d_m^s is negative. For this point, the Gauss–Newton weight update $(-\varepsilon(d_m/d_m^s))$ is positive and leads to an increased error. Therefore, when the second derivative is negative, another method such as the steepest descent

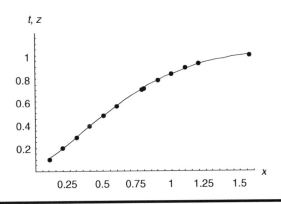

Figure 4.50   Network output from the Gauss–Newton method superimposed on target data.

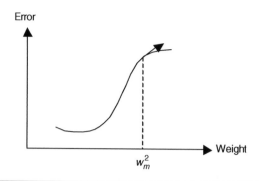

Figure 4.51 **A second derivative of error surface that could lead to increased error.**

must be used for the weight update. Furthermore, a naïve application of second-order methods is risky because numerical instabilities in the inversion of the Hessian matrix can lead to problems [4]. To improve these conditions and guarantee a downhill path on the error surface, the more advanced Levenberg–Marquardt method has been proposed [14].

4.7.4 The Levenberg–Marquardt Method

This method improves the solution to problems that are much harder to solve by only adjusting the learning rate repeatedly, as implied in the Gauss–Newton method that incorporates both first and second derivatives of error. Instead of adjusting ε, the Levenberg–Marquardt (LM) method sets it to unity and a new term e^{λ} is added to the second derivative term, where e is the natural logarithm [7,14]. For example, a network with a single weight w, $\mathbf{R}$, which is the inverse of the second derivative in Equation 4.42, becomes $[1/(d_m^s + e^{\lambda})]$ and the new weight update for epoch m for this weight can be expressed as

$$\Delta w_m = -\frac{d_m}{d_m^s + e^{\lambda}}. \tag{4.57}$$

Because we use the Hessian matrix to denote all second derivatives of error with respect to each weight, the Hessian is modified as

$$\mathbf{H}' = \mathbf{H} + e^{\lambda}\mathbf{I}. \tag{4.58}$$

$\mathbf{R}$ then becomes

$$\mathbf{R} = \frac{1}{(\mathbf{H} + e^{\lambda}\mathbf{I})}, \tag{4.59}$$

where **I** is an identity matrix. The general formula for the weight change of all weights for epoch m can be expressed in matrix form as

$$\Delta w_m = -\frac{d_m}{(\mathbf{H}_m + e^\lambda \mathbf{I})}, \tag{4.60}$$

where d_m is the sum of the first derivative of error and $\mathbf{H}_m$ is the Hessian matrix for epoch m. The term e^λ produces a conditioning effect to the second derivative such that the error never increases in situations where the Gauss–Newton method can result in an increased error (see Section 4.7.3.2), thereby improving the stability of the solution.

In the LM method, λ is chosen automatically until a downhill step is produced for each epoch. Starting with an initial value of λ, the algorithm attempts to decrease its value by increments of $\Delta\lambda$ in each epoch. If the MSE is not reduced, λ is increased repeatedly until a down hill step is produced. When λ is small, the LM method is similar to the Gauss–Newton method in that the second term (conditioning term) in the denominator is small and, therefore, the second derivative plays an important role in the weight update equation. By attempting to reduce λ initially, the LM method essentially attempts to use both the first and second derivatives of error in order to utilize their combined effectiveness, as illustrated in the Gauss–Newton method. The first derivative is already in the numerator.

However, when λ is large, the method is similar to steepest descent in that the conditioning term in the denominator of the weight update equation becomes large. Thus, the effect of the second derivative of error is not significant compared to that of the first derivative, and the error is reduced almost entirely along the direction of the negative error gradient, as illustrated in the steepest descent method. The LM algorithm resorts to this approach and uses a larger λ when the weight change leads to an increased error that is caused by climbing up the error surface due to problems associated with the second derivative of error as previously mentioned. Thus, the LM method is a hybrid algorithm that combines the advantages of the steepest descent and Gauss–Newton methods to produce a more efficient method than either of these two methods. The training terminates prior to the specified number of epochs if the following conditions are met:

$$\lambda > 10\Delta\lambda + \text{Max}[\mathbf{H}],$$

$$\frac{E(w_m) - E(w_{m+1})}{E(w_m)} \le E_{\text{min}}, \tag{4.61}$$

where Max[**H**] is the maximum eigenvalue of the Hessian matrix, which guarantees that a solution is reached in a stable manner down the error curve. (Refer to Chapter 6 for a discussion on eigenvalues.) The first condition in Equation 4.61 specifies the maximum step size $\Delta\lambda$ allowed, guaranteeing the stability of the solution, and the second criterion states that training stops if the proportion of error change between two consecutive epochs becomes less than the minimum specified error. The latter condition is similar to the steepest descent method and is specified by the user.

4.7.4.1 Example: Network Training with LM Method—A Hand Computation

The LM method can be applied to the problem previously started in Section 4.4.1 and continued throughout this chapter. Recall that it involves a one-input, one-output network with one hidden neuron, shown in Figure 4.4a, and used for modeling the data plotted in Figure 4.4b. Start with the same initial weights and the two input–output patterns extracted from the data as given in Table 4.1 in Section 4.4.1. There are two bias weights and two regular weights in the network to be optimized. The two input–output patterns are {0.7853, 0.707} and {1.571, 1.0} and initial weights are $a_0 = 0.3$, $a_1 = 0.2$, $b_0 = -0.1$, and $b_1 = 0.4$.

For the two input–output (x, t) pairs, the modified Hessian matrix $\mathbf{H}'$, which is $(\mathbf{H} + e^{\lambda}\mathbf{I})$, is

$$
\mathbf{H}' = \begin{bmatrix}
0.0011 + e^{\lambda} & 0.0012 & 0.0115 & 0.0072 \\
0.0012 & 0.0016 + e^{\lambda} & 0.0134 & 0.0085 \\
0.0115 & 0.0134 & 0.1235 + e^{\lambda} & 0.0779 \\
0.0072 & 0.0085 & 0.0779 & 0.0491 + e^{\lambda}
\end{bmatrix}. \quad (4.62)
$$

This is similar to **H** for the Gauss–Newton method; however, now there is an e^{λ} term added to the diagonal terms. By denoting the inverse of this Hessian matrix by **R** and inverting it with $\lambda = 5$, as illustrated in *Mathematica* [13], the following **R** is obtained:

$$
\mathbf{R} = \begin{bmatrix}
0.1353 & -0.000022 & -0.0002 & -0.00013 \\
-0.000022 & 0.1353 & 0.00024 & -0.00015 \\
-0.0002 & -0.00024 & 0.133 & -0.0014 \\
-0.00013 & -0.00015 & -0.0014 & 0.1344
\end{bmatrix}. \quad (4.63)
$$

The conditioning effect of the added diagonal term can be ascertained by comparing **R** in Equation 4.63 with **R** in the Gauss–Newton method (Equation 4.55) that does not have the extra diagonal term. The terms in

Equation 4.63 are much smaller and all of the off-diagonal terms are close to zero.

Because the first derivative of error d_m for epoch 1 has been computed already for the two input–output patterns and is found to be $d_m = \{-0.0144, -0.01944, -0.1563, -0.1002\}$ for the four weights (see Section 4.4.1.5), the weight update can be calculated from Equation 4.42 as

$$\Delta w_m = -R d_m$$

$$= - \begin{bmatrix} 0.1353 & -0.000022 & -0.0002 & -0.00013 \\ -0.000022 & 0.1353 & 0.00024 & -0.00015 \\ -0.0002 & -0.00024 & 0.133 & -0.0014 \\ -0.00013 & -0.00015 & -0.0014 & 0.1344 \end{bmatrix} \begin{bmatrix} -0.0144 \\ -0.01944 \\ -0.1563 \\ -0.1002 \end{bmatrix}$$

$$= \begin{bmatrix} 0.0019 \\ 0.0026 \\ 0.0207 \\ 0.0132 \end{bmatrix}.$$

(4.64)

The new weights for the second epoch will be the value of the weight increment added to the respective original weights. The resulting weight vector is $\{0.3019, 0.2026, -0.0793, 0.4132\}$. This results in an MSE of 0.0579, which is smaller than the initial MSE of 0.0602.

4.7.4.2 Network Training with the LM Method—A Computer Experiment

The LM method can be applied to the problem previously started in Section 4.4.1 and continued throughout this chapter up to Section 4.7.4.1. Recall that it involves a one-input, one-output network with one hidden, neuron, shown in Figure 4.4a, and used to model data in Figure 4.4b extracted from the first quarter of the sine wave. The same initial weights are used, as shown in Table 4.1 in Section 4.4.1. They are $a_0 = 0.3$, $a_1 = 0.2$, $b_0 = -0.1$, and $b_1 = 0.4$. Use all of the data in Figure 4.4b to train the network iteratively. There are two bias weights and two regular weights in the network to be optimized.

The training over epochs brings the RMSE down to 0.005 in five epochs, as illustrated in Figure 4.52a. This minimum error is similar to that achieved by the Gauss–Newton method; however, the LM method reaches the minimum error sooner (see also Figure 4.53). The final weights were $a_0 = -0.84$, $a_1 = 2.56$, $b_0 = -0.43$, and $b_1 = 1.5$. These weights are

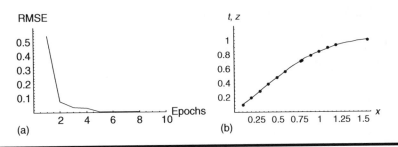

Figure 4.52 Results from the Levenberg–Marquardt method: (a) training performance and (b) network output superimposed on target data.

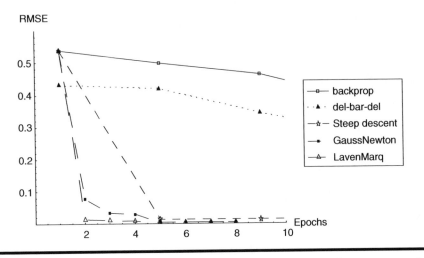

Figure 4.53 A comparison of training performance for Levenberg–Marquardt, Gauss–Newton, steepest descent, delta-bar-delta, and backpropagation methods in mapping the first quarter of the sine function.

comparable to those obtained when using the other methods. The target data are superimposed on the network output in Figure 4.52b, which indicates a perfect fit.

4.7.5 Comparison of the Efficiency of the First-Order and Second-Order Methods in Minimizing Error

How do the second-order methods compare with the first-order methods? The results from backpropagation, delta-bar-delta, the steepest descent method, the Gauss–Newton method, and the LM method are plotted together in Figure 4.53 and are compared. Because the Gauss–Newton and LM methods took less than ten epochs in order to decrease to a minimum error, the performances of the five methods in the first ten iterations are compared in this figure, which reveals that the LM method is the most efficient (largest dashed line) followed by the Gauss–Newton method (second largest dashed line). These two methods are superior to the steepest descent method (third largest dash line), which also brings the error down in almost five epochs. The delta-bar-delta and backpropagation methods (smallest dashed line and solid line) lag behind the first three methods considerably. Therefore, the performance of the training methods used on this simple differs, and the LM, Gauss–Newton, and steepest decent methods are an order of magnitude faster than the delta-bar-delta and backpropagation.

Note that the efficiency of the second-order methods (LM and Gauss–Newton) is gained at a considerable computational cost. This is because computing and inverting the Hessian matrix for large networks trained with a large number of training patterns can be costly computationally and time consuming. Moreover, inverting the Hessian matrix can cause numerical instability problems and the methods may not perform satisfactorily. In these situations, first-order methods such as the steepest descent, backpropagation, and delta-bar-delta can provide effective solutions, although they may take longer to converge. This is because methods such as backpropagation allow a comprehensive search in the parameter (weight) space through their learning parameters.

4.7.6 Comparison of the Convergence Characteristics of First-Order and Second-Order Learning Methods

This section will extend the computer experiment previously addressed in order to compare the search path on the error surface traversed by backpropagation, the steepest descent, the Gauss–Newton, and LM methods. These methods will be used to extract the underlying pattern in another data set. In order to visually evaluate the performance, the data shown in Figure 4.54 will be used so that it can be modeled with a simple single-neuron network with only two weights (bias and a regular weight), as illustrated in Figure 4.55. The activation function of the output neuron is logistic. Error is a two-dimensional surface with the two weights.

The output of this network can be expressed as

$$y = \frac{1}{1 + e^{-(w_0 + w_1 x)}},$$ (4.65)

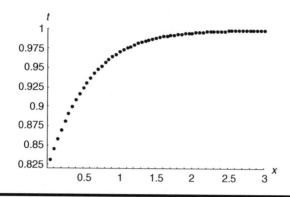

Figure 4.54 Example data to compare convergence characteristics of different learning methods.

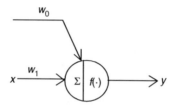

Figure 4.55 One-input, one-output network.

where w_0 and w_1 are bias and input–output weights, respectively. The 50 data points in Figure 4.54 were generated for the values of $w_0 = 1.5$ and $w_1 = -2.0$ in Equation 4.65.

The MSE for this case is

$$E = \frac{1}{N} \sum (t_i - y_i)^2 = \frac{1}{N} \sum \left(t_i - \frac{1}{1 + e^{-(w_0 + w_1 x)}} \right)^2. \qquad (4.66)$$

Note that 2 has been omitted in the denominator to be compatible with the neural networks program used to generate the results. Plot this error surface for a range of w_0 and w_1 values in the neighborhood of the optimum values of 1.5 and -2.0, which were used to generate the data. The error plotted in Figure 4.56 is the RMSE and resembles actual errors.

All four training methods, backpropagation, steepest descent, Gauss–Newton, and LM, should converge to weight values of $w_0 = 1.5$ and $w_1 = -2.0$ when training with random initial values. The underlying pattern in the data is extracted from the four methods and the error

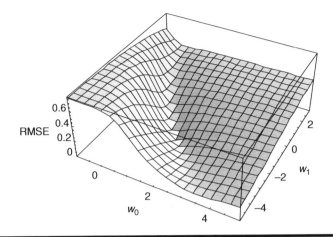

Figure 4.56 Error surface for the weights around the neighborhood of optimum values of (1.5, −2.0).

surface is plotted around these optimum values to find out how efficiently the different methods traverse the error surface down to the optimum values. The results are presented for each method in the next four sections.

4.7.6.1 Backpropagation

The network was first trained using backpropagation training, and the path followed for 200 epochs is shown in Figure 4.57a for a learning rate of $\varepsilon = 0.1$ and momentum $\mu = 0.9$. Figure 4.57b shows the same for $\varepsilon = 0.1$ and $\mu = 0$. In the first case with a high momentum, the iteration path continues from the top of the error surface. The iteration path is not

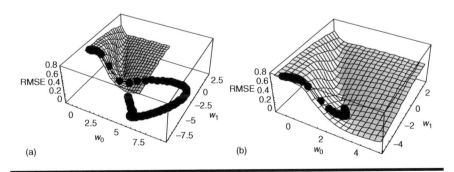

Figure 4.57 Iteration path for backpropagation training: (a) $\varepsilon = 0.1$ and $\mu = 0.9$; (b) $\varepsilon = 0.1$ and $\mu = 0$.

in the most efficient direction to the optimum weights but is from a distance and bypasses it. The path then moves away and later turns around as it tries to approach the optimum toward the end of the 200 specified training epochs. This highlights an interesting point in that the optimum weights would be found eventually. In the latter case where no momentum is used, the path slowly follows the slope of the error surface. Again, the iteration path is not in the most efficient direction but at a distance from the optimum. It then reaches a valley and moves towards the optimum from that point and will eventually reach the optimum weights with more training epochs.

The RMSE for the whole training is shown in Figure 4.58 for the two above cases. Figure 4.58a shows how the error first decreases as weights become closer to the optimum for the high momentum and later increases as the path bypasses it and later reverses the direction. Even if the training were to stop at the point where the error is at a minimum, it would still not produce the optimum weights, as indicated in Figure 4.57. Figure 4.58b shows that error decreases continually when moving towards the optimum when the momentum = 0.

4.7.6.2 Steepest Descent Method

The training path taken in each epoch by the steepest descent method is shown in Figure 4.59. Figure 4.60 shows how the error changes with each epoch.

Figure 4.59 shows that convergence is very slow at the end of the training, but it eventually approaches the true optimum weight values $(1.5, -2.0)$. Figure 4.60 illustrates that it will converge slowly and smoothly to the minimum in about 30 iterations (epochs).

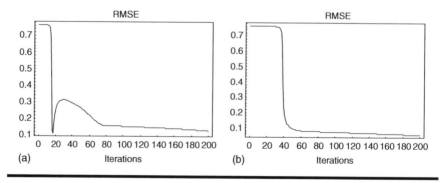

Figure 4.58 Training performance for backpropagation: (a) $\varepsilon = 0.1$ and $\mu = 0.9$; (b) $\varepsilon = 0.1$ and $\mu = 0.$

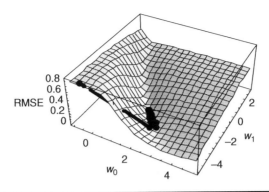

Figure 4.59 **Training path for the the steepest descent algorithm.**

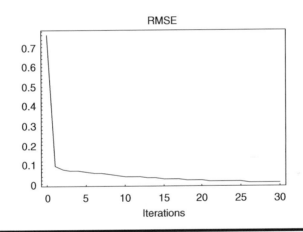

Figure 4.60 **Training performance for the steepest descent algorithm.**

4.7.6.3 Gauss–Newton Method

The performance of the Gauss–Newton algorithm on the same data set is shown in Figure 4.61. Recall that this method uses the second derivative of error and the learning rate is adjusted automatically.

The above figure illustrates that the Gauss–Newton method is more efficient than the steepest descent method and backpropagation in that it traverses the slope of the error surface more efficiently. However, the downhill path it follows is not the most effective. As shown in Figure 4.62, this algorithm also reaches convergence quicker than either of the two previous methods and requires about seven epochs. It also produces a more controlled descent towards the optimum weights.

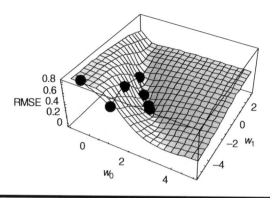

Figure 4.61   Training path followed by the Gauss–Newton algorithm.

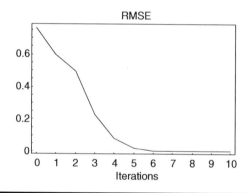

Figure 4.62   Reduction of error with epochs for the Gauss–Newton algorithm.

4.7.6.4   Levenberg–Marquardt Method

Finally, we test the performance of the LM algorithm on the data set. Figure 4.63 shows the training path and Figure 4.64 illustrates how error decreases with each iteration. The training path follows the downhill slope of the error surface almost perfectly, which is a significant improvement over the Gauss–Newton algorithm. The algorithm also converges a little faster than the Gauss–Newton method and requires only four iterations.

This illustration clearly highlights the differences among the four training algorithms. It demonstrates that the LM method is superior to the other three methods. It is faster and more efficient when converging to optimum weights for this simple problem. The Gauss–Newton method is the next best choice in terms of finding an efficient downhill path and the time it takes to converge. Both of these methods that use the first and second derivative of error proved to be superior to the steepest descent

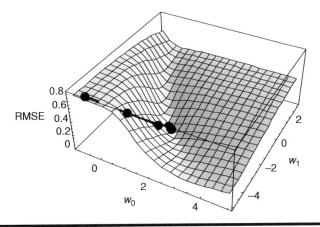

Figure 4.63 Training path for the Levenberg–Marquardt algorithm.

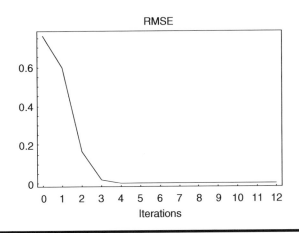

Figure 4.64 Training performance for the Levenberg–Marquardt algorithm.

method and backpropagation that rely only on the first derivative. The third best is the steepest descent algorithm, which performs better than backpropagation. The former three methods do not require trial and error and have yet produced a superior performance to backpropagation. Backpropagation required more training epochs than the 200 that was specified for both cases of learning parameters, one representing a small learning rate with a high momentum and the other a small learning rate without any momentum. Results show the damaging effect of momentum for this example. The fact that the backpropagation algorithm did not converge in this session illustrates that trial and error is indeed needed until appropriate learning parameters that lead to optimum weights sooner are found. Even with appropriate values for learning rate and momentum, the method will find the optimum weights slower than the other methods for the problem presented here.

4.8 Summary

In this chapter, six training methods were discussed: backpropagation, adaptive learning rate (delta-bar-delta), the steepest descent, QuickProp, the Gauss–Newton method, and the Levenberg–Marquardt method. The first three are first-order methods in that they use the first derivative of error (slope) and follow the gradient descent approach. The latter three methods are second-order methods and they rely on both first and second derivative of error (slope and curvature) in the search for the optimum weights. The Gauss–Newton method has the disadvantage of moving in the direction of an error increase in some situations when the second derivative is negative. The LM method addresses this problem by reverting to the steepest descent method and using only the first derivative when the second derivative becomes negative.

The second-order methods provide faster solutions because of the incorporation of an extra second derivative of error information and automatic internal adjustments that are made to the learning parameters. However, this comes at a substantial computational cost of the calculation of the second derivative of error, Hessian, **H**, and inverse of Hessian, **R**, especially for a network with a large number of weights. First-order methods may provide solutions for a variety of problems, yet they can take longer to train, especially during backpropagation training where learning parameters are found by trial and error. For some problems, these may be found relatively easily; however, some other problems may require extensive searching. The first-order methods do give the user the flexibility and direction to improve their analysis when higher order methods may fail to converge because of the numerical instability in the handling of the second derivative.

Problems

1. For the network training example started in Section 4.4.1, calculate error gradients for epoch 2 and verify that the final values and weight adjustments given in Section 4.4.1.5 of the text are correct.
2. Train the example network in Section 4.4.1 with the two input–output patterns given using online (example-by-example) learning. Compare the results for batch learning presented in the text with those from example-by-example learning (refer to Section 4.4.1.6).

3. Explain how momentum adds stability to the search process in error minimization. When is high or low momentum useful?
4. What is the goal of adaptive learning rate or the delta-bar-delta method?
5. What concepts are used in the implementation of the delta-bar-delta method? What are its advantages over backpropagation?
6. In what way are the concepts of momentum and delta-bar-delta similar?
7. What are the main problems associated with backpropagation and delta-bar-delta and how does the steepest descent method address these? What is the advantage of the steepest descent method?
8. Explain the concepts of first-order and second-order error minimization.
9. What are the advantages of second-order methods?
10. Explain the difference between the Gauss–Newton and Levenberg–Marquardt methods. Which method is superior and why?
11. What are some disadvantages of second-order methods?
12. For a problem of choice, implement different learning methods and assess their effectiveness.

References

1. Werbos, P.J. *The Roots of Backpropagation. From Ordered Derivatives to Neural Networks and Political Forecasting*, Wiley Series on Adaptive and Learning Systems for Signal Processing, Communication, and Control, Wiley, New York, 1994.
2. Rumelhart, D.E. and McClelland, J.L. Foundations, *Parallel Distributed Processing—Explorations in the Microstructure of Cognition*, Vol. 1, MIT Press, Cambridge, MA, 1986.
3. Rumelhart, D.E. and McClelland, J.L. Psychological and biological models, *Parallel Distributed Processing—Explorations in the Microstructure of Cognition*, Vol. 2, MIT Press, Cambridge, MA, 1986.
4. Smith, M. *Neural Networks for Statistical Modeling*, International Thompson Computer Press, London, UK; Boston, MA, 1996.
5. Haykin, S. *Neural Networks: A Comprehensive Foundation,* 2nd Ed., Prentice Hall, Upper Saddle River, NJ, 1999.
6. Jacob, R.A. Increased rates of convergence through learning rate adaptation, *Neural Networks*, 1:4, 295, 1988.
7. *Neural Network for Mathematica*, Wolfram Research Inc., Chicago, IL, 2003.
8. Fahlman, S. Faster learning variations of backpropagation: an empirical study. *Proceedings of the 1988 Connectionist Models Summer School*, D.S. Touretzky, G.E. Hinton, and T.J. Sejnowski, eds., Morgan Kaufmann Publishers, Los Altos, CA, 38–51, 1988.
9. Herz, J., Krough, A., and Palmer, R.G. *Introduction to the Theory of Neural Computation*, Addison-Wesley, Reading, MA, 1991.

10. Hassoun, M.H. *Fundamentals of Artificial Neural Networks*, MIT Press, Cambridge, MA, 1995.

11. Dennis, J.E. and Schnabel, R.B. *Numerical Methods for Unconstrained Optimization and Nonlinear Equations*, Prentice Hall, Englewood Cliffs, NJ, 1983.

12. Fletcher, R. *Practical Methods of Optimization*, Wiley-Inter Science, New York, 1987.

13. *Mathematica 5.0*, Wolfram Research Inc., Champaign, IL, 2003.

14. Bishop, C.M., *Neural Networks for Pattern Recognition*, Clarendon Press, Oxford, UK, 1996.

15. Freeman, J.A., *Simulating Neural Networks with Mathematica*, Addison-Wesley, Reading, MA, 1994.

16. *NeuroShell*, Ward Systems Inc., Frederick, MD, 1997.

Chapter 5

Implementation of Neural Network Models for Extracting Reliable Patterns from Data

5.1 Introduction and Overview

Implementing a model to extract patterns from data requires close attention be paid to various aspects of model development such as testing generalization ability, minimizing model complexity, testing robustness of models (i.e., stability of model parameters), and selecting relevant inputs. This chapter focuses on generalization, structure optimization, and robustness of multilayer neural networks; Chapter 6 deals with input selection. Chapter 7 details uncertainty assessment in relation to model parameters, outputs, and network sensitivities to weights and inputs.

First, a model is developed and calibrated to ensure its adequacy. The model is then tested on new data to ensure its generalization ability. A training dataset is used for validation (i.e., calibration by assessing and fine tuning) the model. Generalization means how well a validated model performs on to unseen data and is tested on an independent test dataset. The purpose of validation is to ensure generalization ability of a model.

A model that does not fit the data enough has limited representation, causing lack of fit (bias), and one that fits the data too much models noise as well as leading to overfitting (variance). These occurrences are called bias–variance dilemma, and both situations increase generalization error. Therefore, a tradeoff between these two extremes is sought in reducing generalization error. This bias and variance tradeoff is addressed in Section 5.2. Two approaches for improving generalization, early stopping and regularization, are presented in Section 5.3. The effect of initial random weights, random sampling, and the number of hidden neurons is also presented in Section 5.3; and the nonuniqueness of weights for these cases is addressed. Furthermore, to shed light on the nonuniqueness of weights, a detailed explanation of hidden neuron activation for these cases is presented to illustrate the network's consistent approach toward a solution regardless of the initial random weights, random sampling, and the number of hidden neurons.

Structural complexity is a crucial aspect of model development, meaning that the ideal model has the optimum number of model parameters (i.e., weights). Structural complexity is particularly crucial for neural networks because they tend to have a large number of free parameters that make them very powerful nonlinear processors. A way to address this issue is to prune irrelevant weights and neurons from a network. Section 5.4 details some approaches for pruning multilayer networks. Another important aspect of model development is ensuring that the model parameters are stable and consistent. Section 5.5 addresses trained networks' robustness of weights.

5.2   Bias–Variance Tradeoff

Data collected from many real-world problems or natural systems almost always contains random variations or noise. In many situations, a model that learns to distinguish general trends from the noise in the data is desired. This property is called "generalization ability of a model." Figure 5.1a illustrates a model that generalizes well, captures the required pattern, and reliably predicts unseen data. This generalization ability is crucial if it is to be useful in decision-making where reliable predictions for inputs not seen before by the model are essential.

The generalization is particularly important for neural networks such as multilayer networks. If allowed to be too flexible, the multilayer networks can strictly follow the data, thereby fitting the noise as well resulting in overfitting, as shown in Figure 5.1b. This error introduced by noise is called "variance contribution to the model misfit." The more flexible a model is, the larger the risk of overfitting. This overfitting adversely affects the generalization ability on unseen data. On the other hand, a model with

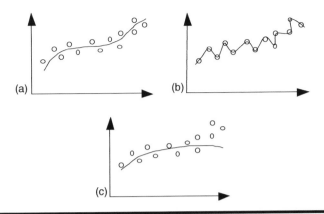

Figure 5.1 Bias–variance dilemma: (a) a model that generalizes well, (b) a model that overfits due to too much flexibility, and (c) a model that underfits due to lack of flexibility.

too little flexibility may not be able to capture essential features in the data, thereby underfitting. A lack of flexibility introduces bias contribution to the misfit of the model. This situation is illustrated in Figure 5.1c where the model has not found the required trend. The correct amount of flexibility is a compromise between these two sources of error and is called bias–variance tradeoff [1,2].

5.3   Improving Generalization of Neural Networks

To avoid overfitting, the flexibility of a neural network must be reduced. Flexibility comes from the hidden neurons, and as the number of hidden neurons increases, the number of network parameters (weights) increases as well. The larger the number of weights, the larger the flexibility. On the other hand, there must be enough neurons to avoid bias or underfitting. There are several ways to handle bias–variance tradeoff and all involve a second validation dataset.

1. *Exhaustive search*: Although more time consuming, the simplest way is to search for the optimum number of neurons is by trial and error. Each time, the performance on a validation dataset must be tested. The one that gives the minimum error on the validation set has the optimum number of neurons.
2. *Early stopping*: The idea here is that a model overfits if it has too much flexibility that is expressed by the number of free parameters (i.e., weights). Weights are called free parameters because they are allowed to change during training, and they define the network's degrees of freedom. If the weights are allowed to grow enough

during training and then stop training at this point, it is possible to restrain the network from overfitting. In this approach, a fixed and large number of neurons is used, and the network is trained while testing its performance on a validation set at regular intervals. Initially, the model error on both the training and validation sets would decrease. At some point, the model generalizes the best. Beyond that point, overfitting sets in. Here, the error on the validation set is the minimum, and it starts increasing with further training, as shown in Figure 5.2. However, an error on the training set would still continue to decrease because overfitting continues to minimize the error on the training set. At the point where overfitting sets in, the weights are taken to be the optimum weights that provide the best generalization on unseen data [2,4].

3. *Regularization*: Another approach to avoid overfitting is regularization. In regularization, a parameter larger than zero is specified, and a regularized performance index is minimized instead of the original mean square error (MSE). This performance index is derived by adding a sum of square weights term to the original MSE term. In neural networks applications, this type of regularization is often called weight decay. The idea is to keep the overall growth of weights to a minimum in such a way that weights are pulled toward zero. In this process, only the important weights are allowed to grow, and others are forced to decay [1–4].

Exhaustive search, although guaranteed to find the optimum number of weights if they exist, is not very practical for all problems. In early stopping and regularization, a network has more weights than are necessary with not all of them optimized as in the first method. Therefore, only a subset that becomes the most sensitive to the output is effectively used. The latter two methods are presented in detail next. A logical extension of early stopping and regularization is to prune irrelevant weights so that the simplest possible network structure remains. This topic is treated later in the chapter.

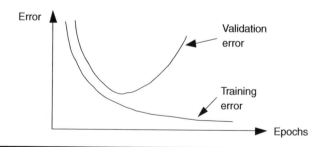

Figure 5.2 Early stopping for improving generalization—training and validation errors.

5.3.1   Illustration of Early Stopping

A simple, one-dimensional example illustrates how the early stopping method works. The data was generated from a cyclical pattern shown along with the data in Figure 5.3; x is the input and t is the target data representing the cyclical pattern. The data shown in the figure was first generated from this pattern, and random noise from a Gaussian distribution with mean 0 and standard deviation of 0.25 was added to each data point to create a noisy dataset of 30 observations. The task of the neural network model is to recognize the pattern depicted by the solid line from the data in the presence of noise.

A two-hidden-neuron model appears to fit the data; however, a multilayer perceptron (MLP) with four hidden neurons with logistic activation (Figure 5.4a) is used to demonstrate the concept of early stopping. The network output z produced by the initial random weights is shown with the original cyclical pattern superimposed on the data in Figure 5.4b, which shows a very poor fit.

To use early stopping, the dataset is randomly divided into two sets. For this example, each set has 15 patterns. The network is trained with the Levenberg–Marquardt method, which is a powerful second-order method discussed in Chapter 4. Training is done with the training set as usual, but the MSE on the validation set is calculated at regular intervals. The training progress in terms of reduction of root mean square error (RMSE) with epochs is shown in Figure 5.5, where the solid line depicts error on the training set and the dashed line shows error on the validation set. In this figure, iteration is an epoch that is one pass of the whole training dataset through the network.

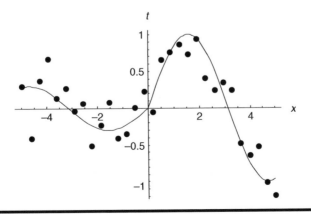

Figure 5.3   Cyclical pattern and noisy data generated from it.

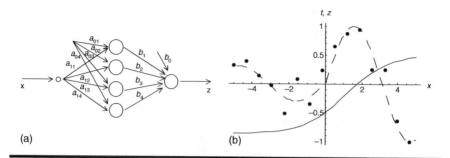

Figure 5.4 Four-hidden-neuron network and initial network output: (a) network configuration, (b) initial network output (solid line) superimposed on training data and target pattern (dashed line).

Figure 5.5 shows that the error on both datasets decreases initially, and after two epochs, validation error increases. The weights at the point where the validation error is minimum are the optimum weights that produce the best model that has optimized the bias–variance tradeoff. The network output for these optimum weights is superimposed on the original pattern and the training data in Figure 5.6.

Figure 5.6 shows that the network output is very close to the true target pattern, indicating that it has recognized the correct trend in the noisy data. The fact that the curve does not rigidly follow the data too closely means that overfitting has been avoided. The fact that the data follows the general trend very closely indicates that the network has enough flexibility to avoid bias (lack of fit).

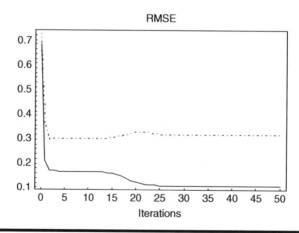

Figure 5.5 Training progress on the training and validation data.

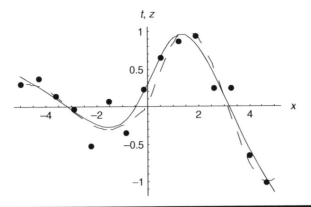

Figure 5.6 Network output for weights obtained from early stopping (solid line) and the original cyclical pattern (dashed line) superimposed on training data.

The MSE on noisy validation data is 0.30 and on the true target pattern is 0.108. The error on the true pattern is calculated from the data generated from the target pattern without using noise. This error indicates the closeness of the model to the true target pattern; however, the error on noisy data highlights the closeness of the predicted pattern to target noisy data, a portion of which was used to model the network. The results indicate that the model output is much closer to the true target pattern uncorrupted by noise than to the noisy validation data. These results show that the network not only has prevented overfitting, but it also has identified the true pattern that was deliberately corrupted with noise.

For comparison, training also continued until error on the training data reached the minimum possible level denoted by the solid line in Figure 5.5. The network output for the weights that produced the minimum error on the training set at the final epoch is shown in Figure 5.7, along with the original true pattern and the noisy training data. It clearly shows a poorer model than the one obtained from early stopping. This network output is overfitted and has poor generalization ability.

Figure 5.7 demonstrates that the solid line goes rigidly through almost all the data points, illustrating overfitting. For the fully trained network, MSE on training data has gone down to 0.107, indicating that the error between the prediction and noisy training data is very small. However, the generalization error on the validation data is now 0.318 and on the true pattern is 0.166. Both are higher than those for the optimum configuration obtained by early stopping (0.30 and 0.108, respectively). The fully trained network error on the true target pattern now is 54 percent, larger than the optimum network. Thus the fully trained model rigidly fits noisy data compared to that obtained from early stopping; therefore, it is too far from the actual pattern that generated data.

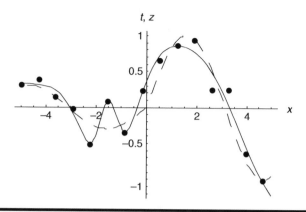

Figure 5.7 **The output of the overfitted network with the weights obtained at the end of the training period (solid line) and the original target pattern (dashed line) superimposed on training data.**

Figure 5.8 shows how weights change in the initial ten epochs of training. The initial weights were small random values closer to zero. The error on the validation set is minimum at epoch 2. The optimum weights are at this point, and they are among the smallest weights compared to those beyond epoch 6 when two hidden-output weights start to increase sharply.

Further training up to 50 epochs dramatically increases these weights, as shown in Figure 5.9. This figure demonstrates that the large weights, with some reaching 20 000, are in deed the cause of overfitting. Thus, early

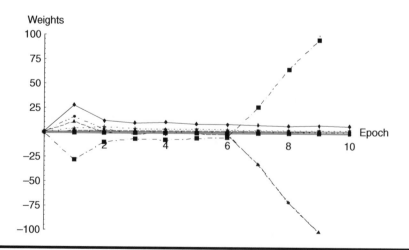

Figure 5.8 **Weight update during first ten epochs (overfitting sets in at epoch 2; two weights that increase drastically at epoch 6 are two hidden-output weights).**

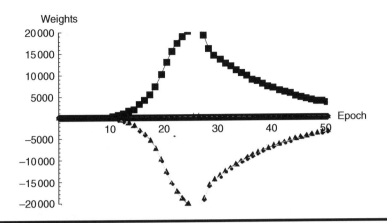

Figure 5.9 Weight change for the whole training period (two weights have become too large and others are comparatively much smaller).

stopping prevents overfitting by keeping weights from growing beyond the point of generalization.

5.3.1.1 Effect of Initial Random Weights

A well-trained network must reach the global absolute minimum error on the error surface. In some cases, there may be several local minima, and the solution can get trapped in these, resulting in suboptimal conditions with larger errors. The same network analyzed in the previous section was trained with three different initial conditions (weights) to study the robustness of the results. Different initial weights put the network at different initial locations on the error surface, and they can help determine if the minimum error and weights achieved are for a local minimum or the global minimum. We have already trained with one set of initial weights for which the results are presented in Figure 5.6 and Figure 5.7 (this will be called initialization 1), so we will do two more trials with different random initializations. In Figure 5.10, the initial network output (a), optimum network output achieved using early stopping (c), and the fully trained network output (d) are shown for the second initial random weight configuration. The training performance is shown in Figure 5.10b.

The RMSE for the optimum model on validation data is 0.301 and on the true target pattern is 0.117. These are very similar to those for the first initialization (0.30 and 0.108, respectively). For the fully trained network, error on training data is 0.118, similar to that for the first initialization (0.107) with both weight initializations leading to overfitting of the fully trained networks. The errors on validation data and true pattern for the fully trained

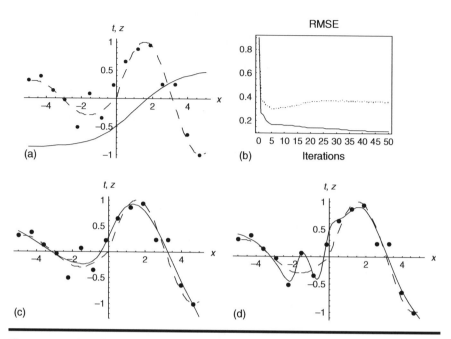

Figure 5.10 **Random weight initialization 2: (a) initial network output (solid line) superimposed on target pattern (dashed line) and training data, (b) root mean square error performance during training (solid line—training, dashed line—validation), (c) optimum network output from early stopping, and (d) fully trained (overfitted) network output.**

network are 0.364 and 0.162, respectively, which are similar to the corresponding values in the first weight initialization (0.318 and 0.166). The fully trained network error on the true pattern now is 38 percent higher than the corresponding optimum network error. Results for the third weight initialization are shown in Figure 5.11.

The third weight initialization produced an optimum network similar to those in the two previous cases (RMSE on validation data and target pattern are 0.299 and 0.111, respectively), but the final network is still very similar to the optimum (i.e., no visible overfitting). This is also evidenced from the training performance shown in Figure 5.11b, where cross validation error does not increase and training error does not decrease with further training as in the other two cases. For the fully trained network, RMSE on both validation and true pattern is 0.306 and 0.1163, respectively, similar to those for the optimum network. The error of the fully trained network on the true target pattern is now only 4.7 percent higher than that for the optimum network. Also, because overfitting has been naturally prevented, the RMSE from the fully trained network on training data is now higher (0.1607) than for the two previous cases where it was 0.107 and 0.118, respectively.

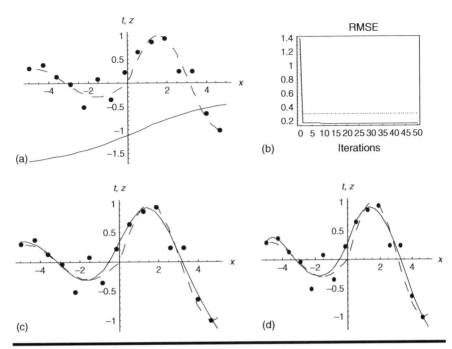

Figure 5.11   Random weight initialization 3: (a) initial network output (solid line), (b) root mean square error change with epochs (solid line—training, dashed line—validation), (c) optimum network output from early stopping, and (d) fully trained network output (all network outputs are superimposed on target pattern (dashed line) and training data).

Therefore, overfitting can be expected in a majority of the cases; however, optimum networks are similar to, and resilient to, random initial weight configurations for this problem.

All three initial conditions have produced similar results in that the optimum networks generalize well. They have similar errors on the validation data and the true pattern indicating that a global minimum has been reached in all three cases. However, in the first two cases, the final fully trained networks seriously overfit the data. In the last case, there is no overfitting at all, indicating that the evolving weight structure has restrained the network and guided toward the global optimum. Certain random initial weights can prevent overfitting, and this situation may correspond to an initial correlated weight structure resembling that of the optimum network. When there are more than enough hidden neurons, the weights become correlated because of redundancy. The correlated structure of weights gives rise to more than one possible set of optimum weights. Some initial weight configurations could be attracted to a possible optimum weight configuration from the beginning, with the result that overfitting is avoided.

5.3.1.2 Weight Structure of the Trained Networks

Table 5.1 lists hidden and output weights for the three cases of weight initialization.

Table 5.1 shows that the final weights for the three initializations are different from each other. This demonstrates that there are many possible combinations of values for the 13 free parameters that can produce the desired outcome. As previously stated, this outcome is attributed to the redundancy because of the correlation of weights. Basically, any set of weights that preserves its correlation structure is a candidate solution. This issue is visited later after further investigation into the network's behavior is conducted. Focus now turns to the behavior of the hidden neurons in the optimum model obtained from cross validation with early stopping for the three cases. (Cross validation refers to calibration of a model built on training data with a validation dataset.) The hidden neuron activations for the three weight initializations are presented in Figure 5.12 for comparison, and these aid in understanding the network's observed behavior.

The first and second initializations have produced similar activations where two sigmoid functions have negative slopes and two have positive slopes. All show their active regions in the input space between -5 and 5. However, in the last plot, all four neurons have negative slopes, and only two are highly active in the input range. This network did not overfit during training. The reason for this is that its flexibility has been dampened by the two neurons that have activations with flatter slopes (contributing less to weight change), leaving two neurons to fit the data. In the first two cases, because all neurons are active, training has to be stopped at the appropriate time to prevent overfitting because of excessive network capability.

A comparison of the three plots in Figure 5.12 reveals that the optimum number of hidden neurons for the data is two. All three networks have found these two neurons that have activations with a negative slope in the three plots. In the first two networks, the two neurons that have activation with a positive slope are the cause of overfitting that is prevented early by the early

Table 5.1 Optimum Weights for Three Random Weight Initializations

Weight Initial	Neuron 1 $(a_{01}, a_{11})\ b_1$	Neuron 2 $(a_{02}, a_{12})\ b_2$	Neuron 3 $(a_{03}, a_{13})\ b_3$	Neuron 4 $(a_{04}, a_{14})\ b_4$	Bias b_0
1	$(-0.73, -0.89)$ -0.46	$(0.91, 0.72)$ 2.59	$(0.31, -0.50)$ 11.47	$(-0.40, 1.3)$ 5.3	-10.1
2	$(1.27, 0.97)$ 2.36	$(0.09, -2.19)$ -1.6	$(1.23, -1.17)$ -3.5	$(-0.6, 0.47)$ -11	6.0
3	$(-0.43, -0.1)$ -66.3	$(0.62, -1.13)$ -3.7	$(0.04, -0.34)$ 41.7	$(-0.33, -0.81)$ -8	11.1

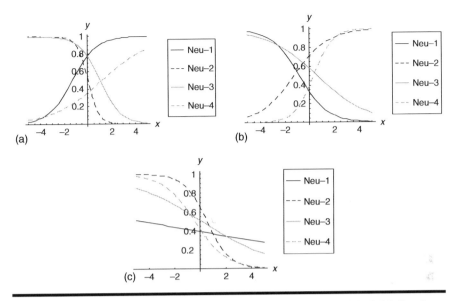

Figure 5.12 Hidden neuron activation for the three random weight initializations: (a) random initialization 1, (b) random initialization 2, (c) random initialization 3 (no overfitting).

stopping method. Because their active regions are in the input space of the data, further training would lead to continuity of activation of these two neurons. In the third network, however, the two redundant neurons weakly contribute to weight change, and they do not cause overfitting.

5.3.1.3   Effect of Random Sampling

The effect of different training samples is demonstrated to address the robustness of networks in relation to different training samples. In many real situations, only a sample from the population of data is available, and the model must give the assurance that results are robust even though only one particular training dataset is used for training. Results have already been shown for one case of training and validation datasets in the two previous sections for the network in Figure 5.4a in Section 5.3.1. The data shown along with the original pattern in Figure 5.3 is called "random sample 1." The first random weight initialization from Figure 5.4b (random weight initialization 1) is used in the following two experiments so that only the datasets change although everything else remains the same.

The two experiments are conducted by repeating the training process twice with data generated from the same true pattern shown in Figure 5.3, but with different random noise values added to the data generated from it. Random sample 2 is shown in Figure 5.13a along with the true pattern. Figure 5.13b–Figure 5.13f presents initial network output, training performance, optimum network output, fully trained overfitted network output, and the weights in the initial epochs of training, respectively, for the training and validation data sampled from random sample 2. The best network is obtained at epoch 2 (Figure 5.13f) where the weights are the smallest beyond what they grow resulting in overfitting. The RMSE on validation data and true pattern for the model obtained from early stopping (Figure 5.13d) are 0.270 and 0.136, respectively, where they are 0.423 and 0.277, respectively, for the fully trained model. Error of the fully trained network on the training data is 0.147.

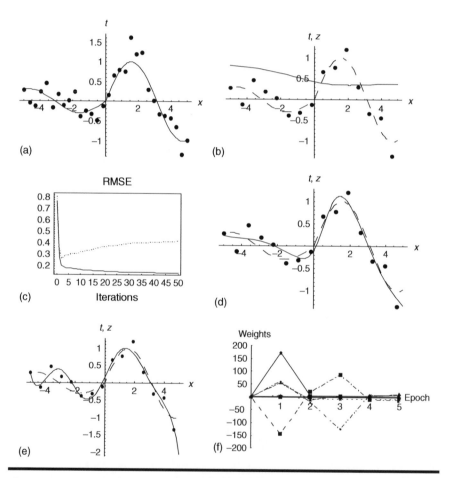

Figure 5.13 Network performance for random sample 2: (a) original pattern and data, (b) initial network performance (solid line) superimposed on target pattern (dashed line) and training data, (c) training performance (solid line—training, dashed line—validation), (d) optimum network output, (e) final overfitted network output, and (f) weights in the initial training epochs (optimum weights occur at epoch 2).

For random sample 3, the optimum network output, training performance, final network output, and the weights in the initial epochs, respectively, are shown in Figure 5.14a–Figure 5.14d. The network behavior on this dataset is interesting. As shown, the network has experienced very little overfitting compared to the first two random samples. The minimum error on the validation set occurs at epoch 7 as shown in Figure 5.14b and Figure 5.14d; Figure 5.14d demonstrates that for this case, initial weights are too small and must grow to a point to capture essential trends in the data. It also shows that weights are stable beyond epoch 7 where they grow in a restrained manner. This again points out that even in a network with a large number of weights, complete training may not lead to severe overfitting in some cases, depending on the manner that the initial weights grow in response to data.

Because in this experiment all three networks had similar initial weights corresponding to weight initialization 1 in the previous section, the effect of random weights has been eliminated. Therefore, the restraint in the growth of weights up to the optimum and not growing beyond this point in this case must have been driven by the data itself. The hidden neuron activation patterns for this network that did not overfit are shown in Figure 5.15a. This activation configuration is similar to that of the networks whose capacity to

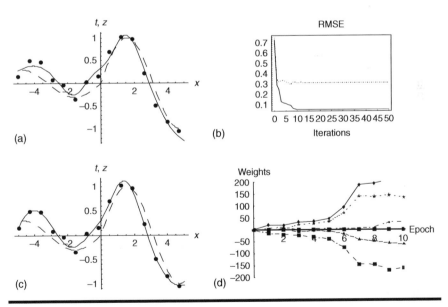

Figure 5.14 Network performance for random sample 3: (a) optimum network output superimposed on training data, (b) training performance (solid line— training, dashed line—validation), (c) final fully trained network output (solid line) superimposed on true pattern (dashed line) and training data, (d) weights in the initial training epochs (optimum weights occur at epoch 7).

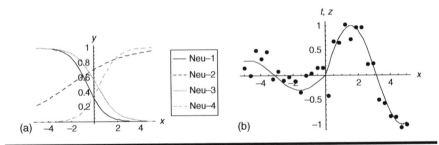

Figure 5.15 Data and hidden neuron activation for the case without overfitting: (a) hidden neuron activation for the optimum network that does not experience overfitting and (b) random data sample 3 that led to the controlled growth of weights to prevent overfitting.

overfit was restrained by early stopping in the previous experiment where the effect of random weight initialization was studied (see Figure 5.12). This indicates that data has regularized the weights by appropriately shifting the active regions of hidden neuron functions to naturally obtain the optimum configuration for them. Such regularization works for data that represents the original pattern uniformly throughout the input space (i.e., the network has no choice but go through the cloud of data). The data in random sample 3 is presented in Figure 5.15b. This concept of controlled growth of weights and subsequent stabilization is used in the method called regularization that helps directly find optimum weights that improve generalization and is shown later in the chapter.

The results from explorations with random sampling indicate that the network is robust in finding the optimum weights with different random training samples. The optimum networks do not follow the data too closely, as do the overfitted networks. However, for some datasets, overfitting may not set in. The optimum network in this case is still reliable. This occurs because for such a sample, some weight regularization has naturally come into effect leading to controlled growth of weights up to the point of best generalization and naturally stopping further growth. It appears that regularization in the network trained with random sample 3 was caused by the data, indicating that the network is sensitive to training data. The next section discusses how to impose regularization from outside as part of model fitting so that regularization takes effect compulsorily, not accidentally.

Generally, the more data obtained, the better the generalization and the less the overfitting. This idea is schematically illustrated in Figure 5.16. In regions of the input space that have many observations, the noise will smear the data patterns, making it difficult for the network to precisely fit individual data points. Thereby, the network is forced to go through the

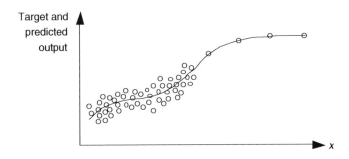

Figure 5.16 Noise in data that helps prevent overfitting and improve generalization.

center of the data cloud preventing overfitting. However, in regions where the data are sparse, it can overfit. To take advantage of this situation, noise can be artificially added to data to improve generalization. For example, if a small random noise is added to each input pattern as it is presented to the network, it will be difficult for the network to follow each data point exactly. In fact, Siestma and Dow [2,5] have shown that, in practice, training with noise leads to improvements in network generalization.

Table 5.2 shows the optimum weights achieved for the three random datasets. The first row shows the results for the first network in the earlier experiment where the effect of random initial weights was tested (see Table 5.1).

The sets of weights in Table 5.1 and Table 5.2 represent different networks. Some networks show some similarity in weights; however, they are not generally similar. Predictions from all the networks are good. As previously explained, the reason for this is that redundancy in weights forms a correlation structure, and this provides latitude for individual weights to change while preserving the correlation structure of the weights. There are many such possibilities. Therefore, it is important to remove such redundancies to obtain models that are robust and consistent and that have optimum complexity.

Table 5.2 Optimum Weights for Random Data Samples

Data Initial	Neuron 1 $(a_{01}, a_{11})\ b_1$	Neuron 2 $(a_{02}, a_{12})\ b_2$	Neuron 3 $(a_{03}, a_{13})\ b_3$	Neuron 4 $(a_{04}, a_{14})\ b_4$	Bias b_0
1	$(-0.73, -0.89)$ -0.46	$(0.91, 0.72)$ 2.59	$(0.31, -0.50)$ 11.47	$(-0.40, 1.3)$ 5.3	-10.1
2	$(-0.22, -0.22)$ -11	$(0.87, 0.1)$ 19.7	$(-1.3, 2.1)$ 4.0	$(-1.2, 0.64)$ -8	19.8
3	$(-0.72, -0.93)$ -9.0	$(0.9, 0.68)$ 0.12	$(0.18, -0.93)$ 32.8	$(-0.42, 1.27)$ 22	-23.3

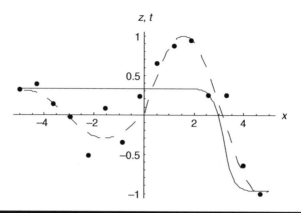

Figure 5.17 One-hidden-neuron network output (solid line) superimposed on true pattern (dashed line) and data.

5.3.1.4 Effect of Model Complexity: Number of Hidden Neurons

Bias–variance dilemma can be resolved by externally controlling model complexity through adjustment of number of hidden neurons. The effect of the number of neurons is examined to evaluate the effectiveness of early stopping.

Single hidden neuron. Figure 5.17 shows the data, initial pattern (dashed line), and a fully trained network output (solid line) for one hidden neuron. With one neuron, the network has captured the trend only in a small region of the pattern.

Two hidden neurons. The improvement of the network output with the addition of another neuron is shown in Figure 5.18. This did not result in

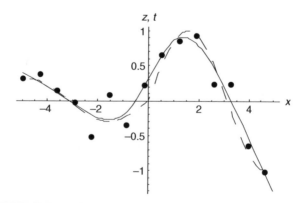

Figure 5.18 Fully trained two-hidden-neuron network output (solid line) super-imposed on actual pattern (dashed line) and training data.

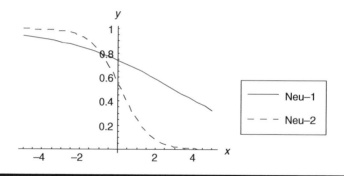

Figure 5.19   Hidden neuron activation for the two-hidden-neuron network.

overfitting either; therefore, a fully trained network is the optimum. This occurs, as Figure 5.18 shows, because a fully trained two-hidden-neuron network output is similar to the one obtained with four hidden neurons trained with early stopping until the validation error starts to increase.

The two-neuron network was trained starting from several initial conditions, and overfitting did not result for any of the conditions, showing that the network is not too rigid. The outputs of the optimum and fully trained networks for all trials were identical. The hidden-neuron activation for a typical two-hidden-neuron network is shown in Figure 5.19. It is evident that this particular combination of neuron activation has been dominant in all the previous networks, regardless of the number of hidden neurons, as illustrated in Figure 5.12 and Figure 5.15a.

Three hidden neurons. Next, the network was trained with three hidden neurons starting from different initial conditions. In four out of five times, overfitting did not set in. Only once did some slight overfitting result. The optimum network prediction was identical to that with two neurons. The hidden-neuron activation for three neurons in a nonoverfitted network is shown in Figure 5.20a. The output of the only overfitted network is shown in Figure 5.20b with the smallest dashes. The output of the network obtained from early stopping (solid line) for this case and the true pattern (intermediate dashes) is also shown. Results indicate that the three-neuron network is on the border line beyond where serious overfitting invariably occurs. Figure 5.20a also points out the dominant trend of the two sigmoid functions with negative slopes that recurred in all previous networks.

5.3.1.5   Summary on Early Stopping

Overfitting is a common outcome of a network with a large number of free parameters. In the preceding sections, the effectiveness of early stopping in obtaining best network weights that produce the best generalization

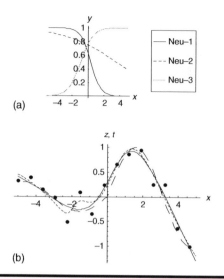

(a)

(b)

Figure 5.20 Three-hidden-neuron network performance: (a) hidden-neuron activation for a nonoverfitted network, (b) optimum network output (solid line) superimposed on its overfitted network output (smallest dashed line), actual pattern (intermediate dashed line) and training data.

(i.e., least error on data not used for training) was investigated. Research found that early stopping effectively finds optimum weights beyond which overfitting sets in. The robustness of early stopping with respect to changes in initial random weights and randomness in the training and validation data was checked. It was revealed that in all cases, early stopping produced a network that generalizes well. In all but two cases, complete training lead to severe overfitting.

Overfitting is a result of weights growing too large. In the two cases where there was no serious overfitting, weights reached stable values even in fully trained networks. The first of these was from trials with random initial weights, indicating that there are regions in the weight space that can evolve naturally into optimum weights and stay around these values resisting change in further training. Such weights would be the ideal set of weights that preserve the correlation structure of weights naturally. However, the initial conditions for these optimum weights are not known a priori; therefore, early stopping provides an effective way to find the set of weight that provides best generalization.

In the second case, similar nonoverfitting conditions were aroused by the quality of the data that led to robust weights resistant to overfitting. Specifically, if noise is present in data in such a way that the network is forced to go through the cloud of data, overfitting is prevented. A similar situation can be artificially created by adding random noise to inputs each time they are presented to the network. Generally, the larger the dataset, the better the generalization; however, typically, the datasets may not be large enough, and it may not be possible to realistically estimate the sample size needed to prevent overfitting, especially when there are many input variables affecting the outcome. In such situations, a method such as early stopping is needed to improve generalization.

In the next section, regularization that prevents the growth of relevant weights beyond the point of best generalization while suppressing the growth of irrelevant weights is examined.

5.3.2 Regularization

The examples used to demonstrate early stopping in the previous sections indicate that overfitting is a result of weights becoming too large. Two network trials there led to controlled growth of weights until the end of training without relying on early stopping. Regularization is a method proposed to do this effectively as part of model fitting by limiting the flexibility of a network [2,4]. It attempts to limit the number of efficient parameters (weights) by minimizing the sum, W, of a regularization term and MSE instead of MSE alone as given in Equation 5.1:

$$W = \text{MSE} + \delta \sum_{j=1}^{m} w_j^2, \tag{5.1}$$

where the second term in the equation is a regularization term, w_j is a weight in the total set of m weights in the network, and δ is a regularization parameter. Basically, regularization keeps the weights from getting large by minimizing the sum of square weights along with sum of square error. It pulls the weights that have only a marginal influence on error toward zero while keeping the weights that efficiently minimize the error. The amount of regularization is controlled by the constant parameter δ, and the larger the δ, the more important the regularization becomes.

In the following example on regularization, the same problem is used that was used to study the early stopping method in Section 5.3.1. The original pattern and the noisy data generated from it using a Gaussian distribution with 0 mean and standard deviation of 0.25 are repeated in Figure 5.21.

Here, the same network, initial weights, and training and validation data are used as those used for illustrating early stopping with the first set of random initial weights in Section 5.3.1. There are four hidden neurons with logistic transfer function in the hidden neurons and a linear output neuron (Figure 5.4a). The network is trained using the Levenberg–Marquardt method, and the validation set is used as a guide, as in the early stopping method. The network was trained with three values of regularization

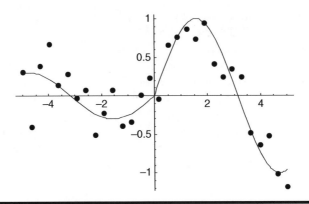

Figure 5.21 Original pattern and noisy data generated from it.

parameter. Figure 5.22 shows the best results obtained (for δ value of 0.0001) after testing several δ values.

Results show that the network training completes in six epochs, and the optimum network is obtained at epoch 4. Weights initially increase, but they

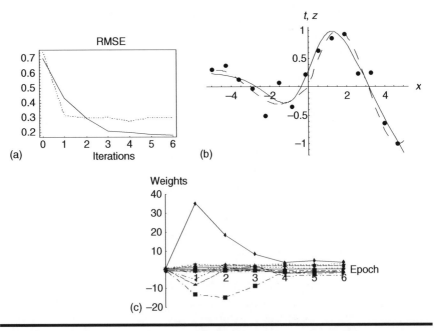

Figure 5.22 Network results for regularization parameter 0.0001: (a) training progress (solid line—training data, dashed line—validation data), (b) optimum network output (solid line) for weights at epoch 4 superimposed on original pattern (dashed line) and noisy data generated from original pattern and (c) network weights for the first six epochs.

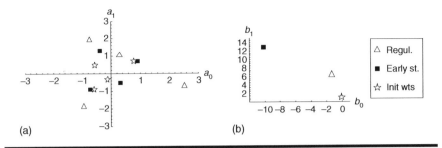

Figure 5.23 The weights from early stopping alone and with regularization ($\delta = 0.0001$) superimposed on initial weights: (a) input-hidden weights and (b) resultant of hidden-output weights plotted against bias weight b_0.

decrease with training and remain constant beyond epoch 4. Training has ended much more quickly (in six epochs) with regularization because the weights do not change after epoch 4, demonstrating the effectiveness of the method. The optimum weights obtained from early stopping alone and with regularization for this case are shown superimposed on initial weights in Figure 5.23a and Figure 5.23b for input-hidden neurons and hidden-output neurons, respectively.

Figure 5.23 shows that the initial input-hidden weights are too small and must grow. Regularization has grown (relaxed) the weights more than early stopping to find the optimum values. This is evidenced by the weights in Figure 5.23a and Table 5.3 that show all but one input-hidden weights are larger for the best regularization parameter of 0.0001 found from trial and error than those for early stopping alone. Results for training with two other values of generalization parameter are also presented in Table 5.3. In contrast, the magnitude of most of the hidden-output weights obtained from regularization is only half that produced by early stopping, as shown in Figure 5.23b where, for clarity, b_0 is plotted against the resultant of all

Table 5.3 Comparison of Optimum Input-Hidden Weights from Early Stopping Alone and with Regularization

Method/Weights	a_{01}	a_{11}	a_{02}	a_{12}	a_{03}	a_{13}	a_{04}	a_{14}
Early stopping	-0.73	-0.89	0.91	0.72	0.31	-0.5	-0.4	1.3
Regularization (0.0001)	-0.97	-1.85	0.29	1.06	2.53	-0.70	-0.76	1.93
Regularization (0.001)	-0.51	-0.93	0.82	0.69	0.68	-0.51	-0.44	1.54
Regularization (0.01)	-1.67	0.59	0.27	-0.4	0.11	-1.48	-0.42	1.77

Table 5.4 Comparison of Optimum Hidden-Output Weights and Network RMSE from Early Stopping and Regularization

Method/Weights	b_0	b_1	b_2	b_3	b_4	Validation RMSE
Early stopping	-10.13	-0.46	2.59	11.4	5.29	0.30
Regularization (0.0001)	-1.35	-2.04	-3.71	3.66	2.90	0.276
Regularization (0.001)	-3.65	-3.08	-1.54	7.47	3.17	0.30
Regularization (0.01)	-0.50	-3.35	2.63	-1.55	1.67	0.285

hidden-output weight values, i.e., $\sqrt{b_1^2 + b_2^2 + b_3^2 + b_4^2}$ from regularization and early stopping conditions. The actual values of hidden-output weights are shown in Table 5.4 along with the RMSE for several values of the regularization parameter. Initial weights used for both layers were small random values closer to zero.

As far as the equivalence of early stopping and generalization is concerned, Table 5.4 shows that a performance (validation RMSE) similar to that obtained from early stopping (RMSE $= 0.3$) is given by the regularization parameter 0.001. For this situation, the optimum weights and validation sample error are shown in Table 5.3 and Table 5.4 and the training performance, network output, and weight evolution are shown in Figure 5.24.

Figure 5.24 indicates that a good generalization with a regularization parameter of 0.001 results. Final input-hidden and hidden-output weights are shown in Figure 5.25.

For the regularization parameter of 0.001, less relaxation on input-hidden weights and hidden-output weights is noted compared to those for optimum δ value of 0.0001. This means that more control is put on keeping the weights small. The input-hidden weights are similar to those obtained from early stopping alone for this regularization parameter. Figure 5.26 shows results for regularization parameter of 0.02. It shows that weights are now pulled toward zero, resulting in poor model predictions because of bias (i.e., lack of fit).

A further increase of δ to 0.1 makes the model completely incapable of finding optimum weights, as shown in Figure 5.27. As the figure shows, when the regularization parameter is too high, more emphasis is put on keeping weights small. Thus, weights are pulled much more strongly toward zero than necessary, to the detriment of model performance causing severe bias.

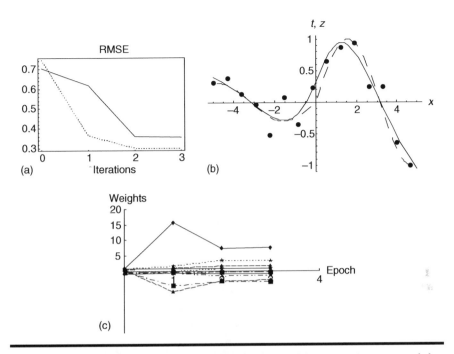

Figure 5.24 Network results for regularization with $\delta = 0.001$: (a) training progress (solid line—training data, dashed line—validation data), (b) optimum network output (solid line) for weights at epoch 2 (found using validation set) superimposed on original pattern (dashed line) and noisy data generated from original pattern and (c) network weights for the first three epochs.

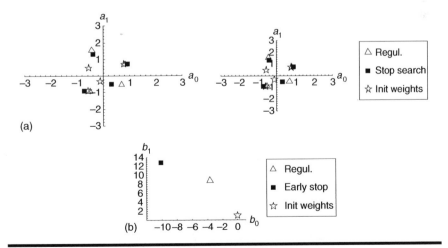

Figure 5.25 The weights from early stopping alone and with regularization ($\delta = 0.001$) superimposed on initial weights: (a) input-hidden weights and (b) resultant of hidden-output weights plotted against bias weight to output neuron.

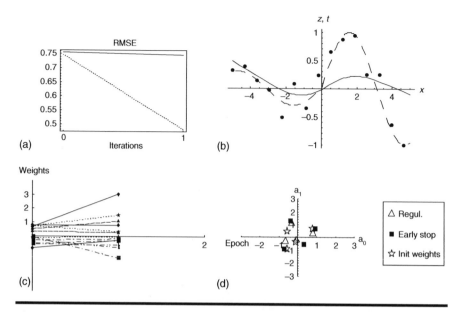

Figure 5.26 Model performance for regularization parameter δ of 0.02: (a) training progress (solid line) and validation performance (dashed line), (b) model predictions (solid line), original pattern (dashed line) superimposed on noisy data generated from the original pattern, (c) weight change during training, and (d) optimized input-hidden weights from early stopping and regularization.

As explained, a good generalization parameter effectively finds optimum weights more quickly than early stopping, but larger values can restrict the growth of weights, severely deteriorating the model performance because of a lack of fit. To put this in perspective, a comparison of regularization's and early stopping's performance is shown in Figure 5.28.

Figure 5.28 shows the effect of regularization parameter on the regularized training criterion, W, and the RMSE on both training and validation data. The top line is for W (regularized criterion), and the next two are for RMSE. Also shown in this figure is the validation error from early stopping shown as a horizontal dashed line drawn for the purpose of comparison. The figure shows that below a certain value of the regularization parameter (about 0.02 in this case), optimizing weights using regularization is superior to that using early stopping.

Because the regularization training criterion (W) has an extra sum of square weight component added to MSE, W is larger than RMSE, as shown in the figure. The effect of regularization on RMSE has an exponential form for this example.

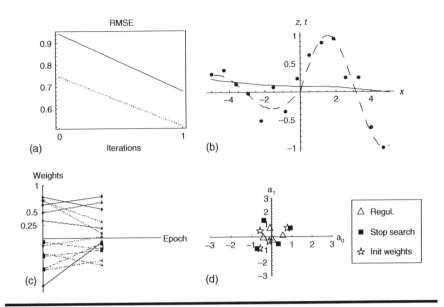

Figure 5.27 **Model performance for regularization parameter of 0.1: (a) training progress (solid line) and validation performance (dashed line), (b) model predictions (solid line), original pattern (dashed line) superimposed on noisy data generated from the original pattern, (c) weight change during training, (d) optimum input-hidden weights from early stopping and regularization superimposed on initial random weights.**

5.4 Reducing Structural Complexity of Networks by Pruning

The previous two sections presented two methods of keeping weights small: early stopping and regularization. In early stopping, training is

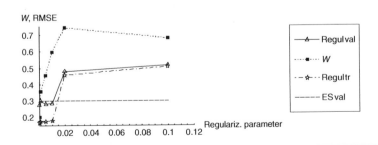

Figure 5.28 **Effect of regularization parameter on the regularized training criterion (*W*) and root mean square error (RMSE) on training (tr), and validation (val) data. Horizontal line depicts the root mean square error for early stopping (ES), drawn for the purpose of comparison with regularization.**

stopped when the error on the validation set starts to increase. Weights at this point are the optimum weights. Further training beyond this point makes weights grow very large. In regularization, a regularization parameter equal to the sum of square of weights is minimized along with MSE. The advantage of this method is that training takes less time, and once the optimum weights are reached, they do not continue to grow. Both these methods use all weights in training and do not reduce the structural complexity of the model. As shown, when more than optimum weights are in the network, different training sessions produce different sets weights. This needs to be resolved to make a model transparent and to make realistic conclusions from the model outcomes. One way of achieving this goal is to reduce the structural complexity of networks so that only the essential weights and neurons remain in the model. An approach to this is network pruning.

Several approaches have been proposed to prune networks. Reed [6] presents the first survey of pruning methods where some simple intuitive methods based on the value of weights and neuron activation values have been proposed. For example, the concept of "goodness of factor" [7] assumes that an important neuron is one that frequently excites and has large weights to other neurons. The concept of "consuming energy" assumes that important units excite neurons in the next layer [7]. The weakness of these two methods is that when a neuron's output is more frequently zero than one, that unit might be removed as unimportant although this may not be the case. Magnitude-based pruning (MBP) is based on the assumption that small weights are irrelevant [5,7]. However, small weights may be important when compared to larger weights because the latter may cause saturation in hidden and output neurons due to their large magnitudes pushing the activation into less active regions of neuron transfer functions. Some later developments are optimal brain damage (OBD) [8] and its variants, optimal brain surgeon (OBS) [9,10] and optimal cell damage (OCD) [11]. These methods perform sensitivity analysis on training error to prune weights. More recently, analysis of variance of sensitivity of error or output to network parameters [12] has been proposed to efficiently prune weights, neurons, and inputs. In the next section, the concepts involved in two methods, OBD and variance analysis of sensitivity, are detailed, and their application to pruning networks to retain a network with the least structural complexity is illustrated.

5.4.1 Optimal Brain Damage

In this method, weights that are not important for input–output mapping are selected and removed. The importance or saliency of a weight is measured based on the cost of setting a weight to zero [3,6]. The saliency of a weight can be computed from the Hessian matrix introduced with the Gauss Newton and Levenberg–Marquardt learning methods in Chapter 4 (see Equation 4.48). Recall that the Hessian is the second derivative of the network error with respect to a pair of weights, w_i, w_j, as repeated in Equation 5.2:

$$H_{ij} = \frac{\partial^2 E}{\partial w_i \partial w_j}. \tag{5.2}$$

This matrix is nonlocal in that it uses the derivative with respect to a pair of weights and for large networks can become computationally costly. A local approximation to this that uses only the diagonal terms involving individual weights can be used as follows to overcome this problem. Then the saliency s_i of a weight w_i can be computed as

$$s_i = \frac{H_{ii} w_i^2}{2}, \tag{5.3}$$

where H_{ii} denotes the diagonal entries of the Hessian matrix, that contain the square of the derivative of network error with respect to each of the individual weights, w_i. Thus, H_{ii} indicates the acceleration of the error with respect to a small perturbation to a weight, w_i. By multiplying H_{ii} by w_i^2, an indication of the total effect of w_i on the error is obtained. The larger the s_i, the larger the influence of w_i on error. The other entries of the Hessian matrix are assumed to be zero; therefore, the second derivative with respect to weights other than itself is ignored. This implies that the weights of the network are independent, which may not be true for a network that has more than the optimum number of weights. To apply this method, a flexible network should be trained in the normal way and saliency computed for each weight. Then, weights with small values of saliency are removed. This may lead to pruning of weights as well as neurons. The reduced network must be trained again with the weights that are kept, starting with their initial values. The trained simplified network should perform as well as the optimum network with larger number of weights. The application of the OBD method is illustrated next.

5.4.1.1 Example of Network Pruning with Optimal Brain Damage

This method is now applied to the previous four hidden neuron optimum network trained with a regularization parameter of 0.0001 using the Levenberg–Marquardt training method to model the noisy data generated from an original pattern. The original network is shown in Figure 5.4a and the true pattern and noisy data generated from it are given in Figure 5.3 and repeated in Figure 5.21. For the trained network, calculated saliency

Table 5.5 Saliency (Importance) of Weights in the Network (the Higher the Saliency, the More Important the Weights Are)

a_{01}	a_{11}	a_{02}	a_{12}	a_{03}	a_{13}	a_{04}	a_{14}	b_0	b_1	b_2	b_3	b_4
0.034	0.08	0.018	0.286	1.66	1.48	0.04	0.13	1.82	1.65	5.99	9.45	3.47

for all 13 weights is given in Table 5.5. Basically, the second derivative of error is calculated for each weight in the trained network, and saliency is computed from Equation 5.3.

Network pruning stage 1 (40 percent of weights pruned). Saliency values in Table 5.5 indicate that all hidden-output weights have high saliencies. However, weights a_{01}, a_{02}, a_{04}, and a_{11} have the smallest saliency values. These four weights are deleted from the network. This removal amounts to deleting all input-hidden weights to neuron 1, eliminating neuron 1, and removing bias weights of hidden neurons 2 and 4. However, when neuron 1 is eliminated, weight b_1 is automatically eliminated. Because b_1 has the smallest saliency of all the hidden-output weights, b_1 is eliminated as well so that neuron 1 is completely eliminated. This amounts to removing five out of 13 (i.e., 40 percent) weights, leaving a total of three hidden neurons and only eight weights. This removal reduces the elasticity or complexity of the network considerably. The reduced network was retrained, starting with the corresponding initial weights from the original training. The network performance is shown in Figure 5.29.

The RMSE on the validation set (Figure 5.29a) for the pruned network is 0.292, comparing well with the performance of the best network that had an RMSE of 0.276 and the second best network that had an RMSE of 0.285 using regularization (Table 5.4). Thus, the selective weight removal based on saliency can produce results comparable to a full network. As indicated in Figure 5.29b, the network has generalized well and is comparable to the best full network.

Figure 5.29c indicates that the remaining weights all achieve convergence in 14 epochs, but training now takes longer. The final weights thus converged are compared to those from early stopping and initial weights in Figure 5.29d. In this figure, the weights that have been eliminated are shown as having a value of zero. It shows that the new weights are quite different from those obtained from early stopping. Recall that the lowest validation RMSE for early stopping was 0.3 (Table 5.4), indicating that the pruned network obtained with regularization performs better. In the pruned network, hidden neurons 2 and 4 have only input-hidden weights active. Only hidden neuron 3 has both bias and input-hidden weights.

Network pruning stage 2 (54 percent weights removed). After the first pruning stage, new saliencies for the remaining weights of the retrained network can be calculated as before, and these, along with the new weights,

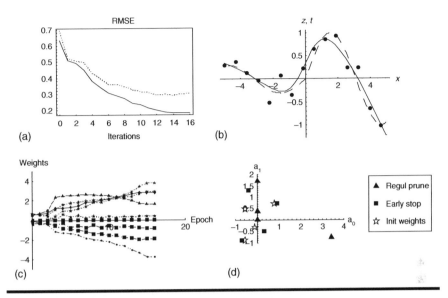

Figure 5.29 **Performance of the pruned network with regularization parameter 0.0001: (a) training progress (solid line) and validation performance (dashed line), (b) model predictions (solid line), original pattern (dashed line) superimposed on noisy training, data generated from the original pattern, (c) weight adaptation during training, and (d) final input-hidden weights of the pruned network superimposed on those from early stopping and initial weights for the full network.**

are presented in Table 5.6. Now weight a_{14} has very low saliency of 0.086. If this weight is removed, neuron 4 is eliminated, and weight b_4 must be eliminated because neuron 4 feeds through b_4 to the output neuron. Now, b_4 saliency (3.16) is the smallest of all values for hidden-output weights.

Neurons 1 and 4 and the links associated with them are eliminated. Altogether seven weights out of 13 (i.e., 54 percent) are eliminated with a total of six remaining (i.e., two weights of hidden neuron 3, one weight of hidden neuron 2, and three output weights including bias). The full network is now reduced to a two-hidden-neuron network. The network was trained with these six weights; and the resulting training performance, target and predicted outcomes, progress of weight adaptation, and final optimum

Table 5.6 **Weights of Pruned Network (40 Percent Weights Removed) and Their Saliency**

	a_{12}	a_{03}	a_{13}	a_{14}	b_0	b_2	b_3	b_4
Weight	0.388	3.42	−0.777	1.71	−2.02	−3.37	2.80	2.67
Saliency	0.358	1.36	1.04	0.086	4.08	3.53	6.22	3.16

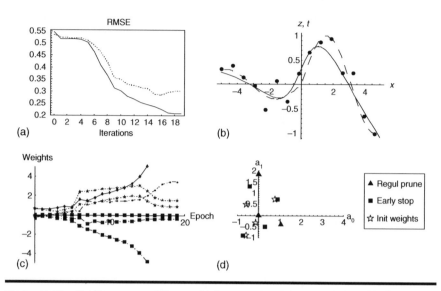

Figure 5.30 **Performance of the pruned network with 54 percent weights (seven weights) eliminated and six weights remaining (regularization parameter 0.0001): (a) training performance, (b) model predictions (solid line), original pattern (dashed line) superimposed on noisy data, (c) weight adaptation during training, and (d) final input-hidden weights of the pruned network superimposed on those from early stopping and initial weights for full network.**

input-hidden weights are shown in Figure 5.30. Results indicate that this simpler network takes even longer to train than the previously pruned network.

The RMSE on validation data is now 0.285, smaller than that of the network with five weights removed (RMSE (validation) = 0.292) yet close to that of the full model with an RMSE of 0.276 on the validation set (see Table 5.4). The training RMSE of 0.208 is, however, slightly higher than that of the full model (0.1805) and the network with five weights removed (RMSE (training) = 0.173). With seven (or 54 percent) weights removed, the reduced network still generalizes well. This two-hidden-neuron network with validation RMSE of 0.285 performs better than the original two-hidden-neuron model trained earlier (validation RMSE = 0.301) in the assessment of the effect of hidden neurons in Section 5.3.1.4 and presented in Figure 5.18. Because the optimum number of neurons is not usually known, a judgment is made. What is known from pruning can be considered as a good approximation.

Network pruning stage 3 (70 percent weights eliminated). To see if the network can be further pruned, the weight saliencies can be calculated on the reduced model for the six remaining weights. These are presented in Table 5.7.

Table 5.7 Weights of Pruned Network after Removing 54 Percent Weights and Their Saliency

Weight	a_{12}	a_{03}	a_{13}	b_0	b_2	b_3
Saliency	0.078	1.28	1.39	28.7	3.10	18.03

Now, the saliency of weight a_{12} (0.078) is very small. If this is dropped, hidden neuron 2 is removed, and b_2 must also be removed. As shown, b_2 has the smallest saliency (3.102) of all hidden-output weights. If these two weights are removed, the result is four weights and one hidden neuron. One neuron is not enough to model the nonlinear function, as illustrated in Figure 5.31. It shows that removing hidden neuron 2 has a severe impact on the network performance. This exercise shows that there is a definite threshold for the architecture of the simplest network that generalizes well before the network performance is severely impacted. It also shows the sensitivity of the network to the removal of an essential neuron. Removal of the last neuron completely deteriorated the performance of the network.

The OBD method systematically removes weights that do not significantly contribute to the network output. To find these weights, it uses the second derivative of the network output with respect to a weight. In this example, the original four neurons can be reduced to two, and the original 13 weights can be reduced to six while maintaining the same level of generalization as the original model that has much greater flexibility.

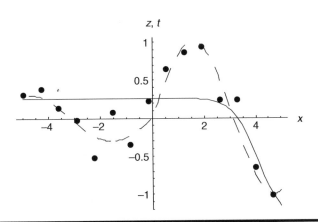

Figure 5.31 Performance of the pruned network with 70 percent (or nine) weights eliminated leaving only four weights associated with one hidden and output neuron (regularization parameter 0.0001). Model predictions (solid line), original pattern (dashed line) superimposed on training data.

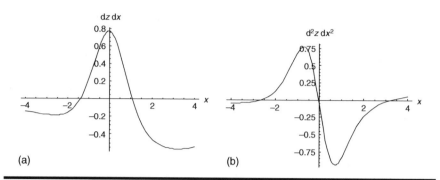

Figure 5.32 Sensitivity of network output to input (a) first derivative (gradient) and (b) second derivative (curvature).

The output of the pruned network with two neurons is

$$z = -5.36 + \frac{2.63}{1 + e^{-1.84x}} + \frac{5.86}{1 + 0.345e^{0.399x}}. \tag{5.4}$$

The sensitivity of the network output to inputs is expressed by the first derivative (gradient) and second derivative (curvature) of network output with respect to inputs. For the pruned network, these can be calculated from Equation 5.4, and the resulting gradient and curvature are shown in Figure 5.32a and Figure 5.32b. In the case of multiple inputs, sensitivities are given by the partial derivative of the output with respect to each input.

Figure 5.32a shows that the output is highly sensitive to the input in the vicinity of $x = 0$, as indicated by the high gradient in this region. The original target function in Figure 5.3 and Figure 5.21 confirm this. Because this is a nonlinear function, the sensitivity of output to input is not constant but situation dependent (i.e., dependent on x). In contrast, for linear models such as linear regression, the sensitivity is a constant that is the coefficient associated with that input in the model, and it is the partial derivative of the output with respect to that input when there are more than one inputs. Therefore, in nonlinear models, situation dependency must be incorporated in the assessment of both the contribution of inputs to output and in selecting relevant inputs. The maximum and minimum of second derivative of a function show the exact locations where the output function changes direction (points of inflexion). These points in Figure 5.32b correspond to the points where the true function changes direction. Thus, the pruned network has captured the intrinsic trend, underlying the true pattern represented by the noisy data. These ideas will be further used to prune irrelevant inputs, hidden neurons, and output neurons as this chapter continues.

This section presented how OBD method uses saliency or importance of weights to remove irrelevant weights. In the process, weights and neurons can be eliminated. In this method, sensitivity analysis is performed with respect to the training error. It results in the use of the second derivative or curvature of the error surface with respect to a weight to compute a saliency measure for each weight that reflects the influence small perturbations to the weight have on the network error. Next, variance analysis of sensitivity (gradient) for network pruning is studied.

5.4.2 Network Pruning Based on Variance of Network Sensitivity

The objective of pruning is to downsize the model to the level of least complexity that still provides the best generalization. In this section, a method that uses the variance analysis of the sensitivity of the output of the network to the perturbation of its parameters, as proposed by Engelbrecht [12], is explored. Sensitivity denotes the derivative or gradient. Hornik et al. [13] and Gallant and White [14] showed that when the network model converges toward the target function, all the derivatives of the network also converge toward the derivatives of the underlying target function. This is also demonstrated in Figure 5.32 for the example in the previous section. Therefore, output sensitivity ($\partial z/\partial \theta$) can be effectively used for assessing the relevance of parameter θ that can be a weight, an input, or a hidden neuron output. Engelbrecht [12] states that the two approaches to sensitivity analysis, using error function and output function, lead to the same results in terms of parameter relevance. However, sensitivity analysis using output function is much simpler.

In variance analysis of sensitivity, a parameter θ can be an input (x), hidden-neuron activation (y) or weight (w). As discussed in Chapter 4, error gradient of a network depends on the inputs meaning that each input pattern results in a unique gradient. The same is true for output (z) sensitivity so each input pattern results in a unique value for parameter sensitivity ($\partial z/\partial w$, $\partial z/\partial y$, $\partial z/\partial x$, etc.). In the variance of network sensitivity approach, it is proposed to compute a "variance nullity measure" that tests whether the variance in parameter sensitivity over all input–output patterns is significantly different from zero. If the variance in parameter sensitivities is not significantly different from zero and the mean sensitivity across all patterns is small, the corresponding parameter has little or no influence on the output of the neural network over all the patterns. This measure is used in a hypothesis test using chi square (χ^2) distribution to test statistically if a parameter should be pruned. The hypothesis simplifies to testing if the expected (mean) value of sensitivity of a parameter over all patterns is zero, as will be demonstrated later. First, an expression for the expected (mean) value must be found. Then, the test statistic following the approach of Engelbrecht [12] must be developed.

If the sensitivity with respect to a parameter is denoted by S_θ, then variance of S_θ denoted by $\sigma^2_{S_\theta}$ can be expressed as

$$\sigma^2_{S_\theta} = \frac{\sum\limits_{i=1}^{N} (S_{\theta i} - \mu_{S_\theta})^2}{N} \tag{5.5}$$

where μ_{S_θ} is the mean sensitivity over the total N input patterns and i is the pattern number. Equation 5.5 can be simplified to

$$\sigma^2_{S_\theta} = \frac{\sum\limits_{i=1}^{N} (S^2_{\theta i} - 2 S_{\theta i} \mu_{S_\theta} + \mu^2_{S_\theta})}{N} = \frac{\sum\limits_{i=1}^{N} S^2_{\theta i}}{N} - 2\mu_{S_\theta} \frac{\sum\limits_{i=1}^{N} S_{\theta i}}{N} + \mu^2_{S_\theta} = \mu_{S^2_\theta} - \mu^2_{S_\theta} \tag{5.6}$$

This yields an expression for the expected value of sensitivity with respect to parameter θ, μ_{S_θ}, in the form of

$$\mu_{S^2_\theta} = \mu^2_{S_\theta} + \sigma^2_{S_\theta}, \tag{5.7}$$

which consists of a bias component, $\mu_{S^2_\theta}$ (i.e., mean of sensitivity square), and a variance component, $\sigma^2_{S_\theta}$. Both must be zero for irrelevance of a parameter. Equation 5.7 states that the mean or expected value of sensitivity square is the square of mean value of sensitivity plus variance of the sensitivity. Because the chosen hypothesis involves both mean and variance, the test of $\mu_{S^2_\theta} = 0$ would encompass testing both $\mu^2_{S_\theta} = 0$ and $\sigma^2_{S_\theta} = 0$. To test the hypothesis, two hypotheses must be tested:

$$(1) \quad H_0: \mu^2_{S_\theta} = 0$$
$$(2) \quad H_0: \sigma^2_{S_\theta} = 0. \tag{5.8}$$

If the first hypothesis is rejected, a parameter is relevant and cannot be pruned. If accepted, then the second hypothesis also must be tested because large positive and negative values of sensitivity can cancel each other and produce a sum close to zero, indicating that the parameter is not significant. In fact, this may not be true. Therefore, testing the second hypothesis is critical and it makes sense to do this first.

For testing the second hypothesis, relevance of a parameter, γ_{S_θ}, is defined in terms of parameter variance nullity. This is the statistical nullity of the variance of the sensitivity of the output to a network parameter calculated over patterns $i = 1, \ldots, N$ and is expressed as

$$\gamma_{S_\theta} = \frac{(N-1)\sigma_{S_\theta}^2}{\sigma_0^2}, \tag{5.9}$$

where $\sigma_{S_\theta}^2$ is the variance of the sensitivity of the network output to perturbation of parameter θ as before and σ_0^2 is a value close to zero. The $\sigma_{S_\theta}^2$ for one output network and N input patterns can be estimated from

$$\sigma_{S\theta}^2 = \frac{\sum_{i=1}^{N}(S_{\theta_i} - \bar{S}_\theta)^2}{N-1}, \tag{5.10}$$

where $\bar{S}_\theta$ is the mean of parameter sensitivity S_θ over the sample of N patterns:

$$\bar{S}_\theta = \frac{\sum_{i=1}^{N} S_{\theta_i}}{N} \tag{5.11}$$

The hypothesis that the variance is close to zero is tested for each parameter θ with the null hypothesis:

$$H_0: \ \sigma_{S_\theta}^2 = \sigma_0^2. \tag{5.12}$$

Since σ_0^2 cannot be made zero in Equation 5.9, the variance, $\sigma_{S_\theta}^2$, cannot be hypothesized as exactly zero. Instead, a small value close to zero is chosen for σ_0^2, and the alternative hypothesis becomes

$$H_1: \ \sigma_{S_\theta}^2 < \sigma_0^2. \tag{5.13}$$

A parameter is pruned if the alternative hypothesis is accepted. Under the null hypothesis, the variance nullity measure in Equation 5.9 follows a $\chi^2(N-1)$ distribution where $N-1$ is the degrees of freedom for N patterns. A lower critical χ^2 value γ_c, obtained from χ^2 distribution tables, is

$$\gamma_c = \chi_{N-1,(1-\alpha/2)}^2,$$

where α is the level of significance. For example, if $\alpha = 0.05$, it means that the acceptable level of incorrectly rejecting the null hypothesis five times out of 100. Smaller α values result in a stricter pruning algorithm. The hypothesis is tested for each parameter. If $\gamma_{S_\theta} \leq \gamma_c$ for a particular parameter, the alternative hypothesis is accepted, and the parameter is pruned. Otherwise, the null hypothesis is accepted, and the parameter is not pruned. The success of pruning depends on the value of σ_0^2. If it is too small, no parameter is pruned. If it is too large, even the relevant parameters will be pruned. Therefore, it is good to start with a small value and increase it if no parameters are pruned. Engelbecht [12] uses an initial value of 0.0001 that is incremented ten-fold in every pruning step.

Concerning the first hypothesis that $\mu_{S_\theta}^2 = 0$, it is not essential to do this step separately because in the test of variance nullity and checking the performance of the pruned network, this step is taken care of automatically. For example, if mean sensitivity and variance are both close to zero for a weight, the variance nullity measure will prune that weight, and the reduced network will perform satisfactorily. If variance is greater than zero but mean is approximately zero, variance nullity would not allow for that weight or parameter to be pruned. In the third case, if variance is approximately zero but mean is larger than zero, a weight should not be pruned. Although variance nullity measure does not specifically test for this case, a part of the pruning process is to test that the pruned network performance is acceptable. If the results are not acceptable, the previous network architecture is restored, preventing the elimination of relevant parameters. Therefore, only hypothesis test 2 can be used where variance nullity measure is tested for eliminating irrelevant parameters.

5.4.2.1 Illustration of Application of Variance Nullity in Pruning Weights

Here, the variance nullity measure is applied to the nonlinear approximation problem studied in the previous sections to see how efficiently the method reduces the complexity of the network. Recall that the problem is one-dimensional, and data was generated from the pattern shown in Figure 5.3 and also in Figure 5.21 by adding a small noise generated from a Gaussian distribution with mean 0 and standard deviation 0.25. The original network had four hidden neurons (Figure 5.4a), and the optimum network obtained from regularization is used here (see Section 5.3.2 and Figure 5.22). Both training and validation datasets had 15 observations each.

Network pruning with variance nullity measure—stage 1. The variance nullity measure requires the calculation of gradient (sensitivity) of the network output to each of the weights. Because gradients were discussed in Chapter 4, the concept is used straightaway and applied to the optimum network shown in Figure 5.22. The combined training and validation datasets (30 observations) is used for this analysis, and the sensitivity of output with respect to each weight for all input patterns resulting in 30 sensitivity values for each of the 13 weights are obtained. The variance of the sensitivity $\sigma_{S_\theta}^2$ for each weight across all patterns is presented in Table 5.8. The critical χ^2 for this case for $\alpha = 0.01$ is $\chi_{(30-1,\ 0.995)}^2 = 13.12$. Initially, a small value of 0.01 for σ_0^2 was used in Equation 5.9 to obtain $\gamma_{s\theta}$, which is the χ^2 test statistic for the 13 weights. This indicated that only b_0 has a χ^2 test statistic (0) smaller than the critical value of 13.2. Therefore, at this level of σ_0^2, null hypothesis is accepted for all weights but rejected for b_0. Therefore, only b_0 can be pruned. Instead, σ_0^2 was increased to 0.1, and the resulting χ^2 test statistic and the difference

Table 5.8 Variance of Sensitivity of Weights and χ^2 Value for $\sigma_0^2 = 0.1$

	a_{01}	a_{11}	a_{02}	a_{12}	a_{03}	a_{13}	a_{04}	a_{14}	b_0	b_1	b_2	b_3	b_4
$\sigma_{s_\theta}^2$	0.025	0.021	0.1	0.255	0.121	2.117	0.05	0.033	0	0.202	0.164	0.054	0.205
χ^2	7.4	6.1	29.0	74.0	35.2	614.1	14.7	9.62	0	58.6	47.7	15.6	59.5
$\chi^2 - \chi_{crit}^2$	−5.7	−7.0	15.9	60.9	22.1	601	1.64	−3.5	−13	45.5	34.6	2.5	46.4

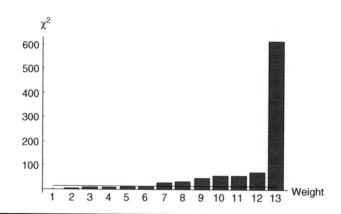

Figure 5.33 Chi-square test statistic for variance of sensitivity of output to weights.

between the test and critical χ^2 are also shown in Table 5.8. The negative values for the difference indicate the weights that can be pruned.

Table 5.8 indicates that a_{01}, a_{11}, a_{14}, and b_0 weights have smaller test values than the critical values. Therefore, the alternative hypothesis (Eq. 5.13) is accepted for these, and weights can be pruned. Note that this result is partially similar to that from the weight saliency measure in OBD in Section 5.4.1. Results from the variance nullity test are graphically presented in Figure 5.33, where bars indicate the test values in ascending order, and the horizontal line represents the critical χ^2 value.

Values below the horizontal solid line are the weights that can be pruned. Neurons 1 and 4 are eliminated (note that the results for a_{04} are on the border line); however, this means that weights b_1 and b_4 must also be eliminated with the two hidden neurons. Because neuron 3 is not deleted, b_3 cannot be deleted; b_0 is left in to compensate for the deleted output weights.

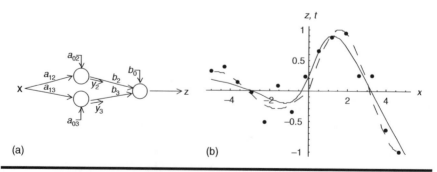

Figure 5.34 Pruned network based on variance nullity measure and its performance: (a) pruned network, (b) trained pruned network output superimposed on target pattern and training data.

Judgment must be used in the decision to prune weights or neurons. The reduced network after pruning six weights (a_{11}, a_{14}, a_{01}, a_{04}, b_1, and b_4) is shown in Figure 5.34a.

The pruned network was retrained with regularization, and the training and validation RMSE for the pruned network are 0.195 and 0.272, respectively, and the validation RMSE is slightly better than that (0.276) for the best full network obtained from regularization in Section 5.3.2. Because this problem requires at least two neurons, further pruning is not necessary. Thus, the variance nullity measure pruned the network more efficiently for this example than did the OBD method using weight saliency.

5.4.2.2 Pruning Hidden Neurons Based on Variance Nullity of Sensitivity

The variance nullity measure can be applied to prune hidden neurons directly based on the sensitivity of network output to hidden-neuron activation. This is an efficient way to reduce the complexity of a network. Recall that for the aforementioned networks, the notation used for this sensitivity is $\partial z/\partial y$, where z is the network output and y is the hidden neuron activation. If linear activation is used in the output neuron, the sensitivity $\partial z/\partial y$ will be equal to the corresponding hidden-output weight (see derivations in Chapter 4) and does not change across all patterns. For this reason, if variance nullity measure for hidden neuron activation is used, a nonlinear activation function for the output neuron must also be used. The arc tan (i.e., $\tan^{-1}$) function for output activation is used in this example for illustrating the concept. The output from the optimum network obtained from training with regularization is shown in Figure 5.35.

How to obtain $\partial z/\partial y$ is discussed in detail in Chapter 4 and therefore the details are skipped here. Presented in Table 5.9 is the variance of the sensitivity values for the four hidden neuron activations (y_1, y_2, y_3, and y_4). Following the same idea as for individual weights, the hypothesis is tested that variance in the sensitivity of output to hidden neuron activation is close to zero. The χ^2 test statistic for the four neurons relative to the critical value is shown in ascending order in Figure 5.36 for $\sigma_0^2 = 0.001$ and 0.01, respectively. These values for the two cases are also presented in Table 5.9. The critical χ^2 value for $\alpha = 0.01$ is $\chi^2_{(30-1,\,0.995)} = 13.12$.

Figure 5.36a shows that neuron 4 can be pruned, and Figure 5.36b shows that both 1 and 4 can be pruned. This leaves neurons 2 and 3 reducing the network to the exact configuration resulted from variance nullity measure on weight sensitivity. This illustrates that the approach is robust. Retraining the reduced network, the results shown in Figure 5.37 are obtained where the network output is superimposed on the target pattern where noisy data shown in the figure was generated.

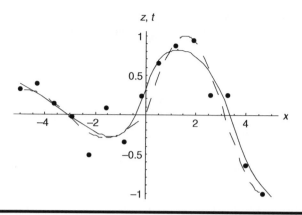

Figure 5.35   Four-hidden-neuron network output with arctan output activation function (solid line) superimposed on target pattern (dashed line) and noisy data generated from the target pattern.

Table 5.9   Variance of Sensitivity of Output to Hidden-Neuron Activation and χ^2 Test Statistic

	y_1	y_2	y_3	y_4
Variance	0.0035	0.007	0.017	0.000193
χ^2 ($\sigma_0^2 = 0.001$)	103	206	503	5.6
χ^2 ($\sigma_0^2 = 0.01$)	10	20	50	0.56

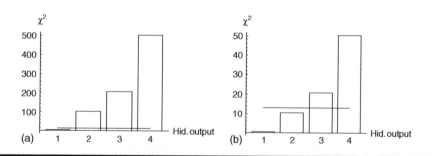

Figure 5.36   Variance nullity measure for sensitivity of output to hidden-neuron activation ($\partial z/\partial y$) for significance level (α) of 0.01: (a) $\sigma_0^2 = 0.001$, which indicates that one neuron has variance of sensitivity close to zero; (b) $\sigma_0^2 = 0.01$, which indicates that two neurons have variance close to zero (critical χ^2 for $\alpha = 0.01$ is $\chi^2_{(30-1, 0.995)} = 13.12$).

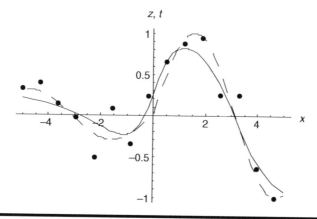

Figure 5.37 Output of the pruned network (hidden neurons 1 and 4 removed) obtained from variance nullity measure on the sensitivity of output to hidden neuron activation, superimposed on target pattern and noisy data.

Figure 5.37 shows a good fit to data. The validation RMSE for the pruned network is 0.277 and for the full network was 0.276 (Section 5.3.2), indicating that generalization of the pruned network is as good as that of the original full model.

The network pruning must be done in stages, and if a pruned network performance is not satisfactory, the original configuration must be restored. If the pruned network performs better or satisfactorily, then further pruning can be done until the performance deteriorates. In this example, at least two neurons are needed so the network will not be further pruned.

The pruning process can also be extended next to removing redundant inputs based on sensitivity of network output to inputs. In the example, there is only one input and, therefore, it cannot be extended to illustrate this point. However, Engelbrecht [12] successfully applied the method to eliminate redundant inputs. Moreover, Engelbrecht [12] tested the variance nullity measure on several real-world examples, including three medical problems, showing that the method is superior to OBD [8], OBS [9,10], and MBP [5,7]. The latter (magnitude based pruning) involves removing weights that have small values; however, this is not a good way to eliminate weights because the network output can still be very sensitive to small weights.

5.5 Robustness of a Network to Perturbation of Weights

Section 5.3.1 illustrates that depending on the randomness in initial weights as well as in sampling of data, different weight values can result. However, the pattern (trend) of hidden neuron activity for these configurations can still be very similar, regardless of the differences in the values of actual weights. It appears that it is not the actual weight values that matter; rather, their correlations [15]. This gives a network with many weights latitude to assume a range of values while producing the desired response. Thus, with redundant weights in a network, there is more than one possible set of weights that preserve the correlation structure. A pruned network would eliminate redundant weights, narrowing down the possible solutions to a set of optimum weights. Even so, inherent random noise in training could cause the weights to fluctuate. A good network can be expected to be stable against perturbation of weights. To test this idea, random noises of increasing magnitude are added to the trained weights of the pruned network in the previous section to see how the network performs.

The weights in the pruned network were perturbed by adding noise from Gaussian distributions with zero mean and a range of standard deviation: 0.01, 0.05, 0.1, and 0.2. Because approximately six standard deviations contain about 99 percent of the observations, these distributions add noise within ± 3 percent, ± 15 percent, ± 30 percent, and ± 60 percent of the individual weight values. For a particular noise distribution, the amount of noise added to each weight is random within the possible range. The network output for these perturbations is shown in Figure 5.38 where the bottom solid line is for a standard deviation of 0.2 providing the maximum perturbation of ± 60 percent. The curves for the other three perturbations are quite close to each other, indicating that the weights are

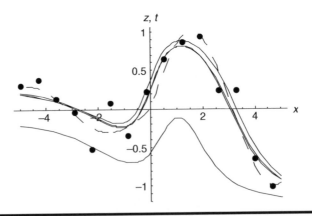

Figure 5.38 Resilience of the pruned network output to perturbation of weights simulated by adding Gaussian noise with zero mean. (Bottom solid line is for noise standard deviation of 0.2 and top three solid lines are for standard deviations of 0.01, 0.05, and 0.1. Dashed line is the target pattern from which noisy data was generated.)

Table 5.10 Training and Validation RMSE for Various Noise Levels Applied to Optimum Weights of Pruned Network

Standard Deviation of Gaussian Noise Distribution	Training RMSE	Validation RMSE
0.01	0.204	0.277
0.05	0.199	0.297
0.1	0.297	0.330
0.2	0.770	0.707

quite robust to random errors up to 30 percent of their mean. The network response shifts while still maintaining the general trend at the noise level of 0.2 standard deviation that perturbs the weights randomly by ±60 percent.

Training and validation RMSE for various noise conditions are given in Table 5.10, showing that network performance is robust against noise up to noise standard deviations of 0.1 causing up to ±30 percent random perturbation to the weights. At the highest standard deviation of 0.2, the error is unacceptable and corresponds to the bottom curve in Figure 5.38.

5.5.1 Confidence Intervals for Weights

Because weights are robust against perturbation with noise levels even up to ±30 percent, confidence intervals for the weights were built [16]. This is illustrated for a noise level of ±15 percent added to weights from a Gaussian distribution with 0.05 standard deviation. Several sets of weights need to be drawn with each representing one network. Ten sets of weights around the optimum weights were generated using the noise distribution. The mean and the standard deviation of the weights are shown in Table 5.11.

From the results in Table 5.11, 95 percent confidence intervals can be constructed using methods of statistical inference based on sampling distribution using

$$(1 - \alpha)\text{CI} = \bar{w} \pm t_{\alpha, n-1} \frac{s_w}{\sqrt{n}}, \tag{5.11a}$$

Table 5.11 Mean and Standard Deviation of Network Weights Perturbed by ±15 Percent Around the Optimum

	a_{02}	a_{12}	a_{03}	a_{13}	b_0	b_2	b_3
Mean	−0.49	2.08	1.44	−0.54	−5.73	2.96	6.02
SD		0.051	0.046	0.053	0.037	0.05	0.033	0.027

Table 5.12 Upper and Lower Confidence Interval (CI) Limits for the Optimum Weights

CI	a_{02}	a_{12}	a_{03}	a_{13}	b_0	b_2	b_3
Upper	−0.455	2.11	1.48	−0.515	−5.7	2.98	6.04
Lower	−0.523	2.05	1.41	−0.566	−5.77	2.94	6.01

where $\bar{w}$ is the mean value of a weight, s_w is the standard deviation of that weight, and n is the sample size. In this case, there are 11 observations. The $t_{\alpha,n-1}$ is the t value from the t distribution for $1-\alpha$ confidence level and degree of freedom (dof) of $n-1$. For $\alpha = 0.05$ and dof $= 10$, t value is 2.228. For the mean and standard deviation given in Table 5.11, confidence intervals for each of the seven weights are constructed and these are presented in Table 5.12.

If the network output is plotted with these lower and upper limits for weights, upper and lower limits for network performance can be constructed. These upper and lower confidence limits are presented in Figure 5.39 along with the training data, target pattern, and network output for the mean (optimum) weights.

The results in Figure 5.39 indicate that mean and confidence limit values of the weights produce output patterns that preserve the trend of the target pattern. The network outputs follow the target pattern more closely than they follow the noisy data, indicating that noise has been eliminated and that the

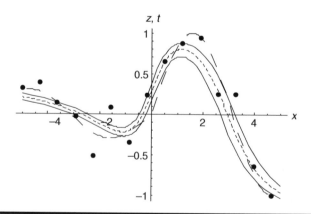

Figure 5.39 Network output for the lower and upper confidence limits for the individual weights. The smaller dashed line is the network performance with mean (optimum) weights and the two solid lines are network output for the upper and lower 95 percent confidence limits for the weights. The medium dashed line represents the target pattern from which noisy data was generated.

networks have found the general trend of the target pattern. The original target pattern in some regions is outside the confidence bands, but the network only sees the noisy data, not the pattern. The network's task is made more difficult by adding a reasonably large noise to the only 15 observations extracted from the target pattern for training and another 15 for cross validation. For a larger sample size or for data with less noise, there is no doubt that the network will approximate the target pattern even better. Recall that a standard deviation of 0.25 was used for the noise distribution when the original data was extracted from the target pattern (see Section 5.3.1).

The perturbation of network weights to obtain a weight distribution for each weight as done here is a simple way to introduce randomness to weights. A random Gaussian noise was added to the optimum weights obtained through training so the perturbation is around a fixed optimum set of weights. In Chapter 7, how Bayesian statistics [2,17] can be used to assess uncertainty of network weights and obtain a probability distribution of weights within a theoretical framework where the optimum network weights become the set that is most plausible (i.e., has the highest probability) is studied. This framework is also used to assess uncertainty of output errors and sensitivity of outputs to inputs in Chapter 7 [18,19].

5.6   Summary

This chapter presents a detailed treatment of network development and assessment of networks' robustness. The first issue dealt with is the bias and variance tradeoff that addresses the issue of having suboptimal models that either underfit (bias) or overfit (variance) the data. A model that does not have adequate flexibility underfits, and one with too much flexibility overfits. What is desired is a model that has the required flexibility and generalizes well for unseen data.

Presented are two methods used for resolving this problem: early stopping and regularization. In early stopping, training stops when error on a validation set starts increasing. In regularization, an additional regularization parameter is added to the square error criterion, and its purpose is to keep weights from growing, which is the cause of overfitting. Through these discussions, the chapter illustrates the effect of initial random weights and random sampling of data. In addition, the problem of nonuniqueness of weights for these cases is addressed. Furthermore, to shed light on the nonuniqueness of weights, the hidden neuron activation for various cases of initial random weights and random sampling is graphically presented to illustrate the consistency of the approach that a network takes toward a solution regardless of the number of hidden neurons, random weights, or random sampling.

Regularization and early stopping produce networks that generalize well, but they are not the optimum in terms of least structural complexity. The simplest possible nonlinear model that generalizes the best is the goal of modeling data. One approach to achieving such optimum models is pruning. This chapter presents approaches to pruning networks (weights and neurons), and it illustrates two pruning methods, optimal brain damage (OBD) and variance nullity measure, in detail. The robustness of the pruned networks is tested by perturbing optimum weights through adding various amounts of random noise to them. This chapter shows that the weights obtained from regularization are robust even up to ± 30 percent noise. Finally, confidence intervals are obtained for weights based on a sample of networks weights extracted from the set of optimum weights by adding noise. From these, upper and lower confidence bounds for network performance are obtained.

Problems

1. Explain the importance of model validation and testing and the specific aspects of model development they address.
2. How do bias and variance affect the generalization ability of a neural network? What are the causes of error in these two situations?
3. What is the basis of the early stopping method used for improving generalization?
4. How can data, and sometimes noise, help optimize model performance? What is the basis for this feature?
5. What is the aim of regularization, and how is it achieved? Explain the difference between early stopping and regularization.
6. Various approaches can be taken to prune networks for optimizing structural complexity. What fundamental concepts are used in pruning? Propose some other potential approaches for improving the efficiency of pruning.
7. Using a dataset of your choice, separate it into training, validation, and testing. Perform early stopping and/or regularization on an initial feedforward MLP network using training and validation data and test the model with test data.
8. If possible, prune the network trained in Problem 7 to obtain the best possible model, and compare the results against actual test data.
9. Extract the optimum weights from the network in Problem 8 or Problem 7 and obtain an analytical expression for the prediction model. Understand how the prediction is affected by the inputs by plotting the model response against inputs.

10. Extract sensitivity of output to inputs $(\partial z/\partial x)$ over the whole database. Comment if this result and trends found in Problem 9 are compatible.

11. Test the robustness of the model in Problem 8 to perturbation of weights by adding random noise to the optimum weights and using these in the model to generate predictions.

12. Generate confidence intervals for the weights, and check if the target response is captured within the confidence bands.

13. Interpret your model in terms of the objectives of the model development task. Are the results meaningful?

14. Compare the model with that obtained from a linear method such as linear regression or any other nonlinear method on the same data. Comment on the advantages or disadvantages of the neural model.

References

1. *Mathematica—Neural Networks*, Wolfram Research Inc., Champaign, IL, 2002.

2. Bishop, C. *Neural Networks for Pattern Recognition*, Clarendon Press, Oxford, 1996.

3. Haykin, S. *Neural Networks: A Comprehensive Foundation*, 2nd Ed., Prentice Hall, Upper Saddle River, NJ, 1999.

4. Hagiwara, K. Regularization learning, early stopping and biased estimator, *Neurocomputing*, 48, 937, 2002.

5. Sietsma, J. and Dow, R.J.F. Creating artificial neural networks that generalize, *Neural Networks*, 4, 67, 1991.

6. Reed, R. Pruning algorithms—A survey, *IEEE Transactions on Neural Networks*, 4, 740, 1993.

7. Hagiwara, M. Removal of hidden units and weights for backpropagation networks, *Proceedings of the International Joint Conference on Neural Networks*, Vol. 1, IEEE, Los Alamitos, CA, 351, 1993.

8. Le Cun, Y., Denker, J.S., and Solla, S.A, Optimal brain damage, *Advances in Neural Information Processing*, Vol. 2, D.S. Touretzky, ed., Morgan Kaufmann Publishers, Los Altos, CA, 598, 1990.

9. Hassibi, B., Stork, D.G., and Wolff, G.J. Optimal brain surgeon and general network pruning, *IEEE International Conference on Neural Networks*, Vol. 1, IEEE, Los Alamitos, CA, 293, 1992.

10. Hassibi, B. and Stork, D.G. Second-order derivatives for network pruning: Optimal brain surgeon, *Advances in Neural Information Processing Systems*, C. Lee Giles, S.J. Hanson, and J.D. Cowan, eds., Vol. 5, IEEE, Los Alamitos, CA, 164.

11. Cibas, T., Fogelman, S., and Raudys, F. Variable selection with neural networks, *Neurocomputing*, 12, 223, 1996.

12. Engelbrecht, A.P. A new pruning heuristic based on variance analysis of sensitivity information, *IEEE Transactions on Neural Networks*, 12, 6, 1386, 2001.

13. Hornik, K., Stinchcombe, M., and White, H. Universal approximation of an unknown mapping and its derivatives using multi-layer feedforward networks, *Neural Networks*, 3, 551, 1990.
14. Gallant, A.R. and White, H. On learning the derivative of an unknown mapping with multilayer feedforward networks, *Neural Networks*, 5, 129, 1992.
15. Aires, F. Neural network uncertainty assessment using Bayesian statistics with application to remote sensing: 1. Network weights, *Journal of Geophysical Research*, 109, D10303, 2004.
16. Rivals, I. and Personnaz, L. Construction of confidence intervals for neural networks based on least squares estimation, *Neural Networks*, 13, 463, 2000.
17. MacKay, D.J.C. A practical Bayesian framework for back-propagation networks, *Neural Computation*, 4, 448, 1992.
18. Aires, F. Neural network uncertainty assessment using Bayesian statistics with application to remote sensing: 2. Output error, *Journal of Geophysical Research*, 109, D10304, 2004.
19. Aires, F. Neural network uncertainty assessment using Bayesian statistics with application to remote sensing: 3. Network Jacobians, *Journal of Geophysical Research*, 109, D10305, 2004.

Chapter 6

Data Exploration, Dimensionality Reduction, and Feature Extraction

6.1 Introduction and Overview

An important first step in modeling is data exploration. This includes data visualization for qualitative assessment of trends and relationships, data cleaning, dimensionality reduction, and extracting relevant inputs and features that help subsequent modeling. Neural networks for prediction and classification, such as multilayer networks including multilayer perceptron (MLP) and radial basis function (RBF) networks, are nonlinear processors that map inputs nonlinearly to outputs. They do this by the use of hidden layer neurons linked with inputs. The greater the number of inputs, the greater the number of weights linking inputs to hidden neurons. This can adversely affect model accuracy. For example, the greater the number of free parameters (i.e., weights), the more demanding is the process of optimizing these weights.

Finding the information most relevant to a problem is generally called feature extraction. What is really necessary is a model that has only the relevant inputs or features of a problem that lead to the least number of free parameters producing the best model with adequate generalization for the given data. Therefore, by removing redundant inputs, for example, the model can be kept more simple. Another problematic aspect of inputs can be multicollinearity where input variables correlate with one another. The correlation structure of inputs leads to nonuniqueness of solution, i.e., the existence of more than one solution for the weights due to interaction between inputs that translate their correlations to weights, thereby complicating the optimization of the free parameters. Furthermore, correlated inputs make a network operate in a dimension reduced from the original. For example, if there are n variables and d variables are highly correlated, then essentially there are $(n-d+1)$ independent variables. The correlated variables can be grouped together, or a representative from this group may be sufficient to incorporate the effect of the whole group on the outcome. A model can be greatly simplified if input variables are independent.

Furthermore, it is not uncommon to encounter practical situations with many variables but few observations because of the nature of the problem or difficulty in collecting data. When a network is trained for such cases with many variables, there will be sparse areas or areas without data in the input space because of the lack of data, causing the network to experience difficulties in adequately estimating the large number of free parameters resulting in a suboptimal solution. In these situations, reducing the number of input variables may be essential.

In many cases, values of some input variables are expressed by large magnitudes whereas others may be quite small. This discrepancy can lead to faulty interpretation by the model because larger weights are adopted for inputs with larger magnitudes, thereby masking the influence of variables with smaller magnitudes. This may require normalization of the input data so that all variables fall in a similar range.

This chapter addresses these important aspects of data preprocessing and illustrates most of the concepts using an example dataset involving thermal conductivity in wood in relation to moisture content, density, and temperature. Histograms, scatter plots, correlation plots, parallel visualizations, and projections of multidimensional data onto two dimensions are presented in Section 6.2 as an aid to understand the character, trends, and relationships in data. Correlation and covariance are used to measure strength of relationships and dispersions in data, respectively, as illustrated in Section 6.3. Section 6.4 presents several approaches to normalization: standardization, range scaling, and whitening. The latter allows normalization of correlated multivariate data.

A great deal of attention is paid to input selection and dimensionality reduction for improving model development and accuracy. The use of statistical tools such as partial correlation, multiple regression, and best subsets regression for input selection are explained in Section 6.5; and principal component analysis and partial least squares regression are presented as suitable methods for dimensionality reduction and feature extraction in Section 6.6. Outlier detection and noise removal in multivariate data are addressed in Section 6.7 and Section 6.8. Various approaches to data preprocessing presented in the chapter are further demonstrated using an example case study involving a real application in Section 6.9. The positive effect of preprocessing on structural complexity and model accuracy is also demonstrated using this case study.

The following example illustrates the methods of data exploration discussed in Section 6.2 through Section 6.7.

6.1.1   Example: Thermal Conductivity of Wood in Relation to Correlated Input Data

The example problem involves exploring thermal conductivity data of wood. In many temperate countries, wooden homes are quite common, and an attractive aspect of wood is that it is a good thermal insulator. The thermal conductivity of a material indicates how much heat is conducted by it, and one with low thermal conductivity is a good insulator. To properly design wooden homes for winter comfort, it is important to know its thermal conductivity accurately. Thermal conductivity of wood depends on density that varies within and across wood species, moisture content, and temperature. Thermal conductivity must be determined from carefully planned experiments, so in our sample there are only 35 observations obtained for a variety of species [1]. In the dataset, the temperature range is between 0 and 100°C; however, most of the observations have been made for a constant temperature of 29°C. The moisture content ranges from 0 to 91.1 percent, and density ranges from 294 to 930 kg m^{-3}. The thermal conductivity for these conditions ranges from 0.0731 to 0.3744 (W m^{-1} K^{-1}). Five input–output pattern vectors extracted from the dataset are shown in Table 6.1.

Table 6.1 A Sample of Five Records from the Dataset for Thermal Conductivity and Related Variables

Species	Temp. (°C)	Moisture (percent)	Density (kg/m³)	Conductivity (W/m K)
Ash white	29	15.6	647	0.1742
Red oak	29	12.4	697	0.1944
Japanese cedar	20	0	294	0.0778
Japanese beech	25	50	800	0.2132
Silver birch	100	0	680	0.25

6.2   Data Visualization

An important but problematic aspect of dealing with multidimensional data is visualization of data. When the dimension is greater than three, the standard methods of visualization become insufficient. However, visualization provides important qualitative clues as to the significance and interaction of the variables. Scatter plots, histograms, parallel visualizations, and projections to reduced dimensions are some popular methods of data visualization. This section explores these visualization methods for the example described in the previous section involving thermal conductivity of wood and its relationship to density, moisture content, and temperature.

6.2.1   Correlation Scatter Plots and Histograms

Visualization of data is an important first step in data analysis and modeling. It can give clues as to the level of nonlinearity, patterns and trends in data and interactions among variables. Figure 6.1 shows plots of all data for the thermal conductivity analysis [2]. The diagonal histograms show the distribution of the individual variables with gray level intensity highlighting correspondence to the value of the output variable, thermal conductivity, denoted by *K*.

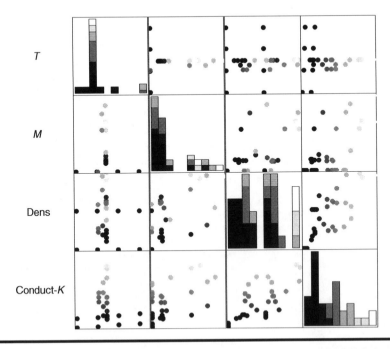

Figure 6.1   Histograms and correlation scatter plots of data highlighting the distribution and the relationship of input variables to themselves and to output variable.

For example, high density (Dens) and high moisture contents (*M*) are associated with high conductivity depicted by lighter shades of gray in histograms. However, the pattern is not so clear for temperature (*T*). Conductivity increases marginally from minimum to maximum temperature, but there is a much larger variation for 29°C represented by the tallest column in the histogram. The off-diagonal scatter plots show how individual variables relate to one another with the gray level highlighting the correspondence to conductivity-*K*. They show that conductivity is positively correlated with density and moisture content. Furthermore, moisture and density are also positively correlated. A dominant feature in the data is the large scatter, which is common in data from biological, ecological, and natural systems.

6.2.2 Parallel Visualization

Parallel visualization is a useful way to visualize multidimensional data. Figure 6.2 shows a parallel plot of data in the thermal conductivity dataset and their interdependencies and highlights how the three variables— temperature, moisture content, and density—together affect conductivity. It shows that high density and high moisture content combination produces high conductivity. The bottom part of the graph shows low conductivities, and the upper part shows high conductivities. Temperature does not have a range of values to cover the whole span, but it can be expected that higher temperature will lead to higher conductivity, especially in combination with high density and moisture content. As can be seen, lack of data along the temperature axis creates sparse data areas for modeling, an effect that can cause difficulties in the parameter estimation process.

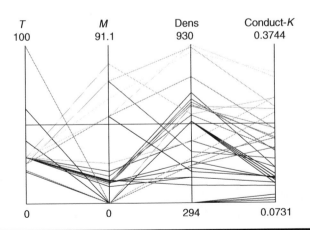

Figure 6.2 Parallel coordinates visualization depicting the range of the variables and their interdependencies highlighted with respect to the value of the output variable (the lighter the shade of the lines, the higher the conductivity).

6.2.3 Projecting Multidimensional Data onto Two-Dimensional Plane

Another powerful visualization method is self-organizing maps (SOMs) where higher-dimensional data is projected onto a two-dimensional (2-D) grid consisting of cells [3]. Chapter 8 is devoted to this topic, so the theoretical issues is not explored here, but the concept is illustrated for the thermal conductivity data discussed in the previous section. Figure 6.3 presents the results of projections of each variable vector, consisting of temperature, moisture content, density, and thermal conductivity, onto a map consisting of 25 (5×5) cells.

These maps are based on the similarity or closeness of the vectors of variables, which are projected so that vectors that are closer together in multidimensional space are also closer together in the 2-D plane, so the projection preserves the spatial correlations (or topology) in the data. Components (variables) of the projected vectors can be visualized separately in component maps as shown in Figure 6.3 where individual cells correspond across maps. For example, the maps in Figure 6.3 can be aligned (or placed on top of one another) for comparison and analysis of correspondence. Gray level highlights the correspondence to thermal conductivity-*K*. These reveal again that high moisture content and high density lead to high conductivity, but temperature has not such a clear

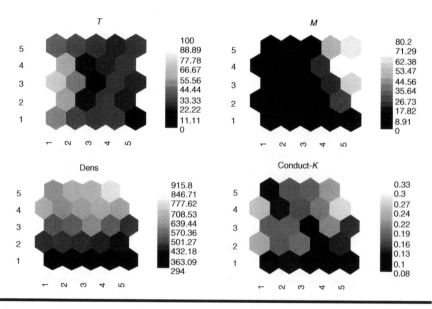

Figure 6.3 Visualization of variables according to their relation to conductivity using self-organizing map projections (lighter shades denote higher conductivity).

relevance to other variables. The maps thus show qualitatively how inputs are related to thermal conductivity throughout their range.

6.3 Correlation and Covariance between Variables

The correlation coefficient r is a measure of the strength of relationship between two variables. The higher the correlation coefficient, the stronger the relationship. The correlation coefficient for two variables x_1 and x_2 with mean $\bar{x}_1$ and $\bar{x}_2$ can be expressed as [4,14,21]

$$r = \frac{\sum_{i=1}^{N}(x_{1i} - \bar{x}_1)(x_{2i} - \bar{x}_2)}{\sqrt{\sum_{i=1}^{N}(x_{1i} - \bar{x}_1)^2 \sum_{i=1}^{N}(x_{2i} - \bar{x}_2)^2}} \tag{6.1}$$

The linear correlation coefficients for the variables in the thermal conductivity dataset are illustrated in the bar chart in Figure 6.4, which shows that conductivity is highly correlated with density (0.775), is reasonably highly related to moisture content (0.647), and has little correlation to temperature (0.172).

The correlation matrix depicting correlation between all four variables is presented in Table 6.2.

The correlation matrix is symmetric with the diagonal values representing the correlation of a variable to itself, which is 1.0. Off-diagonal values are the correlations between pairs of variables denoted by the labels indicated in the first row and column. Moisture content (M) and density (Dens) are correlated at 0.583, but density has very little correlation with temperature (T) (-0.077), and moisture is weakly and negatively related to temperature (-0.221). Thus, judging by the correlation, conductivity is mainly influenced by density and moisture content, and these are reasonably

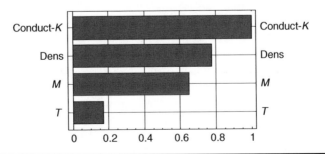

Figure 6.4 Correlation graph for the three independent variables—density, moisture content, and temperature—and the dependent variable, thermal conductivity, K.

Table 6.2 Correlation Matrix for the Input and Output Variables

	T	M	Dens	K
T	1.0	−0.221	−0.077	0.172
M	−0.221	1.0	0.583	0.647
Dens	−0.077	0.583	1.0	0.775
K	0.172	0.647	0.775	1.0

Table 6.3 Covariance Matrix for the Three Input Variables

	T	M	Dens
T	543	−141	−351
M	−141	753	3117
Dens	−351	3117	37 888

highly correlated. Temperature has a weaker relationship to all the variables and is especially weak in its correlation to density.

The covariance of two variables is expressed by [4,14,21]

$$\text{COV} = \frac{1}{N-1} \sum_{i=1}^{N} (x_{1i} - \bar{x}_1)(x_{2i} - \bar{x}_2). \tag{6.2}$$

Basically, the two expressions within parentheses in Equation 6.2 each compute the difference between the value of an input variable and its mean. When two variables coincide (i.e., $x_1 = x_2$), the result is the covariance of one variable with respect to itself, which is its variance. When $x_1 \neq x_2$, the result is the covariance between the two variables. The covariance matrix containing the covariance between each pair of the three input variables is presented in Table 6.3. As with the correlation matrix, the covariance matrix is symmetric. Here, the diagonal values represent the variance of each variable, and off-diagonal values represent the covariance between pairs of variables denoted by the labels in the first row and column. Covariance is a measure of how two variables co-vary in relation to one another. When two variables are not related, their covariance is zero. If two variables move in the same direction, covariance is positive and if they move in opposite directions, it is negative. Table 6.3 indicates a high positive covariance between moisture content (*M*) and density (Dens) and smaller negative covariance between these and temperature (*T*).

Data including spread, trends, relationships, correlations, and covariances helps clarify the problem to determine appropriate strategies for normalization of data and extract relevant inputs and features for model development. These issues are addressed next.

6.4 Normalization of Data

Figure 6.2 indicates that variables in the thermal conductivity dataset have very dissimilar ranges. When variables with large magnitudes are combined with those with small magnitudes, the former can mask the effect of the latter due to the sheer magnitude of the inputs leading to larger weights associated with them. Normalization puts all inputs variables in a similar range so that true influence of variables can be ascertained.

6.4.1 Standardization

There are many ways to normalize data. A simple approach is to standardize the data with respect to mean and the standard deviation using a linear transformation. This transforms all variables into a new variable with zero mean and unit standard deviation. To do this, each input variable is treated separately, and for each variable x_i in the training set, the mean $\bar{x}_i$ and variance σ_i^2 are calculated using

$$\bar{x}_i = \frac{1}{N} \sum_{n=1}^{N} x_i^n$$

$$\sigma_i^2 = \frac{1}{N-1} \sum_{n=1}^{N} \left(x_i^n - \bar{x}_i\right)^2$$

(6.3)

where $n = 1, \ldots, N$ is the pattern number. With the mean and the standard deviation σ_i, each input variable is normalized as

$$x_{Ti}^n = \frac{x_i^n - \bar{x}_i}{\sigma_i}$$

(6.4)

where x_{Ti}^n is the normalized (transformed) value of the nth observation of the variable x_i. The new transformed variable now has zero mean and unit standard deviation. The actual range of the data depends on the original data, but most data fall within $\pm 2\sigma$. For prediction problems, the target output is also normalized using the same procedure for consistency. With the normalization, the inputs and target variables are of the same order; therefore, final weights will also be of order unity. This, furthermore, prevents weights from growing too large and causing training problems in situations where large weights throw the current training into a flat area of the error curve as discussed in Chapter 4.

Table 6.4 Standardized Values for Thermal Conductivity and Related Variables Presented in Table 6.1

Species	Temp	Moisture	Density	Conductivity
White ash	−0.222	−0.204	0.327	−0.083
Red oak	−0.222	−0.321	0.584	0.1611
Japanese cedar	−0.608	−0.772	−1.49	−1.25
Japanese beech	−0.393	1.05	1.11	0.388
Silver birch	2.82	−0.772	0.496	0.833

Look at the thermal conductivity dataset to understand how to do this normalization. The means for the four variables—temperature, moisture content, density, and thermal conductivity—represented in vector form are

$$\bar{\mathbf{x}} = -\{34.2, 1.20, 583, 0.1811\}. \tag{6.5}$$

Similarly, the standard deviations for the same variables represented in vector form are

$$\sigma_{\mathbf{x}} = \{23.29, 27.45, 194.6, 0.0827\}. \tag{6.6}$$

The rescaled data x_{Ti} for those in Table 6.1 using Equation 6.4 is shown in Table 6.4.

Now variables are unit free and have a similar range that varies between ±3 with 0 mean and a standard deviation of 1. The correlations established earlier are not altered by this standardization.

6.4.2 Simple Range Scaling

Another simpler approach is to fix the minimum and maximum values for the normalized variables to 0 and 1 or 1 and −1, respectively. In this case, the mean and the standard deviation of the normalized inputs vary from one input variable to another, but the observations stay in the same range.

A simple linear transformation in the range from 0 to 1 is

$$x_{Ti} = \frac{x_i - x_{imin}}{x_{imax} - x_{imin}} \tag{6.7}$$

where x_{imin} and x_{imax} are the minimum and the maximum values of the variable x_i. A similar transformation can be made for any desired range, e.g., −1 or 1, or any other. For the example thermal conductivity problem, each of the four variables were transformed using Equation 6.7, and the resulting mean and standard deviation for the variables are

$$\bar{x} = \{0.342, 0.233, 0.455, 0.358\}$$

$$\sigma_x = \{0.233, 0.30, 0.306, 0.275\}$$

(6.8)

The above linear transformations are done for each individual variable separately without any consideration given to the correlations among data. The whole set of input variables can be considered together and linear transformations that take into account the correlations among inputs can be done. One such method is called whitening [5]. Table 6.2 shows that the variables in the example thermal conductivity dataset are correlated. The next section examines how whitening transforms these correlated data.

6.4.3 Whitening—Normalization of Correlated Multivariate Data

To illustrate this method, the whole set of input variables k must be considered, so denote the whole group of input variables by vector $\mathbf{x} = \{x_1, x_2, ..., x_k\}$ where x_i is the ith input variable. With this vector arrangement, the mean and variance of each input variable and the covariance between sets of two input variables can be calculated efficiently. Then the mean values can be put into a mean vector $\bar{\mathbf{x}} = \{\bar{x}_1, \bar{x}_2, ..., \bar{x}_k\}$ and the variances and covariances into a covariance matrix (COV) as follows:

$$\bar{\mathbf{x}} = \frac{1}{N} \sum_{n=1}^{N} \mathbf{x}^n$$

(6.9)

$$\text{COV} = \frac{1}{N-1} \sum_{n=1}^{N} (\mathbf{x}^n - \bar{\mathbf{x}})(\mathbf{x}^n - \bar{\mathbf{x}})^T$$

In Equation 6.9, COV is a symmetric matrix of size $k \times k$, where k is the number of input variables. When two variables coincide in the second equation, the result is the variance, and when they are dissimilar, the result is the covariance between the two variables. In the second equation, $(\mathbf{x}^n - \bar{\mathbf{x}})$ is a vector containing the difference between each variable and its mean for the k variables in the nth input pattern. Thus, it is of length k. T denotes transpose, which in this case is the row format of the difference vector. The multiplication of a vector by its transpose is called the outer product and in this case results in a matrix of size $k \times k$ (see the Appendix for an example illustrating transpose of a vector and outer product). This is for one input pattern. If this is done repeatedly for all input patterns, all the corresponding entries are summed, and the sum is divided by $N - 1$, then the operation in Equation 6.9 is performed. The result is the covariance matrix for the dataset. Using vectors, covariance matrix can be calculated more efficiently compared to treating each pair of variables separately as in Equation 6.2.

The COV matrix (see Table 6.3) can be transformed into a new matrix that corresponds to a set of new rescaled variables that have unit variance and are independent of one another. Therefore, covariance between two new variables is zero. This is accomplished by the well-known eigenvalue decomposition method or principal component analysis (PCA), which can be found in many statistical or mathematical software [4–7]. Therefore, mathematical details are kept to a minimum and the concept is emphasized and illustrated using an example. The PCA is represented by

$$\text{COV } u_j = \lambda_j u_j, \tag{6.10}$$

where jth rescaled variable represents the variance (also called the eigenvalue) of the rescaled variable and u_j represents a vector containing the coefficients or the weights indicating the proportion of all the original variables that make up the jth new variable. The value of the jth rescaled variable is obtained from a linear combination of the original variables using these weights. The u_j is called an eigenvector, and there are as many eigenvectors as there are input variables.

Each eigenvector u_j defines the direction of a new axis for the jth rescaled variable called a principal component so that all new axes are perpendicular to each other. This makes the covariance between the rescaled variables zero, meaning that they are uncorrelated. Thus, this process essentially decorrelates the original input variables and creates new variables from them independent of one another.

Figure 6.5 illustrates schematically the original distribution and the whitened distribution for the case of two input variables x_1 and x_2. In the original distribution, x_1 and x_2 are correlated. The new rescaled variables, represented by u_1 and u_2, are perpendicular to each other and uncorrelated. The u_1 is parallel to the major direction of the data in the original

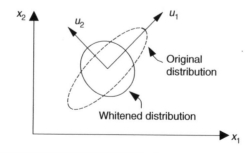

Figure 6.5 **Schematic illustration of the distribution of original correlated data and whitened uncorrelated data.**

Table 6.5 Eigenvectors for the Input Data

	T	M	Dens
u_1	−0.0096	0.0831	0.996
u_2	0.775	−0.629	0.06
u_3	0.632	0.772	−0.058

distribution, thereby capturing most of the variance in the original data. The u_2 is parallel to the minor direction of the original data and captures the variance of the data in that direction. Thus the new rescaled variables represented by u_1 and u_2 capture all the variance in the original data while removing the correlation between x_1 and x_2.

In the case of the thermal conductivity problem, the COV matrix for the three variables of temperature, moisture content, and density is given in Table 6.3. The COV matrix shows that three variables are not independent because covariance (off-diagonal terms) is not zero. This was transformed using Equation 6.10, and the resulting three eigenvectors, u_1, u_2, and u_3, are given in the matrix of eigenvectors in Table 6.5. Each vector consists of the weights or coefficients that transform the original variables to the new coordinate system.

The transformation results in three new variables called principal components (PCs). The first PC is expressed by u_1, the second PC by u_2, and the third by u_3. These are perpendicular to each other and therefore uncorrelated. The variance of the ith new variable (PC) is given by the eigenvalue λ_j, and these were found to be 38151, 630, 402, respectively, for the first, second, and third PCs. The first PC always captures the largest amount of the variance of original data, the second PC captures the largest amount of the residual variance, and so on.

To obtain the transformed variables, the original variables are multiplied by the weights (also called loadings) shown in Table 6.5. The first row defines the weights for the first PC depicted by u_1, and second and third rows contain the weights for the second and third PCs depicted by u_2 and u_3. Prior to this, weights are normalized by dividing by the corresponding standard deviation (or square root of eigenvalue, $\sqrt{\lambda_j}$) of each PC. For example, the first eigenvalue is 38 151; therefore, the standard deviation of the first PC is 195.32. Each loading in the first row in Table 6.5 is divided by 195.32 to normalize the first PC. The normalized coefficients for the three PCs are given in Table 6.6. These are denoted by u'_1, u'_2, and u'_3.

The normalized coefficients are multiplied by the original variables scaled to zero mean to obtain the values for each PC. To transform the original variables to zero mean, simply subtract the corresponding mean value from the values of each of the original variables, (i.e., $x_i - \bar{x}_i$ where the

Table 6.6 Normalized Eigenvectors for the Input Data

	T	M	Dens
u_1'	−0.0000492	0.03086	0.0315
u_2'	0.000425	−0.025	0.038
u_3'	0.0051	0.0024	−0.0029

latter is the mean of the ith input variable). Mean values given in Equation 6.5 are repeated below:

$$\bar{x} = \{34.2, 1.20, 583, 0.1811\}. \tag{6.11}$$

The calculation for the first data vector in the dataset in Table 6.1 will now be performed. The values of the three zero-mean original input variables can now be calculated, and these are $\{-5.2, -14.4, 64\}$ denoted by vector $x_1' = \{x_{11}', x_{21}', x_{31}'\}$. The normalized first eigenvector is given by $u_1' = \{u_{11}', u_{12}', u_{13}'\}$, and from the first row of Table 6.6, its values are $u_1' = \{-0.0000492, 0.0308642, 0.0315057\}$. Thus, the equation for the value of the first component, PC_1, also called the PC score for the first PC, can be written as

$$PC_1 = u_{11}'x_{11}' + u_{12}'x_{21}' + u_{13}'x_{31}'$$

$$= -0.0000492 \times (-5.2) + 0.03086 \times (-14.4) + 0.0315 \times 64 = 1.57.$$

The values of the second and third PCs corresponding to the same input pattern are computed similarly by multiplying the second and third row, respectively, of Table 6.6 by the component of the same input vector. Repeating this process for all the rescaled original input vectors results in the three PCs for each of these input vectors. Now the variance of these PCs is unity, and the mean is zero. Figure 6.6a shows a plot of the distribution of two

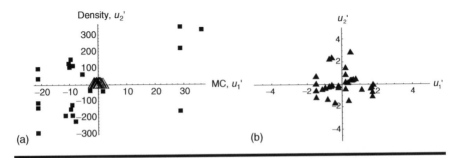

Figure 6.6 Illustration of rescaling of the original input distribution with whitening: (a) original distribution of two variables (moisture content and density) and the whitened distribution with unit variance and zero mean shown with respect to first two PCs and (b) enlarged view of whitened distribution.

original variables (moisture content and density) rescaled to zero mean superimposed on the whitened (transformed) distribution (denoted by triangles shown at the origin) of the first two components represented by u'_1 and u'_2. Because of the scale difference, the form of the whitened distribution cannot be ascertained from this figure; therefore, in Figure 6.6b it is shown separately to demonstrate the form of the distribution that has unit variance and zero mean. All the transformed variables are within a circle.

6.5   Selecting Relevant Inputs

In practical problems, there can be many independent and dependent variables. The first approach considered in many modeling problems is linear regression. This is done using methods such as least squares regression to obtain a mathematical expression relating the output to several input variables. In such regression models, the number of predictor variables determines the number of model parameters and, therefore, the complexity of the model. In neural networks, as presented in Chapter 3, Chapter 4, and Chapter 5, the model complexity is governed by the number of hidden neurons and hidden layers in addition to the number of predictor variables. This gives the modeler greater control in choosing a model with adequate complexity to model the problem. However, overfitting is a problem when a model has too many free parameters (weights). As presented in Chapter 5, overparameterized models can fit the original data well but can yield poor generalization.

In a neural network model, overfitting can be lessened by reducing the number of input variables and hidden neurons [16]. As presented in Chapter 5, early stopping and regularization are two methods used to reduce overfitting using external control over the size of the weights [5,7,8]. This does not reduce the number of free parameters and also does not address redundant or irrelevant variables. Chapter 5 also illustrates several network pruning [9,10,20] methods, and these reduce model complexity by eliminating irrelevant neurons, weights, and inputs. This section examines ways to reduce the number of input variables from the beginning.

In linear regression, for example, too many predictor variables can adversely affect the predicted outcome. Adding a redundant variable to the least squares equation almost always increases the variance of the predicted outcome [13]. Thus, too many variables can make a model very sensitive to noise or small changes in a highly correlated dataset and consequently make it less robust. Therefore, selecting a suitable subset of variables from the original set can be crucial. Some methods that can be used for this purpose are scatter plots, simple and partial correlation coefficients, coefficient of determination, and Mallow's C_p statistic. These topics are discussed next.

6.5.1 Statistical Tools for Variable Selection

6.5.1.1 Partial Correlation

A scatter plot reveals relationships between variables in a dataset, as shown in Figure 6.1. Points lying on a line indicate a linear relationship, a curved set of points denotes a nonlinear relationship, and absence of a pattern indicates that the two variables are uncorrelated. Linear correlation coefficients indicate the strength of the linear relationship between two variables. However, this technique alone is not enough for multivariate data because other variables in the set can affect the correlation of two variables, thereby altering the correlation structure. In such situations, partial correlation can be used to measure the linear association between the two variables while adjusting the effects of other variables by holding them constant [11,14]. The partial correlation is calculated from the matrix of simple correlation coefficients, an example of which is presented in Table 6.2 for the problem of thermal conductivity in relation to density, moisture content, and temperature. Suppose the correlation between two variables x_i and y_j is R_{ij}. The partial correlation, r_{ij}, for the two variables is given by

$$r_{ij} = \frac{-C_{ij}}{\sqrt{C_{ii}C_{jj}}} \qquad (6.12)$$

where C_{ij} is the inverse of the simple correlation coefficient R_{ij} (i.e., $C_{ij} = 1/R_{ij}$).

Returning to the problem on wood thermal conductivity, the inverse of the correlation matrix in Table 6.2 gives the values shown in the diagonal and the top right triangle of Table 6.7 utilizing symmetry. The simple linear correlation coefficients are repeated in the bottom left triangle of Table 6.7, again taking advantage of the symmetry of the correlation matrix.

With the values in Table 6.7, partial correlation can be calculated using Equation 6.12, and these are presented in Table 6.8.

The partial correlation matrix in Table 6.8 has a similar structure to the original correlation matrix, which indicates that in this three-variable case, the simple correlations are not influenced significantly by the other variables. The reason for this is that only moisture and density are significantly related

Table 6.7 Inverse of the Correlation Coefficients C_{ij} (Diagonal and Top Right Triangle) and Simple Linear Correlation Coefficients R_{ij} (Bottom Left Triangle)

	T	M	Dens
T	1.055	0.281	−0.082
M	−0.221	1.59	−0.906
Dens	−0.077	0.583	1.522

Table 6.8 Partial Correlation Coefficients r_{ij} (Diagonal and Top Right Triangle) and Simple Correlation Coefficients R_{ij} (Bottom Left Triangle)

	T	*M*	*Dens*
T	−1.0	−0.216	0.064
M	−0.221	−1.0	0.582
Dens	−0.077	0.583	−1.0

and temperature is weakly related to both variables. For datasets consisting of many variables, the influence of other variables on the correlation between two variables can be significant. This is illustrated in a case study later in the chapter.

6.5.1.2 Multiple Regression and Best-Subsets Regression

Another approach to input selection is multiple regression analysis where a model that linearly fits the output to the input variables is developed through least squares regression. The R^2 or the multiple coefficient of determination represents the portion of the variability of the output explained by the predictor variables. A value of R^2 near 1 indicates a perfect model, and the variables capture all the variance of the outcome. A value near zero indicates a poor model, and the input variables are irrelevant to the outcome. Inputs can be selected based on this approach, but the variance of the predicted output can increase with the inclusion of additional predictor variables. This can cause difficulty in selecting a subset when the number of variables in candidate subsets varies. In such situations, criteria that penalize model complexity are more useful in subset selection. Criteria such as Mallow's C_p statistic [15] have been widely used to evaluate model complexity. This statistic suggests as the criterion the standardized total squared error computed as

$$C_\mathrm{p} = \left(\frac{\mathrm{SS}_{\mathrm{error},\,p}}{\mathrm{SS}_{\mathrm{error},\,\mathrm{total}}} \right) - (n - 2p), \tag{6.13}$$

where $\mathrm{SS}_{\mathrm{error},\,p}$ is the residual error for a multiple linear regression subset model with p inputs, and $\mathrm{SS}_{\mathrm{error},\,\mathrm{total}}$ is the residual error for the model with all n inputs. The correct model has C_p value equal or smaller than p and a wrong model has a C_p value larger than p due to a bias in the parameter estimation. Minimizing C_p over all possible regression can give the best subset model. Good models typically have a (p, C_p) coordinate close to a $45°$ line on a C_p versus p plot.

Regarding the thermal conductivity problem, if models are run with all possible subsets of inputs, the results for the most relevant subsets

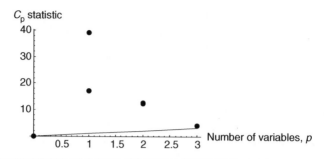

Figure 6.7 C_p **(Mallow's) statistic versus** p **plot from best subset regression for predicting wood thermal conductivity.**

(others had higher C_p values) illustrated in Figure 6.7 show that the best model has all the variables in the model. This case is denoted by the dot lying near the 45° line (note that the scales of the two axes are different).

6.6 Dimensionality Reduction and Feature Extraction

6.6.1 Multicollinearity

In previous sections, several methods of data rescaling have been explored. The objective of rescaling and normalizing is to make the range of all the variables similar. In whitening, normalization is done for correlated variables by transforming them into new uncorrelated variables with zero mean and unit variance. These methods primarily focus on rescaling. It is very common to encounter collinearity (correlation) between measurements in data. An example of this is illustrated in previous sections where a simple analysis of correlation and covariance on multivariate data is presented.

Using collinear measurements in inferential modeling can potentially lead to high prediction variance and ill conditioning [17]. Furthermore, as discussed previously, highly correlated data provides redundant input dimensions to the network causing it to operate, in effect, in a reduced space while having to do redundant computations with redundant weights and neurons. Therefore, it is useful to reduce dimensionality so that the essential correlations of the data are preserved while lower dimensional features characteristic of data are extracted from the original data. Basically, original correlated variables are transformed to fewer uncorrelated variables where the original correlations are embedded. An approach used to address multicollinearity in multivariate data is principal component analysis (PCA), which was already used in data rescaling with whitening. There, the focus is not on dimensionality reduction, but on data transformation. However, the basic idea still applies.

6.6.2 *Principal Component Analysis (PCA)*

In PCA, correlated data is linearly combined to form new variables (PCs) that are uncorrelated and ordered according to the portion of the total variance in the data accounted for by the PCs. The first few components retain the variation in all original variables. In dimensionality reduction, an additional step is used to select the required number of PCs and use these uncorrelated variables in the model instead of the original correlated inputs [23]. Thus data is compressed so that only the essential information is retained in the new variables. In this process of dimensionality reduction, the portion of variance accounted for by each PC is examined.

As presented in Section 6.4.3, PCA involves decomposition of the covariance matrix of the original dataset so that the original coordinate system of correlated variables is transformed to a new set of uncorrelated variables called PCs. The actual transformation is done through eigenvectors also called loadings or weights. Each eigenvector contains loadings for each of the original variables that transform them to the new variables (PCs). The values of the new variables are called scores (or PC scores) that are obtained by projecting the original inputs onto the eigenvectors (i.e., multiplying original inputs with corresponding loadings).

Eigenvalues denote the variance of the new variables (PCs), and they are in descending order. Therefore, the first PC represented by the first eigenvector captures the largest amount of variation in the original data, each subsequent PC captures the largest amount of remaining variance, and so on. The amount of variation captured by each PC is given by their corresponding eigenvalues. Theoretically, there are as many PCs as there are input variables, but because the first few PCs capture most of the variance, a threshold maximum variance can be defined as a suitable cutoff point (90 percent, 95 percent, etc.) taking into account the noise and variance in the data to select an adequate number of PCs that sufficiently represent the original data while discarding the rest. Recall that each PC is a linear combination of all the input variables. The selected PCs can then be used as inputs to a neural network as an alternative to using the original variables. This approach will result in a network with less complexity as the number of inputs to the model is substantially reduced. Using uncorrelated PCs also helps prevent overfitting in neural networks while original inputs are appropriately represented in the network. This is demonstrated later in the chapter through a case study.

The method will now be applied to the example problem involving wood thermal conductivity affected by three variables: temperature, moisture content, and density. However, the standardized inputs will be used with zero mean and unit standard deviation to do the PCA because the variables have dissimilar ranges. If variables with largely dissimilar ranges are used without standardizing, variables with large magnitudes will overshadow the effect of those with smaller magnitudes, thereby misrepresenting the real effects of multicollinearity. The standardized data is shown in Table 6.4, and it is obtained by subtracting the mean from the data and dividing by the standard deviation. The mean and the standard deviation of the data are presented in Equation 6.5 and Equation 6.6. The PCA is done on the covariance matrix of the standardized variables [16]. For the standardized variables, the covariance matrix is the same as the correlation matrix, and the PCs are extracted from this matrix. The resulting eigenvalues are

$$\{1.65, 0.95, 0.40\},$$

which account for (55 percent, 31 percent, and 13.3 percent, respectively) variance across the whole input dataset. The eigenvectors representing principal components u_i and their loadings thus obtained are presented in Table 6.9.

According to Table 6.9, the first PC (represented by the first row) strongly features both moisture content and density, which is realistic. This suggests that density and moisture content are correlated so that one represents the other, or alternatively they can be combined, along with temperature, into a PC using the loadings. The second component strongly features temperature, indicating that it correlates much less with the other variables. The total variance accounted for by the first two components is 86 percent. Whether this accuracy is sufficient depends on the problem, and if it is not sufficient, more components need to be added. In this example, it amounts to using all three components, which is the same as the number of original variables. However, the new variables are uncorrelated. The third PC adds the last 13.3 percent variance, and it again strongly features moisture content and density which account for the residual variance after the variance of the first two components are taken into account. The eigenvalue plot (scree plot) for the components shown in Figure 6.8 illustrates the variance captured by each component.

Table 6.9 Principal Components (u_i') and Their Loadings Extracted from the COV Matrix of the Standardized Variables

	T	M	Dens
u_1	0.311	−0.69	−0.65
u_2	0.93	0.093	0.345
u_3	−0.176	−0.718	0.673

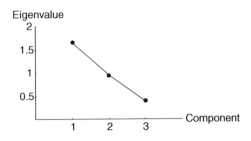

Figure 6.8   Eigenvalue plot (scree graph) from PCA analysis based on COV matrix of standardized variables.

Figure 6.9a provides an illustration of the loadings depicting the direction of the original variables with respect to the first two principal directions, which shows that the density and moisture content axes are closer to the first principal direction. The closeness of moisture content and density axes reflects the correlation between the two variables. The first component is dominated by these two. The temperature axis is close to the second principal direction and almost perpendicular to the direction of density and moisture content, indicating low correlation between temperature and the other two variables. Temperature dominates the second component.

Figure 6.9b presents the principal scores obtained by transforming all original input variables using the corresponding component loadings. Recall that these scores are computed by multiplying the loadings by the corresponding input variables (e.g., $PC_1 = u_{11}x_1 + u_{12}x_2 + u_{13}x_3$, where u_{1i} is the ith component of the first eigenvector, and x_1, x_2, and x_3 are the three original variables). It is clear that most of the data lies along the general direction indicated by moisture and density showing that they account for most of the data variation. Temperature is almost perpendicular to it, as indicated by its having been featured strongly in the second component.

In many PCs, it is not uncommon to encounter tens, hundreds, or sometimes thousands of input variables, and correlation between many of the variables is inevitable. In such situations, there are as many PCs as there are input variables, yet because of the correlations between variables, few components will capture the total variance in the data so that these will be retained and others discarded.

How many PCs must be selected mainly follows rule of thumb and is ad hoc, and the justification is that they are intuitively plausible and work. For example, the variables that are highly correlated with the output must be in the PCs chosen. A method that is commonly used to select the number of PCs is to define a threshold cumulative percentage variance that must be reached by the set of PCs. This threshold is problem dependent and can range from 75 to 90 percent.

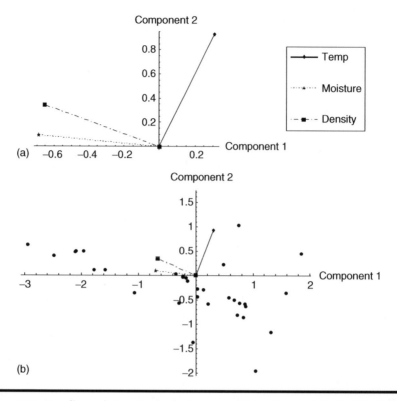

Figure 6.9 Loadings plot and principal scores. (a) Loadings plot of original input variables with respect to first two principal directions and (b) principal scores (transformed values) of original data superimposed on the loadings plot with respect to first two principal directions.

The cumulative variance of all PCs is equal to the total (combined) variance of all original variables. Thus

$$\sum_{j=1}^{p} l_j = \sum_{j=1}^{p} S_{jj} \tag{6.14}$$

where the left-hand expression is the sum of variance l of all p PCs and the right-hand expression is the sum of variances S of all p original variables. The percentage of variance accounted for by the first k PCs can be expressed as

$$t_k = 100\frac{\sum_{j=1}^{k} l_j}{\sum_{j=1}^{p} S_{jj}} = 100\frac{\sum_{j=1}^{k} l_j}{\sum_{j=1}^{p} l_j} \tag{6.15}$$

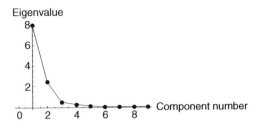

Figure 6.10 Scree graph from a PCA analysis with nine input variables. (From Warne, K., Prasad G., Rezvani S. and Maguire L., *Engineering Applications of Artificial Intelligence*, 17, 871, 2004. With permission from Elsevier.)

By choosing a threshold value for t_k, for example in the range between 70 and 95 percent, k number of PCs that contain most of the information in the original variables set can be retained. The threshold value generally decreases as the number of original variables increases or the number of observations increases [12].

Another rule for selecting an adequate number of PCs is based on the scree graph [18]. The scree graph is a plot of eigenvalues versus the PC number as shown in Figure 6.8. The rule suggests looking for the point beyond which the scree graph is more or less straight, not necessarily horizontal. The number of PCs in the example thermal conductivity problem is not adequate for illustrating this point; therefore, a new plot with more inputs and more PCs relevant to another problem is shown in Figure 6.10 for demonstration purposes.

The curve in Figure 6.10 approaches a straight line, and the first point in this straight line is the last component to be retained. For the above plot, the third, fourth, and fifth components account for 96.2, 98.1, and 99 percent, respectively, of the total variance. Figure 6.10 indicates that PCs beyond three or four components lie on a more or less straight line. Although this method is less subjective than using a threshold variance, some judgment is still needed because the first point on a straight line may not be clearly discerned from the scree graph. A way to overcome this problem and boost confidence may be to use an approach similar to that of cross validation. The number of PCs in the prediction of the output is increased successively until the overall prediction does not improve with the addition of an extra PC. The number of PCs is then the minimum number required for optimal prediction. The optimum number chosen must be the one that satisfies all the criteria discussed here; hence, these can be used together to facilitate the best choice.

6.6.3 *Partial Least-Squares Regression*

Multivariate statistical projection methods such as partial least squares (PLS) overcome the overfitting problem by performing regression on a

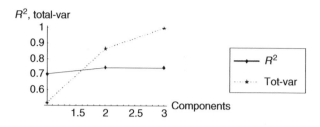

Figure 6.11 Results from partial least square regression on wood thermal conductivity data.

smaller number of orthogonal latent predictor variables that are linear combinations of the original variables. This results in reliable predictions based on well-conditioned parameter estimates [4,11,21]. These can also be useful in preliminary analysis, preprocessing of input data, or feature extraction. The orthogonal latent predictors are similar to PCs, and PLS performs analysis of variance (ANOVA) on these predictors. The results for the example problem relating to wood thermal conductivity are presented graphically in Figure 6.11, which shows that regression on two components captures 87 percent of the variance. The last component, although accounting for 13 percent of the total variance, does not improve the R^2 (Figure 6.11) value of the model.

6.7 Outlier Detection

In data preprocessing, variables are usually scaled so that important variables with small magnitudes are not overshadowed by those with larger magnitudes. Then the data is usually tested for outliers, which may be caused by a measurement error or a genuine observation that differs from the rest of the data for unknown reasons. Outliers severely distort the results if they are included in model development. Therefore, outliers need to be identified and eliminated from the training set before modeling.

Trimming and Winsorizing are two simple approaches for removing outliers from single variables. They involve sorting each variable and removing or modifying a small percentage (typically 1–5 percent for large datasets) of extreme values of the variable [22]. Only the extreme values of one variable of the input vector are modified at a time. In trimming, most extreme values are set to missing. In Winsorizing, most extreme values are given a value closer to the mean, often three standard deviations from the mean, or the last good value for the variable in the dataset [22]. Replacing all values larger than the 99th percentile or smaller than the first percentile with the values of those limits is also done. Trimming or Winsorizing can remove a large majority of trivial outliers caused by erroneous measurements, badly transcripted data, nonworking instruments, etc. However, there may be real and interesting extreme values that deviate less severely but still need to be considered as outliers and, therefore, scrutinized and kept from further modeling. However, this requires methods for outlier detection in multivariate data [22].

Methods for detecting outliers in multivariate data are less extensive compared to those for single variables [11]. PCA as discussed earlier is one approach. Although outliers must be detected early on during preprocessing, the discussion was delayed until PCA was thoroughly explained. As presented in Section 6.4.3 and Section 6.6.2, PCA involves a transformation of the original data to an orthogonal coordinate system represented by PCs where each successive component accounts for a decreasing amount of variance in the original data. The first few PCs capture the largest amount of variation, whereas the last few PCs refer to the directions associated with small variance. The outliers inflate variance and covariance in the data, so the first few PCs can highlight these if they are present in the data. Specifically, the outliers that are detectable from a plot of the first few PCs are those that inflate variances and covariances [11]. Joliffe [12] proposes the following test statistics for detecting these outliers:

$$d_{1i}^2 = \sum_{k=p-q+1}^{p} z_{ik}^2$$

$$d_{2i}^2 = \sum_{k=p-q+1}^{p} \frac{z_{ik}^2}{l_k} \qquad (6.16)$$

$$d_{2i}^2 = \sum_{k=p-q+1}^{p} l_k z_{ik}^2$$

where z_{ik} is the kth PC score for the ith observation, p is the number of variables, q represents the number of low variance PCs (for example, variance less than one from PCA based on the COV matrix of standardized variables), and l_k is the variance of the kth PC. The first two statistics detect observations that do not conform to the correlation structure of the data, and the last statistic detects those observations that inflate the variance of the data. As can be seen, the test statistics are calculated for each original observation i, and those observations whose test statistic deviates by more than three standard deviations from the mean statistic are considered outliers. For the example relating to wood thermal conductivity, the plot of the first two PCs shown in Figure 6.9b does not indicate any outliers; however, a case study in Section 6.9 shows how this feature is used to remove outliers.

6.8 Noise

In most datasets almost all variables contain noise. Thus there is a relevant part and a noise part to a variable. In modeling, the relevant parts of variables are expected to interact forming the model. Noise in data should be attenuated as much as possible. As a first step, a linear filter can be used on both independent and dependent variables to remove drift and high-frequency disturbances that adversely affect model development. Removing excessive noise from the variables therefore helps subsequent model development. Latent variable models such as PCA and PLS regression estimate the relevant part and noise of each variable in a set of variables and, therefore, are suitable for noise removal in multivariate data.

One source of noise can be due to error of measurement or experimental reproducibility. When such error comprises the majority of the noise, mildly weighted (partial) least squares, after trimming and Winsorizing, may be adequate [22]. However, measurement error is usually just a small part of noise. Additional correlated noise over observations and variables is common. Then unweighted (partial) least squares will be less risky than most other approaches [22].

Using the methods discussed in this section, a suitable subset of inputs or features can be selected with consideration given to the adequacy of the selected inputs or features, outlier removal, and attenuation of noise. The other component of model adequacy is the model itself represented by the network type and architecture. The next section presents a case study involving multivariate inputs and illustrates the use of most of the preprocessing methods discussed so far in the chapter.

6.9 Case Study: Illustrating Input Selection and Dimensionality Reduction for a Practical Problem

Estimating the quality of substrates used in medical packaging products. Some of the techniques for input selection and dimensionality reduction discussed in the previous sections will now be applied to a real-world problem. The study described here has been reported by Warne et al. [11]. It presents data conditioning, input selection, PCA-based dimensionality reduction, and neural network model development systematically.

The problem domain for this case study is the coating industry, which supplies medical packaging products. Packaging requires that the substrate such as paper or plastic be coated. A water-based adhesive is used to coat the substrate, and the wet-coated material called web is dried in an oven furnace that has three drying zones. The purpose of drying is to control the quality of the adhesion measured by an industry standard quality measure called anchorage. It is crucial for the industry that the best possible anchorage consistent with the industry standard is produced by the coating process. The substrate, guided by rollers, moves in sheet form throughout the process of coating, drying, chilling, inspecting, and rewinding onto rolls. Thirteen variables affect anchorage. Data has been generated for the full range of variables that involved perturbing the normal operating values by ±5 percent. Several hours of data consisting of 600 observations have been collected.

6.9.1 Data Preprocessing and Preliminary Modeling

Outlier detection. Using the statistics given in Equation 6.16, 27 observations whose test statistic deviated by more than three standard deviations from the mean have been deleted reducing the dataset to 573 observations.

Influential input selection. The desired model is the one that estimates anchorage, which is the quality measure of the coating. Using process knowledge, two variables have been eliminated as irrelevant. For the remaining set, scatter plots and simple and partial correlations were obtained for each of the dependent and independent variable combinations. The simple and partial correlation between the variables and anchorage is shown in Table 6.10. Some of these variables relate to the oven, some to coating, and the rest to the web itself. It shows that simple correlation coefficients are very high for some variables but partial

Table 6.10 Variables Influencing Anchorage as Identified by Correlation

Variable Number	Predictor Variable	Correlation Coefficient	Partial Correlation Coefficient
1	Zone 3 temperature	−0.935	0.670
2	Zone 2 temperature	−0.909	0.822
3	Zone 1 temperature	−0.905	0.673
4	Web temperature	−0.770	0.436
5	Rewind tension	−0.465	0.295
6	Oven tension	−0.462	0.372
7	Coat weight	−0.418	0.305
8	Unwind tension	−0.268	0.293
9	Applicator speed	−0.002	0.317
10	Coating tension	0.146	0.327
11	Web (line) speed	0.148	0.301

Source: From Warne, K., Prasad G., Rezvani S., and Maguire L., *Engineering Applications of Artificial Intelligence*, 17, 871, 2004. With permission from Elsevier.

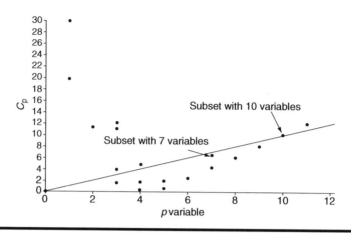

Figure 6.12 C_p **versus** p **plot. (From Warne, K., Prasad G., Rezvani S., and Maguire L., *Engineering Applications of Artificial Intelligence*, 17, 871, 2004. With permission from Elsevier.)**

correlation coefficients are lower, indicating the existence of multi-collinearity. Furthermore, partial correlation coefficients indicate that the 2nd and 3rd variables are more influential than the first one that has the largest simple correlation coefficient.

The variables in Table 6.10 were then subjected to C_p statistical analysis to test their suitability for inclusion in a smaller subset of variables that estimate anchorage. The C_p statistic was calculated for all possible combinations of variables in Table 6.10. The variables considered in each subset are the set that produced the smallest C_p statistic out of all possible combinations for that number of inputs in the subset. The plot of C_p versus p (the number of predictor variables) is shown in Figure 6.12 for some relevant subsets. The sets with C_p statistic on or close to the 45° degree line are indicators of a good model that accounts for most of the variation in anchorage.

According to Figure 6.12, there are several subsets with the C_p statistic close to the line with the ten-variable subset being the best. However, the subset with seven variables ($C_p = 6.4$) was selected because it involves fewer variables, and the extra three variables in the ten-variable subset are weakly correlated with the output. The seven variables are the first seven in Table 6.10.

Model development. A three-layer neural network was trained using backpropagation with the seven variables as input and seven input neurons in a single hidden layer. The training performance of the network is illustrated in Table 6.11 which shows the training mean square error (MSE), model accuracy (percent), validation MSE, and prediction variance at various stages

Table 6.11 Performance Measures for Neural Network

Iterations	Training Accuracy		Validation	
	MSE	Percent	MSE	Prediction Variance
50	0.0291	76.38	0.54	1.1271
100	0.0248	76.9	0.487	0.677
400	0.01818	80	0.456	0.468
700	0.0131	73.07	0.467	0.519
1000	0.0131	73.46	0.553	0.521
1500	0.0091	71.65	0.62	0.561
5000	0.0039	69.23	0.633	0.698
10 000	0.00292	65.99	0.635	1.1248

Source: From Warne, K., Prasad G., Rezvani S., and Maguire L., *Engineering Applications of Artificial Intelligence*, 17, 871, 2004. With permission from Elsevier.

of training. As can be seen, the training MSE decreases throughout the training, but overfitting sets in after 400 iterations when model accuracy is highest (80 percent) and validation MSE (0.456) and prediction variance (0.468) are lowest.

The predicted outcome of the best model is shown in Figure 6.13 along with the laboratory measured anchorage for the times the data was collected. The prediction accuracy is 80 percent with error accounting for

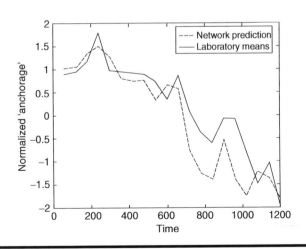

Figure 6.13 The predicted outcome of the best model using original input variables. (From Warne, K., Prasad G., Rezvani S., and Maguire L., *Engineering Applications of Artificial Intelligence*, 17, 871, 2004. With permission from Elsevier.)

Table 6.12 Result of PCA

| PC No. | Percent Variance Captured by PCA Model | | |
	Eigenvalue	Variance (percent)	Total Variance (percent)
1	8.297	75.431	75.431
2	1.935	17.591	93.022
3	0.349	3.181	96.203
4	0.216	1.963	98.167
5	0.093	0.843	99.010
6	0.057	0.523	99.533
7	0.027	0.253	99.786
8	0.011	0.096	99.883
9	0.007	0.063	99.946
10	0.005	0.049	99.996
11	0.0004	0.004	100

Source: From Warne, K., Prasad G., Rezvani S., and Maguire L., *Engineering Applications of Artificial Intelligence*, 17, 871, 2004. With permission from Elsevier.

the high discrepancy between the prediction and experimental values for some observations.

Multicollinearity and dimensionality reduction. In the previous analysis, the most influential variables are selected and used in the model. However, multicollinearity in data is not explicitly addressed. Serious multicollinearity effects can lead to suboptimal models. To test this, PCA on the set of input variables has been performed. The results from the PCA are presented in Table 6.12. It shows 11 PCs and their corresponding eigenvalues (variance), percent total variance accounted for by each PC, and the total variance accumulated by the PCs up to that PC in the list. It shows that five PCs take into account 99.01 percent of the total variance.

The scree graph of variance versus component number is shown in Figure 6.14. As discussed previously, how many PCs must be selected is mainly determined by rule of thumb and ad hoc, and the justification is that they are intuitively plausible and work. For example, the variables that are highly correlated with the output must be in the PCs selected. A method commonly used to select the number of PCs is to define a threshold cumulative percentage variance that must be reached by the set of PCs. This threshold is problem dependent and can range from 75 to 95 percent. Another approach is to check the scree graph and select the point where the graph becomes more or less horizontal. A third approach to determine the number of PCs required is cross validation where the number of PCs that produces the least validation error is found through testing.

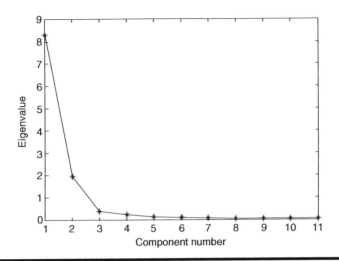

Figure 6.14 **The scree graph of variance versus component number. (From Warne, K., Prasad G., Rezvani S., and Maguire L.,** *Engineering Applications of Artificial Intelligence,* **17, 871, 2004. With permission from Elsevier.)**

In this case study, all three methods have been used. The first two rules were used to select a subset of PCs, and the selected subset was then subjected to cross validation to select the number of PCs to produce optimal results. Catell [18] suggests 70 percent as the threshold variance, and if it is adopted for this case, only one component is retained. Others have suggested cut-off points ranging from 70 to 95 percent [11], which significantly affects the number of PCs retained. The scree graph in Figure 6.14 indicates that beyond three or four PCs, the graph is more or less a straight line. Table 6.12 points out that beyond four PCs, the difference in variation between the two successive components is fairly constant. Both three and four PCs were retained, and cross validation was conducted to determine the best number to provide the least validation error.

6.9.2 PCA-Based Neural Network Modeling

After the desired number of PCs is retained, neural networks can be trained with these PCs as input instead of the original variables. This architecture is illustrated in Figure 6.15.

The smaller number of PCs results in a less complex network, and the orthogonal transformed variables (PCs) that are linear combinations of original variables overcome the problem of overfitting commonly encountered in neural networks. The network has three or four inputs

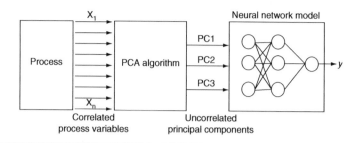

Figure 6.15   The PCA-based neural network architecture. (From Warne, K., Prasad G., Rezvani S., and Maguire L., *Engineering Applications of Artificial Intelligence,* **17, 871, 2004. With permission from Elsevier.)**

corresponding to the number of PCs, two hidden neurons, and one neuron in the output layer; and this structure remains fixed to compare different input selection techniques discussed previously. Training was done using backpropagation. The results for the network with three inputs are shown in Figure 6.16a along with the laboratory-measured mean value for anchorage. The comparison of network performance with three and four PCs is illustrated in Figure 6.16b. The mean squared error for training and validation data using the PCs is shown in Table 6.13.

The advantage of using the PCs in a network is that they remove the effect of multicollinearity that could lead to ill conditioning (numerical instability in generating a solution) and high prediction variance. Now that

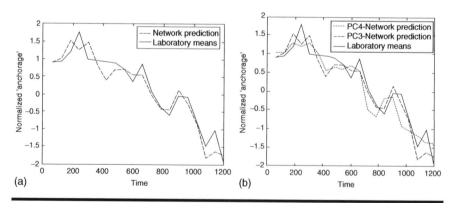

Figure 6.16   Comparison of network performance with laboratory-measured anchorage: (a) performance with three PCs, (b) performance with three and four PCs superimposed on laboratory measurements. (From Warne, K., Prasad G., Rezvani S., and Maguire L., *Engineering Applications of Artificial Intelligence,* **17, 871, 2004. With permission from Elsevier.)**

Table 6.13 Performance Measures for PCA-Based Neural Networks Using Different Number of PCs as Inputs

| | PC-3 | | | | PC-4 | | | |
| | | | Validation | | | | Validation | |
Iterations	Training MSE	MSE	Accuracy (percent)	Prediction Variance	Training MSE	MSE	Accuracy (percent)	Prediction Variance
50	0.073441	0.538	73.95	0.604	0.089401	0.145	85.50	0.401
100	0.06225	0.4805	77.80	0.467	0.088209	0.144	85.98	0.372
400	0.0625	0.435	78.95	0.458	0.06969	0.146	85.27	0.378
700	0.05499	0.356	81.99	0.429	0.058081	0.144	85.34	0.387
1000	0.0522	0.33	82.10	0.448	0.063504	0.1305	86.10	0.359
1500	0.052	0.19	84.20	0.38	0.045796	0.14	85.44	0.388
5000	0.05095	0.126	87.80	0.362	0.044524	0.142	85.40	0.401
10 000	0.0384	0.111	88.90	0.344	0.042849	0.156	83.34	0.404

Source: From Warne, K., Prasad G., Rezvani S., and Maguire L., *Engineering Applications of Artificial Intelligence*, 17, 871, 2004. With permission from Elsevier.

network training with an input set with collinear variables (original best predictors) as well as a set of orthogonal variables (PCs) has been carried out in this case study, the effect of multicollinearity can be studied. Table 6.11 and Table 6.13 show remarkable differences for the non-PCA- and PCA-based methods. For instance, Table 6.11 clearly shows that for the non-PCA approach training error continues to decrease, but the validation error decreases initially and then starts increasing after 400 iterations (epochs). This indicates that the network starts overfitting at this point, as shown in Figure 6.17 which compares the training and validation for non-PCA- and PCA-based approaches.

The PCA-based approach, in contrast, almost eliminates overfitting. It can be seen that both the training and validation errors decrease and level off to a constant value which is particularly evident for the network with three PCs. Another important observation relates to prediction variance. The PCA-based networks show an improvement in prediction variance, indicating that collinearity exists within the data and must be addressed in the development of robust inferential models. The results show that PCA is suitable for dealing with this issue.

With regard to how many PCs are optimal, Table 6.13 indicates clearly that three PCs as inputs are the best because for this case, both MSE and prediction variance are the smallest. This illustrates that the choice of the number of PCs is important, and although more PCs may capture more of the variance in the original data, this may not always result in an improved model.

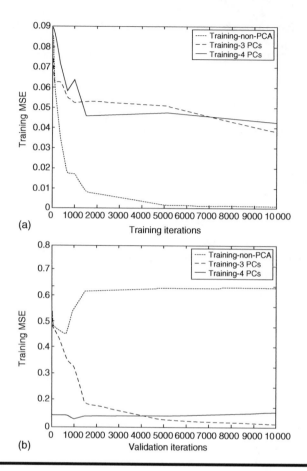

(a)

(b)

Figure 6.17 (a) Training and (b) validation errors for non-PCA- and PCA-based modeling approaches. (From Warne, K., Prasad G., Rezvani S., and Maguire L., *Engineering Applications of Artificial Intelligence,* **17, 871, 2004. With permission from Elsevier.)**

6.9.3 Effect of Hidden Neurons for Non-PCA- and PCA-Based Approaches

So far in this case study, the number of neurons has been fixed to compare the effect of different approaches to input selection and feature extraction. The effect of the number of hidden neurons and layers has also been investigated, and the prediction (validation) error for networks with 2, 3, 4, 5, 6, 7, and 8 neurons for the non-PCA- and PCA-based approaches are shown in Figure 6.18.

Figure 6.18 highlights some interesting observations. For the non-PCA-based approach, prediction error decreases as the number of neurons

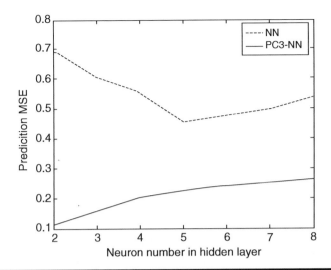

Figure 6.18   Effect of number of hidden neurons on the prediction error for non-PCA- and PCA-based approaches. (From Warne, K., Prasad G., Rezvani S., and Maguire L., *Engineering Applications of Artificial Intelligence,* **17, 871, 2004. With permission from Elsevier.)**

increases from two to five and increases thereafter. This means that five neurons are optimal; beyond five neurons gives the network too much flexibility, and further increasing the number of hidden neurons makes them mimic and memorize the data, thereby eroding prediction accuracy further. However with PCs as inputs, two hidden neurons are optimal, as opposed to five in a non-PCA case. Furthermore, the prediction MSE increases only slightly and plateaus quickly with further increase in the number of hidden neurons indicating the resilience of the PCA-based networks. It was also found that increasing the number of hidden layers increases the MSE for both approaches.

6.9.4   Case Study Summary

This case study presents a systematic framework for the development of inferential models from correlated data. The problem involves developing a robust model to assess a quality measure (anchorage) of adhesive coated substrates used for medical packaging from 11 influential variables. The results show clearly that neural network performance can be significantly improved by incorporating a PCA initialization model to the inferential model. In this particular case, PCA reduced the dimensionality from eleven to three, which accounted for 96 percent of the variation in the original process variables and produced the optimum network performance. Moreover, the PCA addressed the issue of multicollinearity within data and eliminated overfitting. The optimum number of hidden neurons for the PCA-based network is fewer than that required for a network with original variables, and further increases beyond the optimum number of neurons result in only a small effect on prediction error. Because the number of inputs is significantly reduced, the network architecture is less complex and training time is reduced.

6.10   Summary

This chapter addresses some important aspects of data preprocessing and illustrates the concepts using examples. The positive effect of preprocessing on structural complexity and model accuracy is also demonstrated using a case study. Histograms, scatter plots, correlation plots, parallel visualizations, and projections of multidimensional data onto two dimensions are presented as an aid in understanding the character, trends, and relationships in data. Correlation and covariance are used to measure strength of relationships and dispersions in data. Data normalization is helpful in putting the values of all variables in a similar range so that the influence of those with smaller values is not masked by that of the variables with higher values. This chapter presents several approaches to normalization: standardization, range scaling, and whitening. The latter allows normalization of correlated multivariate data.

A great deal of attention is paid to input selection and dimensionality reduction for improving model development and accuracy. The use of statistical tools such as partial correlation, multiple regression, and best subsets regression for input selection are explained, and PCA and partial least squares regression are presented as suitable methods for dimensionality reduction and feature extraction. Outlier detection and noise removal in multivariate data are also addressed in this chapter. Various approaches to data preprocessing presented in the chapter are further demonstrated using an example case study involving a real application. The case study illustrates that dimensionality reduction leads to less complex models with higher generalization ability than those using original correlated input data. The validity of this approach for a more complex problem [20] is illustrated in Chapter 7.

This chapter deals with linear approaches to input selection and feature extraction (except for self-organization maps). These have been widely used in data preprocessing. However, in many biological and natural systems, correlations can be highly nonlinear and approaches that capture these nonlinear trends are advantageous in input selection in feature extraction. Some of the emerging techniques for nonlinear data preprocessing—Partial Mutual Information, Self-Organizing Maps, Generalized Regression Neural Networks and Genetic Algorithms—are presented in detail in Chapter 9, Section 9.10.1.

Problems

1. What is correlation, and what does it measure?
2. What information does covariance convey about multivariate data?
3. For a multivariate dataset of choice, obtain histograms and scatter plots and learn as much as possible about the statistical nature (mean, standard deviation, etc.) and trends in data.
4. For the data in Problem 3 above, determine correlation coefficients and covariances, and learn as much as possible about the variables that are influential in predicting the output and those that are strongly correlated.
5. What is the purpose of normalization of data?
6. Ascertain if any normalization is useful for the data in Problem 3 and use all or suitable normalization methods discussed in Section 6.4 on the data. Would one method be more advantageous than the others? Try any other methods that you are familiar with.
7. Apply the input selection methods presented in Section 6.5 (partial correlation, multiple regression, best subsets regression) or any other method that you know, and select the inputs that influence the output most.
8. Train a multilayer network with original inputs and the best subset of inputs, and scrutinize the difference. Is there any advantage in input selection?
9. What is dimensionality reduction, and how is it useful in modeling?
10. Explain purpose of principal component analysis and the basic idea behind PCA.
11. For the data in Problem 3, perform PCA, and select the appropriate number of PCs.
12. Train a new model with the selected PCs, and compare the results with that based on the best subset of influential variables.

References

1. Avramidis, S. and Iliadis, L. Predicting wood thermal conductivity using artificial neural networks, *Wood and Fiber Science*, 37(4), 682, 2005.
2. *Machine Learning Framework for Mathematica: Multi-method System for Creating Understandable Computational Models from Data*, uni software plus, Linz, Austria, 2002.
3. Kohonen, T. *Self-Organizing Maps*, 3rd Ed., Springer-Verlag, Berlin, 2001.
4. Hair, J.F. et al. *Multivariate Data Analysis*, 5th Ed., Prentice Hall, Upper Saddle River, NJ, 1998.
5. Bishop, C. *Neural Networks for Pattern Recognition*, Clarendon Press, Oxford, 1996.
6. *Minitab 14*, Minitab Inc., State College, PA, USA, www.minitab.com, 2004.

7. *Mathematica*, Wolfram Research Inc., Champaign, IL, 2002.

8. Haykin, S. *Neural Networks: A Comprehensive Foundation*, 2nd Ed., Prentice Hall, Upper Saddle River, NJ, 1999.

9. Le Cun Y., Denker J.S., and Solla, S.A. Optimal brain damage, *Advances in Neural Information Processsing*, Vol. 2, D.S. Touretzky, ed., Morgan Kaufmann Publishers, Los Altos, CA, pp. 598–605, 1990.

10. Engelbrecht, A.P. A new pruning heuristic based on variance analysis of sensitivity information, *IEEE Transactions on Neural Networks*, IEEE, Los Alamitos, CA, 12, 1386, 2001.

11. Warne, K., Prasad, G., Rezvani, S., and Maguire, L. Statistical and computational intelligence techniques for inferential model development: A comparative evaluation and a novel proposition for fusion, *Engineering Applications of Artificial Intelligence*, 17, 871, 2004. (Figures reprinted with permission from Elsevier.)

12. Jolliffe, I.T. *Principal Component Analysis*, Springer, New York, 1986.

13. Walls, R. A note on the variance of a predicted response in regression, *American Statistician*, 23, 24, 1969.

14. Steel, R.G. *Principles and Procedures of Statistics: A Biometric Approach*, 2nd Ed., McGraw-Hill, New York, 1980.

15. Mallow, C. Some comments on C_p, *Technometrics*, 15, 661, 1973.

16. Smith, M. *Neural Networks for Statistical Modeling*, International Thompson Computer Press, London, UK; Boston, MA, 1996.

17. Qin, S.J. et al. Self-validating inferential sensors with application to air emission monitoring, *Industrial Engineering and Chemical Research*, 35, 1675, 1997.

18. Catell, R.B. *The Scientific Use of Factor Analysis in Behavioural and Life Sciences*, Plenum Press, New York, 1966.

19. Aires, F. Neural network uncertainty assessment using Bayesian statistics with application to remote sensing: 3. Network Jacobians, *Journal of Geophysical Research*, 109, D10305, 2004.

20. Hassibi, B., Stork, D.G., and Wolff, G.J. Optimal brain surgeon and general network pruning, *IEEE International Conference on Neural Networks*, Vol. 1, IEEE, Los Alamitos, CA, p. 293, 1992.

21. Johnson, R. and Wichern, D. *Applied Multivariate Statistical Methods*, 3rd Ed., Prentice Hall, Upper Saddle River, NJ, 1992.

22. Kettaneh, N., Berglund, A., and Wold, S. PCA and PLS with very large data sets, *Computational Statistics and Data Analysis*, 48, 69, 2005.

23. Chandraratne, M.R., Samarasinghe, S., Kulasiri, D., Frampton, C., and Bickerstaffe, R. Determination of lamb grades using texture analysis and neural networks. *3rd IASTED International Conference on Visualization, Imaging and Image Processing* (VIIP 2003), M.H. Hamza (ed.), Acta Press, Calgary, Canada, p. 396, 2003.

Chapter 7

Assessment of Uncertainty of Neural Network Models Using Bayesian Statistics

7.1 Introduction and Overview

A rigorous statistical approach requires not only a good estimation of the model outputs but also an uncertainty estimate of the model parameters (i.e., weights). This assessment must include individual uncertainties and the correlation structure of these uncertainties. This is the common approach to investigating the reliability of a model. In addition, uncertainty of model predictions (i.e., output errors) must be evaluated for assessing the reliability of predictions, especially if they are to be used in decision making or as input to other models, which is common. Moreover, rigorous estimates of the sensitivities, such as sensitivity of output(s) to inputs can provide useful information about relevant inputs and an estimate of their contribution to the model predictions.

Although neural networks have been used to model complex phenomena, until now tools for assessing the uncertainty of neural network statistical models have been limited. In this chapter, uncertainty estimate tools for neural networks in real-world applications based on Bayesian statistics are discussed. The uncertainty assessment tools provided here can be used for a variety of probabilistic quantities related to the overall uncertainty of neural network models, i.e., weights, output errors, sensitivities for "automatic relevance" detection in selecting more relevant inputs, and "novelty detection" to monitor outliers.

This chapter presents a detailed treatment of uncertainty assessment of neural network models. These methods for neural networks are in their infancy. In Section 7.2.1, the standard training criterion involving square error and weight regularization used in network training (see Chapter 5) is shown to be a special case of a more realistic case where variance of noise and initial weights are used to regularize the training criterion. That standard error minimization is equivalent to minimizing negative log likelihood in statistical parameter estimation is also demonstrated. Using these concepts, the optimum weights obtained from regular network training are put in a Bayesian context to obtain a posteriori probability distribution functions (PDFs) for the final weights of a network in Section 7.2.2 to Section 7.2.4. In this formulation, the optimum weights obtained from network training are the most probable or maximum a posteriori weights and the weight distribution represents the uncertainty in weights.

In Section 7.2.5, the derivation of the PDF of weights using a case study involving estimation of geophysical parameters that drive Earth's atmospheric radiative transfer phenomena from satellite data is illustrated. A sample of weights extracted from the PDF is presented to illustrate their behavior and correlation structure. The weight distribution is used to generate uncertainty estimates for output error that is divided into model error and intrinsic noise. These two aspects are treated in detail in Section 7.3.1 and illustrated using the case study.

The last section of the chapter (Section 7.4) involves uncertainty estimation of network input–output sensitivities for the purpose of selecting inputs that significantly influence the outputs. Approaches to determine the influence of inputs to outputs of a feedforward network are presented in Section 7.4.1 with examples in Section 7.4.2. A theoretical treatment of uncertainty of sensitivities is given in Section 7.4.3 and illustrated through an example case study in Section 7.4.4. Specifically, a detailed treatment of the effects of multicollinearity in inputs and outputs on network training and outcomes is presented in Section 7.4.4.1 and it is shown that the principal component (PC) decomposition of inputs and outputs produces networks that have more stable sensitivities than those using regular inputs and outputs that are correlated (Section 7.4.4.2 to Section 7.4.4.5). The sensitivities obtained from principal component analysis (PCA)-based networks also have smaller standard deviations indicating less uncertainty in their values. The PCA approach produces a much clearer view of how inputs are linked to outputs and therefore relevant inputs can be identified with more confidence. A summary of the chapter is presented in Section 7.5.

7.2 Estimating Weight Uncertainty Using Bayesian Statistics

In neural network training, some quality criterion, such as mean square error or an additional regularization term, is minimized. This finds a single set of values for the network weights. The Bayesian approach considers a PDF in weight space representing the relative degree of belief in different values for the weight vector [1]. This weight PDF is initially set to some prior distribution $P(W)$, as shown in Figure 7.1. When no prior information is available, it is set to a uniform distribution. After the data (D) has been observed, it can be converted to a posterior distribution ($P(W|D)$) using Bayes' theorem. The optimum network weights found from minimizing the quality criterion are the most probable weights represented by W_{MP} in Figure 7.1. The posterior distribution of weights can then be used to evaluate uncertainties of predictions of the trained network for new values of inputs. The weight distribution can also be used to assess other uncertainties such as network sensitivities.

To demonstrate the relevance of Bayesian statistics, the usual approach to training neural networks is revisited and then Bayesian concepts are introduced appropriately.

7.2.1 Quality Criterion

As discussed in Chapter 4, the square error is the most commonly used error criterion. A regularization term (i.e., sum of square weights) has been added to the square error term in the "weight decay" scheme so that the weights are kept small; large weights are often the cause of learning instability and they lead to poor generalization. The combined error criterion is repeated in a

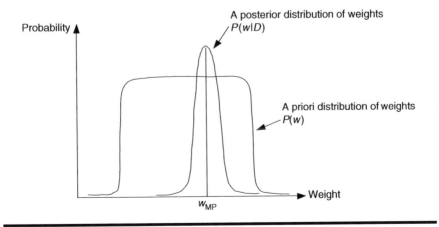

Figure 7.1 Initial and posterior weight distributions.

compact form in Equation 7.1:

$$E(w) = \beta E_D(w) + \alpha E_r(w), \tag{7.1}$$

where $E_D(w)$ is the square error term and $E_r(w)$ is the weight decay term expressed as

$$E_D(w) = \frac{1}{2}\sum_{n=1}^{N}\sum_{k=1}^{K}(t_k^n - y_k^n)^2$$

$$E_r(w) = \frac{1}{2}\sum_{i=1}^{W}w_i^2, \tag{7.2}$$

where t_k^n is the kth desired output component and y_k^n is the kth component of the neural network output vector for the nth input–output pattern. N is the total number of observations and K is the number of output components (desired outputs). The term β in Equation 7.1 is the relative weight given to the square error term and α is that of the weight decay. However, the real meaning of β is that it represents the inverse of the observation noise variance for all outputs and α is linked to the a priori general variance of the weights.

The expression for $E(w)$ in Equation 7.2 is a simplification of a more realistic expression. For example, consider the case of a single output. The error ε_y measured as the difference between the target t and predicted y is often supposed to follow a Gaussian distribution with a mean of zero and variance σ^2. The ideal variance for the error distribution is called "intrinsic noise" or the "natural variability" of the output variable y. If σ^2 is known, its inverse a (i.e., $a = 1/\sigma^2$) can be incorporated into the square error quality criterion $E_D(w)$ as a normalization term:

$$E_D(w) = \frac{1}{2}\sum_{n=1}^{N}\frac{1}{\sigma^2}(\varepsilon_y^n)^2 = \frac{1}{2}\sum_{n=1}^{N}a(\varepsilon_y^n)^2, \tag{7.3}$$

where ε_y^n is the error for the nth input pattern, i.e., $\varepsilon_y^n = (t^n - y^n)$. When no information is available on σ^2 prior to training, it will be dropped and this error criterion becomes the first expression in Equation 7.2, which is simply the sum of square error for the difference between target and predicted outputs. When there is more than one output, error distribution is multivariate with a separate error variance for each output. In this case, the error covariance among pairs of outputs in the error term must be considered and this is efficiently accomplished by using the covariance matrix. The ideal covariance matrix $\mathbf{C}_{in}$ for this multivariate error distribution is the intrinsic noise or the natural variability of the variable vector y. If $\mathbf{C}_{in}$ is known, its inverse $\mathbf{C}_{in}^{-1} = \mathbf{A}_{in}$ can be incorporated into the square error quality criterion $E_D(w)$ as

$$E_D(w) = \frac{1}{2} \sum_{n=1}^{N} (\varepsilon_y^n)^T \cdot A_{in} \cdot \varepsilon_y^n \tag{7.4}$$

where A_{in} is a matrix containing the inverse of the noise variance for each output. The ε_y^n is the error vector for the nth input–output pair and $(\varepsilon_y^n)^T$ is the transpose of the error vector organized in a row for easier manipulation of the error vectors to obtain the square error. As explained for the one-output case, when no information is available on C_{in} prior to training, it will be dropped and the error criterion becomes the first expression in Equation 7.2.

As far as the regularization using weight decay is concerned, a similar expression can be developed. Weight decay implies that one choose a Gaussian a priori weight distribution for weight uncertainty [2]. This means that each weight has an a priori Gaussian weight distribution. Consider the case of a single weight w. If the a priori variance σ_r^2 of the weight is known, its inverse a_r (i.e., $a_r = 1/\sigma_r^2$) can be introduced as a normalization term into the weight decay term $E_r(w)$ of the error criterion in Equation 7.2 in a similar manner to that used for incorporating intrinsic variance into the square error criterion:

$$E_r(w) = \frac{1}{2} \frac{w^2}{\sigma_r^2} = \frac{1}{2} a_r w^2 \tag{7.5}$$

If the prior weight distribution $P(W)$ is unknown, a uniform distribution is assumed, as illustrated in Figure 7.1 for the case of a single weight. Then, a_r is dropped from Equation 7.5 and it is reduced to the commonly used weight decay term with sum of square weights in Equation 7.2.

Many weights involve a multivariate weight distribution. If the covariance matrix of the a priori distribution of network weights is C_r, then its inverse C_r^{-1} (i.e., $C_r^{-1} = A_r$) can be used in the weight decay term as

$$E_r(w) = \frac{1}{2} w^T \cdot A_r \cdot w, \tag{7.6}$$

where w is the weight and T denotes transpose. The expression on the right hand basically repeats the modified decay term shown in Equation 7.5 efficiently for each of the weights in the network using the covariance for all the weights. The covariance matrix C_r of a priori distribution of weights has a different variance for each weight w_i and describes a structure of correlation between them. If the prior weight distribution $p(W)$ is unknown, a uniform distribution is assumed for each weight as illustrated in Figure 7.1 for the case of a single weight. Then, A_r is dropped from Equation 7.6 and it is reduced to the commonly used weight decay term with sum of square weights in Equation 7.2. Figure 7.1 and Equation 7.5 illustrate for a single weight that learning in a general sense involves transforming the initial weight distribution to a posterior distribution of the final weight, where the W_{MP} is the most probable (MP) weight that is determined from the usual network training methods. Equation 7.6 extends this concept for the multiple output case.

The parameters $\mathbf{A}_{in}$ and $\mathbf{A}_r$, representing the inverse of output error (observation noise) variance and inverse of the a priori variance of weights, respectively, are called "hyperparameters." As can be seen, the α and β terms are simplified forms of the inverse of the relevant covariance matrices, i.e., α is a weight for the regularization term and is linked to the a priori general variance of the weights and β represents the inverse of the observation noise variance for all outputs. This is obviously a poorer and less general formulation than the more realistic matrix formulation; however, it is difficult to estimate hyperparameters $\mathbf{A}_{in}$ and $\mathbf{A}_r$ and a method to estimate these very important parameters is described later in this chapter.

7.2.2 Incorporating Bayesian Statistics to Estimate Weight Uncertainty

The classical neural network theory can now be linked with Bayesian statistics. In statistical parameter estimation, two approaches are commonly used: maximum likelihood and Bayesian parameter estimation methods. In fact, it can be shown that the mean square error criterion used thus far in network training is motivated from the principle of maximum likelihood. The goal of network training is to model the underlying generator of data. The most complete description of the generator of data is the expression of the joint probability $p(x,t)$ in the joint target–input space. The joint distribution using Bayes' rule can be decomposed as [1,2]

$$p(x, t) = p(t|x)p(x), \tag{7.7}$$

where $p(t|x)$ is the probability density of t given that x takes a particular value and $p(x)$ is simply the probability density of x. The goal is to estimate $p(x,t)$ from data. Assuming that each data point is drawn independently and probabilities can therefore be multiplied, the likelihood for a training pattern $\{x^n, t^n\}$ can be expressed as

$$L = \prod_{n=1}^{N} p(t^n|x^n)p(x^n), \tag{7.8}$$

where $\prod$ denotes multiplication over the N training patterns. In maximum likelihood, it is attempted to maximize L. However, instead of maximizing the likelihood, it is more convenient to minimize the negative logarithm of the likelihood

$$E = -\ln L = -\sum_n \ln p(t^n|x^n) - \sum_n \ln p(x^n), \qquad (7.9)$$

where E is called the error function. The second term in Equation 7.9 does not depend on the network weights and it therefore can be dropped so that the error function becomes

$$E = -\sum_n \ln p(t^n|x^n). \qquad (7.10)$$

In the error function, an error term $(-\ln p(t^n|x^n))$ for each pattern n is summed over all patterns. The goal is to minimize E and this constitutes the likelihood approach for parameter estimation. The t is a continuous variable in the case of prediction and a class label in the case of classification. That this is equivalent to sum of square error criterion will now be demonstrated.

7.2.2.1 Square Error

In parameter estimation, it is assumed that the target variable t can be expressed by a deterministic function $f(x)$ with added Gaussian noise ε. The deterministic function can be obtained by any means and here it is treated as a neural network regression model. For a single output case this can be expressed as

$$t = f(x) + \varepsilon \qquad (7.11)$$

and illustrated in Figure 7.2a. As shown in the figure, the error, which is the difference between the predicted outcome from $f(x)$ and target t, is assumed to follow a normal distribution with zero mean and standard deviation σ that does not depend on x (i.e., uniform variance or homoscedasticity).

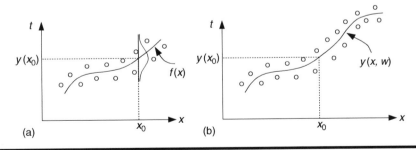

Figure 7.2 Target function and network approximations: (a) target function $y=f(x)$ to be approximated from data and (b) neural network approximation $y(x,w)$ of target function.

Thus the error distribution can be expressed as a normal distribution with zero mean and standard deviation σ in the usual way:

$$p(\varepsilon) = \frac{1}{\sqrt{2\pi\sigma^2}} e^{-\frac{(\varepsilon)^2}{2\sigma^2}} \tag{7.12}$$

Assume that the unknown function $f(x)$ is approximated with a neural network model $y(x,w)$ where x is input and w is weight. A trained network function must go through the mean of the target for any values of x, as shown in Figure 7.2b. For example, the network output for a given value of x_0 is the mean of the target values for x_0. Due to noise, there can be many target values for a given value of x.

The distribution of the target values for a given value of x is the conditional distribution of target values for that x depicted by $p(t|x)$. This distribution is essentially the noise distribution around the mean of the target values for a given value of x, as illustrated in Figure 7.3 for a particular value of x_0. For the case of random noise and optimum model, this distribution is Gaussian with standard deviation equal to noise standard deviation σ. The mean is the mean of the targets approximated by the network output. In regression it is assumed that variance of this distribution is constant across all value of x when the dataset becomes large.

The mean of the target for any value of x can be expressed as

$$f(x) = y(x, w) = t - \varepsilon. \tag{7.13}$$

Thus, the conditional distribution of targets $p(t|x)$ can be expressed as a normal distribution with the mean as given in Equation 7.13 and standard deviation σ in the usual way:

$$p(t|x) = \frac{1}{\sqrt{2\pi\sigma^2}} e^{-\frac{(t-y)^2}{2\sigma^2}} = \frac{1}{\sqrt{2\pi\sigma^2}} e^{-\frac{(\varepsilon)^2}{2\sigma^2}} \tag{7.14}$$

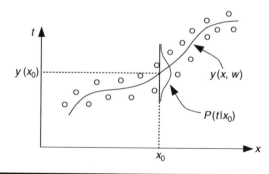

Figure 7.3 Conditional probability distribution of target data for a given value of x.

where $y(x,w)$ is expressed as y for clarity. Thus, with the optimal regression model, the conditional distribution of targets for a given value of x is the noise distribution. Now, taking the logarithm of Equation 7.14:

$$\ln p(t|x) = -\frac{(t-y)^2}{2\sigma^2} - \frac{1}{2}\ln 2\pi - \ln \sigma. \tag{7.15}$$

This expression can be substituted into the negative log likelihood expression in Equation 7.10 to obtain

$$E = \frac{1}{2\sigma^2} \sum_{n=1}^{N} (t^n - y^n)^2 + N \ln \sigma + \frac{N}{2} \ln 2\pi. \tag{7.16}$$

The second and third terms in Equation 7.16 do not depend on weights and can therefore be dropped in error minimization with respect to weights. Thus, the negative log likelihood function becomes

$$E = \frac{1}{2\sigma^2} \sum_{n=1}^{N} (t^n - y^n)^2. \tag{7.17}$$

Equation 7.17 can be further simplified by omitting $1/\sigma^2$, which is constant. The standard square error minimization equation

$$E = \frac{1}{2} \sum_{n=1}^{N} (t^n - y^n)^2 \tag{7.18}$$

is then obtained. Therefore, the mean square error criterion is a simplified form of negative log likelihood. Therefore, minimizing square error is equivalent to minimizing negative log likelihood (or maximizing likelihood).

Having found w^*, one can also find the optimum value for the error variance σ^2 by minimizing E in Equation 7.16. This can be performed analytically to obtain

$$\frac{dE}{d\sigma} = -\frac{1}{\sigma^3} \sum_{n=1}^{N} \{t^n - y(x, w^*)^n\}^2 + N\frac{1}{\sigma} = 0,$$

$$\sigma^2 = \frac{1}{N} \sum_{n=1}^{N} \{y(x, w^*)^n - t^n\}^2, \tag{7.19}$$

which states that the optimum value for σ^2 is the residual value of the mean square error function at its minimum.

After a network is trained, weights are fixed at the optimum values obtained from the learning process. To assess uncertainty of weights, a weight distribution for the trained network must be obtained. This distribution is called an a posteriori weight distribution and is illustrated in Figure 7.1 as $p(w|D)$ for the case of a single weight, where w_{MP} is the optimum weight obtained from error minimization. Bayesian statistics provides a framework for generating these distributions. To do this, the intrinsic uncertainty of targets will first be examined and the concepts derived there will be used to generate a posterior PDF for weights.

7.2.3 Intrinsic Uncertainty of Targets for Multivariate Output

In model prediction there is usually a discrepancy between the target t and prediction y. As presented in Section 7.2.2, the conditional probability $P(t|x,w)$ represents the variability of target t for a given input x and weight w. This variability results from a variety of sources, including the error in the model linking x to t (i.e., error in weights) or the observational noise on x and t. If the trained neural network fits the data well, the intrinsic variability of the target is evaluated by comparing the target output t for each input x with the predicted output y. The expression for the case of a single output was given in Equation 7.14, where it is represented by a conditional PDF with a mean of zero and variance of σ^2. For the multivariate output case, this distribution can generally be approximated by a Gaussian distribution with zero mean and covariance $\mathbf{C}_{in}$ that measures the covariance between errors for different output components. Thus, the variability of targets or the output conditional probability for a given x in the case of multiple outputs can be expressed as a multivariate normal distribution (analogous to Equation 7.14) using the multivariate error $E_D(w)$ represented in Equation 7.4:

$$P(t|x, w) = \frac{1}{Z} e^{-\frac{1}{2}\varepsilon_y^T \cdot \mathbf{A}_{in} \cdot \varepsilon_y} \tag{7.20}$$

where ε_y is the error vector for the prediction y, ε_y^T is the transpose of ε_y. $\mathbf{A}_{in}$ is the inverse of covariance matrix $\mathbf{C}_{in}$ of the intrinsic error for different output components, as given in Equation 7.4. The Z is used to denote the constant normalization factor in the expression for a normal distribution which now becomes $1/(2\pi)^{w/2}|\mathbf{C}_{in}|^{1/2}$ where $\mathbf{C}_{in}$ is the covariance matrix. The likelihood of the model, i.e., the likelihood of the model adequately representing the data, can be expressed by evaluating the output conditional probability (Equation 7.20) over the entire training database that includes target output vector t^n for each input pattern x^n. If the whole set of t^n is denoted by D, then the likelihood of the model $P(D|x,w)$ is the multiplication of the conditional probabilities for each of the target patterns t^n, assuming that they are independent:

$$p(D|x, w) = \prod_{n=1}^{N} p(t^n|x^n, w) \tag{7.21}$$

where Π denotes multiplication. This provides the probability distribution for the output variability for the entire dataset. Substituting Equation 7.20 into Equation 7.21 and simplifying:

$$p(D|x, w) = \prod_{n=1}^{N} \frac{1}{Z} e^{-\frac{1}{2} \varepsilon_y^{(n)^T} \cdot A_{in} \cdot \varepsilon_y^{(n)}} = \frac{1}{Z^N} e^{-\frac{1}{2} \sum_{n=1}^{N} \varepsilon_y^{(n)^T} \cdot A_{in} \cdot \varepsilon_y^{(n)}}, \qquad (7.22)$$

and substituting the expression in Equation 7.4:

$$P(D|x, w) = \frac{1}{Z^N} e^{-E_D(w)}. \qquad (7.23)$$

This simply states that the smaller the square error criterion E_D, the more likely that the output data set D is generated from the model (i.e., closer all predictions y are to target t). The likelihood of the model depends on inputs x, but because the weights are of interest and the distribution of x is not, x is dropped from the subsequent expression. Focus will now be on $p(D|w)$, which is the likelihood of the model given the weights. With this information, the probability distribution of weights can now be derived.

7.2.4 Probability Density Function of Weights

In neural network training, a point estimate of the model parameters w is searched. This is the optimum weight vector. In the Bayesian context, uncertainty of the weights is described by a probability distribution. This is called the a posteriori distribution of weights, or conditional probability of weights given data $p(w|D)$, as shown in Figure 7.1, and expressed using Bayes' rule as

$$p(w|D) = \frac{p(D|w)p(w)}{p(D)} \qquad (7.24)$$

where $p(D|w)$ is the likelihood of the model already derived in Equation 7.23 and $p(w)$ is the a priori probability of weights. When no prior information is available, a uniform distribution is used for $p(w)$, as illustrated in Figure 7.1. The $p(D)$ is the data probability that does not depend on weights. Therefore, the latter two components can be considered as constant normalization factors and the expression for $p(D|x,w)$ in Equation 7.23 can be used to get $p(w|D)$ as

$$p(w|D) = \frac{1}{Z_1^N} e^{-E_D(w)} \qquad (7.25)$$

where Z_1 is a new constant normalization term including those for $p(D)$ and $p(w)$. This is an expression for the a posteriori weight distribution containing the square error criterion $E_D(w)$. To obtain $E_D(w)$, an approximation is used.

Because w^* is known after training, the error criterion around the optimum weights $E_D(w^*)$ can be expanded using a second-order Taylor expansion, as illustrated graphically in Figure 7.4.

This gives an expression for the square error term for weights w in the vicinity of w^* as

$$E_D(w) = E_D(w^*) + b^T \cdot \Delta w + \frac{1}{2} \Delta w^T \cdot \mathbf{H} \cdot \Delta w \qquad (7.26)$$

where $E_D(w^*)$ is the minimum of MSE for the optimum weights, $\Delta w = w - w^*$, b is the Jacobian (first derivative) of $E_D(w)$ with respect to weights, i.e., gradient $(\partial E_D(w))/\partial w$, for each weight. The second term on the right hand side of Equation 7.26, $b^T \cdot \Delta w$, is the amount of error increase based on the gradient for a small increment in weight by Δw around w^*. The T is the transpose. Because gradients are zero at the optimum weights, the linear term $b^T \cdot \Delta w$ drops from Equation 7.26. The third component in the equation is the amount of error increase due to a small increment in weight Δw based on the curvature expressed by $\mathbf{H}$, the Hessian matrix, which is the second derivative of error term with respect to weights, i.e., the curvature $(\partial^2 E_D(w))/(\partial w_i \partial w_j)$ with respect to a pair of weights w_i and w_j. (A full treatment of gradients and the Hessian is given in Chapter 4.) Thus, a second-order approximation to the weight distribution incorporating only the curvature effects can be obtained by substituting Equation 7.26 into Equation 7.25 as

$$p(w|D) = \frac{1}{Z_1} e^{-E_D(w^*) - \frac{1}{2}\Delta w^T \cdot \mathbf{H} \cdot \Delta w}$$
$$\propto e^{-\frac{1}{2}\Delta w^T \cdot \mathbf{H} \cdot \Delta w}. \qquad (7.27)$$

In the last expression, $\propto$ denotes proportionality and this simplification is possible because $E_D(w^*)$ is constant at the optimum weights. This means that the a posteriori distribution of weights follows a Gaussian distribution with

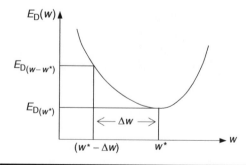

Figure 7.4 Taylor series expansion of $E_D(w)$ around optimum weight w^*.

mean w^* (note that $\Delta w = w - w^*$) and covariance matrix $\mathbf{H}^{-1}$. With the mean and variance expressed this way, the probability distribution for the weights can be generated with the w^* obtained from training and the Hessian matrix determined as discussed in Chapter 4. This probability represents the plausibility for the weight w, not the probability of obtaining w from the learning algorithm. Because the calculation of the Hessian is addressed in Chapter 4, it will not be discussed further here.

7.2.5 Example Illustrating Generation of Probability Distribution of Weights

7.2.5.1 Estimation of Geophysical Parameters from Remote Sensing: A Case Study

To illustrate the use of the concepts described for obtaining probability distribution for the weights, a case study reported by Aires [2] is presented. The study involves estimation of geophysical parameters that drive Earth's atmospheric radiative transfer phenomena from the data collected through remote sensing. These parameters are Earth surface skin temperature T_s, the integrated water vapor content WV, and microwave surface emissivities E_m. They are the required parameters for models representing the radiative transfer phenomenon. To make good atmospheric and climatological predictions relating to this phenomenon, reliable estimates for these geophysical parameters are needed. Direct estimation is not easy due to the complexity of the physical process; therefore, an indirect method must be used.

The goal of the study is to apply neural networks for remote sensing of these surface and atmospheric parameters where a nonlinear multivariate neural network regression model represents the inverse radiative transfer function. This type of problem is called "inverse parameter estimation," where the parameters of a process are estimated from the known outcomes of the process [15–17]. In direct methods, outcomes are predicted from known parameters. The task for the neural network model is to retrieve T_s, WV, and E_m between 19 and 85 GHz frequencies from the remotely measured microwave brightness temperatures (TB) obtained from satellite data. Thus, the parameters are retrieved from indirect radiative measurements obtained from satellite data. The emissivities (E_m) are described by seven measurements made by seven different frequency/polarization channels, making the total number of network outputs equal to nine variables (T_s, WV, and seven E_m components). The inputs are TB measured at the same seven frequencies for which surface emissivities are retrieved and surface temperature ($Tlay$) (i.e., eight indirect inputs).

The database has been produced from global clear-sky data collected from satellite images with 1 239 187 pixels repeatedly sampled every 3 hours, from July 1992 to June 1993 over land between 60°S and 80°N. Basically, from the satellite images, microwave brightness temperature (*TB*) has been determined and the coincident geophysical parameters have been matched to the *TB* from various available sources to complete the database consisting of eight inputs and nine outputs. When a model is developed from these data, it would relate the inputs to a parameter space that allows estimation of the geophysical properties for a particular combination of inputs.

Neural network development. The learning database consisted of 20 000 randomly selected observations from the very large database. Sample size was reduced because the computation of Hessian is time consuming. Of the 20 000 observations, 15 000 were used for training and model calibration and 5000 for testing generalization behavior. The number of neurons in the hidden layer was estimated by monitoring generalization error for different choices and the optimal number was kept.

The network was multiplayer perceptron (MLP) with 17 inputs; these were the seven original input variables plus the first guesses for the nine parameters to be retrieved. Using experiential knowledge, initial guesses can be estimated and using these as inputs can help a model narrow its search space for the optimum weights. The hidden layer had 30 neurons and the output layer had nine neurons corresponding to the nine geophysical parameters to be retrieved. The channels from which the seven inputs were obtained have a Gaussian instrumental noise of 0.6°K standard deviation. The inputs and outputs were first centered and then normalized. The weights of the network were initialized prior to training using a uniform distribution between -1 and 1. The network was trained with conjugate gradient descent, a second-order learning method with improved search direction. Training was done using training data and the separate test dataset was used for assessing generalization behavior. The trained network with optimum weights w^* statistically represents the inverse radiative transfer equation. Because the database was constructed in such a way that it was representative of the whole domain, the network is valid for all observations.

Results. Figure 7.5 presents plots of the first guesses and retrieved parameters against target data for three parameters: T_s, WV, and E_m for 19 GHz vertical polarization (E_m19V). The figure shows that for each of these parameters, the network retrievals are concentrated along the diagonal, indicating that they are closer to the target than the first guesses. A similar observation was made for the other six parameters.

Posterior distribution of network weights. A rigorous model parameter estimation should be followed by sensitivity analysis. A model can only be trusted when its sensitivities to all the employed hypotheses are known.

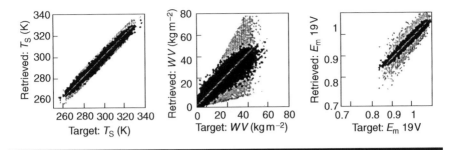

Figure 7.5 Scatterplots of network output for three geophysical parameters: first guess (gray) and retrieved (black) parameters against actual parameters. (From Aires, F., *Journal of Geophysical Research*, 109, D10303, 2004.)

Here, the sensitivity analysis is performed by estimating the uncertainty of the network weights. To derive the PDF of network weights, the Hessian must be estimated, which when inverted gives the covariance matrix of weights. The nature of the Hessian matrix is related to the structure of network weights. For example, by definition, the Hessian is the second derivative of error with respect to weights, as treated in detail in Chapter 4. For two different weights w_i and w_j, the Hessian is a depiction of how error gradient with respect to w_i is sensitive to w_j, i.e., how independent (or correlated) the weights are. If two weights are independent, the Hessian will be zero (or small) and if they are related, the Hessian will be large. Therefore, the nondiagonal entries of the Hessian matrix, representing two nonidentical weights w_i and w_j, indicate whether the weights in the network are independent.

Ideally, weights should be independent and therefore nondiagonal entries should be zero (or minimum). The diagonal terms that represent rate of change of a gradient with respect to the same weight can be small or large. The Hessian for this problem indicated that the weights between input and hidden layer are more related than those between hidden and output layers. After **H** is inverted, it becomes the covariance matrix of network weights. For this structure of Hessian, the uncertainty of weights between hidden-output layers is larger than that between the input-hidden layers.

By using the Hessian, the PDF of weights was obtained from Equation 7.27. The w^* with plus or minus two standard deviations are shown in Figure 7.6 for the first 100 weights in the top figure and for all the 821 weights in the bottom figure. Weights from 510 to 819 are for hidden-output layer connections and these are obviously more variable than the input-hidden weights. This is because the first stage of processing at the hidden layer is high-level processing that encompasses the non-linearity of the network, whereas the second stage of processing at the

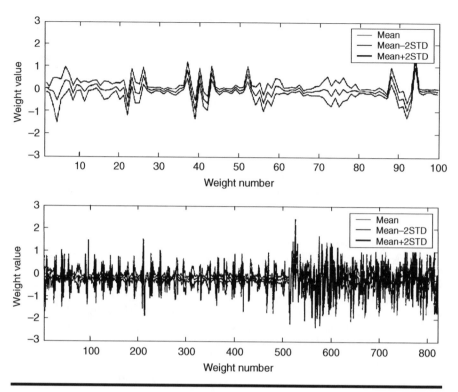

Figure 7.6 The mean network weights w*, and w* plus or minus two standard deviations of weights for the first 100 weights corresponding to input-hidden layer (top figure), and for all 821 weights with 510–819 indicating hidden-output layer weights. (From Aires, F., *Journal of Geophysical Research*, 109, D10303, 2004.)

output layer is low level, involving just a linear postprocessing of the hidden-layer outputs.

Interpretation of weight uncertainty. From the distribution of weights, sets of weights can be drawn for analyzing uncertainty. Each set represents a particular network. Together, the sets of weights represent the uncertainty of all network weights. Four sets of weights (mean plus three sets) extracted from the weight distribution are shown in Figure 7.7.

In Figure 7.7, one sample is one network and it shows that the weights of the three networks follow each other closely. Even if the sets of weights are included within the large variability of the two standard deviations envelope, the correlation constraints prevent any random oscillations due to noise from imposing a structure on them. In other words, the weights have considerable latitude to change but their correlations restrict them to following a strong dependency structure. This explains why different weight configurations can produce the same outputs.

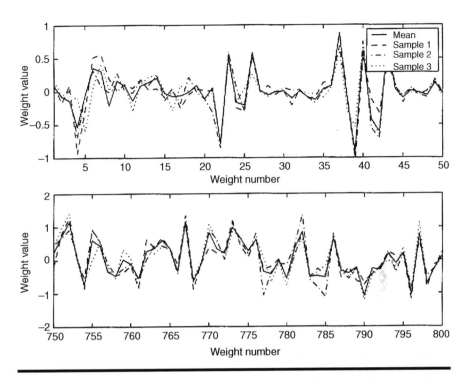

Figure 7.7 Mean (optimum) network weights and three samples of weights simulated from the a posteriori distribution of weights for the first 50 input-hidden weights (top) and for 50 weights corresponding to hidden-output layer (bottom). (From Aires, F., *Journal of Geophysical Research*, 109, D10303, 2004.)

When there are many weights (in this study the number is 819, but it can be thousands), an efficient sampling strategy must be used to sample weights. The four sets of weights (mean plus three sets) shown in Figure 7.7 were extracted from the weight distribution using eigen-decomposition-based sampling (i.e., PC). In eigen-decomposition, principal components of weights are derived so that a lesser number of PCs than the actual number of weights are used while preserving the weights and their correlation structure in the PCs. From these, sets of weights can be extracted. Refer to Chapter 6 for a detailed treatment of PCA.

The pattern of the weights depicted in Figure 7.6 and Figure 7.7 indicates that the structure of the weight correlations, not the actual values, is the most important for the processing in the network. For example, if the difference between two inputs is a good predictor, then as long as the two weights linked to these inputs perform the difference, the absolute value of the weights is not essential. Another source of uncertainty for the weights arises from the fact that some permutations of neurons have no impact on the network. For example, if two hidden neurons in the network are permuted, the network output would not change. This is because the sigmoid transfer function used in the hidden neurons is saturated for weighted sum of inputs entering it that are too high or too low; therefore, change in weight going to such a neuron would have a negligible effect on its output.

Some of the reasons why network weights can vary and still provide a good model that generalizes well have been explained. Results indicate that the variability of the network weights can be considered as a natural variability that is inherent to the neural technique. From the user's point of view, the uncertainty that this variability produces in the network outputs (or even more complex quantities such as sensitivities of output to inputs) is more important than the variability of weights.

The case study described involved the development of tools to provide insights into how the neural network model actually works and how the network outputs are estimated. These novel developments draw neural network techniques closer to better understood classical regression methods in which it is standard practice to estimate uncertainties of the parameters and it is completely mandatory before using the model. The availability of similar statistical tools for investigation of internal structure puts neural networks on a stronger theoretical and practical base and presents them as a natural alternative to some traditional regression methods with the advantage of nonlinear modeling.

The a posteriori PDF of the network weights derived here is useful in investigating many types of uncertainties. In the next case study, the probability distribution of the network sensitivities are investigated on the basis of the PDF of weights described here. Another important application of the PDF of network weights is the comparison of different network models in light of the observations [3].

7.3 Assessing Uncertainty of Neural Network Outputs Using Bayesian Statistics

A technique to estimate the uncertainty of network weights as proposed and demonstrated by Aires [1] was presented in the previous section. In this section, these weight uncertainty estimates will be used to compute uncertainties in the network outputs (i.e., correlation structure of the output errors and error bars). A rigorous model development requires not only good quality outputs but also an uncertainty estimate of the outputs, such as error bars and correlation structure of the error. This is especially important where output accuracy is paramount in subsequent decision making, or where the outputs are subsequently used as inputs into other models that use the estimated uncertainties, such as in meteorological and climatological models.

The reliability of network predictions is important for any application. For classical linear regression, the method of confidence intervals (CIs) is well established. For nonlinear models, such approaches are more recent and for neural networks they are rare. For neural networks, only root mean square error (RMSE) of the generalization error is used, but this single quantity is an average estimate and not situation dependent (i.e., does not vary with inputs). The RMSE is good for linear problems. Situation dependency, however, is more realistic for complex nonlinear relationships. Bootstrap techniques have been used to estimate CIs for neural networks, but they require a large number of computations. Rivals and Personnaz [4,5] introduced CIs based on least-squares estimation. Here, a Bayesian statistics approach to estimate errors for multiple outputs is presented. Earlier, the Bayesian approach to estimating uncertainties in network weights characterized by the PDF of weights was demonstrated. This PDF of weights can be used to provide a framework for the characterization and analysis of various sources of network errors, such as output errors and sensitivities. These concepts are presented by extending the case study described in Section 7.2.5.1.

7.3.1 Example Illustrating Uncertainty Assessment of Output Errors

Estimation of geophysical parameters from remote sensing: a case study.
The application of the approach for assessing uncertainty in network outputs using the same remote sensing problem described previously in Section 7.2.5.1 is demonstrated. It involves the retrieval of geophysical parameters that drive the Earth's atmospheric radiative transfer phenomena from satellite data collected over land. These parameters are T_s, WV, and E_m. A case study is presented here as reported by Aires et al. [6].

7.3.1.1 Total Network Output Errors

The network error can be divided into two sources: (1) errors due to network weight uncertainty (i.e., model uncertainty), and (2) error from all remaining sources (i.e., intrinsic noise that includes random error due to measurement noise, error due to finite resolution of the observation system, etc.). The model uncertainty itself results from imperfections in data, nonoptimum network structure and nonoptimum learning algorithms. These two uncertainty quantities for the model and intrinsic noise can be estimated. First, however, the total network output error will be examined.

After learning, a network with good generalization behavior is obtained. Model parameters are the optimum denoted by w^*. The network output errors can be computed as $\varepsilon_y = t - y$ over the database, where t is the target and y is the network output; the output error follows a Gaussian distribution with zero mean.

7.3.1.2 Error Correlation and Covariance Matrices

For each output variable, the mean and variance of error can be calculated along with covariance of error between outputs. Thus, the total output error covariance matrix, $\mathbf{C}_0$, can be obtained from the model using the dataset and this covariance matrix of error describes the PDF of error. The diagonal terms of this matrix are the error variance of each output variable and off-diagonal terms are the covariance of output errors for pairs of output variables. If the output variables are independent (i.e., not correlated), the off-diagonal terms are zero.

In the remote sensing problem described here, the nine outputs of the network correspond to T_s, WV, and E_m for the seven frequency/polarization channels. The frequency ranges from 19 to 85 GHz and polarization can be either vertical or horizontal. For the trained network, the output error covariance matrix, along with the correlations, are shown in Table 7.1 for the nine output variables. The right/top triangle in the table denotes error correlations and the left/bottom triangle presents error covariances. The diagonal values are the variance of the error for individual output parameters. The table clearly indicates that some errors are highly correlated. For this reason, it is a mistake to monitor the error bars although they are easier to understand because correlation structure can distort individual output patterns. Aires et al. [6] state that the correlation of errors in Table 7.1 demonstrates the expected behavior among the parameters. For example, errors in T_s are negatively correlated with the other errors and errors in WV are weakly correlated with other errors. Correlations between emissivity errors are always of the same sign and are high for the same polarization (vertical or horizontal) and decreases as the difference in frequency increases.

7.3.1.3 Statistical Analysis of Error Covariance

The correlations present in Table 7.1 make it necessary to understand uncertainty in multidimensional output space. This is more challenging than determining individual error bars but is also much more informative [6]. To statistically analyze $\mathbf{C}_0$, PCA was used that decomposes the covariance matrix into eigenvectors (PCs) that are decorrelated (orthogonal). This means that the set of eigenvectors constitutes a set of error patterns such

Table 7.1 Covariance Matrix C_0 and Correlation Matrix of Network Output Error Estimated over the Database

	T_s	WV	E_m19V	E_m19H	E_m22V	E_m37V	E_m37H	E_m85V	E_m85H
T_s	2.138910	−0.24	**−0.87**	**−0.72**	**−0.76**	**−0.84**	**−0.72**	**−0.49**	**−0.32**
WV	−1.392113	14.708836	0.16	−0.06	0.14	0.05	−0.15	−0.18	**−0.37**
E_m19V	−0.006294	0.003179	0.000024	**0.77**	**0.88**	**0.89**	**0.74**	**0.60**	**0.42**
E_m19H	−0.005261	−0.001143	0.000019	0.000024	**0.72**	**0.73**	**0.81**	**0.60**	**0.56**
E_m22V	−0.006274	0.003140	0.000024	0.000020	0.000031	**0.84**	**0.71**	**0.71**	**0.54**
E_m37V	−0.006121	0.001049	0.000021	0.000018	0.000023	0.000024	**0.81**	**0.70**	**0.50**
E_m37H	−0.005290	−0.002954	0.000018	0.000020	0.000020	0.000020	0.000025	**0.65**	**0.67**
E_m85V	−0.004895	−0.004945	0.000020	0.000020	0.000027	0.000023	0.000022	0.000046	**0.79**
E_m85H	−0.003906	−0.011933	0.000017	0.000022	0.000024	0.000020	0.000027	0.000044	0.000067

The right/top triangle is for correlation and left/bottom triangle is for covariance; the diagonal gives the variance. Correlations with an absolute value higher than 0.3 are in bold. (*Source*: Aires, F., Prigent, C., and Rossow, W.B., *Journal of Geophysical Research*, 109, D10304, 2004.)

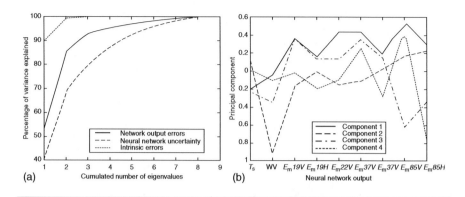

Figure 7.8 Principal component decomposition of total error covariance matrix: (a) Variance of the original output error explained by the PCs; (b) loadings of the first four PCs with respect to errors of actual output variables. (From Aires, F., Prigent, C., and Rossow, W.B., *Journal of Geophysical Research*, 109, D10304, 2004.)

that the contribution of each of these patterns to the total error is decorrelated [6,7]. Each pattern represents a proportion of the total error variance. For example, the curve denoted by "network output errors" in Figure 7.8a presents the cumulative percentage of variance explained by the PCs of C_0. The components 1 and 2 explain 55 percent and 30 percent, respectively, of the total error variance which means that the errors are concentrated in the first two components. The other curves of this figure will be explained shortly.

Figure 7.8b illustrates the factor loadings (i.e., weights or coefficients) of the first four PCs for the actual output variable errors. Recall that these weights indicate the relative importance of the variable to the PC. For example, the first component is mainly related to T_s and emissivities with vertical polarization. Negative weight for T_s and positive weights for emissivities are consistent with the correlations in Table 7.1 that indicate that T_s and emissivities are anticorrelated. *WV* dominates the second component, along with emissivities for channels that are sensitive to *WV*, specifically 19 GHz and 85 GHz with horizontal polarization.

7.3.1.4 Decomposition of Total Output Error into Model Error and Intrinsic Noise

Network output errors due to model parameter uncertainty. The output error described in the previous section is the total error consisting of model error and intrinsic noise. To obtain the covariance of the output error due to weight (model) uncertainty, Bayesian statistics can be used. For example, the probability of outputs can be expressed as

$$p(t|x, D) = \int p(t|x, w) \cdot p(w|D) dw \tag{7.28}$$

where D is the set of target outputs t in the database corresponding to inputs x, i.e., dataset containing pairs of $\{x^n, t^n\}$. The expression in Equation 7.28 represents the output uncertainty taking into account the weight uncertainty (distribution). The first component on the right hand is the conditional probability distribution of the targets and represents the noise distribution as given in Equation 7.20 that has the form

$$p(t|x, w) = \frac{1}{Z} e^{-\frac{1}{2}(t - y(x,w))^{\mathrm{T}} \cdot \mathbf{A}_{\mathrm{in}} \cdot (t - y(x,w))} \tag{7.29}$$

where $y(x,w)$ is the network function. The second component in Equation 7.28 is the a posteriori PDF of weights expressed in Equation 7.27 and repeated here:

$$p(w|D) = \frac{1}{Z^N} e^{-E_D(w^*) - \frac{1}{2}\Delta w^{\mathrm{T}} \cdot \mathbf{H} \cdot \Delta w} \tag{7.30}$$

Substituting Equation 7.29 and Equation 7.30 into Equation 7.28:

$$p(t|x, D) = \frac{1}{z} \int e^{-\frac{1}{2}(t - y(x,w))^{\mathrm{T}} \cdot \mathbf{A}_{\mathrm{in}} \cdot (t - y(x,w))} \cdot e^{-\frac{1}{2}\Delta w^{\mathrm{T}} \cdot \mathbf{H} \cdot \Delta w} dw \tag{7.31}$$

where the constant term $e^{E_D(w^*)}$, which contains the square error term for optimum weights, has been put together with the other constant normalization factors in z. Introducing a first-order expansion of the neural network function, $y(x,w)$, about w^* gives

$$y(x, w) = y(x, w) + \mathbf{G}^{\mathrm{T}} \cdot \Delta w, \tag{7.32}$$

where $\mathbf{G}$ is the first derivative of network output with respect to weights at the optimum weights w^*, i.e. $G = (\partial y(x, w) / (\partial w_{w=w^*})$, which is a $W \times M$ matrix where W is the total number of weights and M is total number of outputs. Substituting Equation 7.32 into Equation 7.31 and simplifying (details not shown) gives

$$p(t|x, D) \propto e^{-\frac{1}{2}\varepsilon_y^{\mathrm{T}} \cdot \mathbf{C}_0 \cdot \varepsilon_y}, \tag{7.33}$$

where

$$\mathbf{C}_0 = \mathbf{C}_{\mathrm{in}} + \mathbf{G}^T \cdot \mathbf{H}^{-1} \cdot \mathbf{G}. \tag{7.34}$$

The $\mathbf{C}_0$ in Equation 7.34 is the total output error covariance matrix already presented in Table 7.1 and the above derivation has decomposed it into two components: (1) intrinsic error covariance matrix $\mathbf{C}_{\mathrm{in}}$, which in a single output case is the variance of the output error σ_{in}^2 that was explained in Section 7.2.3, and (2) covariance matrix for the model errors $\mathbf{C}_{\mathrm{m}}$

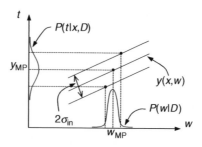

Figure 7.9 **Uncertainty of network outputs dominated by variance of intrinsic noise compared to uncertainty of network weights.**

expressed by $\mathbf{G}^{\mathrm{T}} \cdot \mathbf{H}^{-1} \cdot \mathbf{G}$. The latter is the error variance due to network uncertainty (i.e., uncertainty of weights). Recall that $\mathbf{H}^{-1}$ is the covariance matrix of weight uncertainty explained under "Posterior distribution of network weights" in Section 7.2.5.1, where the example case study was first introduced. For the case of one weight and one output, the covariance of the model error simplifies to

$$\mathbf{C}_{\mathrm{m}} = \mathbf{G}^{\mathrm{T}} \cdot \mathbf{H}^{-1} \cdot \mathbf{G} = \frac{\left(\frac{\partial y}{\partial w}\right)^2}{\frac{\partial^2 y}{\partial w^2}}. \tag{7.35}$$

For a single weight w and single input–output case, the effect of the intrinsic noise and weight uncertainty on the uncertainty of network output, which is $P(t|x,D)$ where D is the target data in the data set, is schematically illustrated in Figure 7.9 and Figure 7.10 [1]. The mean of the predicted output y_{MP} is from the optimum weight that is the most probable weight w_{MP} in Bayesian formalism. Figure 7.9 depicts a case where the noise

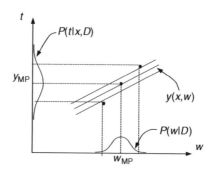

Figure 7.10 **Uncertainty of network outputs dominated by uncertainty of network weights compared to variance of intrinsic noise.**

contribution dominates the output uncertainty compared to model error. In Figure 7.9, $y(x,w)$ is model prediction and the two lines depict plus or minus 1 standard deviation of the noise (σ_{in}). It also shows a posteriori distribution of weights $P(w|D)$, which is very narrow in relation to noise distribution. Therefore, in this case, width of the distribution of network outputs is primarily determined by the noise.

Figure 7.10 depicts a case where model error contribution dominates output uncertainty compared to intrinsic noise. In this case, the a posteriori distribution of weights is larger than the noise distribution and the width of the distribution of network outputs is dominated by the distribution of network weights.

The two error components for the example case study can now be analyzed using the concepts presented here.

Network output errors due to model uncertainty, C_m. Because the optimum network structure is known, the gradient with respect to weights **G** is easily calculated. As presented in Section 7.2.5.1, the inverse Hessian, $\mathbf{H}^{-1}$, is the covariance of the PDF of network weights. Recall that the Hessian is a matrix containing the second derivative of network error with respect to a pair of weights. Table 7.2 represents the covariance matrix of error associated with uncertainty of weights, $\mathbf{G}^T \cdot \mathbf{H}^{-1} \cdot \mathbf{G}$, computed for the whole database. The top right triangle presents the correlation of output errors due to model uncertainty.

Although some of the bottom left values representing covariance matrix are close to zero (artifact due to ranges of variables being different to each other), the correlation structure is still apparent from the top right portion where the correlation coefficients are presented. The correlation of errors due to model uncertainty has a relatively small magnitude (maximum of 0.55). However, the structure of these is similar to the correlation structure for the global error covariance matrix in Table 7.1, with the same signs for correlation and similar relative values between variables.

This covariance matrix was decomposed using PCA to find error patterns (PCs) involved in this component of error. The curve depicted by neural network uncertainty in Figure 7.8a shows the cumulative percentage error variance due to model uncertainty explained by the PCs, indicating that first, second, and third components account for 40, 30, and 10 percent, respectively, of variance, indicating that the model error variance is spread across several significant components. Figure 7.11a shows the behavior (loadings) of the first four PCs.

The overall behavior of the first component is rather similar to that of the global error covariance matrix in Figure 7.8b, except for the sign. For example, the first PC is related to T_s and emissivities in vertical polarization with positive weight for T_s and negative weight for E_m. The behavior of component 2 is very similar to that of the global error covariance matrix and

Table 7.2 Covariance Matrix of Output Error Due to Model Uncertainty ($G^T \cdot H^{-1} \cdot G$) Averaged over the Whole Database

	T_s	WV	E_m19V	E_m19H	E_m22V	E_m37V	E_m37H	E_m85V	E_m85H
T_s	0.493615	−0.14	−0.28	−0.14	−0.25	**−0.32**	−0.16	−0.19	−0.06
WV	−0.106484	1.063071	0.10	−0.02	0.09	0.02	−0.07	−0.15	−0.25
E_m19V	−0.000325	0.000167	0.000002	**0.33**	**0.55**	**0.55**	0.28	0.27	0.08
E_m19H	−0.000255	−0.000060	0.000001	0.000006	0.26	0.22	0.29	0.10	0.13
E_m22V	−0.000268	0.000152	0.000001	0.000001	0.000002	**0.50**	0.26	0.28	0.12
E_m37V	−0.000330	0.000033	0.000001	0.000000	0.000001	0.000002	**0.34**	**0.38**	0.14
E_m37H	−0.000270	−0.000183	0.000001	0.000001	0.000000	0.000001	0.000005	0.16	0.26
E_m85V	−0.000231	−0.000282	0.000000	0.000000	0.000000	0.000000	0.000000	0.000002	**0.43**
E_m85H	−0.000128	−0.000681	0.000000	0.000000	0.000000	0.000000	0.000001	0.000001	0.000006

The right/top triangle is for correlation and left/bottom triangle is for covariance; the diagonal gives the variance. Correlations with absolute value higher than 0.3 are in bold. (*Source:* Aires, F., Prigent, C., and Rossow, W.B., *Journal of Geophysical Research,* 109, D10304, 2004.)

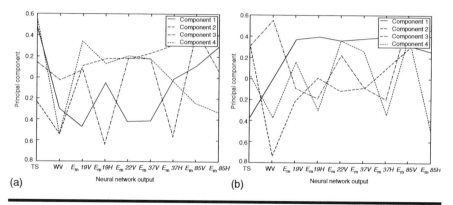

Figure 7.11   Principal component decomposition of model error and intrinsic noise covariance matrices: (a) loadings of the first four PCs of model error covariance matrix with respect to errors of actual output variables, (b) loadings of the first four PCs of intrinsic noise covariance matrix with respect to errors of actual output variables. (From Aires, F., Prigent, C., and Rossow, W.B., *Journal of Geophysical Research,* **109, D10304, 2004.)**

features *WV* and emissivities, especially the higher frequency channels that are more sensitive to water vapor. The third component is mostly related to low frequency emissivities in horizontal polarization. The second and third components together account for model error variance stemming from *WV* and related emissivities in horizontal polarization; the first component mainly accounts for T_s and emissivities for vertical polarization, especially the low-frequency ones.

Network output errors due to intrinsic noise of observations. Because the total error covariance matrix, C_0, and that for the error due to model uncertainty, C_m, are known, the covariance matrix for the intrinsic error can be found by subtracting the latter from the former:

$$C_{in} = C_0 - C_m \qquad (7.36)$$

averaged over the whole database. This covariance matrix and correlations among intrinsic noise in the nine output variables are presented in Table 7.3, where the top right portion shows the correlation and the bottom left portion presents the covariance structure.

Table 7.3 indicates that intrinsic error correlations can be very large, reaching 0.99. The structure of the correlation matrix is interestingly similar to that of the global error matrix, C_0, except for the larger values. The PCA shows that most of the error variability (90 percent) is captured by the first PC, as shown in the graph denoted by intrinsic errors in Figure 7.8a, meaning that the number of degrees of freedom in the intrinsic output error variability is limited. Figure 7.11b shows that the first

Table 7.3 Covariance Matrix of Intrinsic Noise Errors, C_{in}, Estimated over the Whole Database

	T_s	WV	E_m19V	E_m19H	E_m22V	E_m37V	E_m37H	E_m85V	E_m85H
T_s	1.645294	−0.27	−0.99	−0.92	−0.86	−0.95	−0.88	−0.55	−0.37
WV	−1.285629	13.645765	0.17	−0.06	0.14	0.05	−0.16	−0.19	−0.39
E_m19V	−0.005968	0.003011	0.000021	**0.89**	**0.91**	**0.92**	**0.83**	**0.63**	**0.46**
E_m19H	−0.005006	−0.001083	0.000017	0.000017	**0.83**	**0.86**	**0.98**	**0.71**	**0.66**
E_m22V	−0.006005	0.002988	0.000023	0.000019	0.000029	**0.87**	**0.80**	**0.75**	**0.58**
E_m37V	−0.005790	0.001015	0.000020	0.000017	0.000022	0.000022	**0.90**	**0.72**	**0.54**
E_m37H	−0.005019	−0.002770	0.000017	0.000018	0.000019	0.000019	0.000019	**0.74**	**0.76**
E_m85V	−0.004663	−0.004662	0.000019	0.000019	0.000026	0.000022	0.000021	0.000043	**0.82**
E_m85H	−0.003777	−0.011251	0.000016	0.000021	0.000024	0.000019	0.000026	0.000042	0.000060

The right/top triangle is for correlation and left/bottom triangle is for covariance; the diagonal gives the variance. Correlations with absolute value higher than 0.3 are in bold. (*Source:* Aires, F., Prigent, C., and Rossow, W.B., *Journal of Geophysical Research,* 109, D10304, 2004.)

component is mostly related to T_s with a negative loading and to emissivities with similar loadings. Thus, most of the intrinsic noise variance is due to these variables. The second component shows a remarkable similarity to the second component of the total error covariance matrix C_0. It is mainly related to WV and the emissivities at high frequencies. These, however, account for only about 10 percent of the total intrinsic noise variance.

7.4 Assessing the Sensitivity of Network Outputs to Inputs

Neural networks nonlinearly relate several inputs to one or more outputs. How can the influence of individual variables to outputs be rated? For this purpose, the sensitivity of output(s) to inputs provides vital information. The sensitivity of a network can be easily expressed as the partial derivative of the network output with respect to inputs. They are very important in that they allow statistical validation of how a trained neural network model derives the outputs from inputs. These in essence reflect the trends captured by the model and highlight the internal working of a model that correctly captures essential relationships in the data. From a validated model, the output sensitivity can be used to determine the most influential variables. Moreover, these sensitivities are useful in identifying nonrobust models and are therefore useful in model selection.

7.4.1 Approaches to Determine the Influence of Inputs on Outputs in Feedforward Networks

The existing approaches to determining the importance of input variables to the output of a feedforward network can be grouped into two classes. One is based on the magnitude of weights and the other is based on sensitivity analysis [8].

7.4.1.1 Methods Based on Magnitude of Weights

Methods based on the magnitude of weights group together those approaches that are exclusively based on the values of the weights to estimate the relative influence of each input variable on each of the network outputs. These methods involve the calculation of the product of the weighs (a_{ij}) between input i and hidden neuron j, and b_{jk} between hidden neuron j and output neuron k, for each of the hidden neurons and then summation of the products. A representative of this type is given by

$$Q_{ik} = \frac{\displaystyle\sum_{j=1}^{m}\left(\frac{a_{ij}}{\sum_{r=1}^{n} a_{rj}} b_{jk}\right)}{\displaystyle\sum_{i=1}^{n}\sum_{j=1}^{m}\left(\frac{a_{ij}}{\sum_{r=1}^{n} a_{rj}} b_{jk}\right)}, \tag{7.37}$$

where m is the number of hidden neurons and n is the number of inputs [9]. The $\sum_{r=1}^{n} a_{rj}$ is the sum of the weights between n inputs and neuron j. The Q_{ik} is the relative influence of input variable x_i on output y_k in relation to the rest of the input variables. The sum of relative influence of all inputs must be equal to 1.0, and thus it represents the percentage contribution of inputs to outputs.

A variant of the weight-based approach is the weight product [10] that incorporates the ratio of the value of input variable x_i and value of output y_k to the sum of the weight products as

$$WP_{ik} = \frac{x_i}{y_k} \sum_{j=1}^{m} a_{ij} b_{jk}, \tag{7.38}$$

where WP_{ik} is the influence of the input variable x_i on the output y_k. However, analysis based on weights is not the most effective method for determining the influence of variables [8].

7.4.1.2 Sensitivity Analysis

Sensitivity analysis is based on the effect observed in the output y_k due to a small change in input x_i. The greater the observed effect, the greater the sensitivity of output to that input (i.e., the greater the influence of the input). The input sensitivity, also called the Jacobian, is obtained by the partial differentiation of the output with respect to each of the inputs. Because different input patterns can provide different sensitivity values, the mean value is used to represent the overall sensitivity of an output to an input.

Sensitivity analysis can also be carried out with respect to the effect observed in error due to changes in input x_i. A common approach to this sensitivity analysis is to clamp the input variable of interest x_i to a fixed value, usually the mean, and compare the effect of this on the output error to the original error when the input variable is not clamped. The greater the effect of clamping an input on output error, the greater the importance of that input to the output.

7.4.2 Example: Comparison of Methods to Assess the Influence of Inputs on Outputs

Montano and Palmer [8] generated data for three input variables (X_1, X_2, and X_3) and the output Y from the following expression for comparative study on different methods for assessing the contribution of input variables:

$$Y = 0.0183e^{(4X_2)} + \tanh(X_3) + \varepsilon N(0, 0.01) \qquad (7.39)$$

where the function between Y and X_2 is exponential with a range between 0 and 1, and the function between Y and X_3 is a hyperbolic tangent with a range between -1 and 1. A random error ε is added to the above expression from a normal distribution with mean zero and standard deviation of 0.1. The output variable was rescaled to a range between 0 and 1. The X_1 is not featured in the expression and therefore does not make any contribution to the output.

The datasets consist of 500 patterns in the training set, 250 in the validation set, and 250 in the test set. The network has three inputs, two hidden neurons, and one output. Hyperbolic activation is used in the hidden neurons and linear activation in the output neuron. The network was trained with backpropagation with a learning rate of 0.25 and momentum of 0.8. The validation set was used to obtain the best network and the test set was used to assess the contribution of variables. To use the weight-based methods, weights were extracted from the trained network. To determine the weight product, WP_i, the mean over all input patterns in the test dataset was used; this is because each input pattern gives a different WP_{ik} value. The sensitivity analysis requires the partial derivative of network output(s) to inputs (Jacobians). (How to obtain the partial derivatives is explained in Chapter 4.) The results from the three approaches (percentage influence Q_i, weight product WP_i, and sensitivity analysis) are presented in Table 7.4. For the latter two approaches, means and standard deviations are presented. Similar to WP_{ik}, sensitivity also varies with input patterns and therefore mean sensitivity must be used to assess the influence of an input.

Table 7.4 Comparison of Three Methods for Assessing the Contribution of Inputs to Output

	Q_i (percent)	WP_i		Sensitivity Analysis	
		Mean	*SD*	*Mean*	*SD*
X_1	1.395	0.009	0.025	0.005	0.003
X_2	23.27	0.374	0.366	0.199	0.038
X_3	75.34	3.194	3.906	1.376	0.665

Results in Table 7.4 show that the three methods have correctly established the significance of input variables identifying the hierarchy among the inputs, with X_3 being the most influential, followed by X_2 and a negligible influence of X_1. For example, percentage contribution (Q_i) shows that the X_3 contribution is 75.34 percent, X_2 contribution is 23.27 percent and X_1 contribution is only 1.395 percent. The weight product and sensitivity analysis methods have also captured the correct order of importance of variables; however, with these two methods, mean values can take any value between $-\alpha$ and $+\alpha$. In terms of reliability, sensitivity analysis is more accurate than the weight-product method because the standard deviation of sensitivities is much smaller than that from weight product. In the sensitivity analysis method, the more randomness there is in function Y, the larger the standard deviation of the sensitivity. In the next section, the discussion on sensitivity analysis is extended to assess the uncertainty of sensitivities. First, however, the issues surrounding inputs and outputs in general in relation to network structure are examined to shed light on the uncertainty of network output sensitivity to inputs.

7.4.3 Uncertainty of Sensitivities

There are several concerns in relation to neural network model structure in assessing uncertainty. These stem from the fact that models are trained to obtain a good statistical fit of inputs to output(s), but no constraints are generally applied to the internal structure. Statistical inference is often considered to be an ill-posed inverse problem [9,10] and consequently, many solutions can be found for the network parameters (weights) that provide satisfactory outputs. One of the reasons for the nonunique solution is that multicollinearity can exist among the variables. Such relationships are also a major problem in linear regression and can lead to very unstable regression parameters that drastically vary from one trial to another [9]. Partial derivatives are the equivalent of the linear regression parameters, so it can be expected that in a neural network the partial derivatives (sensitivity) can be highly variable and unreliable in the presence of multicollinearity although output statistics are very good.

The solution to multicollinearity and all robustness problems in general is to use some form of regularization [13]. Many regularization methods exist that reduce the complexity (degrees of freedom) of a model resulting from multicollinearity or excessive model flexibility. One method of regularization involving weight decay that controls the magnitude of the weights, and weight-pruning methods that reduce the degrees of freedom of a model were discussed in Chapter 5. Reducing the number of inputs is one direct method of regularization that can also be used as a model selection tool. However, Aires et al. [11] states that the introduction of redundant information in the input to a network can be useful in reducing observational noise, as long as the network is regularized in some way. One efficient method of eliminating multicollinearity while keeping most of the variables is PCA, which linearly combines the inputs to produce new inputs that are not correlated to each other. The same approach can be applied to remove correlations among output variables in the case of multiple outputs. In this representation, where correlations among inputs and outputs are suppressed, the solution is expected to be unique, meaning that the sensitivities should be more reliable and physically more meaningful. Furthermore, the PCA representation suppresses part of the noise during data compression, such that initial PCs represent the real variability of the output and the remaining PCs are more likely to be related to Gaussian noise of the instrument (or measurement) error, or to very minor variability (i.e., unimportant information) [11].

7.4.4 Example Illustrating Uncertainty Assessment of Network Sensitivity to Inputs

Estimation of geophysical parameters from remote sensing: a case study. The use of the concepts of network sensitivities for estimating the relevance of inputs and related network uncertainties in conjunction with PCA will be demonstrated. The work presented here was reported by Aires et al. [11] and is a continuation of the case study started in Section 7.2.5.1 and continued in Section 7.3.1. Recall that the objective is to retrieve T_s, WV, and E_m from satellite data. The emissivities are described by seven measurements made by seven different frequency/polarization channels, thereby making the total number of outputs equal to nine variables (T_s, WV, and seven E_m components). The inputs are microwave brightness temperature (TB) measured at the same seven frequencies for which surface emissivities are retrieved and $Tlay$ (i.e., eight indirect inputs). As noted in Section 7.2.5.1, it has been possible to come up with first guesses for the nine outputs so that the search for the optimum model can be made efficient. These first guesses were also used as inputs, thus the total inputs were 17. The problem is to retrieve the nine parameters (outputs) from 17 inputs.

7.4.4.1 PCA Decomposition of Inputs and Outputs

In PCA, new components are derived from original variables by multiplying them by eigenvectors that contain factor loadings (i.e., coefficients) for each of the original variables. If the original set of inputs and output variable vectors are denoted by x and y, respectively, and the corresponding set of means are denoted by $\bar{x}$, $\bar{y}$, then the expression for the jth PC denoted by vector x'_j can be expressed as

$$x'_j = u_j^T \left(\frac{x - \bar{x}}{s_x} \right)$$

$$y'_j = v_j^T \left(\frac{y - \bar{y}}{s_y} \right),$$

(7.40)

where s_x and s_y are standard deviation vectors of inputs and output variables, respectively, and u_j and v_j are the jth eigenvector for the inputs and outputs, respectively. The T denotes transpose. For a dataset with N input and output patterns, there are N values for each x'_j and y'_j. As can be seen from Equation 7.40, each original variable is standardized by subtracting its mean and dividing by its standard deviation. The u_j and v_j are derived from the covariance matrix of the standardized original variables.

The dataset contained 20 000 input and output vectors. Input and output variables showed multicollinearity as illustrated in Section 7.3.1 and therefore PCA was appropriate. The results of PCA for the inputs and outputs are shown in Table 7.5, where the first column indicates the number

Table 7.5 Cumulative Percentage Variance of Inputs and Outputs Accounted for by the Number of PCs

Number of PCA Components Used	Cumulative Explained Variance for Inputs (percent)	Cumulative Explained Variance for Outputs (percent)
1	42.50	57.6844
2	68.23	94.1111
3	81.94	98.1895
4	86.38	99.4994
5	90.56	99.9346
6	92.64	99.9675
7	94.47	99.9910
8	96.09	99.9974
9	97.49	100.0000
10	98.35	–
11	99.01	–
12	99.46	–
13	99.78	–
14	99.87	–
15	99.94	–
16	99.98	–
17	100.00	–

of PCs and the second and third columns represent the cumulative explained variance (percent) for inputs and outputs, respectively.

Figure 7.12 shows the first four significant PCs (loadings) for the inputs. In this study it was assumed that PCs that are at the bottom of the list (higher-order) describe instrumental noise and unimportant information.

The first PC, which explains 42.5 percent of the total variance according to Table 7.5, is dominated by the first guess of T_s and regular inputs of TB with very similar weights for all frequencies. The differences between the information carried by horizontal and vertical polarization channels are represented by the second component. The first guess for WV only dominates in the fourth component.

Figure 7.13 shows the first four PC basis functions for the output data. The first component explains more than half of the variance of the original set of outputs (see Table 7.5) and is dominated by T_s and E_m with similar weights for all E_m. The WV, as well as the differences in E_m polarizations, are represented in the second component. These two PCs together account for 94 percent of the variance of the outputs.

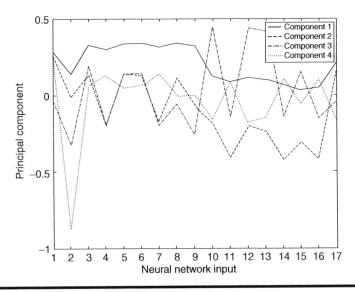

Figure 7.12 Loadings for the 17 original input variables in the first four principal components of input covariance matrix. (The inputs are, respectively, T_s, WV, $TB19V$, $TB19H$, $TB22V$, $TB22H$, $TB37V$, $TB37H$, $TB85V$, $TB85H$, and first guesses for the parameters of E_m19V, E_m19H, E_m22V, E_m37V, E_m37H, E_m85V, E_m85H, and $Tlay$.) (From Aires, F., Prigent, C., and Rossow, W.B., *Journal of Geophysical Research*, 109, D10305, 2004.)

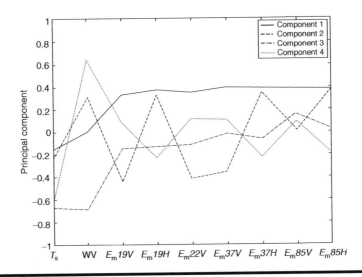

Figure 7.13 Loadings for the nine original output variables in the first four principal components of output covariance matrix. (From Aires, F., Prigent, C., and Rossow, W.B., *Journal of Geophysical Research*, 109, D10305, 2004.)

Aires et al. [9] state that even if the PCA is only optimal for datasets that follow Gaussian distributions, it can still be used for more complex distributions with satisfactory compression levels. Applying PCA to non-Gaussian data results in non-Gaussian distributions for the PCA components. Figure 7.14 shows the first four output PC distributions, which indicate that some distributions are skewed (C and D) and may have positive kurtosis, i.e., platykurtic (Figure 7.14d), or negative kurtosis, i.e., leptokurtic (Figure 7.14c). This makes the use of a nonlinear model, such as neural networks, even more important. Dealing with non-Gaussian distributed data requires a model that is able to incorporate the complex and nonlinear dependencies in the data. Furthermore, extreme events can occur that are represented by very strong absolute values for some PCs. If these are considered outliers, they can be removed using multivariate outlier detection methods discussed in Chapter 6.

To check if the PCs of outputs are consistent with physical data, the original data was projected, as shown in Figure 7.15, onto the first two PCs that represent most (94 percent) of the variability. In the figure, clouds of points are represented by one-sigma (standard deviation) contour lines for clarity, meaning that the mean and the standard deviation of the represented Gaussians are the mean and the standard deviation of the cloud of data points. As can be seen, the projection onto principal space differentiates

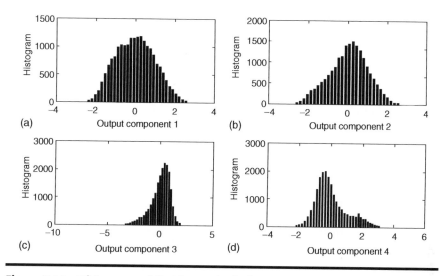

Figure 7.14 Histograms of first four principal components of the outputs. (From Aires, F., Prigent, C., and Rossow, W.B., *Journal of Geophysical Research*, 109, D10305, 2004.)

different land surface types in a set of Gaussian modes that are well separated and physically consistent.

PC maps, such as that shown in Figure 7.15, can therefore be used for clustering and classification because they provide physically meaningful interpretations. For example, the negative first component values, such as those for rainforest, mean that for this vegetation type, T_s' (earth surface skin temperature), is above mean value. Recall that principal scores are calculated by multiplying the PC loadings (see Figure 7.13) by the corresponding value for the original standardized input variables. (Recall also that the original variables are standardized by subtracting the mean and dividing by the standard deviation.) The loading on the T_s' in the first component is negative in Figure 7.13. Therefore, for the scores shown in Figure 7.15 to be negative for the rainforest, the standardized values of the output T_s' must be positive in this case. This can happen only if T_s' is above the mean for rainforest.

In contrast, tundra has a positive first component, indicating that T_s' is below the mean value. Similarly, the second component is highly positive for the rainforest, indicating that *WV* is higher than the mean in the equatorial regions, as expected. This is because the loading on *WV* in the second component is positive. It is known that surface types represent a large part of variability in the parameters (outputs); the fact that PCs can coherently separate different surfaces demonstrates the significance of the PCA representation. This is particularly important because in the next

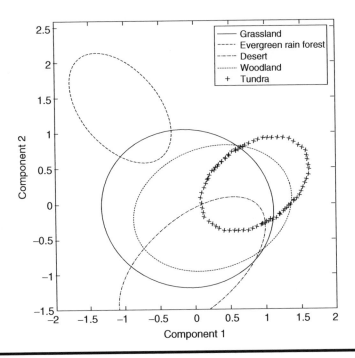

Figure 7.15 One-sigma contour lines of distributions of output parameters projected onto the first two principal components. (From Aires, F., Prigent, C., and Rossow, W.B., *Journal of Geophysical Research*, 109, D10305, 2004.)

section, PCs are used to regularize network learning. The patterns that are found by PCs will distribute the contribution of each input and each output for a given sensitivity and it is essential that these patterns have a physical meaning [11].

7.4.4.2 PCA-Based Neural Network Regression

As discussed in Chapter 6, the use of PCs instead of raw inputs has several benefits: it makes the network faster to train because of the reduced dimensionality and reduced noise level in the observations [11,12]. The network is also less complex with fewer inputs and outputs and there are therefore fewer parameters to be estimated. Consequently, the variance in the determination of the actual values of weights is also reduced. Reduced dimensions also make training simpler because the inputs are decorrelated. As discussed earlier, correlated inputs, or multicollinearities, are known to cause problems in model fitting; therefore, suppressing these makes the minimization of the error criterion more efficient because it is easier to minimize with less probability of getting trapped in a local minimum. It has the general effect of suppressing the uncertainty in the determination of weights.

Selecting the number of PCs. Several methods for selecting the number of components are discussed in Chapter 6. For neural networks, experience shows that if the network is well regularized, once sufficient information has been provided as input, adding more PCs as input does not have a large effect on the retrieved results with the disadvantage of the extra effort expended on training. In this study, a conservative approach has been taken by keeping more components than the denoising optimum would indicate so that all possibly useful information is kept.

Apart from the benefits of denoising and compression, reduction of the number of PCs has other effects on output retrieval quality. During the learning stage, the network is able to relate each output to the inputs that help predict it and disregard those that vary randomly. In some cases, the number of inputs is so large (few thousands) that compression is essential for meaningful interpretation of the network outputs in terms of inputs. In this study, 12 components accounting for 99.46 percent of the total variance (see Table 7.5) have been selected as inputs to the PC-based network.

The number of PCs for network output is related to the retrieval error magnitude for a nonregularized (non-PCA based) network. If the compression error due to PCA is minimal compared to retrieval error, the number of output components used is satisfactory [11]. It is not practical to retrieve something that is, in essence, noise. Furthermore, doing so could lead to numerical problems and interfere with the retrieval of other more important components. In this study, five output PCs have been used, accounting for 99.93 percent of the total variance of the outputs (see Table 7.5). The number of input and output components selected has other consequences, too. As presented in Chapter 4, the Hessian, which by definition is the second derivative of error, is used in second-order error minimization methods. The final form of the Hessian simplifies to the second derivative of network outputs with respect to weights. In some situations, the Hessian becomes ill-conditioned, leading to numerical instability problems. This problem is intimately related to the number of inputs and outputs selected for the network [11].

Postprocessing of data after PC decomposition. Data normalization (i.e., standardization) is performed before PC decomposition, but a post-PC normalization is also required. This is needed for both input and output PCs because they have different dynamic ranges. For a particular input PC denoted by $\mathbf{x}'_j$ and output PC denoted by $\mathbf{y}'_j$, the normalized PCs ($\mathbf{x}''_j$ and $\mathbf{y}''_j$) are calculated as

$$\mathbf{x}_j'' = \frac{\mathbf{x}_j' - \bar{\mathbf{x}}_j'}{\mathbf{s}_{\mathbf{x}_j'}}$$

$$\mathbf{y}_j'' = \frac{\mathbf{y}_j' - \bar{\mathbf{y}}_j'}{\mathbf{s}_{\mathbf{y}_j'}}$$

$$\tag{7.41}$$

where $\bar{\mathbf{x}}_j', \bar{\mathbf{y}}_j'$ are the mean of input PC $\mathbf{x}_j'$ and output PC $\mathbf{y}_j'$, respectively, and $\mathbf{s}_{\mathbf{x}_j'}$ and $\mathbf{s}_{\mathbf{y}_j'}$ are the corresponding standard deviations.

Furthermore, the importance of each of the output components is not equal; the first component represents 52.68 percent of the total variance, whereas the fifth component represents only 0.43 percent. Therefore, giving the same weight to each of these components during learning could be misleading. To correct this, a weighting equal to the standard deviation of the component has been given to each component in the square error criterion, i.e., multiplying square error for each component by its standard deviation. This is equivalent to using Equation 7.4 for the error criterion $E_D(w)$, where $\mathbf{A}_{in}$ is a diagonal matrix with diagonal terms equal to the standard deviation of the PCs. The off-diagonal terms are zero because there is no correlation between the PCs. Thus, the first component is 50 times more important than the fifth component.

Retrieval of results. The PCA-regularized network structure had 12 inputs, 30 hidden neurons and 5 outputs, whereas the nonregularized network architecture had 17, 30, and 9, respectively. The root mean square (RMS) retrieval error for the network with PC-based inputs and outputs was slightly higher than that for the non-PCA based network. For example, the RMS error for T_s was 1.53 compared to 1.46 for the non-PCA network. This is expected because PCA representation reduces overfitting (variance) and therefore, increases RMS error (bias), as illustrated in Chapter 6. The difference in RMS error in this case is negligible.

The evolution of learning statistics for the first three output components is illustrated in Figure 7.16, which indicates how the RMS error of the retrieval of each output component decreases with learning iterations. It shows that learning is unstable for some outputs with large initial oscillations due to the complex mixing of the components that the network tries to retrieve, meaning that each component mixes variability from each of the nine original variables. The network decreases error in one component, and then to reduce error in another component makes compromises that cause a sudden hike in RMS error in another component. However, when these error curves are translated back to the physical variables, more stable error curves that steadily decrease with iterations are observed, as presented in Figure 7.17 for three original outputs T_s, WV, and E_m19H. The character of the RMS curves for the other six original outputs is similar.

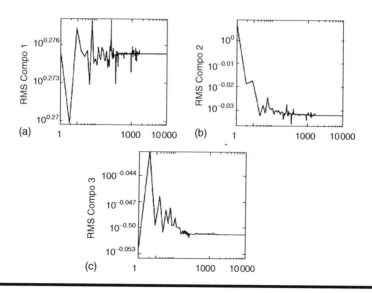

Figure 7.16 RMS error curves for the first three network output principal components during learning. (From Aires, F., Prigent, C., and Rossow, W.B., *Journal of Geophysical Research,* **109, D10305, 2004.)**

7.4.4.3 Neural Network Sensitivities

A trained network is a statistical model relating inputs to outputs. It also provides an efficient calculation of the network derivative with respect to inputs, which is called a network Jacobian. The Jacobian concept is powerful in that it allows for a statistical estimation of the multivariate and nonlinear sensitivities connecting the inputs and outputs in a given model. Specifically, the Jacobian gives the global mean sensitivities for each retrieved output parameter thereby indicating the relative contribution of

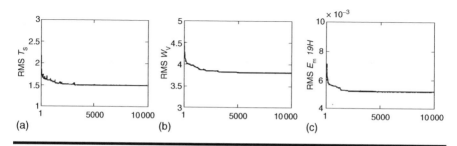

Figure 7.17 RMS error curves translated for three original network outputs during learning (curves for the other six variables are similar). (From Aires, F., Prigent, C., and Rossow, W.B., *Journal of Geophysical Research,* **109, D10305, 2004.)**

each input to the retrieval of that output parameter. For a nonlinear model, the Jacobian is situation dependent, meaning that it is not constant, but depends on the input vector; the mean is therefore given over the whole batch of input vectors.

Because the inputs are normalized, the network Jacobian would also be normalized quantities. The Jacobian for the original physical quantities $(\partial y_j/\partial x_i)$ (the sensitivity of output y_j to input x_i) can be obtained using Equation 7.40 and Equation 7.41 with the chain rule of differentiation:

$$\frac{\partial y}{\partial x} = \frac{\partial y}{\partial y'} \frac{\partial y'}{\partial y''} \frac{\partial y''}{\partial x''} \frac{\partial x''}{\partial x'} \frac{\partial x'}{\partial x}, \tag{7.42}$$

where $\mathbf{x}$ and $\mathbf{y}$ are original input and output vectors, $\mathbf{x}'$ and $\mathbf{y}'$ are the PCs and $\mathbf{x}''$ and $\mathbf{y}''$ are standardized PCs. From Equation 7.40, by putting all eigenvectors for inputs together in a matrix $\mathbf{U}$ and those for outputs in $\mathbf{V}$, each of the derivatives in Equation 7.42 can be derived as

$$\frac{\partial y}{\partial y'} = \frac{\mathbf{s_y}}{\mathbf{V}^{\mathrm{T}}}; \quad \frac{\partial y'}{\partial y''} = \mathbf{s_{y'}}; \quad \frac{\partial x''}{\partial x'} = \frac{1}{\mathbf{s_{x'}}}; \quad \frac{\partial x'}{\partial x} = \frac{\mathbf{U}^{\mathrm{T}}}{\mathbf{s_x}}. \tag{7.43}$$

Substituting the derivatives in Equation 7.43 into Equation 7.42:

$$\frac{\partial y}{\partial x} = \mathbf{s_y} \cdot [\mathbf{V}^{\mathrm{T}}]^{-1} \mathbf{s_{y'}} \cdot \frac{\partial y''}{\partial x''} \cdot \mathbf{s_{x'}^{-1}} \cdot \mathbf{U}^{\mathrm{T}} \cdot \mathbf{s_x^{-1}}, \tag{7.44}$$

which translates the derivatives $(\partial y''/\partial x'')$ based on standardized PCs to those with respect to original variables, $(\partial y/\partial x)$. In Equation 7.44, $\mathbf{x}$ is the vector of n original inputs $\{x_1, x_2,..., x_n\}$ and $\mathbf{Y}$ is the vector of k original outputs $\{y_1, y_2,..., y_k\}$ and $\mathbf{s_x}$ and $\mathbf{s_y}$ are the vectors of standard deviations of original inputs and outputs. The $\mathbf{U}$ and $\mathbf{V}$ are the matrices of eigenvectors (sets of $\mathbf{u_i}$ and $\mathbf{v_j}$ given in Equation 7.40) derived from the covariance matrices of original inputs and original outputs, respectively. Recall that $\mathbf{u_i}$ contains loadings for each original input in the ith input PC and $\mathbf{v_j}$ contains loadings for each original output in the jth output PC. The left side of Equation 7.44 is a matrix containing the sensitivity of each original output y_j to each of the original input variables x_i. To compare sensitivities between variables with different variation characteristics, their standard deviations $(\mathbf{s_x}, \mathbf{s_y})$ in Equation 7.44 can be suppressed so that for each input and output variable, normalization by their standard deviation is used. The resulting nonlinear Jacobians indicate the relative contribution of each input to the retrieval of a given output variable.

7.4.4.4   Uncertainty of Input Sensitivity

In an investigation of the sensitivity (Jacobian) of the network, its uncertainty or variance must be considered. Uncertainty is expressed as PDFs and PDFs of the Jacobians are therefore necessary. There is no direct way to estimate PDFs of Jacobians, so simulations must be used to indirectly obtain these quantities using the network weight uncertainties. After training, the maximum a posteriori (MAP) weights w^* are obtained, which are also the most probable weights. With these weights, the mean or MAP Jacobian corresponding to the most probable weights can be obtained. However, the mean Jacobian is not sufficient for real sensitivity analysis and uncertainty in this estimate is needed.

Even if uncertainty estimates are not of interest, using only w^* to directly estimate other dependent quantities may not be optimal. This is because in a high dimensional space, the densest areas of the weight distribution (location where the probability is higher) can be far from the most probable state, which is the MAP state. In fact, the high dimensions make the masses of PDFs more on the periphery of the density domain and less at the center [11]. Nonlinearity can also distort the distribution of estimated quantities such as weights. This is another reason why it is better to use a sample of weights from the weights PDF. Weight PDFs have already been derived in Section 7.2.4 using Bayesian statistics and can therefore be used for assessing uncertainty in network sensitivities.

As mentioned previously, without a priori information, the internal regularities of the network have no constraints, which can lead to high variability of the Jacobians that must be assessed. To estimate these uncertainties, 1000 samples of weights were extracted from the weights PDF using a Monte Carlo simulation that is designed to sample mostly the significant part of the weight space. Sampling a PDF in high dimensional weight space (819 weights in this study) can be very time consuming, and methods such as Monte Carlo simulations are needed for efficient sampling. One sample is one set of values for the 819 weights in the network (i.e., one network). For each weight sample w^r, the mean Jacobian over the entire dataset was calculated. With 1000 samples, there were 1000 mean Jacobian values for each output–input combination ($\partial y_j/\partial x_i$), which comprises the Jacobian matrix ($\partial \mathbf{y}/\partial \mathbf{x}$). These are used to obtain the PDF for each individual component of the Jacobian in ($\partial \mathbf{y}/\partial \mathbf{x}$) (i.e., all ($\partial y_j/\partial x_i$) components). From these PDFs, the mean and standard deviation of each Jacobian can be obtained; these are shown in Table 7.6 for the non-PCA-based network. These values indicate the relative contribution of each input to the retrieval of a given output parameter. In the table, rows are inputs and columns are outputs. The latter half of the input variables in the first column represents the first guesses for the output variables.

Table 7.6 Global Mean Nonregularized Sensitivities of Outputs to Inputs $(\partial y_j/\partial x_i)$

	T_s	WV	E_m19V	E_m19H	E_m22V	E_m37V	E_m37H	E_m85V	E_m85H
TB19V	0.26±0.19	0.04±0.23	**0.91±0.18***	-0.20±0.23	0.57±0.22*	0.02±0.19	-0.29±0.15	-0.17±0.21	-0.12±0.19
TB19H	0.08±0.19	**0.42±0.27**	-0.16±0.24	**1.26±0.40***	**-0.46±0.36**	**-0.54±0.23***	0.03±0.18	**-0.30±0.23**	**-0.43±0.27**
TB22V	0.11±0.19	**-0.79±0.27***	0.17±0.21	-0.14±0.25	**0.59±0.28***	-0.15±0.21	-0.09±0.17	**-0.77±0.21***	-0.26±0.22
TB37V	0.20±0.18	-0.16±0.21	0.19±0.18	-0.25±0.21	0.25±0.21	**0.12±0.19***	0.05±0.15	**0.63±0.20***	0.01±0.19
TB37H	0.15±0.18	**-0.67±0.23***	-0.28±0.17	-0.00±0.22	-0.13±0.21	0.18±0.19	**0.84±0.15***	-0.20±0.20	**0.61±0.21***
TB85V	0.24±0.16	-0.05±0.20	**-0.54±0.17***	-0.14±0.19	**-0.61±0.23***	-0.29±0.18	**-0.33±0.14***	**1.06±0.18***	-0.15±0.20
TB85H	-0.13±0.15	**1.60±0.18***	0.05±0.15	-0.16±0.17	0.09±0.18	-0.12±0.16	0.02±0.13	-0.14±0.16	**0.45±0.17***
T_s	0.18±0.08*	-0.15±0.11	-0.27±0.08*	-0.14±0.09	-0.26±0.10*	**-0.31±0.08***	-0.12±0.07	-0.26±0.09*	-0.07±0.09
WV	-0.04±0.05	**0.33±0.07***	0.03±0.06	0.04±0.08	-0.01±0.09	0.03±0.06	-0.04±0.05	-0.06±0.07	-0.15±0.08
E_m19V	-0.07±0.04	0.07±0.06	0.12±0.05*	0.11±0.08	0.09±0.07	0.15±0.05*	0.06±0.04	0.16±0.05*	0.03±0.05
E_m19H	-0.11±0.08	-0.03±0.10	0.22±0.09*	-0.04±0.15	0.29±0.14*	0.19±0.09*	0.09±0.07	0.16±0.10	0.18±0.10
E_m22V	-0.06±0.04	0.04±0.05	0.11±0.04*	0.04±0.04	0.15±0.04*	0.13±0.04*	0.05±0.03	0.13±0.04*	0.06±0.04
E_m37V	-0.07±0.04	0.02±0.05	0.11±0.04*	0.07±0.05	0.13±0.05*	0.16±0.04*	0.07±0.03*	0.15±0.04*	0.07±0.04
E_m37H	-0.08±0.06	-0.07±0.07	0.14±0.06*	0.09±0.07	0.15±0.07*	0.18±0.06*	0.11±0.05*	0.20±0.06*	0.15±0.06*
E_m85V	-0.04±0.04	-0.05±0.06	0.07±0.04	0.07±0.05	0.11±0.05*	0.10±0.05	0.04±0.04	0.20±0.05*	0.10±0.05*
E_m85H	-0.03±0.07	-0.18±0.09	0.12±0.09	-0.04±0.11	0.17±0.10	0.12±0.08	0.05±0.06	0.21±0.08*	0.21±0.07*
Tlay	-0.03±0.06	0.13±0.09	-0.04±0.08	0.01±0.09	-0.07±0.09	-0.06±0.08	-0.03±0.06	-0.13±0.08	-0.04±0.07

Columns are network outputs, y, and rows are network inputs, x. Sensitivities with absolute value higher than 0.3 are in bold and positive 5 percent significance tests are indicated by a star. The first part of the table is for satellite observations (TBs), the second part correspond to first guesses for outputs. (*Source:* Aires, F., Prigent, C., and Rossow, W.B., *Journal of Geophysical Research*, 109, D10305, 2004.)

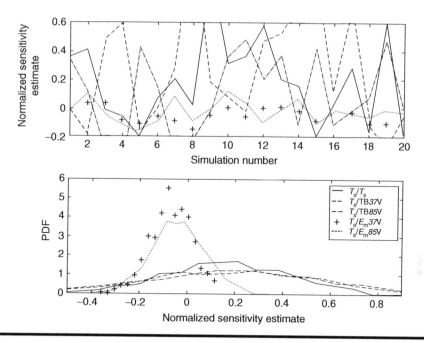

Figure 7.18 **Twenty samples of five sensitivities from non-PCA regularized network** $\left(\frac{\partial T_s}{\partial T_s} \; \frac{\partial T_s}{\partial TB37V} \; \frac{\partial T_s}{\partial TB85V} \; \frac{\partial T_s}{\partial E_m 37V} \; \frac{\partial T_s}{\partial E_m 85V} \right)$ **(top) and their PDFs (bottom). (From Aires, F., Prigent, C., and Rossow, W.B.,** *Journal of Geophysical Research,* **109, D10305, 2004.)**

Table 7.6 indicates that the variability of the Jacobians is large in that they can be up to several times the mean value. For most cases, Jacobian value is not in the confidence interval (the symbol * above a number indicates the sensitivities with positive 5 percent significance tests), meaning that the actual value is not significant. In the table, sensitivities higher than 0.3 are in bold. In linear regression, nonsignificant parameters are often an indication that multicollinearity exists.

Twenty samples of five sensitivities (normalized) extracted from the sensitivity PDFs, along with the distribution of these sensitivities, are shown in Figure 7.18. They confirm that sensitivities are highly variable and the distributions of Jacobian show that most of them are not statistically significant. The reason for such uncertainty could be the interference of multicollinearity during the learning process resulting from the introduction of compensation phenomena. For example, if two correlated variables are used to predict an outcome, then learning has some indeterminacy, meaning that it can give more emphasis to the first input and compensate for it by underallocation in the second correlated variable. This can be reciprocated from one epoch to another and therefore the two corresponding sensitivities can be highly variable from epoch to epoch. The output prediction could be just as good for both cases, but the internal structure (e.g., Jacobians) of the model would be different. Because these structures are of interest in assessing relative contribution of inputs, the problem must be addressed.

First, to see if multicollinearity is at the root of such large uncertainty in Jacobians, correlation between the sensitivities were determined and given in Table 7.7.

Table 7.7 shows that some sensitivities are indeed significantly correlated, indicating that correlated sensitivities are related following the compensation principle. For example, correlation between $(\partial T_s / \partial TB\ 19V)$ and $(\partial T_s / \partial TB22V)$ is larger in absolute value than T_s to high frequency TB. The negative sign of this relationship can be explained by the fact that $TB19V$ and $TB22V$ are highly correlated. For example, a large sensitivity of T_s to $TB19V$ is compensated for in the network by a low sensitivity to $TB22V$, giving rise to a negative correlation between the two sensitivities. Although the absolute values of correlation are not high, when added, these correlations define quite a complex and strong dependency structure among sensitivities. This is a sign that multicollinearities exist and consequent compensations are occurring in the network.

7.4.4.5 PCA-Regularized Jacobians

Now that the structure of sensitivities of the nonregularized network that uses original inputs is known, it can be compared with the sensitivity structure for PCA-regularized network. The mean and standard deviations for sensitivity of output PCs to input PCs in the PCA-based network described in Section 7.4.4.2 are given in Table 7.8. These results are much more stable and satisfactory than those for the non-PCA network. Compare the standard deviations in Table 7.8 and Table 7.6.

Table 7.8 shows some high sensitivities and they were all significant to the 5 percent confidence level. This sensitivity matrix also highlights how the network relates inputs to outputs. For example, the first output component is highly related to the first input component with a sensitivity of 0.81, but also to the third component with a sensitivity of 0.51. This indicates that the network has to nonlinearly transform input components to retrieve the output components. As the number of the important output components increases, the number of input components increases, too; however, higher order input components (i.e., higher than fifth) have limited impact. It must be noted that even if the mean sensitivity of an input component is low for the major output components, an input component can have an impact on the retrieval of output components in

Table 7.7 Correlation Matrix for a Sample of Neural Network Output–Input Sensitivities

	$\frac{\partial T_s}{\partial T_s}$	$\frac{\partial T_s}{\partial TB19V}$	$\frac{\partial T_s}{\partial TB19H}$	$\frac{\partial T_s}{\partial TB22V}$	$\frac{\partial T_s}{\partial TB37H}$	$\frac{\partial T_s}{\partial TB85V}$	$\frac{\partial T_s}{\partial TB85H}$	$\frac{\partial T_s}{\partial E_m19V}$	$\frac{\partial T_s}{\partial E_m19H}$	$\frac{\partial T_s}{\partial E_m85H}$
$\frac{\partial T_s}{\partial T_s}$	1.00	–	–	–	–	–	–	–	–	–
$\frac{\partial T_s}{\partial TB19V}$	−0.19	1.00	–	–	–	–	–	–	–	–
$\frac{\partial T_s}{\partial TB19H}$	−0.15	−0.18	1.00	–	–	–	–	–	–	–
$\frac{\partial T_s}{\partial TB22V}$	−0.05	**−0.44**	−0.16	1.00	–	–	–	–	–	–
$\frac{\partial T_s}{\partial TB37H}$	0.13	−0.00	**−0.59**	−0.01	1.00	–	–	–	–	–
$\frac{\partial T_s}{\partial TB85V}$	−0.08	−0.04	0.12	−0.18	−0.03	1.00	–	–	–	–
$\frac{\partial T_s}{\partial TB85H}$	−0.01	0.18	−0.05	−0.08	**−0.38**	**−0.45**	1.00	–	–	–
$\frac{\partial T_s}{\partial E_m19V}$	0.14	−0.17	0.25	−0.16	−0.06	0.15	0.04	1.00	–	–
$\frac{\partial T_s}{\partial E_m19H}$	−0.00	0.14	**−0.44**	0.09	0.03	−0.03	0.01	**−0.41**	1.00	–
$\frac{\partial T_s}{\partial E_m85H}$	−0.01	0.18	**−0.41**	0.17	0.06	0.03	−0.12	**−0.31**	0.26	1.00

Correlations with absolute value higher than 0.3 are in bold. (*Source:* Aires, F., Prigent, C., and Rossow, W.B., *Journal of Geophysical Research,* 109, D10305, 2004.)

Table 7.8 Global Mean Regularized Neural Sensitivities, y_j''/x_i'', of Output Principal Components to Input Components

NN inputs	Compo 1[a]	Compo 2	Compo 3	Compo 4	Compo 5
			NN outputs		
Compo 1	-0.81 ± 0.01	-0.25 ± 0.01	0.53 ± 0.01	-0.05 ± 0.01	-0.03 ± 0.01
Compo 2	-0.21 ± 0.01	-0.69 ± 0.01	-0.62 ± 0.01	0.16 ± 0.01	0.10 ± 0.02
Compo 3	0.51 ± 0.01	-0.65 ± 0.01	0.41 ± 0.01	0.47 ± 0.01	0.02 ± 0.01
Compo 4	0.17 ± 0.01	-0.46 ± 0.01	0.05 ± 0.01	-0.44 ± 0.01	-0.07 ± 0.01
Compo 5	-0.06 ± 0.01	0.04 ± 0.01	-0.00 ± 0.01	-0.02 ± 0.01	0.77 ± 0.01
Compo 6	-0.01 ± 0.01	0.02 ± 0.01	0.01 ± 0.01	0.02 ± 0.01	0.07 ± 0.01
Compo 7	-0.01 ± 0.01	0.02 ± 0.01	-0.01 ± 0.01	-0.02 ± 0.01	0.10 ± 0.01
Compo 8	-0.03 ± 0.01	-0.10 ± 0.01	0.01 ± 0.01	-0.04 ± 0.01	-0.05 ± 0.01
Compo 9	0.01 ± 0.01	0.02 ± 0.01	-0.01 ± 0.01	-0.00 ± 0.01	-0.25 ± 0.01
Compo 10	-0.01 ± 0.01	0.03 ± 0.01	0.00 ± 0.01	0.02 ± 0.01	0.12 ± 0.01
Compo 11	-0.14 ± 0.01	0.22 ± 0.01	-0.01 ± 0.01	0.05 ± 0.01	0.25 ± 0.01
Compo 12	0.10 ± 0.01	-0.18 ± 0.01	0.10 ± 0.01	0.08 ± 0.01	-0.14 ± 0.01

Columns are network outputs, y_j'', and rows are network inputs, x_i''. Sensitivities with absolute value higher than 0.3 are in bold. [a]Compo refers to PCA component. (*Source:* Aires, F., Prigent, C., and Rossow, W.B., *Journal of Geophysical Research*, 109, D10305, 2004.)

some situations, such as shown in bold for the output component 5. This is because the nonlinearity of a network makes it situation dependent (dependence on the input vectors) and therefore an input component can be valuable in some situations.

The Jacobians in Table 7.8 are in the PC space and therefore cannot be compared directly with nonregularized sensitivities in Table 7.6. Therefore, PCA-based sensitivities were transformed to those for actual physical variables using Equation 7.44 and presented in Table 7.9 for comparison with Table 7.6. In Table 7.9, rows are the original input variables and columns are the original output variables. The latter half of the variables in the first column represents the first guesses for the retrieved output variables.

The uncertainty of mean sensitivities is extremely low now, and most of the mean sensitivities are significant to the 5 percent level. Compare again the standard deviations in Table 7.9 and Table 7.6. These results illustrate that PCA has resolved the problem of multicollinearity by suppressing them in the network. Because the interferences among variables are suppressed, the standard deviation of each sensitivity is very small compared to those found for a non-PCA network (Table 7.6). Furthermore, these sensitivities are physically more meaningful. For example,

Table 7.9 Global Mean Regularized Neural Sensitivities for Original Variables, $\partial y_j/\partial x_i$, Translated from PCA-Based Sensitivities (Columns Are Network Outputs y_j and Rows Are Inputs x_i)

	T_s	WV	E_m19V	E_m19H	E_m22V	E_m37V	E_m37H	E_m85V	E_m85H
TB19V	0.23 ± 0.02	$\mathbf{-0.52\pm0.02}$	0.06 ± 0.00	-0.00 ± 0.00	0.06 ± 0.00	0.04 ± 0.00	-0.01 ± 0.00	-0.02 ± 0.00	-0.03 ± 0.00
TB19H	0.06 ± 0.001	0.14 ± 0.01	0.00 ± 0.00	0.03 ± 0.00	-0.00 ± 0.00	-0.01 ± 0.00	0.03 ± 0.00	-0.01 ± 0.00	0.02 ± 0.00
TB22V	0.21 ± 0.01	$\mathbf{-0.34\pm0.01}$	0.05 ± 0.00	-0.01 ± 0.00	0.04 ± 0.00	0.03 ± 0.00	-0.01 ± 0.00	-0.01 ± 0.00	-0.02 ± 0.00
TB37V	0.21 ± 0.01	$\mathbf{-0.27\pm0.01}$	0.04 ± 0.00	-0.01 ± 0.00	0.04 ± 0.00	0.03 ± 0.00	-0.01 ± 0.00	0.00 ± 0.00	-0.02 ± 0.00
TB37H	-0.06 ± 0.01	-0.28 ± 0.01	-0.02 ± 0.00	0.02 ± 0.00	-0.01 ± 0.00	-0.01 ± 0.00	0.02 ± 0.00	0.01 ± 0.00	0.02 ± 0.00
TB85V	-0.12 ± 0.01	$\mathbf{0.39\pm0.01}$	-0.02 ± 0.00	-0.01 ± 0.00	-0.02 ± 0.00	-0.00 ± 0.00	-0.01 ± 0.00	0.04 ± 0.00	0.01 ± 0.00
TB85H	-0.01 ± 0.02	$\mathbf{0.80\pm0.02}$	-0.06 ± 0.00	0.01 ± 0.00	-0.05 ± 0.00	-0.03 ± 0.00	0.01 ± 0.00	0.04 ± 0.00	0.04 ± 0.00
T_s	-0.20 ± 0.02	-0.18 ± 0.02	-0.04 ± 0.00	-0.01 ± 0.00	-0.05 ± 0.00	-0.05 ± 0.00	-0.01 ± 0.00	-0.04 ± 0.00	-0.01 ± 0.00
WV	-0.06 ± 0.01	$\mathbf{0.42\pm0.01}$	0.01 ± 0.00	0.00 ± 0.00	0.01 ± 0.00	-0.00 ± 0.00	-0.01 ± 0.00	-0.02 ± 0.00	-0.01 ± 0.00
E_m19V	-0.09 ± 0.01	0.09 ± 0.01	0.03 ± 0.00	0.01 ± 0.00	0.03 ± 0.00	0.02 ± 0.00	0.01 ± 0.00	0.01 ± 0.00	0.01 ± 0.00
E_m19H	-0.07 ± 0.02	-0.08 ± 0.02	0.03 ± 0.00	0.05 ± 0.00	0.02 ± 0.00	0.00 ± 0.00	0.04 ± 0.00	-0.04 ± 0.00	0.01 ± 0.00
E_m22V	-0.07 ± 0.01	0.05 ± 0.01	0.02 ± 0.00	0.01 ± 0.00	0.02 ± 0.00	0.02 ± 0.00	0.01 ± 0.00	0.02 ± 0.00	0.01 ± 0.00
E_m37V	-0.08 ± 0.01	0.02 ± 0.01	0.02 ± 0.00	0.00 ± 0.00	0.02 ± 0.00	0.03 ± 0.00	0.01 ± 0.00	0.03 ± 0.00	0.01 ± 0.00
E_m37H	-0.08 ± 0.02	-0.13 ± 0.02	0.02 ± 0.00	0.03 ± 0.00	0.02 ± 0.00	0.02 ± 0.00	0.03 ± 0.00	0.00 ± 0.00	0.02 ± 0.00
E_m85V	-0.05 ± 0.01	-0.06 ± 0.01	0.01 ± 0.00	-0.00 ± 0.00	0.01 ± 0.00	0.03 ± 0.00	0.00 ± 0.00	0.05 ± 0.00	0.02 ± 0.00
E_m85H	-0.05 ± 0.02	-0.16 ± 0.02	0.01 ± 0.00	0.01 ± 0.00	0.01 ± 0.00	0.02 ± 0.00	0.02 ± 0.00	0.04 ± 0.00	0.03 ± 0.00
Tlay	-0.03 ± 0.02	0.11 ± 0.02	-0.00 ± 0.00	-0.01 ± 0.00	-0.00 ± 0.00	-0.01 ± 0.00	-0.01 ± 0.00	-0.01 ± 0.00	-0.01 ± 0.00

Sensitivities with absolute value larger than 0.3 are in bold. The first column of the table is for observed inputs and the latter part of it is for first guesses. (*Source:* Aires, F., Prigent, C., and Rossow, W.B., *Journal of Geophysical Research*, 109, D10305, 2004.)

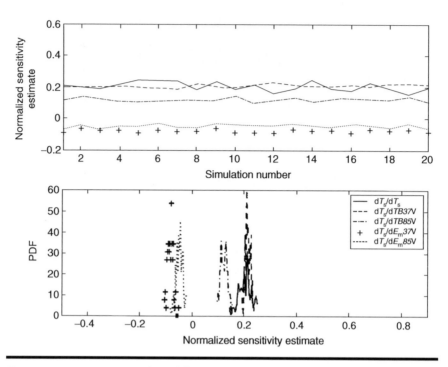

Figure 7.19 **Twenty samples of five sensitivities from PCA regularized network** $\left(\frac{\partial T_s}{\partial T_s}\ \frac{\partial T_s}{\partial TB37V}\ \frac{\partial T_s}{\partial TB85V}\ \frac{\partial T_s}{\partial E_m 37V}\ \frac{\partial T_s}{\partial E_m 85V}\right)$ **(top) and their PDFs (bottom). (From Aires, F., Prigent, C., and Rossow, W.B., *Journal of Geophysical Research*, 109, D10305, 2004.)**

retrieved T_s is sensitive to *TB* at vertical polarization for the low frequencies and to the first guess of T_s. The sensitivity of retrieved T_s to the first guess emissivities is weak regardless of the polarization and frequency.

Information on *WV* clearly comes from the 85 GHz horizontal polarization channel. The sensitivity of *WV* to *TB85H* is almost twice as large as that to first guess *WV*, indicating that the most relevant information for the retrieval of *WV* is extracted from this channel. The sensitivity of the retrieved emissivities depends on the polarization. For example, vertical polarization emissivities are more directly related to first guess of T_s and *TBV* whereas those for horizontal polarization are dominated by their first guess emissivity. Thus, the sensitivity matrix clearly illustrates how the network extracts information from the inputs in the retrieval of outputs.

A sample of these sensitivities and their PDFs are presented in Figure 7.19 and the comparison of the corresponding figure for non-PCA network in

Figure 7.18 confirms the robustness and stability of the Jacobians from the PCA regularized network. Aires et al. [11] states that such PCA regularized neural networks have robust Jacobians even if the network architecture, such as number of hidden neurons, is changed. This illustrates the significant impact of PCA regularization on the reliability and robustness of network Jacobians and the network model. In addition to helping understand how the network links inputs to outputs, the sensitivity matrix can help refine the model by revealing inputs that are insignificant for the prediction of an outcome.

7.4.4.6   Case Study Summary

This section presented a framework for the characterization, analysis, and interpretation of Jacobians and their uncertainties in any neural network-based parameter estimation and illustrated the concept using a case study. The Jacobian of a nonlinear model is a powerful concept [14] and the study illustrated its use to understand the sensitivities of network outputs to inputs and investigate the relative contribution of each input to a given output. PDFs of sensitivities were developed by Monte Carlo simulations involving sampling (1000 times) of sets of weights from the weight PDFs derived from Bayesian statistics and analyzing the corresponding network sensitivities. The results showed that without PCA regularization, sensitivities are highly variable and exhibit multicollinearity. PCA regularization makes them very stable and robust and helps extract physically meaningful relationships from sensitivities. This was made possible by the suppression of multicollinearities among original input and output variables in the PCA representation of both inputs and outputs. This representation also made it easy to explain the variability of the output parameters caused by different land cover types. The PCA-based approach makes both the learning process more stable and Jacobians more reliable and physically meaningful.

7.5   Summary

This chapter presents a detailed treatment of uncertainty assessment of neural network models. These methods for neural networks are in their infancy. First, the standard error criterion used in regular network training is shown to be a simplified version of a more realistic case where variance of noise and initial weights are used to regularize error minimization. It was also demonstrated that the standard error minimization is a special case of minimizing negative log likelihood in statistical parameter estimation. Based on these developments, the optimum weights obtained from regular network training were placed in Bayesian context to obtain a priori PDFs for the final weights of a network. In this formulation, the optimum weights obtained from network training are the most probable or maximum a posteriori weights, and the distribution represents the uncertainty in weights.

The derivation of PDF of weights is illustrated using a case study involving estimation of geophysical parameters that drive Earth's radiative transfer phenomena from satellite data. A sample of weights extracted from the PDFs are presented to illustrate their behavior and correlation structure. The weight distribution is used to generate uncertainty estimates for output error that is divided in to model error and intrinsic noise. These two aspects are treated in detail in the chapter and illustrated in the case study.

The latter part of the chapter involves uncertainty estimation of network input–output sensitivities so that significant inputs can be identified. A detailed treatment of the effect of multicollinearity in inputs and outputs is presented and it is shown that the PC decomposition of inputs and outputs produces networks that have more stable sensitivities than those using regular inputs and outputs that are correlated. The sensitivities obtained from PCA-based networks also have smaller standard deviation, indicating less uncertainty in their values. The PCA approach produces a much clearer view of how inputs are linked to outputs and therefore how relevant inputs can be identified with more confidence.

Problems

1. Explain the reasons for nonunique solution for weights in feedforward networks.
2. What quantities do β and α in the regularized training criterion in Equation 7.1 approximate? Explain the meaning and significance of these quantities.
3. What is intrinsic error of targets and what are the sources of it? What can be learned from error covariance matrix for a multiple output situation? What is the structure of an ideal error covariance matrix?
4. What is an a priori distribution of weights and a priori weight covariance? What can be learned from a weight covariance matrix?
5. Show that the square error minimization is a special case of minimizing negative log likelihood.
6. What are the statistical characteristics of intrinsic uncertainty of targets for multivariate outputs?
7. What are the mean and covariance of an a posteriori distribution of weights. How can they be obtained?
8. What information does the a posteriori weight covariance matrix convey about the relationship between weights? What is an ideal weight covariance matrix?

9. For a multivariate problem of choice, train an appropriate multilayer network and obtain the maximum a priori weights. Compute the Hessian matrix for the network and assess the correlation structure of the weights using the Hessian.

10. For the network in Problem 9, obtain weight covariance matrix by inverting the Hessian and then determine the PDF of weights. Assess the behavior of a selected subset of network weights. One set represents one network. What can you say about the variability of weights?

11. For the data in problem 9, use the PC decomposition on inputs and outputs and select an appropriate number of input and output components. Train a network and obtain a new probability distribution for the weights. Compare the weight distributions from the PCA-based network and the non-PCA network in Problem 9.

12. On what basis can the prediction error from a model be decomposed into model error and intrinsic error? How can these be minimized?

13. Obtain estimates for model error and intrinsic error for the models in Problems 9 and 11. Compare the results.

14. What concepts are used in assessing the influence of inputs on the outputs and selecting relevant inputs?

15. Derive the expression for network sensitivity (Jacobian) in Equation 7.44. Simplify it to one input, one output, and one hidden neuron network.

16. Obtain sensitivities from the non-PCA- and PCA-based networks in Problems 9 and 11, respectively, and compare the results in terms of stability and uncertainty. Select the most significant inputs from the sensitivities.

17. What are some advantages of PCA-based networks?

18. What are the advantages of uncertainty assessment of networks? How can uncertainty be minimized?

References

1. Bishop, C. *Neural Networks for Pattern Recognition*, Clarendon Press, Oxford, 1996.

2. Aires, F. Neural network uncertainty assessment using Bayesian statistics with application to remote sensing: 1 Network weights, *Journal of Geophysical Research*, 109, D10303, 2004. (Figures reproduced by permission of American Geophysical Union.)

3. MacKay, D.J.C. A practical Bayesian framework for back-propagation networks, *Neural Computation*, 4, 448, 1992.

4. Rivals, I. and Personnaz, L. Construction of confidence intervals for neural networks based on least squares estimation, *Neural Networks*, 13, 463, 2000.

5. Rivals, I. and Personnaz, L. MLP (mono-layer polynomials and multi-layer perceptron) for nonlinear modelling, *Journal of Machine Learning Research*, 3, 1383, 2003.
6. Aires, F., Prigent, C., and Rossow, W.B. Neural network uncertainty assessment using Bayesian statistics with application to remote sensing: 2 Output error, *Journal of Geophysical Research*, 109, D10304, 2004. (Figures and tables reproduced by permission of American Geophysical Union.)
7. Rodgers, C.D. Characterization and error analysis of profiles retrieved from remote sounding measurements, *Journal of Geophysical Research*, 95, 5587, 1990.
8. Montano, J.J. and Palmer, A. Numeric sensitivity analysis applied to feedforward networks, *Neural Computing and Applications*, 12, 119, 2003.
9. Garson, G.D. Interpreting neural-network connection weights, *AI Expert*, 6, 47, 1991.
10. Tchaban, T., Taylor, M.J., and Griffin, A. Establishing impacts of the inputs in a feedforward network, *Neural Computing and Applications*, 7, 309, 1998.
11. Aires, F., Prigent, C., and Rossow, W.B. Neural network uncertainty assessment using Bayesian statistics with application to remote sensing: 3 Network Jacobians, *Journal of Geophysical Research*, 109, D10305, 2004. (Figures and tables reproduced by permission of American Geophysical Union.)
12. Warne, K., Prasad, G., Rezvani, S., and Maguire, L. Statistical and computational intelligence techniques for inferential model development: A comparative evaluation and a novel proposition for fusion, *Engineering Applications of Artificial Intelligence*, 17, 871, 2004.
13. Vapnik, V. *The Nature of Statistical Learning Theory*, Springer-Verlag, New York, 1997.
14. Werbos, P.J. *The Roots of Backpropagation. From Ordered Derivatives to Neural Networks and Political Forecasting*, Wiley Series on Adaptive and Learning Systems for Signal Processing, Communication, and Control, Wiley, New York, 1994.
15. Rajanayaka. C., Kulasiri, D., and Samarasinghe, S. A comparative study of parameter estimation in hydrology modelling: Artificial neural networks and curve fitting approaches, *Proceedings of International Congress on Modelling and Simulation (MODSIM'03)*, D.A. Post, ed., Vol. 2, Modelling and Simulation Society of Australia and New Zealand, Townsville, Australia, p. 843, 2003.
16. Rajanayake, C., Samarasinghe, S., and Kulasiri, D. Solving the inverse problem in stochastic groundwater modelling with artificial neural networks, *Proceedings of the 1st Biennial Congress of the International Environmental Modelling and Software Society*, A.E. Rizzoli and A.J. Jakeman, eds., Vol. 2, Servizi Editorial Association, Lugano, Switzerland, p. 154, 2002.
17. Rajanayake, C., Samarasinghe, S., and Kulasiri, D. *A Hybrid Artificial Neural Networks Approach to Solve Inverse Problem in Advection-Dispersion Models*, Applied Computing Mathematics and Statistics Publication Series, Division of Applied Computing and Management, Lincoln University, Christchurch, New Zealand, ISSN 1174-6696, Serial QA75.5 Res. no. 2002/04, 2002.

Chapter 8

Discovering Unknown Clusters in Data with Self-Organizing Maps

8.1 Introduction and Overview

Unsupervised networks are used to find structures in complex data. For example, they can locate natural clusters and one- or two-dimensional relationships in data in a natural way without using an externally provided target output. This is useful because there are many real-life phenomena in which the data is multidimensional and its structure and relationships are unknown a priori; in these situations, the data must be analyzed to reveal the patterns inherent in it. The reason for this is that there are many complex processes in biology, ecology, and the environment that are understood either partially or not at all. However, it is possible to observe these processes and to gather data with relative ease, but to make sense of this data it must be synthesized into coherent and meaningful structures.

For example, what weather patterns drive certain ecological processes? What is the underlying amino acid sequence that produced the specific structure of a protein? What types of species—insects, fish, plants, and so on—assemble together, and what are the conditions under which they form assemblages? All these and many other open-ended problems in the natural world do not have desired outputs; the data must be explored deeply to find meaning in it and to extract answers from it. Unsupervised networks do this by projecting high-dimensional input data onto one- or two-dimensional space to represent it in a compact form so that its inherent structure and patterns can be interpreted meaningfully and validated visually [1]. The task of unsupervised networks presented in this chapter is to perform this projection. Recall that in feedforward networks such as the multilayer perceptron (MLP), each input vector must accompany a desired or target output. In unsupervised networks, targets are not involved; unsupervised networks bear that name because only the inputs are used.

This chapter presents a detailed discussion of unsupervised networks with an emphasis on self-organizing map (SOM) networks, highlighting their internal workings, practical examples, and new developments in the field. Section 8.2 presents the structure of unsupervised competitive networks; Section 8.3 introduces learning in these networks. Implementation of competitive learning is illustrated with examples in Section 8.4. A range of topics pertaining to SOMs are formally introduced in Section 8.5. Specifically, Section 8.5.1 will address learning in SOMs to illustrate SOM training using neighborhood operations for topology preservation in one-dimensional networks, with examples and a real-life case study. The discussion is extended to two-dimensional SOMs in Section 8.5.2, which presents a variety of topics, including map training, preservation of the spatial proximity of data on the map through neighborhood features, quantization (or map error), and the distance matrix (U-matrix) using examples of two-dimensional data.

The SOM concepts are further extended and illustrated in Section 8.5.4, with an application to multidimensional data; this section addresses topics including: map training and analysis of final map structure (Section 8.5.4.2 and Section 8.5.4.3); the quality of representation of input probability density of data by a map; the projection of input data onto a trained map (Section 8.5.4.5); and the quality of retrieval of inputs (Section 8.5.4.6 and Section 8.6.1.1). The chapter then continues to cover classification with maps; cluster formation on the map; the optimization of clusters (Section 8.5.5); and map validation (Section 8.5.6), using examples. In Section 8.6, SOM concepts are further expanded to include evolving SOMs, a strategy which is proposed to evolve a map to a desired size and complexity; this model is able to address the limitations of the regular SOM with a fixed-size map. Several approaches to evolving maps are presented, with examples.

8.2 Structure of Unsupervised Networks

An unsupervised network usually has two layers of neurons: an input layer and an output layer. The input layer represents the input variables, x_1, x_2, …, x_n, for the case of n inputs. The output layer may consist of neurons arranged in a single line (one-dimensional) or a two-dimensional grid, forming a two-dimensional layer. These two forms are shown in Figure 8.1.

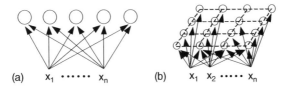

Figure 8.1 Unsupervised network structures: (a) one-dimensional and (b) two-dimensional network configurations.

The main feature of an unsupervised network is the weights that link the inputs to the output neurons. These weights are the free parameters of a network; learning involves adapting these weights. Each output neuron receives inputs through the weights that link it to the inputs, so the weight vector has the same dimensionality as the input vectors. The output of each neuron is its activation, which is the weighted sum of inputs (i.e., linear neuron activiation). The objective of learning is to project high-dimensional data onto one- or two-dimensional output neurons on the basis of their activation in such a way that each output neuron incrementally learns to represent a cluster of data. The weights are adjusted incrementally; the final weights of neurons representing the input clusters are called codebook vectors or weights. These weights are found by unsupervised learning mechanisms.

8.3   Learning in Unsupervised Networks

Unsupervised learning is a central part of our daily life: every day, the human brain naturally implements unsupervised learning. Humans are accustomed to synthesizing a myriad of information and organizing it into compact forms, such as perception, recognition, and categorization, that are meaningful to our lives. One key question addressed by early researchers of neural networks was how the neurons in the human brain facilitate this natural self-organization of information. An important contributor in this area was Frank Rosenblatt [2], who introduced the concepts of unsupervised or competitive learning as a possible learning mechanism in the brain, as presented in Chapter 2. These and other of his pioneering ideas led to the development of the perceptron and the first implementation of learning in neural networks, as discussed in Chapter 2.

To explain how the human brain recognizes similar patterns and distinguishes them from dissimilar patterns, Rosenblatt [2] proposed a model of competitive learning between neurons. In his model that attempts to mimic this brain function, neurons inhibit each other by sending their activation as inhibitory signals, the goal being to win a competition for the maximum activation corresponding to an input pattern. The neuron with the maximum activation then represents the input pattern that led to its activation. This neuron alone becomes the winner and is allowed to adjust its weight vector by moving it closer to that input vector; however, the neurons that lose the competition by succumbing to the inhibition are not allowed to change their weights. Another neuron may become the winner for another input pattern; this neuron gets to adjust its weights, moving them closer to the input pattern for which it was the winner. Over time, the individual output neurons learn to specialize, responding to a specific set of inputs. This idea of competitive learning is presented in Figure 2.9 of Chapter 2, in which the original perceptron hypothesis is applied to vision.

8.4 Implementation of Competitive Learning

The implementation of competitive learning in a simple network will now be presented. First, the number of output neurons must be determined. In many cases, the number of data clusters is unknown; it is therefore necessary to use a reasonable estimate based on the current understanding of the problem. When there is uncertainty, it is better to have a larger number of output neurons than the possible number of clusters because redundant neurons can be eliminated. The problem determines the dimensionality of the input vector. As with feedforward networks such as MLP, the larger the number of input variables, the larger the number of weights and hence the higher the complexity of the network, which is undesirable. After the number of input variables and output neurons has been set, the next step is to initialize the weights. These may be set to small random values, as was done in the MLP networks. Another possibility is to randomly choose some input vectors and use their values for the weights. This has the potential to speed up learning. Now that the structure of the networks has been defined, learning can commence.

8.4.1 Winner Selection Based on Neuron Activation

In competitive learning, an input is presented to the network and the winner is selected based on the neuron activation (i.e., net input into a neuron). This involves the presentation of an input vector $\mathbf{x}$ (with components $\mathbf{x}_i$) to the network and the computation of the net input ($\mathbf{u}$), which is the weighted sum of inputs, into each of the neurons. Thus

$$\mathbf{u}_j = \sum_{i=1}^{n} \mathbf{w}_{ij}\mathbf{x}_i, \tag{8.1}$$

where $\mathbf{x}_i$ is the ith input variable, $\mathbf{w}_{ij}$ is the weight from input $\mathbf{x}_i$ to output neuron j, and n is the dimensionality of the input. Once each output neuron has computed its activation, competition can begin. There are several ways this can happen; a simple way is for each neuron to send its signal in an inhibitory manner, with an opposite sign to other neurons. Once each neuron has received signals from the others, each neuron can compute its net activation by simply summing the incoming inhibitory signals and its own activation. If the activation drops below a threshold (or zero), that neuron drops out of the competition. As long as more than one neuron remains, the cycle of inhibition continues until one winner emerges; its output is set to one. This neuron is declared the winner because it has the highest activation and it alone represents the input vector.

8.4.2   Winner Selection Based on Distance to Input Vector

The competition described above can be implemented much more simply by using the concept of distance between an input and a weight vector. By scaling the weights and inputs so that their relative lengths ($\|\mathbf{x}\|$ and $\|\mathbf{w}\|$) are one, then it can be shown that a weight that is closer to an input vector would cause a larger activation than one that is far away from the vector. This is because the net input in Equation 8.1 can also be presented as $\|\mathbf{x}\|\|\mathbf{w}\| \cos \theta$, where θ is the angle between the input vector $\mathbf{x}$ and the weight vector $\mathbf{w}$. For $\|\mathbf{w}\| = 1$, $\|\mathbf{x}\| = 1$, the inputs that are closer to $\mathbf{w}$ will make a smaller angle, leading to higher value of $\cos \theta$ and consequently a higher net input. This is illustrated in Figure 8.2. In this case, the input and weight vectors are normalized to unit length by dividing each component of the vector by its length. For example, if the two components representing two input variables of an input vector are 3.0 and 4.0, then the length of the vector is $\sqrt{(3^2 + 4^2)} = 5$ and, therefore, the normalized vector components are 3/5 and 4/5. The length of the normalized vector is 1.0.

Thus, the competition will eventually be won by the neuron associated with the weight that is closest to an input, which will consequently have the highest activation. A simple measure of the closeness of a weight to an input vector is the Euclidean distance between them; this is defined as

$$d_j = \mathbf{x} - \mathbf{w}_j = \sqrt{\sum_{i=1}^{n} (\mathbf{x}_i - \mathbf{w}_{ij})^2}, \qquad (8.2)$$

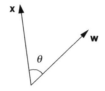

Figure 8.2   Closeness between an input vector x and weight vector w.

where d_j is the distance between the input vector **x** and the weight vector $\mathbf{w}_j$ associated with the *j*th output neuron. This is illustrated in Figure 8.3, in which the subscript *j* has been dropped for clarity. Once the distance between an input vector and all the weights has been found, the neuron with the smallest distance to the input vector is chosen as the winner, and its weights are updated so that it moves closer to the input vector, as

$$\Delta\mathbf{w}_j = \beta(\mathbf{x} - \mathbf{w}_j) = \beta d_j \qquad (8.3)$$

where β is the step length (or learning rate), which indicates what portion of the distance between the two vectors (input and weight) the weight vector must cross towards the input vector. The updated weight $\mathbf{w}'$ (dashed line) is also shown in Figure 8.3, in which $\mathbf{w}^0$ is the weight of a winning neuron before the update. The other weights remain unchanged. When the inputs are processed and the weights adjusted based on a distance measure such as the expression in Equation 8.3, the clustering resulting from competitive learning is analogous to cluster analysis in statistics. The weights associated with each neuron represent a cluster center representing the inputs that are closest to it. For example, if an input vector closest to a cluster center is presented to the trained network, the winning neuron will be the one with weights representing that cluster center.

8.4.2.1 Other Distance Measures

The most widely used distance measure is Euclidean distance. Other related measures include correlation, direction cosines, and city block distance. Correlation is a measure of the similarity between two vectors; in the case of the similarity between an input vector and a weight vector, correlation simply is the weighted sum of the inputs. The higher the weighted sum, the higher the correlation, and the more similar the two vectors. The direction cosine is the angle between two vectors, and as shown in Figure 8.2, when the two vectors have unit lengths, the larger the cosine angle, the closer the two vectors are to each other. City block distance is the sum of the absolute values of the difference between corresponding vector components (i.e., $d_j = \sum_{i=1}^{n} |\mathbf{x}_i - \mathbf{w}_{ij}|$).

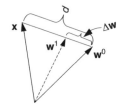

Figure 8.3 Illustration of weight update in competitive learning.

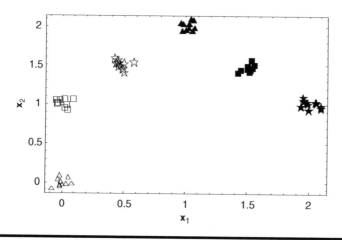

Figure 8.4 Two-dimensional input data with six distinct clusters. (Data sourced from *Mathematica—Neural Networks*, Wolfram Research, Inc., Champaign, IL.)

8.4.3 Competitive Learning Example

This section will present a computer experiment to promote understanding of unsupervised learning. Figure 8.4 depicts a two-dimensional input dataset with six distinct clusters, which was extracted from *Mathematica— Neural Networks* [3]. Two inputs are denoted x_1 and x_2. In this case, the cluster structure is obvious and six neurons are needed in the output layer. Thus, the network structure has two inputs and six output neurons, as shown in Figure 8.5.

There are six sets of weights associated with the six output neurons. Before training, these need to be initialized to random values. Figure 8.6 shows the initial weight vectors superimposed on the data. It can be seen that the position of these vectors is far from the cluster centers.

The objective of training is to evolve weights so that they each assume the center position of a cluster. Consequently, each neuron with its respective weights is more sensitive to inputs in its own cluster than to other inputs and

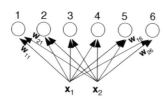

Figure 8.5 Unsupervised one-dimensional network configuration for the data in Figure 8.4.

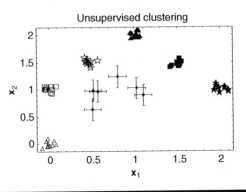

Figure 8.6 Initial random weight vectors (crosses) superimposed on data.

thus becomes the winner for inputs from its cluster. For this to happen, each weight vector should represent a cluster whose inputs are closer to the corresponding weight vector than to the weight vectors representing other clusters. Unsupervised or competitive learning facilitates such movement of weights to respective cluster centers through either recursive or batch learning.

8.4.3.1 Recursive Versus Batch Learning

Unsupervised networks can be trained in either recursive or batch mode. In recursive mode, the weights of the winning neurons are updated after each presentation of an input vector, whereas in batch mode, the weight adjustment for each neuron is made after the entire batch of inputs has been processed. In the latter case, the weight update for each input vector is noted, but the weights are not changed until all the input patterns have been presented. After an epoch (i.e., one pass of the whole training dataset through the network), the average weight adjustment for each neuron is computed and the weights are changed by this amount. In this method, the weights are adjusted such that the distance between a representative weight vector and the inputs in the cluster it represents decreases in an average sense. Conceptually, this is similar to the batch learning in multiple layer networks presented in Chapter 3 and Chapter 4. Training terminates when the mean distance between the winning neurons and the inputs they represent is at a minimum across the entire set of clusters, or when this distance stops changing.

8.4.3.2 Illustration of the Calculations Involved in Winner Selection

Before examining the results from training this network, a small hand calculation will be performed to determine the distance, find the winner neuron, and adjust the weights recursively for two randomly selected input patterns from the data shown in Figure 8.4. The two patterns are

Input 1: $(-0.035, 0.030)$; Input 2: $(-0.033, 1.013)$.

The random initial weight vectors extracted from the neural network program (*Mathematica—Neural Networks* [3]) for the six output neurons are

$$
\begin{bmatrix}
0.805 & 1.234 \\
0.520 & 0.977 \\
0.574 & 0.963 \\
1.027 & 1.023 \\
1.106 & 0.893 \\
0.514 & 0.626
\end{bmatrix}.
$$

The distance between an input and a weight vector is calculated by computing the difference between their respective vector components, summing the square of these differences, and finding the square root of the sum. Thus, the distance d_1 between the first input vector and the weight vector of the first output neuron is

$$d_1 = \sqrt{(-0.035 - 0.805)^2 + (0.03 - 1.234)^2} = 1.468.$$

The distance d_2 between the first input vector and the second output neuron weight vector is

$$d_2 = \sqrt{(-0.035 - 0.52)^2 + (0.03 - 0.977)^2} = 1.098.$$

Similarly, the distance from the same input vector to the other four output neuron weight vectors, d_3, d_4, d_5, and d_6, can be calculated; they are 1.115, 1.455, 1.431, and 0.810, respectively. Thus, the first input pattern is closest to the weight vector of output neuron 6, which will be the winner.

The weight update for this neuron, using Equation 8.3 for a learning rate or step length β of 0.1, is

$$
\begin{aligned}
\Delta \mathbf{w}_6 &= \beta(\mathbf{x} - \mathbf{w}_6) \\
&= 0.1 \times [(-0.035, 0.03) - (0.514, 0.626)] \\
&= 0.1 \times [(-0.035 - 0.514), (0.03 - 0.626)] \\
&= (-0.055, -0.059).
\end{aligned}
$$

Following the same procedure for the second input pattern yields the following distances between the second input vector and each of the six output neuron weight vectors: 0.866, 0.554, 0.609, 1.06, 1.146, and 0.665. The second output neuron, with the minimum distance of 0.554 from the input vector, is the winner. Its weight vector update is

$$\Delta\mathbf{w}_2 = \beta(\mathbf{x} - \mathbf{w}_2)$$
$$= 0.1 \times [(-0.033 - 0.520), (1.013 - 0.977)]$$
$$= (-0.055, -0.0036).$$

The training criterion is the mean distance (the sum of the squared distance) between all the inputs and their respective winning neuron weights which represent the cluster centers. For this small dataset with two input vectors, the training criterion is

$$\text{mean dist} = 0.810^2 + 0.554^2 = 0.963.$$

The objective of training is to minimize the mean distance over iterations.

This is an illustration of the training process in recursive mode with only two input patterns. Training with the whole dataset in batch mode involves finding the winning neuron for each input pattern and calculating the weight change for the winning neuron. After the whole dataset has been presented once (i.e., one epoch), the total increment for each weight is determined and the weights are updated. As training progresses, the mean distance will decrease because during each epoch, as the weights are moved closer to the inputs in the cluster they represent. The mean distance will eventually reach a minimum, at which point training stops. The mean distance D between all inputs and their respective cluster centers represented by the weight vector of each of the winning neurons, can be expressed as

$$D = \sum_{i=1}^{k} \sum_{n \in c_i} (\mathbf{x}^n - \mathbf{w}_i)^2, \tag{8.4}$$

where $\mathbf{x}^n$ is the nth input vector belonging to cluster c_i, whose center is represented by $\mathbf{w}_i$, i.e., the weight vector of the winning neuron representing cluster c_i. There are k clusters. The two summations mean that the distance is computed over all clusters and all input patterns in clusters. This distance D is minimized over the learning epochs. Turning to the example problem, it will now be demonstrated how the network performs on the data.

8.4.3.3 Network Training

Using *Mathematica—Neural Networks* [3], the network in Figure 8.5 was trained with random initial weights until the mean distance was minimized. Figure 8.7 shows the training performance, indicated by how quickly the mean distance decreases with each epoch. In this figure, one iteration represents an epoch. It shows that training was complete in seven epochs, reaching a minimum mean distance of 0.065.

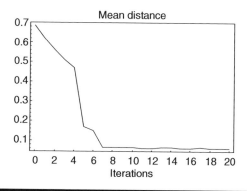

Figure 8.7 Reduction in mean distance with training epochs during network training.

It is now possible to see the position of the new weights and how well they represent the data clusters. In Figure 8.8, the final weights (denoted by crosses) are superimposed on the data, and the weights have indeed assumed the position of cluster centers, indicating that training was successful.

Now each of the six output neurons is sensitive to inputs from its cluster, so for any new input presented to the network, the neuron representing the cluster that the input belongs to will be highly active and declare itself the winner. In this manner, it is possible to determine to what cluster any unknown input vector belongs. Figure 8.9 presents the same training results as Figure 8.8, but in a slightly different way. Here also, the final weights are superimposed on the data, but with a label indicating the class or category label for each cluster. These labels denote the output neurons in the map in Figure 8.5, where the leftmost neuron is labeled 1 and the rightmost neuron 6.

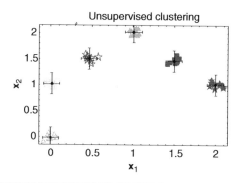

Figure 8.8 Final output neuron weights of the trained network superimposed on data.

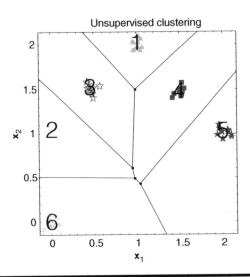

Figure 8.9 Classification regions, called Voronoi cells, that define classification boundaries for each output neuron, superimposed on data; the numbers denote the label of an output neuron in the network in Figure 8.5 and indicate the position of final weights that represent cluster centers.

It can be seen that the neurons in the trained network represent data clusters randomly, i.e., neurons that are closer in the physical network in Figure 8.5 do not represent clusters that are closer together in input space in Figure 8.9. More importantly, Figure 8.9 shows the classification (or influence) region of each output neuron. These regions are called Voronoi cells.

Voronoi cells are enlarged regions around input clusters, and are like input catchments for the output neurons representing clusters. If an input falls within a Voronoi cell, the corresponding output neuron will become active, indicating that the input belongs to that cluster. The Voronoi cells are determined by the distances between weight vectors. Specifically, if the positions of two winning neurons in Figure 8.9 are joined by a line, its perpendicular bisector meets other bisectors to form regions resembling a honeycomb. This division of the input space is called Voronoi tessellation, and the individual regions are the resulting Voronoi cells [1].

It is interesting to observe how, during training, the weights evolve from the initial random positions to cluster centers. This can be seen from the intermediate results of training as the weights gradually change their positions; as the weight vector for each of the output neurons moves in search of a cluster center, its path is denoted by a dashed line, as shown by Figure 8.10. It can be seen that each weight can follow either a linear or a nonlinear path to its destination.

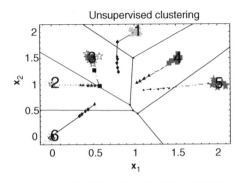

Figure 8.10 Evolution of weights from initial positions to cluster centers during training (lines trace the position of weights at intermediate stages of training).

8.5   Self-Organizing Feature Maps

It has been accepted that competition among neurons is the mechanism responsible for self-organization in the brain [5]. It is known that the human brain is a self-organized entity, in which different regions correspond to specific aspects of human activity; moreover, these regions are organized such that tasks of a similar nature, such as speech and vision, are controlled by regions that are in spatial proximity to each other [5]. This preservation of spatial organization is called topology preservation, and was incorporated into competitive learning in artificial neural networks during the early 1980s by Kohonen [4,25], who invented self-organizing feature maps (SOFMs or SOMs, also called Kohonen's maps). In SOFMs, not only the winner neuron but also neurons in the neighborhood of the winner adjust their weights together so that a neighborhood of neurons becomes sensitive to a specific input. This neighborhood feature helps to preserve topological characteristics of inputs. This is an important aspect in the implementation of self-organization, as many phenomena in the natural world are driven by spatially correlated processes or attributes. Therefore, inputs that are spatially closer together must be represented in close proximity in the output layer or map of a network.

8.5.1   Learning in Self-Organizing Map Networks

In SOFM learning, not only the winner but also the neighboring neurons adjust their weights. Neurons closer to the winner adjust weights more than those that are far from it. Thus, we need to define the size of the neighborhood as well as by how much the neighbor neurons must adjust their weights.

8.5.1.1   Selection of Neighborhood Geometry

There are several ways to define a neighborhood. Linear, square, and hexagonal arrangements shown in Figure 8.11 are the most common. If only

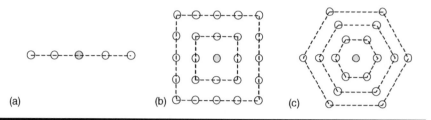

Figure 8.11 Neighborhood definitions: (a) linear (b) square, and (c) hexagonal neighborhood surrounding a winning neuron (solid circle denotes winner and empty circles denote neighbors).

the most immediate neighbors of the winner are considered, the distance, also called radius r, is 1. If two levels of adjacent neighbors are considered, then the radius is 2. For example, in the linear case, a radius of 1 includes one neighbor to the right and one to the left of the winner. A radius of 2 would include two neighbor neurons each to the left and right of winner, making a total of four in the neighborhood. In the case of a square map, a radius of 1 includes all neurons separated by one step from the winner and includes eight neurons as shown in Figure 8.11b. A hexagonal neighborhood is associated with a map where neurons are arranged in a hexagonal grid. For a radius of 1, this includes six neurons; a radius of 2 will encompass another layer of neurons located an additional step away.

8.5.1.2   Training of Self-Organizing Maps

Training an SOM follows in a similar manner to the standard winner-takes-all competitive learning. However, a new rule is adopted for weight changes. Suppose that for a random n-dimensional input vector $\mathbf{x}$ with components $\{x_1, x_2, \ldots, x_n\}$, the position of the closest codebook vector of the winning neuron is identified and indexed as $\{i_{win}, j_{win}\}$ on the map. Then all the codebook vectors $\mathbf{w}_j$ of the winner and neighbors are adjusted to $\mathbf{w}_j'$ according to

$$\mathbf{w}_j' = \mathbf{w}_j + \beta\, NS[\mathbf{x} - \mathbf{w}_j], \tag{8.5}$$

where NS is the neighbor strength that varies with the distance to a neighbor neuron from the winner and β is the learning rate as described previously. Neighbor strength defines the strength of weight adjustment of the neighbors with respect to that of the winner as presented in the next section.

8.5.1.3   Neighbor Strength

With the neighbor feature, all neighbor codebook vectors are shifted towards the input vector; however, the winning neuron update is the most

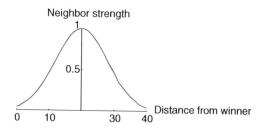

Figure 8.12   Gaussian neighbor strength function.

pronounced and the farther away a neighbor neuron is, the less its weight update. The NS function determines how the weight adjustment decays with distance from the winner. There are several possibilities for this function and some commonly used functions are linear, Gaussian, and exponential.

The simplest form of NS function is the linear decay function, where the strength decreases linearly with distance from the winning neuron. The Gaussian form of the NS function makes the weight adjustments decay smoothly with distance, as shown in Figure 8.12, and is given by

$$NS = Exp\left[\frac{-d_{i,j}^2}{2\sigma^2}\right],\tag{8.6}$$

where $d_{i,j}$ is the distance between the winning neuron i and any other neuron j, and σ is the width of the Gaussian. This width is usually defined in terms of the radius of the neighborhood and the width of the function shown in Figure 8.12 is 20. The strength is maximum (1.0) at the winning neuron, which is positioned at the center of Figure 8.12. The distance from the winner must be extracted accordingly.

The exponential decay NS function is given by

$$NS = Exp[-k\,d_{i,j}],\tag{8.7}$$

where k is a constant. For $k = 0.1$, the form of the function is shown in Figure 8.13 where the strength is maximum at the winning neuron which is positioned at the center of the figure; the distance from the winner must be extracted accordingly.

8.5.1.4   *Example: Training Self-Organizing Networks with a Neighbor Feature*

To gain confidence in the method, the six-cluster problem that was solved with competitive learning in Section 8.4.3 will be revisited. The data is presented in Figure 8.4 and the network in Figure 8.5. Assume the same initial weights and one of the input patterns used previously in

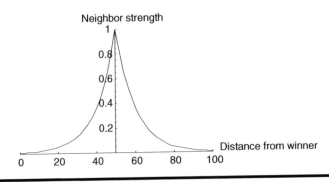

Figure 8.13 Exponential decay neighbor strength function.

Section 8.4.3.2 in this exercise. The only differences are that a neighbor-topology-defining distance d from a neighbor to winner, and neighbor strength NS that depends on distance d, are introduced. For this example, the exponential NS function given in Equation 8.8 is used. There are six codebook vectors, and for convenience in a hand calculation, an initial neighborhood size of two (i.e., two neighbors each to the left and right of winner) will be used.

$$NS = Exp[-0.1d]. \tag{8.8}$$

The neighbor strength function in Equation 8.8 behaves as in Figure 8.14 for $d = 2$.

The initial codebook vectors repeated from Section 8.4.3.2 are

$$\begin{bmatrix} 0.805 & 1.234 \\ 0.520 & 0.977 \\ 0.574 & 0.963 \\ 1.027 & 1.023 \\ 1.106 & 0.893 \\ 0.514 & 0.626 \end{bmatrix} \tag{8.9}$$

The input pattern will now be processed and the weights will be adjusted using the neighbor feature.

Iteration 1: Input 1: $(-0.035, 0.030)$. Previously, in Section 8.4.3.2, the codebook vector 6 was the winner for this input pattern and only its weights were adjusted. Here, however, the neighbor codebook vectors will also be adjusted according to NS. The neighborhood size is two, but neuron 6 has only two neighbors to the left and none to the right (see Figure 8.5). Therefore, only neurons 4 and 5 adjust their weights. Assume that the learning rate is 0.5.

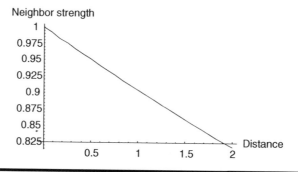

Figure 8.14 **Exponential neighbor strength function for neighborhood size of 2.**

Weight update for neuron 6:

From Equation 8.8 and Figure 8.14: $NS = 1.0$; $\beta = 0.5$;

$$\mathbf{x} = (-0.035, 0.030); \quad \mathbf{w}_6(0) = (0.514, 0.626).$$

From the weight update equation (Equation 8.5):

$$\mathbf{w}'_j = \mathbf{w}_j + \beta\, NS(\mathbf{x} - \mathbf{w}_j) \tag{8.10}$$

$$\mathbf{w}_6(1) = (0.514, 0.626) + 0.5 \times 1.0\big[(-0.035, 0.03) - (0.514, 0.626)\big]$$

$$= (0.514, 0.626) + (-0.2745, -0.2979) = (0.239, 0.328).$$

Weight update for neuron 5:

From Equation 8.8 and Figure 8.14, for distance $d = 1$, $NS = 0.9048$; $\beta = 0.5$;

$$\mathbf{x} = (-0.035, 0.030); \quad \mathbf{w}_5(0) = (1.106, 0.893);$$

$$\mathbf{w}_5(1) = (1.106, 0.893) + 0.5 \times 0.9048\big[(-0.035, 0.030) - (1.106, 0.893)\big]$$

$$= (1.106, 0.893) + (-0.51658, -0.390403) = (0.590, 0.5026)$$

Weight update for neuron 4:

From Equation 8.8 and Figure 8.14, for distance $d = 2$, $NS = 0.8187$; $\beta = 0.5$:

$$\mathbf{x} = (-0.035, 0.030); \quad \mathbf{w}_4(0) = (1.0275, 1.0235);$$

$$\mathbf{w}_4(1) = (1.0275, 1.0235) + 0.5 \times 0.8187[(-0.035, 0.030) - (1.0275, 1.0235)]$$

$$= (1.0275, 1.0235) + (-0.435053, -0.406696) = (0.5924, 0.6168)$$

This completes the weight adjustment of the winner and the neighbors in a large map for input 1. The process is repeated for all the inputs until the mean distance from inputs to representative cluster centers (codebook vectors) does not change or becomes acceptable. Now that the method for adjusting weights in a neighborhood has been demonstrated, an efficient method to obtain distance to the winner from neighbors in a large map will be discussed.

8.5.1.5 Neighbor Matrix and Distance to Neighbors from the Winner

When the map is large, an efficient method is required to determine the distance of a neighbor from the winner to compute neighbor strength. As shown in Figure 8.11, linear, square, and hexagonal neighborhoods are common. From these neighbor configurations, the distance can be determined. For example, the distance to the immediate neighbor is one, and to the next immediate neighbor is two, etc. To compute these distances, a neighbor matrix (**NM**), also called distance matrix, can be designed that defines the organization and size of the neighborhood of neurons around the winner and identifies the distance between the winning neuron and the neighbor neurons.

Suppose that a map consists of a single layer of four neurons. A neighbor matrix for this map that considers all neurons on the map as neighbors can be designed as

$$\mathbf{NM} = [3, 2, 1, 0, 1, 2, 3], \qquad (8.11)$$

in which a distance of zero is given to the position c of the winner in the neighbor matrix. In this example with three neurons, $c = 4$ (i.e., fourth position in the neighbor matrix). The numbers in the matrix indicate the distance from the winner. For example, if the winner is the leftmost neuron on the map, all three adjacent neurons to the right of it are neighbors and the distance to the rightmost neuron is 3. Similarly, if the winner is the rightmost neuron, all three adjacent neurons to the left of it are neighbors and the distance to the leftmost neuron is 3. Basically, the neighbor matrix translates this information to a template for efficient extraction of the distance from the winner to all neighbors in the neighborhood. The neighbor matrix's use will now be demonstrated.

Suppose that the coordinate (position) of the winner on the actual map is indicated by i_{win}. Then the distance between the winner and any neighbor neuron at position i on the map is found by matching the coordinates of the winner i_{win} with position c, designated for the winner in the neighbor matrix template, indicated by zero. In other words, the neighbor template is matched with the map so that the position of the winner on the map coincides with the designated center position of the winner on the template.

For example, if the winner is the very first of the four neurons (leftmost) on the map, then its coordinate $i_{win} = 1$ and the neighbor template is positioned so that zero distance is matched with the first neuron. It is found that there are three neighbors to the right of the first neuron. Where there are no neurons physically on the map to match the positions on the template, these positions are ignored. The distance d between the winner i_{win} and any neuron i can be found from

$$d = \mathbf{NM}[c - i_{win} + i]. \tag{8.12}$$

Thus, for the leftmost winning neuron, the distance to the immediate neighbor at position $i = 2$ on the map is

$$d = \mathbf{NM}[4 - 1 + 2] = \mathbf{NM}[5] = 1,$$

where $\mathbf{NM}[5]$ is the number in the fifth location in the neighbor matrix template, which is 1. Similarly, for the neighbor at position $i = 3$ on the map:

$$d = \mathbf{NM}[4 - 1 + 3] = \mathbf{NM}[6] = 2.$$

Two-dimensional maps will be formally studied in Section 8.5.2, but the design of a neighbor matrix for a two-dimensional map will be introduced here to complete the discussion. For a map of 12 neurons arranged in three rows and four columns, as shown in Figure 8.15, the neighbor (or distance) matrix for a rectangular neighborhood is

$$\mathbf{NM} = \begin{bmatrix} 3 & 2 & 2 & 2 & 2 & 2 & 3 \\ 3 & 2 & 1 & 1 & 1 & 2 & 3 \\ 3 & 2 & 1 & 0 & 1 & 2 & 3 \\ 3 & 2 & 1 & 1 & 1 & 2 & 3 \\ 3 & 2 & 2 & 2 & 2 & 2 & 3 \end{bmatrix} \tag{8.13}$$

where zero indicates the position of the winning neuron in the template.

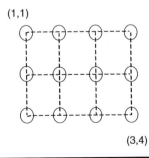

(1,1)

(3,4)

Figure 8.15 **A map of 12 neurons arranged in three rows and four columns.**

The numbers indicate the distance from the winner. This template is used to find the distance between the winning neuron and its neighbors by matching the coordinates of the winner on the map with its designated position in the neighbor matrix (i.e., center). This is basically superimposing the template on the two-dimensional map so that its center coincides with the winning neuron.

Suppose that the horizontal and vertical coordinates of the winner neuron on the two-dimensional map in Figure 8.15 are indicated by (i_{win}, j_{win}). Then the distance between the winner and any neighbor neuron at position (i, j) is found in a manner similar to that used for a linear neighborhood as

$$d = \mathbf{NM}\big[[c_1 - i_{win} + i, c_2 - j_{win} + j]\big], \tag{8.14}$$

where $\{c_1, c_2\}$ is the position of the winner in the neighbor matrix **NM** (Equation 8.13), which indicates the neighborhood organization for the 3×4 map. For this case, $c_1 = 3$ and $c_2 = 4$.

To illustrate how the distance to a neuron is determined, suppose that the winner is found to be the neuron at the position $(1, 1)$ (i.e., top left corner) on the map in Figure 8.15. Thus, $i_{win} = 1$ and $j_{win} = 1$. Now, if the neighbor matrix template is superimposed so that its center coincides with the winning neuron, it can be seen that the bottom right quarter of the matrix spans the whole map, indicating that for this size **NM**, all neurons on the map are neighbors. With the known position of the center, that is $\{c_1, c_2\} = \{3, 4\}$, we can extract the distance to all neighbors. For example, the distance to the neuron at the location of $\{1, 4\}$ on the map, (i.e., $i = 1$ and $j = 4$, which is the last neuron on the first row of the map) from the winner located at $i_{win} = 1$ and $j_{win} = 1$, is

$$d = \mathbf{NM}[3 - 1 + 1, 4 - 1 + 4] = \mathbf{NM}[3, 7] = 3,$$

where **NM**[3,7] refers to the position indicated by the third row and seventh column in the neighbor matrix in Equation 8.13, which indicates that $d = 3$. This way, the distance to all neighbors from the winner can be obtained efficiently by matching the template with the winning neuron on the physical map so that the position indicating zero distance coincides with the winner.

In training SOMs, several other features are used for efficient convergence and fine tuning of maps. One is that the neighborhood size shrinks during learning so that only the winner or its immediate neighbors remain in the neighborhood. The other is that the learning rate also decays during training so that the accuracy of representation of the input space by the map becomes more refined. These topics will be discussed in the next two sections and the application of these features in training of SOMs will be illustrated through an example case study.

8.5.1.6 Shrinking Neighborhood Size with Iterations

A large initial neighborhood guarantees proper ordering and placement of neurons in the initial stages of training to broadly represent spatial organization of input data. However, subsequent shrinking of neighborhood is required to further refine the representation of the input probability distribution by the map. A larger initial neighborhood is necessary because smaller initial neighborhoods can lead to metastable states corresponding to local minima [1,6]. Therefore, the size of a large starting neighborhood is reduced with iterations and there are several forms that can be used for shrinking the neighborhood size. Equation 8.15 shows a linear function commonly used for this purpose:

$$\sigma_t = \sigma_0(1 - t/T), \tag{8.15}$$

where σ_0 is the initial neighborhood size, σ_t is the neighborhood size at iteration t, and T is the total number of iterations that would bring σ_t to zero (i.e., only the winner) or few desired number of neurons. The T can be adjusted to reach the desired neighborhood size. For example, Equation 8.15 with $T = 1000$ produces a linear decay of the neighborhood size with iterations as shown in Figure 8.16, where the initial size of 50 neurons in the neighborhood is reduced to zero, leaving only the winning neuron after 1000 iterations.

Exponential decay is another form used for adjusting neighborhood size with iterations, as given by

$$\sigma_t = \sigma_0 \text{Exp}[-t/T], \tag{8.16}$$

where σ_t is the width of the neighborhood at iteration t, σ_0 is the initial width of the neighborhood, and T is a constant that allows exponential function to decay to zero with iterations. This is illustrated in Figure 8.17, where initial neighborhood of size 10 decays to the winner after 300 iterations for $T = 50$.

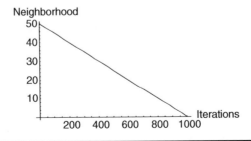

Figure 8.16 Linear decay of neighborhood size with iterations.

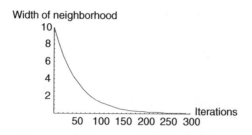

Figure 8.17 Exponential decay of neighborhood width with iterations.

The σ_t in Equation 8.16 can be substituted into Equation 8.6 so that the decay in neighborhood size is integrated into the NS function as

$$NS(d,t) = \text{Exp}\left[-d_{i,j}^2/2\sigma_t^2\right] = \text{Exp}\left[-d_{i,j}^2/2\{\sigma_0\text{Exp}(-t/T)\}^2\right], \quad (8.17)$$

where $NS(d,t)$ indicates that the neighbor strength changes not only with distance d, but also with the neighborhood size, which shrinks with iterations t. An illustration of how the neighborhood shrinks with iterations is given in Figure 8.18. Specifically, an initial neighbor strength function with $\sigma_0 = 20$, as shown in Figure 8.18a, shrinks to that shown in Figure 8.18b after 1500 iterations for a value of $T = 800$ in Equation 8.17.

In the next section, learning rate decay is addressed and both learning rate and neighborhood decay are integrated into the original weight update formula presented in Equation 8.5.

8.5.1.7 Learning Rate Decay

The step length, or the learning rate β, is also reduced with iterations in self-organizing learning and a common form of this function is the linear decay,

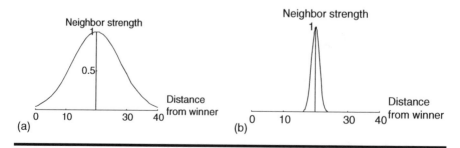

Figure 8.18 Neighbor strength decay during training: (a) initial neighbor strength and (b) final strength indicating that only the winner or few neighbors are involved in the weight update.

given by

$$\beta_t = \beta_0(1 - t/T), \qquad (8.18)$$

where β_0 and β_t are the initial learning rate and that at iteration t, respectively. T is a constant that brings the learning rate to zero or a very small value at the end of the specified number of t iterations. Another form is the exponential decay of the learning rate given by

$$\beta_t = \beta_0 \text{Exp}[-t/T], \qquad (8.19)$$

where T is a time constant that brings the learning rate to a very small value with iterations.

A general guide is to start with a relatively high learning rate and let it decrease gradually but remain above 0.01. For $\beta_0 = 0.1$ and $T_2 = 1000$, for example, the learning rate decay from Equation 8.19 is shown in Figure 8.19.

8.5.1.8 Weight Update Incorporating Learning Rate and Neighborhood Decay

Learning rate and neighborhood decay can now be integrated into the original weight update formula in Equation 8.5. Thus, the weight update after presenting an input vector **x** to a SOM incorporating both neighborhood size and learning rate that decrease with the number of iterations can be expressed as

$$\mathbf{w}_j(t) = \mathbf{w}_j(t-1) + \beta(t)\,\text{NS}(d,t)\,[\mathbf{x}(t) - \mathbf{w}_j(t-1)], \qquad (8.20)$$

where $\mathbf{w}_j(t)$ is the weight update after t iterations, $\mathbf{w}_j(t-1)$ is the update after the previous iteration, $\beta(t)$ is the learning rate variation with iterations t, and $\text{NS}(d,t)$ is the neighbor strength as a function of distance d from the winner to a neighbor neuron at iteration t. The $\mathbf{x}(t)$ is the input vector presented at the tth iteration.

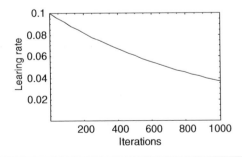

Figure 8.19 Exponential decay of earning rate.

8.5.1.9 Recursive and Batch Training and Relation to K-Means Clustering

Similar to competitive learning, training in SOM networks can be done in recursive or batch mode. In recursive mode, weights are adjusted after each input pattern has been presented and in batch mode weight adjustments are accumulated and one adjustment is made after an epoch. In batch mode, the unsupervised algorithm without neighbor feature becomes equivalent to K-means clustering. When the neighbor feature is incorporated, it allows nonlinear projection of the data as well as the very attractive feature of topology preservation, by which regions closer in input space are represented by neurons that are closer in the map. For this reason it is called a feature map.

8.5.1.10 Two Phases of Self-Organizing Map Training

Training is usually performed in two phases: ordering and convergence. In the ordering phase, learning rate and neighborhood size are reduced with iterations until the winner or a few neighbors around the winner remain. In this phase, topological ordering of the weight vectors takes place. Depending on the iterations required, careful consideration must be given to the choice of learning rate and neighborhood function. A recommendation is that the learning rate parameter should begin with a relatively high value and should thereafter gradually decrease, but must remain above 0.01. The neighborhood size should initially cover almost all neurons in the network when centered on a winning neuron and then shrink slowly with iterations. Depending on the problem, the ordering phase may take few to thousands of iterations, during which, neighborhood size is allowed to reduce to a few neurons around the winning neurons or just the winner itself. For a two-dimensional map of neurons, therefore, the initial neighborhood size σ_0 in Equation 8.16, for example, can be set to the radius of the map. The time constant T in Equation 8.16 must be chosen accordingly.

In the convergence phase, the feature map is fine tuned with the shrunk neighborhood so that it produces an accurate representation of the input space [6]. This phase may also run from a few to hundreds or thousands of iterations. In this phase, learning rate is maintained at a small value, on the order of 0.01, to achieve convergence with good statistical accuracy. Haykin [6] states that the learning rate must not become zero because the network can get stuck in a metastable state that corresponds to a feature map configuration with a topological defect. The exponential decay learning rate (Equation 8.19) prevents the network from getting stuck in metastable states because the learning rate never becomes equal to zero, thereby allowing the map to slowly approach convergence. With linear learning rate decay in Equation 8.18, however, this is not the case. The NS function should contain only the nearest neighbors of the winning neuron and may slowly reduce to one or zero neighbors (i.e., only the winner remains).

8.5.1.11 Example: Illustrating Self-Organizing Map Learning with a Hand Calculation

The six-cluster problem used previously for illustration of competitive learning (Section 8.4.3.2) and self-organization (Section 8.5.1.4) is now revisited, but a neighbor strength that depends on the topology (i.e., distance to winner) that decays with iterations is introduced along with a learning rate that also decays with iterations. The data is presented in Figure 8.4 and the network is presented in Figure 8.5. The same two inputs and same initial weights will be used that were introduced in Section 8.4.3.2. There are six codebook vectors; for convenience in a hand calculation, an initial neighborhood of size 2 (i.e., two neighbors each to the left and right of winner) will be used, as will three iterations in which neighbor size becomes zero and only the winner remains. The same exponential NS function used in the previous calculation with self-organization in Section 8.5.1.4 (Equation 8.8 repeated here) will be used here:

$$NS = Exp[-0.1d],\qquad(8.21)$$

where d is the distance. This function behaves as in Figure 8.20.

A simple learning rate formula will now be introduced and adjusted according to

$$\beta_t = \frac{2}{3+t},\qquad(8.22)$$

where t is the iteration number [3]. For three iterations, learning rate drops, as shown in Figure 8.21, from 0.5 to 0.333.

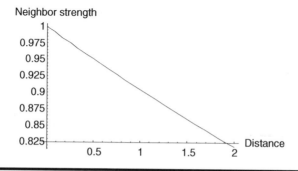

Figure 8.20 **Exponential neighbor strength function for an initial neighborhood size of 2.**

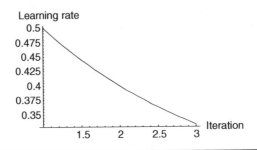

Figure 8.21 Learning rate decay in three iterations.

Lastly, the neighborhood size variation with iteration σ_t must be set. A linear neighborhood decay function will be used here:

$$\sigma_t = \sigma_0(1 - t/T), \qquad (8.23)$$

where σ_0 is the initial neighborhood size, T is a constant, and t is current iteration. This function behaves as shown in Figure 8.22 for three iterations with $T = 3$. In three iterations, neighbor size drops to zero and only the winner remains.

All necessary parameters have now been attained and training with the initial codebook vectors can continue with the two input vectors. The initial codebook vectors are repeated here:

$$\begin{bmatrix} 0.805 & 1.234 \\ 0.520 & 0.977 \\ 0.574 & 0.963 \\ 1.027 & 1.023 \\ 1.106 & 0.893 \\ 0.514 & 0.626 \end{bmatrix} \qquad (8.24)$$

Iteration 1: Input 1: (−0.035, 0.030): As previously demonstrated in Section 8.5.1.4, codebook vector 6 was the winner for this input pattern and the winner and neighbors adjusted their weights. Here, however, the neighbors' codebook vectors will be adjusted according to the NS and learning rate decay functions selected for this purpose. Initial neighborhood size from Equation 8.23 and Figure 8.22 is two, but neuron 6 has only two neighbors to the left and none to the right. Therefore, only neurons 4 and 5 adjust their weights.

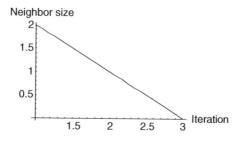

Figure 8.22   Neighborhood size decay in three iterations.

Weight update for neuron 6:
   From Equation 8.21 and Figure 8.20, NS = 1.0;
   from Equation 8.22 and Figure 8.21, $\beta = 0.5$:

$$\mathbf{x} = (-0.035, 0.030); \quad \mathbf{w}_6(0) = (0.514, 0.626).$$

From the weight update equation (Equation 8.20):

$$\mathbf{w}_j(t) = \mathbf{w}_j(t-1) + \beta(t)\text{NS}(d,t)[\mathbf{x}(t) - \mathbf{w}_j(t-1)] \qquad (8.25)$$

$$\mathbf{w}_6(1) = \big(0.514, 0.626\big) + 0.5 \times 1.0\big[(-0.035, 0.03) - \big(0.514, 0.626\big)\big]$$

$$= \big(0.514, 0.626\big) + (-0.2745, -0.2979) = (0.239, 0.328).$$

Weight update for neuron 5:
   From Equation 8.21 and Figure 8.20, for distance = 1, $NS = 0.9048$;
$\beta = 0.5$;

$$\mathbf{x} = (-0.035, 0.030); \quad \mathbf{w}_5(0) = (1.106, 0.893);$$

$$\mathbf{w}_5(1) = \big(1.106, 0.893\big) + 0.5 \times 0.9048\big[(-0.035, 0.030) - \big(1.106, 0.893\big)\big]$$

$$= \big(1.106, 0.893\big) + \big(-0.51658, -0.390403\big) = \big(0.590, 0.5026\big).$$

Weight update for neuron 4:
   From Equation 8.21 and Figure 8.20, for distance $d = 2$, NS = 0.8187;
$\beta = 0.5$:

$$\mathbf{x} = (-0.035, 0.030); \quad \mathbf{w}_4(0) = (1.0275, 1.0235);$$

$$\mathbf{w}_4(1) = (1.0275, 1.0235) + 0.5 \times 0.8187[(-0.035, 0.030) - (1.0275, 1.0235)]$$

$$= (1.0275, 1.0235) + \big(-0.435053, -0.406696\big) = \big(0.5924, 0.6168\big)$$

This completes the weight adjustment of the winner and the neighbors for input 1. Input pattern 2 is now presented.

Iteration 2: Input 2: (−0.033, 1.013): Omitting the details, the distances from input to the six codebook vectors calculated from Equation 8.2 are presented.

$$\text{Distance} = (0.8664, 0.5544, 0.6092, 0.7406, 0.8057, 0.7376)$$

$$\text{Winner} = \text{neuron 2}$$

Because the neighbor size changes with iterations, for $t = 2$, $d = 1$ from Equation 8.21 and Figure 8.22. Only the nearest neighbors update weights. Neuron 1 is to the left and neuron 3 to the right of neuron 2. Therefore, these get updated. Since the learning rate decreases with iterations, the new learning rate β from Equation 8.22 and Figure 8.21 is 0.4.

Weight update for neuron 2:
From Equation 8.21 and Figure 8.20, NS = 1.0;
from Equation 8.22 and Figure 8.21, $\beta = 0.4$:

$$\mathbf{x} = (-0.033, 1.013); \quad \mathbf{w}_2(0) = (0.5203, 0.9774);$$

$$\mathbf{w}_2(1) = (0.5203, 0.9774) + 0.4 \times 1.0[(-0.033, 1.013) - (0.5203, 0.9774)]$$

$$= (0.5203, 0.9774) + (-0.2213, 0.01447) = (0.2990, 0.9919)$$

Weight update for neuron 1:
NS = 0.9048 for $d = 1.0$; $\beta = 0.4$:

$$\mathbf{x} = (-0.033, 1.013); \quad \mathbf{w}_1(0) = (0.805, 1.234);$$

$$\mathbf{w}_1(1) = (0.805, 1.234) + 0.4 \times 0.9048 [(-0.033, 1.013) - (0.805, 1.234)]$$

$$= (0.805, 1.234) + (-0.3033, -0.0798) = (0.5017, 1.154).$$

Weight update for neuron 3:
NS = 0.9048 for $d = 1.0$; $\beta = 0.4$:

$$\mathbf{x} = (-0.033, 1.013); \quad \mathbf{w}_3(0) = (0.5742, 0.9634);$$

$$\mathbf{w}_3(1) = (0.5742, 0.9634) + 0.4 \times 0.9048[(-0.033, 1.013) - (0.5742, 0.9634)]$$

$$= (0.5742, 0.9634) + (-0.2198, 0.01815) = (0.354, 0.9815).$$

The two input patterns have now been presented and the weights have been updated recursively. The last iteration with input pattern 1 will now be performed, followed by an examination of the results to see if self-organization has begun.

Iteration $t = 3$: Input 1: $(-0.035, 0.030)$:

$$\text{Distance} = (1.24595, 1.01825, 1.02825, 0.859255, 0.783708, 0.405178)$$

$$\text{Winner} = \text{neuron } 6$$

From Figure 8.22, neighborhood size $\sigma_t = 0$ for $t = 3$; only the winner updates weights.

Weight update for neuron 6:
From Equation 8.22 and Figure 8.21, $\beta = 0.333$:

$$\text{NS} = 1.0;$$

$$\mathbf{w}_6(1) = (0.239, 0.328)$$

$$\mathbf{w}_6(1) = (0.239, 0.328) + 0.333 \times 1.0\,[(-0.035, 0.03), (0.239, 0.328)]$$

$$= (0.239, 0.328) + (-0.0915, -0.0993) = (0.1478, 0.2287).$$

The codebook vectors after three training iterations are

$$\begin{bmatrix} 0.501 & 1.154 \\ 0.299 & 0.992 \\ 0.354 & 0.981 \\ 0.592 & 0.617 \\ 0.590 & 0.502 \\ 0.148 & 0.229 \end{bmatrix}$$

It can now be determined if the codebook vectors have begun ordering themselves. Figure 8.23 shows a plot of these codebook (weight) vectors superimposed on the original data. They indeed appear to have ordered themselves even with such a small number of iterations and only two input patterns, as indicated by the fact that all codebook vectors have experienced neighborhood operations resulting in the ordering of class labels. Compare this result with Figure 8.9 that shows the position of codebook vectors when the neighborhood feature is not incorporated. There, class labels are not in sequential order, indicating the absence of preservation of neighbor relations.

Interestingly, topology preservation quickly takes effect in self-organizing learning. If more inputs and a larger number of iterations are used, codebook vectors would move towards cluster centers while preserving topological relationships (i.e., order of classes that are neighbors). To illustrate this, the training is completed with all the data

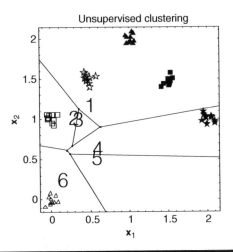

Figure 8.23 Ordering of codebook vectors for only two input patterns and a total of three iterations during which neighborhood size shrinks from 2 to 0.

and the map is trained until convergence. The performance during computer training is shown in Figure 8.24, which shows that the network achieves convergence in about 40 epochs, reaching a minimum mean distance of 0.15. In the figure, iterations denote epochs.

The final codebook vectors superimposed on the data are shown in Figure 8.25. It shows that except for one, all vectors have found cluster centers. The codebook vectors have been ordered such that the rightmost cluster is cluster 1 and the number increases towards the left. For this reason, the codebook vectors form an orderly topology when they are connected, indicating that the clusters that are closer in the input space are also closer on the map.

In situations such as the one presented here, where one cluster is not properly represented, it is beneficial to have more output neurons than

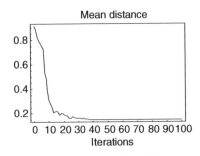

Figure 8.24 Complete training performance with all data.

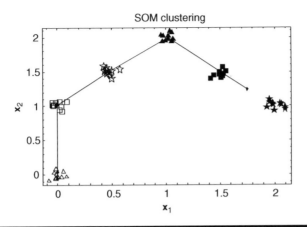

Figure 8.25 Final output neuron codebook vectors representing cluster centers superimposed on data (output neurons are connected in an orderly fashion indicating topology preservation on the map).

clusters. This helps the codebook vectors find all possible centers and unused vectors can be easily removed. The same data was trained with eight neurons on the map grid and the mean distance decreased to 0.05 in about 40 epochs. The new vectors superimposed on data are shown in Figure 8.26a.

Figure 8.26a shows that with eight vectors, there is perfect clustering. Once again, the vectors have been ordered such that the rightmost cluster is cluster 1 and the number increases towards the left. Cluster 6, at the left bottom corner, is the last cluster. There is one vector between clusters 1 and

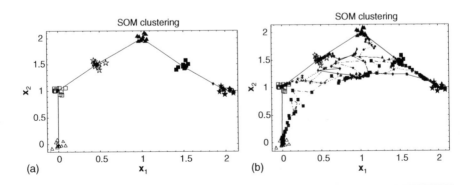

Figure 8.26 Training SOM with eight codebook vectors: (a) position of final codebook vectors and map topology, (b) evolution of codebook vectors during training as indicated by their position (initial codebook vectors are at the center of the figure).

2 and another one just beyond cluster 1. These can be removed so that there are six vectors organized in a topology-preserving manner.

The movement of codebook vectors towards clusters during training is shown in Figure 8.26b, which illustrates the position of codebook vectors at some selected training iterations from the beginning to the end. The initial random vectors are at the center of the figure. Careful examination of Figure 8.26b reveals the neighborhood operations in action, especially compared to Figure 8.10, where the network was trained without neighborhood considerations.

8.5.1.12 SOM Case Study: Determination of Mastitis Health Status of Dairy Herd from Combined Milk Traits

The application of SOM is illustrated in this section with a case study conducted at Lincoln University in New Zealand. It involves the determination of the health status of dairy cows in terms of the state of progression of mastitis—a bacterial infection in the udder—from milk traits. Mastitis is probably the most important disease affecting the dairy industry. It not only affects the yield and composition of milk, but it also affects the welfare of cows due to increased use of antibiotics and physical damage to the udder. Therefore, accurate detection of mastitis in the early stages of infection is important [26,27]. Mastitis is known to affect milk in various ways. These include changes in electrical conductivity (EC), somatic cell populations, and other traits such as fat and protein percentages in milk. As the bacterial infection progresses, the number of somatic cells that counteract the infection increases, leading to high somatic cell count (SCC), and electrical conductivity increases due to chemical changes in milk and elevated temperature resulting from bacterial activity. The SCC and EC are the most widely used indicators in the diagnosis of mastitis in the dairy industry. However, it is believed that better insights can be gained of the mastitic status of a cow at a specific time using several traits related to milk composition.

In a study over a period of 14 weeks, 107 cows from a farm were assessed for several of these milk traits to study the incidence of mastitis in the herd from the beginning until the middle of lactation [26]. Quarter (udder) milk samples from all four quarters of each cow were collected weekly from each cow, resulting in a total of 6848 quarter milk observations. These were analyzed for EC, fat percentage (FP), protein percentage (PP), SCC, and microbiological profile.

To reduce the number of inputs needed for the SOFM, hierarchical cluster analysis was employed. As a result, two separate groups were formed: EC, and the clustering of the other traits and their interactions.

The EC group was represented by a conductivity index (CI) that redefined the EC of a quarter in relation to its ratio to the sum total EC over all quarters. This index is derived from

$$CI = 2 + [(EC/100) - IQR],$$

where

$$IQR = EC/\sum_{i=1}^{4} EC_i,$$

which is the ratio of the EC of a quarter to the sum total of conductivity of all four quarters. The second group was represented by a composite milk index (CMI) that was the sum of all the other traits of FP, PP, SCC, and their interactions (products). The details of these derivations are excluded to keep the focus on the central issue of cluster formation and self-organization in maps and the reader is advised to refer to the original reference for more details.

These two traits, CI and CMI, were selected as inputs to a one-dimensional SOM and four arbitrary health categories were defined as outputs (i.e., healthy ($H0$), moderately ill ($H1$), ill ($H2$), and severely ill ($H3$)). Thus, the network has two inputs and four outputs, and the inputs were normalized in the range 0–1. Clustering into health categories was made using NeuroShell2™ [30].

Several variations of NS and learning rate functions were tested to select the best combination and assess the overall robustness of results. The networks were robust and the results obtained from training with a Gaussian neighbor strength function and a linear learning rate decay function are presented here. Initial neighborhood size included all neurons and decayed to the winner at the end of the training.

The health categories discovered by the SOM are shown in Figure 8.27, which illustrates the character of the four clusters in relation to CMI, CI, and SCC. Although SCC was used only indirectly as an input to SOM in the CMI, it was included in the plot because SCC is used in industry as an indicator of mastitis. However, its effectiveness as an indicator of mastitis is not always high due to the variable immune response of cows. From the health clusters, it was found that about 80 percent of the quarters were either in $H1$ or $H2$ category, while the percentage of $H3$ animals was maintained at less than 5 percent and the remaining cases were considered as $H0$. The SOM has clustered the data into statistically significant ($P<0.01$) mastitis categories. K-means clustering was also performed for comparison and an overall correlation of 0.89 ($P<0.01$) was observed between the clustering results from SOM and those from the K-means method. Thus, the clusters formed by the map were reasonable.

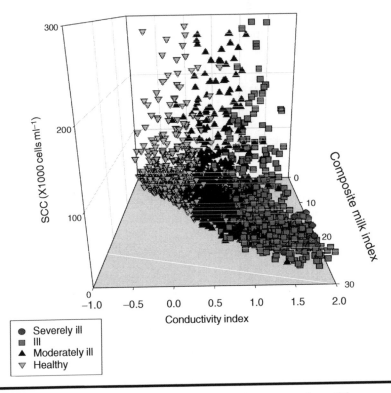

Figure 8.27 Clusters formed by the SOM indicating status of mastitis.

The mean and the standard deviation of each of the milk traits in the clusters found by SOM are shown in Table 8.1. After clustering, statistical tests (ANOVA and LSD) were conducted to compare the means between the health categories on SPSS statistical software [29] and results are shown in the last column of Table 8.1. It indicates that all the health categories are statistically significant based on the mean values of CI and CMI, the two inputs used to train the map. The same is valid for the FP as well. For SCC and PP, however, three groups had similar values.

The SOM clustered health categories in Figure 8.27 were assessed to determine if they are physically meaningful; they were found to appropriately represent the health status of a cow. For example, the results indicate that the health status of an animal deteriorates progressively from healthy to seriously ill as the CMI and CI values increased. It is generally accepted that the higher the electrical conductivity, the more serious the state of health and the network has detected this trend. The effect of the CMI can also be explained by the traits that make up the CMI.

The SCC was not the best indicator of mastitis compared with other traits such as CI because its values were spread out among the categories

Table 8.1 Mean and Standard Deviation of the Milk Traits in the SOM Clustered Mastitis Health Categories and the Statistical Significance of the Means between the Categories

Trait	Health Category				LSD (P<0.05)
	H0	H1	H2	H3	
PP	3.61 ± 0.01	3.52 ± 0.01	3.51 ± 0.01	3.51 ± 0.02	H0 vs H1, H2, H3
FP	4.01 ± 0.06	3.33 ± 0.04	2.97 ± 0.04	2.64 ± 0.13	All
SCC	51 ± 4.7	53 ± 4.2	83 ± 14.9	1237 ± 221	H3 vs H0, H1, H2
CI	-0.08 ± 0.01	0.49 ± 0.01	1.03 ± 0.01	2.24 ± 0.10	All
CMI	11.91 ± 0.15	16.38 ± 0.11	20.66 ± 0.15	44.90 ± 2.80	All

(Figure 8.27) and there was not a clear boundary of distinction to determine thresholds. This is because animals that would previously be considered as ill could have actually been in a healthy category. Figure 8.27 demonstrates that a low SCC does not necessarily mean a low incidence of mastitis. This supports the hypothesis that it is the speed at which cells can be mobilized into the udder that determines the degree of illness. Cows that had higher than acceptable SCC levels and were in the *H*0 (healthy) category had normal or desirable values in the other traits. This could indicate that the cow may have suffered an infection, but it quickly recovered from it because its immune system was effective against the pathogen.

8.5.2 Example of Two-Dimensional Self-Organizing Maps: Clustering Canadian and Alaskan Salmon Based on the Diameter of Growth Rings of the Scales

To illustrate the concept of self-organization in two-dimensional maps, a small two-dimensional classification problem studied in Section 2.5.3.2 is revisited. Two-dimensional problems allow easier graphical interpretation of the concepts of SOMs. The problem involved identifying the breed of salmon, whether Alaskan or Canadian, from the growth-ring diameter of the scales in freshwater and ring diameter of scales in seawater [7]. The data is shown in Figure 8.28. The problem stems from the concerns of environmental authorities about the depletion of salmon stocks; to address these concerns, authorities have decided to regulate the catches. To do so, it is necessary to identify whether the fish is of Alaskan or Canadian origin. Fifty fish from each place of origin were caught and the growth-ring diameter of scales was measured for the time they lived in freshwater and for the subsequent time they lived in saltwater. The aim is to identify the origin of fish from the growth-ring diameter in freshwater and saltwater.

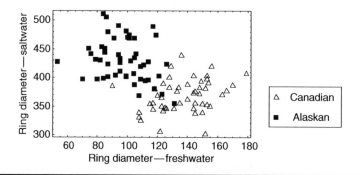

Figure 8.28   Ring diameter of scales for Canadian and Alaskan salmon in saltwater and freshwater.

Although the categories are known in this case, we use the input data to illustrate how a two-dimensional map represents the input space.

8.5.2.1   Map Structure and Initialization

In this example, an SOM consisting of a square grid of size 5×5 was trained for clustering the data in Figure 8.28. The coordinates of the position of the bottom left neuron on the map are {1, 1}; those of the top right neuron are {5, 5}, as shown in Figure 8.29a. The network has 25 output neurons and 2 inputs: ring diameter in freshwater and ring diameter in saltwater. The weights associated with the 25 neurons were randomly initialized and they are shown by the gray dot superimposed on the data in Figure 8.29b. All 25 initial weight values are centered in the original data.

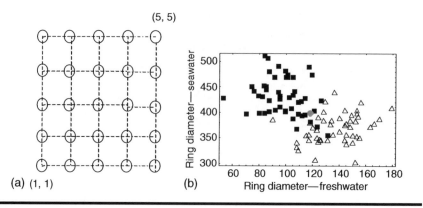

Figure 8.29   The map structure and initial weights: (a) 25-neuron map of size 5×5 and (b) initial weights of the map (gray dot) superimposed on data.

8.5.2.2 Map Training

The map was trained using a square neighborhood with learning rate β expressed as

$$\beta = 0.01 \quad \{t < 5$$
$$= \frac{2}{3 + t} \quad \{t > 5 \tag{8.26}$$

Learning rate is a small constant value in the first four iterations so that the codebook vectors find a good orientation (this is not always done but is used here to highlight some interesting features of map orientation). Then it is increased to 0.25 at the fifth iteration to speed up the convergence. From this high value, the step length is slowly decreased until the map converges, which can only be achieved if the learning rate converges towards zero. The learning rate given in Equation 8.26 decays to 0.025 in 100 iterations, as shown in Figure 8.30.

The neighbour strength function used was

$$\text{NS} = \text{Exp}[-0.1d] \qquad \text{if } t < 5$$
$$= \text{Exp}\left[-\frac{(t-4)}{10}d\right] \quad \text{otherwise} \tag{8.27}$$

During the first four iterations, all neurons on the map are neighbors of a winning neuron and all neighbors are strongly influenced. The influence on the nearer neighbors, however, is greater than that on the distant ones, as shown in Figure 8.31a. The stronger influence on the neighbors in the initial iterations makes the network conform to a nice structure and avoids knots. From the fifth iteration, the influence on neighbors decreases with iterations,

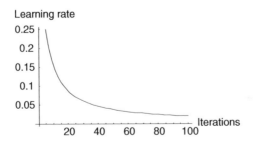

Figure 8.30 Learning rate decay with iterations.

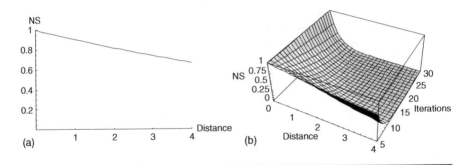

Figure 8.31 Neighborhood strength function in relation to distance from winner and iterations: (a) constant neighborhood influence in the first four iterations, and (b) neighborhood decay after five iterations.

as shown in Figure 8.31b, which shows that at iteration 30 only the winner and the nearest neighbors update weights.

The map was trained in two phases: ordering phase and convergence phase. In the ordering phase, the map is trained; in the convergence phase, it is finetuned. These two phases will now be discussed in detail.

Ordering phase. The map was trained using recursive update where codebook vectors are updated after each presentation of an input pattern. The training performance is shown in Figure 8.32 that indicates how the mean distance between the codebook vectors and input vectors in the corresponding clusters decreases with iterations. The network has reached stability in about 30 iterations and at this stage only the winner and the nearest neighbors are active, as shown in Figure 8.33.

The trained map is superimposed on the data in Figure 8.34, which shows that the initial weights have traversed the input space so that the map represents the input data distribution well. The dots on the map indicate the position of a codebook vector that represents a cluster of data surrounding it. As can be seen, there are no knots in the map, indicating that neighboring

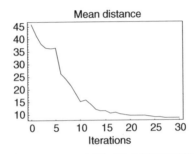

Figure 8.32 Training performance of the two-dimensional map.

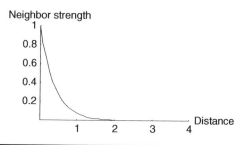

Figure 8.33 Neighbor strength in relation to distance from winner after 30 iterations.

neurons represent data that is in close proximity in the input data space. This is possible due to the neighbor feature that attempts to preserve the topological structure of the data on the map grid.

Figure 8.34 shows that the initial contracted map has unraveled and unfolded during training to traverse the whole input space. Each junction of the trained map represents a neuron; its codebook vector components are the corresponding ring diameter in saltwater and freshwater in the plot in Figure 8.34. The fact that there are no knots in the unfolded map indicates that the weights have traversed freely after finding a good orientation in the first four iterations. This process can be observed in Figure 8.35, where the progress of unraveling of the initial map during training is shown. Careful examination of the path of map evolution indicates that the map has indeed found a good orientation in the first four iterations where the neighbor

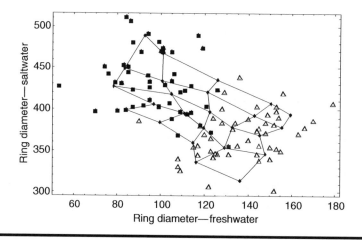

Figure 8.34 Codebook vectors of the trained map at the completion of the ordering phase superimposed on input data.

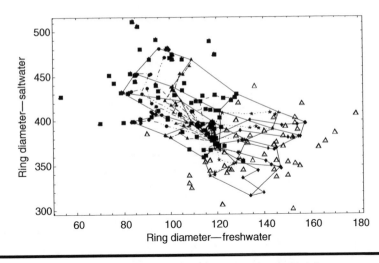

Figure 8.35   Evolution of the map codebook vectors during training; map traverses the input space freely after having found a good orientation in the first four iterations.

influence is larger, thereby making the subsequent traversal in the input space free of entanglement.

An even clearer picture of the unfolding of the map can be seen in Figure 8.36, where the configuration of the map in the first 15 iterations is presented. It shows the good orientation found by the map in the first five iterations and from that point on it simply expands to cover the input space. It is noteworthy that in the first five iterations, the map has orientated itself along the main principal direction of data. Recall that the main principal direction indicates the direction in which the variation in the original data is the largest, and clearly in this example, this direction is very much along the diagonal in the plot of the original data. The other dimension of the map therefore assumes the second principal direction that is perpendicular to the first principal direction. Thus, the SOM procedure has found the principal directions of the data relatively quickly.

Convergence phase. To finetune and make sure that the map has converged, the trained map was trained further in batch mode for ten epochs. Because the network in this case appears to have approached convergence, the learning rate has been set to 1.0 because there will be only small or straightforward adjustments to the position of the codebook vectors with further training. The neighbor strength is limited to the winning neuron. The training performance with respect to reduction in mean distance is presented in Figure 8.37, which indicates that the map has now converged. If, however, the network has not approached convergence in the ordering phase, further training with a smaller constant learning rate on

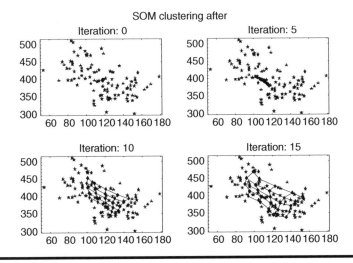

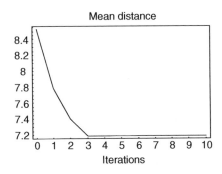

Figure 8.36 **Map evolution superimposed on data at selected iterations of 0, 5, 10, and 15 (the second map indicates that it has orientated itself along the main principal direction of the data after five iterations).**

the order of 0.01 may be appropriate. The neighbor strength then may be limited to a few neighbors and decrease to the winner or the nearest neighbors towards the end of training.

The final map superimposed on the data in Figure 8.38 indicates that it has reached more data in the outlying regions compared to the map formed at the end of the ordering phase (Figure 8.34).

The final converged map and the map obtained at the completion of the ordering phase, along with the input data, are shown in Figure 8.39, where the outer map is the fine-tuned map. There has been a shift in the position of

Figure 8.37 **Final fine tuning of the trained map in batch mode in convergence phase (iterations denote epochs).**

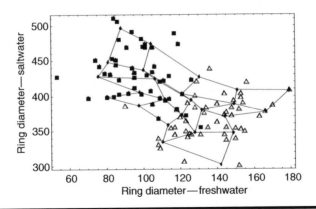

Figure 8.38 Final converged map formed at the end of convergence phase superimposed on data.

the map in the final fine-tuning phase. Figure 8.39 reveals that the difference in the two maps before and after fine-tuning is significant. The final map is much better in terms of representing the input data distribution. It has expanded to better reach the data and it has assumed a more centered position in the data space.

The converged map in Figure 8.38 and Figure 8.39 is a representation of the input space defined by the two classes. Each codebook vector represents a subregion in the input space, and as such, the set of codebook vectors more compactly represents the input space. When an input vector is

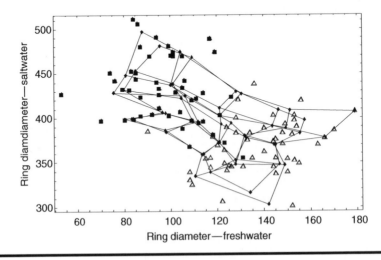

Figure 8.39 Comparison of the final converged map and the map obtained at the completion of ordering phase in relation to input data (outer map is the fine-tuned map).

presented, the appropriate codebook vector will be activated, indicating the region of the input space where the new input vector belongs. Because there are two classes of salmon (Canadian and Alaskan), a cluster of codebook vectors, not a single vector, defines each class. This gives the map its ability to form nonlinear cluster boundaries. This cluster structure can be used to discover unknown clusters in data. The map can also be used for subsequent supervised classification. For example, when class labels are known, the codebook vectors that represent corresponding input vectors can be used as input to train a feedforward classification network to obtain the class to which a particular unknown input vector belongs. This is called learning vector quantization. However, it is not demonstrated here in favor of continuing to explore the map.

Figure 8.40 presents the number of input vectors from each class represented by each of the 25 codebook vectors. The numbers from 1 to 5 denote the row and column positions of the neurons on the map and each of these neurons are represented by a cell in Figure 8.40. The cell at the bottom right corner denoted by coordinates {1, 5} represents the neuron at the bottom right corner of the original map in Figure 8.29a. Figure 8.40 shows that the map has clustered the data such that most of the Canadian salmon (class-0) are mapped to the right side and the Alaskan salmon are mapped to the left side, as they are in the input space in Figure 8.28. In the middle of the map, few neurons have clustered data belonging to both classes and there is consequently some overlap between the classes near the apparent boundary in the original data.

Each codebook vector represents the center of gravity of a cluster of inputs that it represents and therefore approximates the average or point density of the original distribution in a small cluster region. Therefore, the magnitude (length) of the codebook vectors should reflect this. A properly ordered map should show evenly varying length of the codebook vectors

	1	2	3	4	**5**
5	2×1	4×1	2×1 1×0	8×0	5×0
4	2×1	6×1	1×0	6×0	4×0
3	4×1	2×1	2×1	3×0	4×0
2	7×1	6×1 1×0	2×1 3×0	3×0	5×0
1	3×1	4×1	3×1 1×0	1×1 1×0	4×0

Figure 8.40 The number of input data vectors clustered by the 25 codebook vectors according to class definition (0—Canadian; 1—Alaskan).

on the map. The magnitude of the codebook vectors of each neuron is shown in Figure 8.41.

In Figure 8.41, darker cells represent small magnitudes and lighter cells represent larger values. It shows that the map progressively becomes lighter as it extends up and to the right, indicating that it has preserved the topological character of the data. The cell at the bottom right corner in Figure 8.41 represents the neuron at the bottom right corner of the map in Figure 8.29a (recall that the map has orientated in the principal direction of the data while maintaining the relative position of the neurons on the map grid due to topology preserving quality of self-organization). Thus the map can be considered as a discontinuous, smoothed approximation of the probability distribution of the input data. Each codebook vector approximates the average of the inputs in its catchment or Voronoi region.

8.5.2.3 U-Matrix

The distance between the neighboring codebook vectors can highlight different cluster regions in the map and can be a useful visualization tool. The distance for each neuron or cell is the average distance between neighboring codebook vectors. Neurons at the corners and edges have fewer neighbors. If two vectors are denoted by $\mathbf{w}_1 = \{\mathbf{w}_{11}, \mathbf{w}_{21}, ..., \mathbf{w}_{n1}\}$ and $\mathbf{w}_2 = \{\mathbf{w}_{12}, \mathbf{w}_{22}, ..., \mathbf{w}_{n2}\}$, where n is the number of input dimensions, then the Euclidean distance d_{12} between the two vectors is calculated in the usual way as

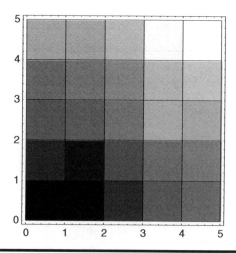

Figure 8.41 Magnitude (length) of codebook vectors (dark—small, light—large: smallest is 334.0 mm, largest is 505.0 mm).

$$d_{12} = \sqrt{(\mathbf{w}_{11} - \mathbf{w}_{12})^2 + (\mathbf{w}_{21} - \mathbf{w}_{22})^2 + \ldots + (\mathbf{w}_{n1} - \mathbf{w}_{n2})^2}. \qquad (8.28)$$

The average of the distance to the nearest neighbors is called unified distance, and the matrix of these values for all neurons is called the U-matrix [1,8]. The U-matrix for the map shown in Figure 8.38, Figure 8.39, and Figure 8.40 for the example data in Figure 8.28 is shown in Figure 8.42.

In Figure 8.42, lighter regions represent larger distances between the neighboring vectors and darker regions represent smaller distances between them. It shows that in the middle region along the diagonal, distances are smaller, indicating a concentration of neurons in close proximity in this region, whereas towards the top right and bottom left corners, the map distances are larger, indicating sparse neurons. There is an explanation for this result. In Figure 8.28, where original data is presented, there is a large concentration of data along the diagonal. The map has taken this into account by allocating closely spaced neurons to represent these dense regions. Thus the map has not only orientated itself in the principal directions of the data, but has also learned to represent the density distribution of the input data.

Although the U-matrix can indicate the cluster structures in data, it cannot do so explicitly in this example because the boundary between the two classes is too close to the data and the vector distances near the boundary are not appreciably different from those away from the boundary. In well-separated clusters, the distances between neurons in the boundary region are

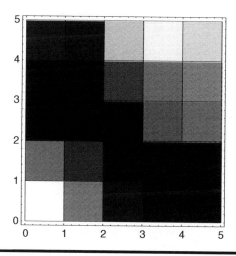

Figure 8.42 U-matrix for salmon data. Darker colors indicate smaller distances and lighter colors indicate larger distances between neighboring neurons; (minimum (black) average distance is 20.0 and maximum (white) average distance is 45.0 mm).

much larger than those within clusters thereby enabling the discovery of unknown clusters. The discovery of cluster structures will be illustrated in another example where clusters are quite separate in Section 8.5.4.

8.5.3 Map Initialization

Weights or codebook vectors of a map can be initialized in various ways. Random initialization clusters the initial vectors near the center of gravity of inputs and assigns random values in this center region. The center of gravity of the map coincides with the center of gravity of data; therefore, this initialization would allow the network to unfold from its own center and the center of gravity of data. This form of random initialization is used in the example previously presented in Section 8.5.2.

Deterministic initialization is another approach, where some input vectors from the dataset are used as initial vectors. This can accelerate map training. Yet another approach is to train a map with random initialization for a few iterations and use the resulting vectors as initial vectors (random-deterministic). There are other approaches as well. Recall that in the initial few iterations, the map orientates so that the map direction coincides with the data distribution. In the above example, it has orientated itself such that its axes are parallel to main directions of variation in the data, resembling principal directions. Therefore, another possible approach to initialization is to find the first two principal directions of data using principal component analysis and use these two directions for map directions [1]. With this approach, the initial map will already be orientated in the direction of data before training. Visual inspection of data may be very useful, not only in initialization of the map, but also in deciding the dimensions of the map.

8.5.4 Example: Training Two-Dimensional Maps on Multidimensional Data

An example is studied in this section in which the input dimension is larger than two. The example discussed here has been widely used in literature for illustrating classification and clustering properties of neural networks. It involves classification of iris flowers into three classes of species depending on four attributes: petal length, petal width, sepal length, and sepal width. The dataset contains 150 iris flowers belonging to the species of *Iris setosa, Iris versicolor,* and *Iris virginica.* The dataset is available at the UCI machine learning repository at http://www.ics.uci.edu/~mlearn/MLRepository.html [9]. There are four numerical attributes; the last variable is the class to which a flower with a particular set of attributes belongs and has the values 1, 2, and 3. Because data has more than two dimensions, all data cannot be visualized in a meaningful way. The SOMs can be used not only to cluster input data, but also to explore the relationship between different attributes of input data. The following analysis was conducted on *Machine Learning Framework for Mathematica* [10].

8.5.4.1 Data Visualization

It is useful to visualize the data to obtain a view of the cluster structure. For this, all input data can be plotted, indicating their relationship to the species class, as shown in Figure 8.43.

Figure 8.43 shows the cluster structures in all combinations of two-dimensional plots. It reveals that one cluster is quite distinct in relation to all input dimensions, whereas the other two are not quite separable from each other. This is not uncommon in multidimensional data. The histograms along the diagonal are for each attribute and they are colored according to the species that a particular attribute value is associated with. The histograms indicate that the species *Iris virginica* has higher values for the attributes of sepal length, petal length, and petal width, and medium values for sepal width. Thus, this species is at the high end of most of the attributes. Species *Iris versicolor* is at the low end of the attribute range for sepal length, petal length, and petal width, and at the high end of the range for sepal width. The species *Iris setosa* is in the mid-range for all variables except for sepal width, where it is positioned at the lower end of the range. This species overlaps with species *Iris virginica* at all attribute boundaries as indicated by the scatter plots. The goal is to organize and project high dimensional data onto a two-dimensional map with a limited number of neurons where realistic cluster structures are formed. These clusters would then be useful in deciding the cluster to which a particular unknown flower belongs.

8.5.4.2 Map Structure and Training

An SOM of size 8×8 with 64 neurons was trained using the four input attributes and an exponential decay learning rate with an initial learning rate of 0.5. In the learning rate decay formula, the maximum number of iterations T is set at 20 times the total number of training patterns, which is $20 \times 150 = 3000$. Neighbor strength function is Gaussian with hexagonal neighbor geometry. Neighbor size decreases to one at the end of the iterations. Figure 8.44 shows the resulting map where neurons are numbered from 1 to 64.

In Figure 8.44, maps have been separated for each attribute and can be thought of as panels that can be overlaid on top of each other. Each neuron

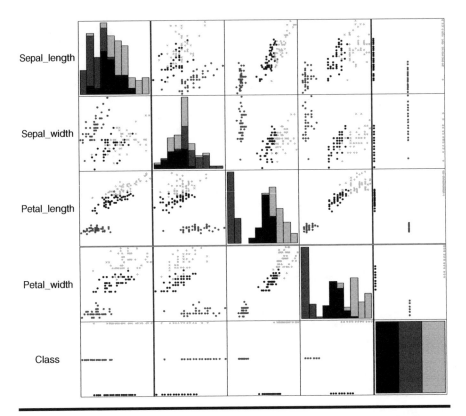

● Class_Is_Iris-setosa
● Class_Is_Iris-versicol
◉ Class_Is_Iris-virginica

Figure 8.43 Scatter plots and histograms of input attributes of iris flowers in relation to species class.

is depicted by a codebook vector that approximates the center of a cluster of inputs and therefore has attribute values that reflect the average of those inputs in that cluster. These values are color-coded in the map to show their magnitude. The last panel denotes species class and each cell in it presents the average of the class value for all the original inputs represented by the codebook vector of that cell. Although the species class was not used in map training, a separate panel can be drawn because the original data contains a class label for each input vector and it was depicted by numbers 1, 2, and 3.

The position of neurons in each panel coincides with each other and by following the same neuron position in each panel, the values of attributes of a flower and the class it belongs to can be traced. Furthermore, the relation of attributes to a class can also be discerned from the panels. For example,

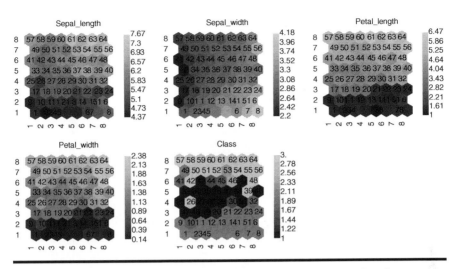

Figure 8.44 Mapping of 4 dimensional data onto a two-dimensional SOM. The first four plots present the input attributes and the last plot shows the species class.

petal length and petal width have a structure similar to the class structure. Low petal length correlates with low petal width (located at the bottom of the respective panels) and these strongly relate to class 2, which is *Iris versicolor*; this agrees with the relations indicated by the plots in Figure 8.43. Sepal length is similar to the petal length and petal width, but the sepal width has a significantly different spread of its values in the map, which is also supported by the plots in Figure 8.43.

The relationships between data clusters and neurons can now be observed by mapping the neuron onto data. This is presented in Figure 8.45 where data clusters and corresponding neurons that have clustered together to represent the data clusters can be clearly seen in a plot of two input attributes of petal length and sepal length. In the figure, smaller solid circles denote input data with the color indicating the species. The larger gray circles with numbers are the neurons; their position indicates the codebook vector and the number indicates the label given to it in the map in Figure 8.44. The gray lines indicate connectivity between neurons.

There is one distinct cluster at the bottom of the figure and neurons 1–21 associated with this cluster are grouped together at the bottom of the class panel in Figure 8.44. This group of neurons represents *Iris versicolor*. The top cluster is represented by neurons 48–64 in Figure 8.45; these are located in the top region in the class panel in Figure 8.44. This group of neurons represents *Iris virginica*. The middle cluster overlaps with the top cluster as shown in Figure 8.45 and is represented by neurons 17 – 46 that are located in the middle region of the map. This group of neurons represents *Iris setosa*.

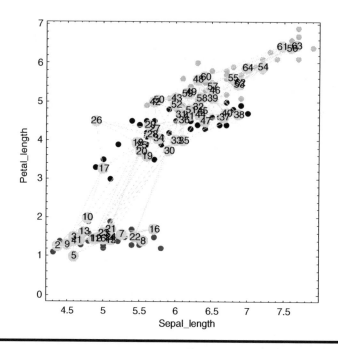

Figure 8.45 Clusters of neurons formed on the map in relation to two attributes of original data (numbered circles are neurons and smaller dots are inputs with the color indicating class; gray lines represent neuron connectivity).

Neuron 26, belonging to the top cluster, is located in the middle region of the map among those representing the middle cluster because it has input attributes similar to those in the second cluster. See Figure 8.43 and Figure 8.45 to locate this input data point.

In the boundary between the neurons that appear to represent the top cluster and middle cluster in the map, some neurons have a class value that is between those for the two clusters, as indicated in the last panel in Figure 8.44. These represent the regions in the data space where clusters overlap. These neurons represent input vectors from both classes that fall in their catchment and therefore may produce errors in classification if the map is used for this purpose.

The clusters of neurons formed on the map are highlighted in Figure 8.46 where the input data is denoted by very small dots in the background. The black lines show the connectivity between neurons. The fact that there are no entanglements in the lines indicates that neurons that are closer in the map represent input data that are closer in the input space. Thus, the trained map has preserved the topology of data.

From the trained map, various statistics can be obtained. For example, since we have class labels (i.e., 1, 2, 3) that have been used to create panel 5

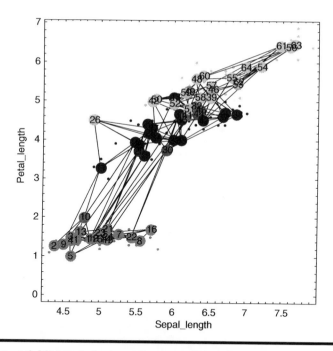

Figure 8.46 Highlight of clusters of neurons formed on the map (small dots are data points and black lines are connections between neurons).

in Figure 8.44, the classification accuracy of the map can be assessed from the performance of the map on the data. We can also identify the data points that have been misclassified. To perform this classification, the input patterns must be presented to the network individually, and the codebook vector nearest to an input vector must first be found. The class to which the input belongs is the class label associated with this nearest codebook vector. This label is the average of all class labels of the input vectors belonging to that codebook vector, as depicted in panel 5 in Figure 8.44. This method of extracting class labels is for neighborhood size of zero, meaning that only the winner decides the classification.

The data set was passed through the map and these tasks were performed; the resulting percentage classification accuracy is presented in the form of a confusion matrix and a bar chart in Figure 8.47a. It indicates that one class has been classified with 100 percent accuracy; this is the bottom cluster in Figure 8.43, representing *Iris versicolor*. The first cluster, *Iris setosa*, has been classified with 86 percent accuracy and the third cluster, *Iris virginica*, with 88 percent accuracy. These are the clusters that overlap at the boundary. The total classification accuracy of the map is 91.3 percent.

A neighbor size other than zero can also be specified to include some neighbors of the winner, such that several neurons participate in decision

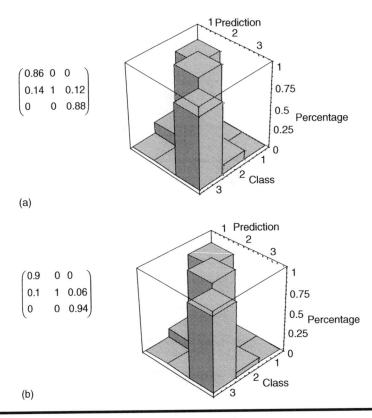

$$\begin{pmatrix} 0.86 & 0 & 0 \\ 0.14 & 1 & 0.12 \\ 0 & 0 & 0.88 \end{pmatrix}$$

(a)

$$\begin{pmatrix} 0.9 & 0 & 0 \\ 0.1 & 1 & 0.06 \\ 0 & 0 & 0.94 \end{pmatrix}$$

(b)

Figure 8.47 Classification performance of trained map on data: (a) confusion matrix and bar chart depicting the percentage classification accuracy of the trained map for neighborhood size of 0 where only the winner decides classification, and (b) confusion matrix and bar chart depicting the percentage classification accuracy of the trained map for neighborhood of size 2 where winner and neurons in a neighborhood decides classification.

making. In this case, the weighting given to the neighbors must be decided. By giving equal weighting and using a neighborhood size of 2, a much improved classification results, as shown in Figure 8.47b. Basically, once the winner has been selected for an input pattern, the class label associated with all neurons in a neighborhood of size 2 is extracted and averaged to obtain a final classification.

Figure 8.47b indicates that one class has been classified with 100 percent accuracy; this is the bottom cluster (*Iris versicolor*) in Figure 8.43. The classification accuracy of the first cluster (*Iris setosa*) has increased to 90 percent from 86 percent and that of the third cluster (*Iris virginica*) has increased to 94 percent from 88 percent, when only the winner decided the classification. The neighborhood operation has increased the total classification accuracy to 94.7 percent compared to 91.3 percent when only the winner decided the classification. All together, eight input patterns have been misclassified; these can be extracted from the results and plotted, as shown in Figure 8.48, where misclassified patterns are shown as bigger gray circles around the original data depicted by solid circles with the color indicating species class. The map has misclassified three patterns belonging to *Iris setosa* (middle cluster represented by black dots) and five belonging to *Iris virginica* (top cluster represented by lighter shades of gray).

8.5.4.3   U-Matrix

From the trained map, average distance between a neuron and its neighbors can be obtained and, as explained earlier, this distance is easily expressed using the U-matrix. The U-matrix for the example iris problem is shown in Figure 8.49. The larger the distance between neurons, the larger the U value and more separated the clusters. In the U-matrix map shown in Figure 8.49a, the lighter the color, the larger the U value. As can be seen, the cluster at the bottom of the map (bottom cluster of *Iris versicolor* in data plot in Figure 8.43) is at a distance from that represented in the middle region. The latter cluster is the middle cluster depicting *Iris setosa* in the data plot. Because there is a large distance between these two clusters in the data plot, the difference is preserved in the map. Large distances between codebook vectors indicate a sharp boundary between the clusters. A different view of the distance between neurons is presented in the surface graph shown in Figure 8.49b, where the height represents the distance. In

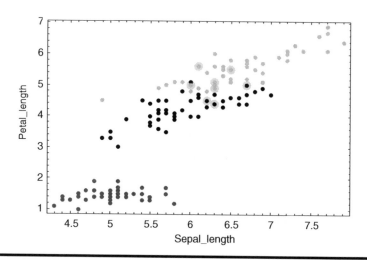

Figure 8.48   The eight misclassified input vectors shown as gray circles around original data depicted by solid circles with the color indicating species class.

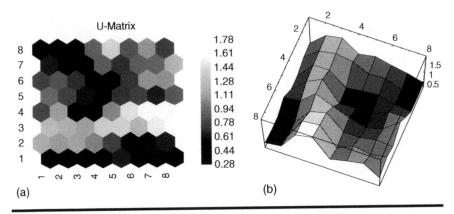

(a)   (b)

Figure 8.49 **Two views of the U-matrix representing average distance between codebook vectors of neighboring neurons: (a) density map and (b) three-dimensional surface graph.**

this figure, coordinate {0, 0} coincides with the bottom left corner of map in Figure 8.49a.

The large height in the third row in Figure 8.49b indicates a separation between the bottom and middle clusters. The same cannot be said about the other two clusters. The clusters in the middle and top regions do not have a distinct boundary due to the overlap of the original data at the cluster boundary. This is clearly shown on the map. There is no large difference in neighbor distance in the middle to top region, indicating that progressively placed neurons in this region cannot find any distinct boundary anywhere in the input space because the two clusters overlap. Thus the map preserves the features of the data and the map has captured the cluster relationships well.

8.5.4.4 Point Estimates of Probability Density of Inputs Captured by the Map

From the trained map, we can also determine the number of input vectors represented by each neuron. Two views of this are shown in Figure 8.50. Each neuron represents the local probability density of inputs. In the left panel, the lighter the color, the larger the number of inputs falling onto that neuron. In the right panel, the larger the circle and the darker the hexagon in which it is located, the larger the number of inputs falling onto that neuron.

Figure 8.50 illustrates that some neurons have captured more inputs than others. Ideally, the map should preserve the probability density of the input data so that the map is a faithful preservation of the data distribution. In Figure 8.51, the codebook vectors are superimposed on the data so that

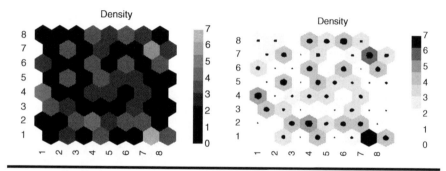

Figure 8.50 **Two views of local probability density of inputs captured by the map.**

the distribution of codebook vectors and the input data distribution can be compared. Four different pairs of input attributes have been plotted to show different views of the multidimensional cluster relationships.

In Figure 8.51, codebook vectors are denoted by smaller dots, and input data by larger dots with a different shade for each cluster. It shows that the map has followed the distribution of data. Neurons have attempted to assume central positions in local data clusters and reached out to data that are not too close to a cluster.

8.5.4.5 Quantization Error

The distribution of mapping error can also be determined throughout the map. In training, the objective is to get the codebook vectors as close as possible to the input vectors that are closer to them than to the others. Quantization error is a measure of the distance between codebook vectors and inputs. If for an input vector $\mathbf{x}$, the winner's weights vector is $\mathbf{w}_c$, then the quantization error can be described as a distortion error, e, expressed as [1]

$$e = d(\mathbf{x}, \mathbf{w}_c),$$ (8.29a)

which is the distance from the input to the closest codebook vector. It may be more appropriate to define the distortion error in terms of neighborhood function because the neighbor feature is central to SOM. With the neighbor feature, the distortion error of the map for an input vector $\mathbf{x}$ becomes

$$e = \sum_i NS_{ci} d(\mathbf{x}, \mathbf{w}_i),$$ (8.29b)

where NS_{ci} is the neighbor strength, c is the index of the winning neuron closest to input vector $\mathbf{x}$, and i is any neuron in the neighborhood of the winner, including the winner. The $d(\mathbf{x}, \mathbf{w}_i)$ is the distance between the input vector $\mathbf{x}$ and a codebook vector $\mathbf{w}_i$ in the neighborhood of the winning

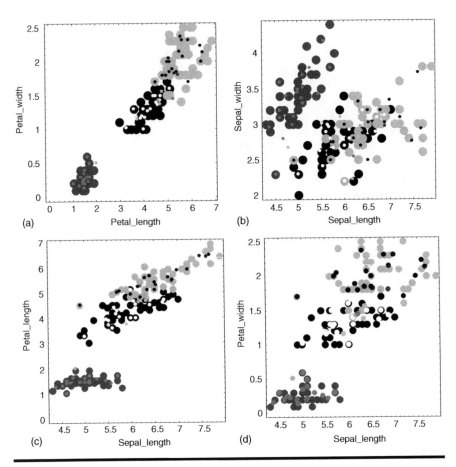

Figure 8.51 Codebook vectors representing local cluster centers superimposed on data to ascertain the preservation of the local probability distribution of input data by the map (smaller dots represent codebook vectors and larger dots represent input data with a different shade for each of the three species clusters).

neuron responding to input vector $\mathbf{x}$. Computing the distortion measure for all input vectors in the input space, the average distortion error E for the map can be calculated from

$$E = \frac{1}{N} \sum_{n} \sum_{i} \text{NS}_{ci} \; d(\mathbf{x}^n, \mathbf{w}_i). \qquad (8.30a)$$

When the neighbor feature is not used, Equation 8.30a simplifies to

$$E = \frac{1}{N} \sum_{n} d(\mathbf{x}^n, \mathbf{w}_i). \qquad (8.30b)$$

In Equation 8.30, $\mathbf{x}^n$ is the nth input vector, N is the total number of input vectors and c denotes a neighborhood. Thus the goal of SOM can alternatively be expressed as finding the set of codebook vectors $\mathbf{w}_i$ that globally minimizes the average map distortion error E.

For the iris problem under study, the average distortion measure for each neuron without the neighbor feature, which indicates the average distance to the inputs represented by the neuron, is presented in Figure 8.52 in the form of a density graph (left) and a three-dimensional plot (right) where the height indicates the distortion error.

Figure 8.52 indicates that the distortion error is more uniform in the interior region of the map and higher in some outer regions. High distortion error indicates areas where the codebook vector is relatively far from the inputs. The distortion error is a function of the input distribution, size of the map, and distribution of codebook vectors in the input space. High distortion error is an indication that either the data in this region is far from the representative codebook vector or that there is a lack of representation by neurons in the region. Such information can be used to refine the map to obtain a more uniform distortion error measure if a more faithful reproduction of the input distribution from the map is desired.

8.5.4.6 *Accuracy of Retrieval of Input Data from the Map*

A recall operation can now be used to demonstrate how accurately the map covers the input data space. If the dataset is sent through the map, it identifies the best matching codebook vector. The resulting codebook vector can be thought of as the retrieved input because it is the closest to

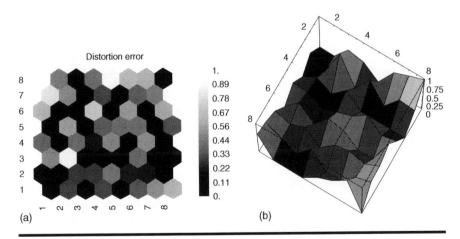

Figure 8.52 Two views of map distortion error: (a) density plot and (b) surface graph (the origin of the graph coincides with the bottom left corner of the map in (a)).

that input. All codebook vectors representing a dataset then become the retrieved dataset. These codebook vectors are superimposed on data in Figure 8.53a. They represent the retrieved input data distribution in compact form because each vector is the center of a local cluster of original inputs. The figure shows that the vectors represent the data well.

If a neighborhood of neurons is used in the retrieval, more than one codebook vector can be activated and these codebook vectors can be interpolated to obtain a recalled match of the input to the map. Then the retrieved inputs are not the codebook vectors, but fall between them due to interpolation. This was done with a neighborhood size 2 and the resulting interpolated codebook vectors are superimposed on the actual data in Figure 8.53b.

In Figure 8.53, the distribution of retrieved inputs with $N = 2$ has a retrieval error of 0.154; for $N = 0$, the error is slightly higher at 0.159. The significance of this difference depends on the required retrieval accuracy for a particular problem. This error is the average distance between the actual data vectors and their corresponding interpolated codebook vectors defining the best position for those input vectors in the trained map. Thus, a neighborhood provides a better approximation to this input distribution than a single codebook vector.

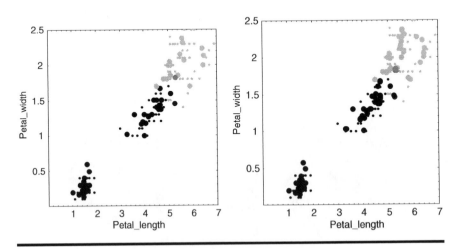

Figure 8.53 The position of retrieved inputs (i.e., codebook vectors) superimposed on original data for: (a) neighborhood size $N = 0$ and (b) $N = 2$ involving interpolation of codebook vectors within a neighborhood of two neurons (small dots denote original data, large dots denote the position of input data retrieved from the map).

8.5.5 Forming Clusters on the Map

In many practical situations, the number of clusters is unknown. Ideally, the neurons in the map should naturally form distinct clusters; however, it is not always very clear from the map exactly how many distinct clusters are present. Therefore, it will be useful if the map can be used to find realistic and distinct clusters. Because each neuron in the map is represented by a codebook vector that approximates the center of gravity of the inputs that are closest to it, the distance between the codebook vectors (U-matrix) can be used to find borders between data, as discussed in Section 8.5.4.3. The larger the distance, the more likely a cluster boundary exists between the vectors. Because a neighborhood topology is used in training, cluster boundaries may be blurred, but neurons on the boundary may receive few or no inputs and can indicate possible cluster boundaries.

After a map has been trained, the total inputs distribution has been replaced with a smaller number of codebook vectors; thus, the dataset has been compressed. In the process, noise has been removed, as have outliers because the map only contains codebook vectors—not the original input data—and each of these vectors represents a sample of input data. Outliers or inputs with noise map to a more central codebook vector representing the cluster to which these data belong.

Although in unsupervised clustering the number of clusters is not known a priori, because the topological structure of the data is preserved in the map, the neurons (i.e. codebook vectors) that are closer together can be clustered to form the most likely number of clusters. These clusters of codebook vectors would represent clusters in the original data. This idea is presented schematically in Figure 8.54, which illustrates two levels of data reduction [11]. The first occurs when the input data is projected onto a map where they are represented by trained codebook vectors. The second level of reduction occurs where the codebook vectors are further grouped into clusters of codebook vectors. For a new input vector, this approach will first find the best matching codebook vector. Then the cluster to which the best

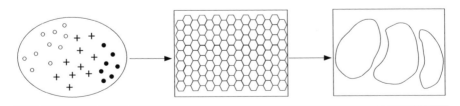

Figure 8.54 Illustration of the clustering of a trained map based on two levels of data reduction: map formation with trained codebook vectors at the first level and clustering of codebook vectors at the second level.

matching vector belongs will determine the cluster to which the original input vector belongs.

8.5.5.1 Approaches to Clustering

Any established clustering method can be used for clustering the codebook vectors. Generally, there are two main approaches to clustering data: hierarchical and partitive approaches [11,13,28]. In hierarchical clustering, a clustering tree called a dendrogram is produced in a hierarchical fashion. This approach can be further divided into agglomerative and divisive algorithms, depending on whether top-down or bottom-up approaches are used to build the hierarchical clustering tree [11].

Agglomerative methods are more commonly used than divisive methods and the basic mechanism of agglomerative clustering is as follows. At the beginning, each input vector forms its own cluster. Then the distance between all clusters is computed and the two clusters that are closest to each other are merged. This process is repeated until there is one cluster left, which is the total dataset. Basically, in this method, data points are merged together to successively form a clustering tree that finally consists of one cluster, which is the whole dataset. This tree can be used to interpret data structure and determine the number of clusters.

In partitive clustering, a dataset is divided into a number of clusters while minimizing some criterion or error function. An example is the commonly used K-means clustering where a dataset is divided into k clusters. The number of clusters is usually predetermined, but it can also be made part of the error function [12]. The basic approach to partitive clustering with predetermined number of clusters is as follows. The number of clusters is first determined and cluster centers are initialized. Then the clusters are updated with the presentation of inputs by bringing the best matching cluster centers closer to the corresponding input vectors. This is repeated until there is no change in the updates.

If the number of clusters is unknown, a partitive algorithm can be repeated for a set of different number of clusters, typically from two to $\sqrt{N}$ where N is the number of input vectors. The K-means clustering for example, minimizes the error function E:

$$E = \sum_{k=1}^{C} \sum_{x \in \text{Clust } k} \|\mathbf{x} - c_k\|^2, \tag{8.31}$$

where C is number of clusters, $\mathbf{x}$ is an input vector belonging to cluster k (Clust k), and c_k is the center of cluster k. Basically, the distance between a codebook vector representing a cluster center and the inputs represented by it is accumulated over the total number of clusters. Partitive clustering does not depend on previously found clusters; it is therefore better in this respect. However, these methods make implicit assumptions about the shape of clusters. For example, K-means clustering attempts to find spherical clusters.

8.5.5.2 Example Illustrating Clustering on a Trained Map

Clustering on the map will be illustrated by continuing the iris example presented in Section 8.5.4. Level one of abstraction, depicted in Figure 8.54, has already been completed by training a map. The second level involves clustering the codebook vectors. The effects of hierarchical clustering on the trained map will now be demonstrated and we specifically choose agglomerative clustering in this example. In clustering the SOM, our inputs to the clustering algorithm are the codebook vectors, not the original input data. There are 64 neurons on the map, so there are 64 corresponding codebook vectors. These neurons are shown on trained map component panels in Figure 8.44 and are repeated in Figure 8.55.

The dendrogram of these 64 trained vectors are shown in Figure 8.56, which demonstrates how clusters are formed on the trained map. At the bottom of the map are the individual neurons making separate clusters; the numbers indicate the labels given to the neurons in the map in Figure 8.55. Horizontal lines and their height from the bottom indicate the levels of clustering. By following the number on the neurons in the dendrogram, the formation of clusters can be observed. For example, at the lowest level, it has first clustered neurons 3 and 4 that have the minimum distance. Then it has clustered neurons 11 and 12, which are later formed into a bigger

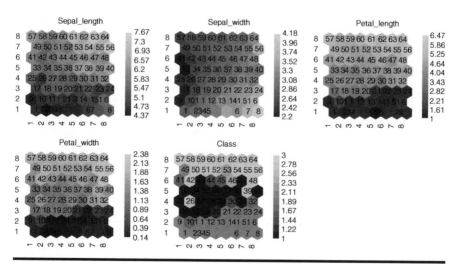

Figure 8.55 Trained SOM network with respect to input attributes and species class labels.

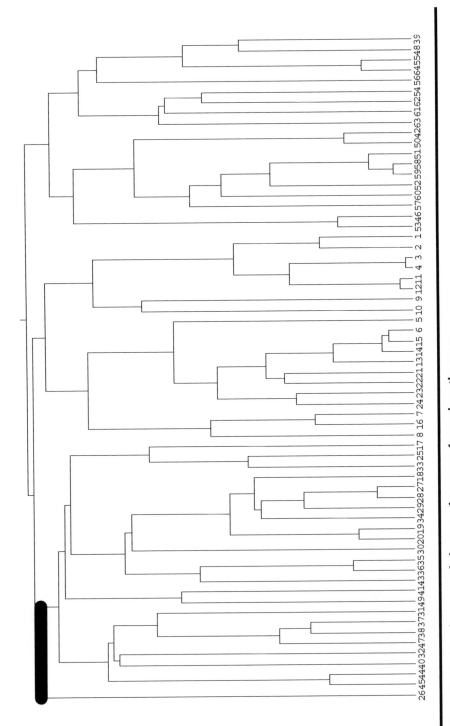

Figure 8.56 Dendrogram of clusters of neurons formed on the map.

cluster with 3 and 4. If one follows these numbers in the map in Figure 8.55, it can be seen that these four neurons are closer together in the bottom two rows of the map. Next it has clustered neurons 58 and 59 and these are at the top row of the map. By following this hierarchical structure, the process of clustering at various levels can be observed.

At the beginning of the formation of the tree, there are as many clusters as there are codebook vectors and the two closest vectors are merged based on a distance measure (Euclidean distance is used here). After these initial clusters are formed, the distance between clusters must be computed to determine which clusters must be merged next. There are several methods for such clustering, including single linkage, complete linkage, and Ward method [13,28]. In the single linkage (nearest-neighbor) method, the nearest neighbors across clusters are used to determine the distance between clusters [12]. Clusters with minimum single linkage are merged. In the complete linkage method, the distance between clusters is determined by the greatest distance between any two data points (farthest neighbors). Clusters with minimum farthest-neighbor distance are merged. This method is effective if the clusters naturally form "clumps" [13]. However, the method is inappropriate for clusters that are somewhat elongated, or are of a "chain type." There are many other methods of forming clusters. The Ward method [14] is used in this example to determine which clusters to merge in the dendrogram in Figure 8.56.

Invented in 1963, Ward is an agglomerative clustering method that is very efficient when the dataset is not too large [14]. It is the most widely used agglomerative technique. The method uses an analysis of variance approach to evaluate distances between clusters. Specifically, the method attempts to minimize the within-group sum of squares distance as a result of joining two possible (hypothetical) clusters. The within-group sum of squares is defined as the sum of the squares of the distance between all objects in the cluster and the centroid of the cluster. Two clusters that produce the least sum of square distance (or variance, which increases with each step) are merged in each step of the clustering process. The distance measure is called the Ward distance (d_{ward}) and is expressed as

$$d_{ward} = \frac{(n_r \times n_s)}{(n_r + n_s)} \|x_r - x_s\|^2 \tag{8.32}$$

where r and s denote two specific clusters, n_r and n_s denote the number of data points in the two clusters and x_r and x_s are the center of gravity of the two clusters. $\|x_r - x_s\|$ is the Euclidean norm (i.e., magnitude of the distance between the two cluster centers). As the number of data points in the two clusters increase and the distance between the cluster centers increases, the Ward distance measure increases. The two clusters that produce the least distance measure are merged at each step of clustering. The mean of the two merged clusters is computed from

$$\mathbf{x}_{r(new)} = \frac{1}{n_r + n_s}(n_r \times \mathbf{x}_r + n_s \times \mathbf{x}_s) \tag{8.33}$$

and the new cluster size is updated to $n_r + n_s$.

The Ward method tends to produce compact groups of similar and spherical size clusters and it can be sensitive to outliers. When used on a trained SOM, the homogeneity of clusters within the map is achieved by merging only neurons and clusters that are neighbors in the map.

The dendrogram still does not provide the optimum partitioning of the data as required by the user. Partitioning can be achieved by cutting the dendrogram at certain levels. For example, when cut across a horizontal line, the clusters that drop are the clusters formed at this level. This works because more and more neurons and neuron clusters are grouped together as the distance from the bottom increases. However, this still does not indicate the best possible clustering of the data. In this example, we use the Ward method [14] that was used to merge clusters in the dendrogram. It helps find the optimum cluster structure, depicted by the thick horizontal line in Figure 8.56, that indicates the best level of clustering for this problem. Basically, the method computes the likelihood of various numbers of clusters from which the most appropriate number of clusters can be obtained based on a likelihood index.

The Ward likelihood index is defined as

$$\text{Ward Index} = \frac{1}{NC}\left(\frac{d_t - d_{t-1}}{d_{t-1} - d_{t-2}}\right) = \frac{1}{NC}\left(\frac{\Delta d_t}{\Delta d_{t-1}}\right), \tag{8.34}$$

where d_t is the distance between the centers of the two cluster to be merged in the current step, and d_{t-1} and d_{t-2} are the distance between merged clusters in the last step and the step prior to the last. NC is the number of clusters left. Thus the Ward likelihood index is a measure of the difference in distance between the two clusters to be merged at the current step and the two clusters merged at the last step relative to the difference in distance in the last step, normalized by the number of clusters left. The clusters that are far apart have a higher denominator and therefore a higher Ward index; this can be used as an indication of the separability of clusters.

The likelihood of 13 clusters for the map shown in Figure 8.55 is presented in Figure 8.57. It shows that three clusters have the maximum Ward likelihood index, which in this case is optimum because we know that there are three clusters in the data. The larger the index, the more likely the clusters separate more distinctly.

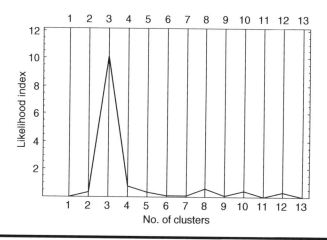

Figure 8.57   Ward likelihood index for various number of clusters of map neurons.

8.5.5.3   Finding Optimum Clusters on the Map with the Ward Method

To select the best partitioning, the maximum Ward likelihood index is used as a guide. As indicated in Figure 8.57, three clusters have the largest likelihood index of 10 and the other clusters have a much lower likelihood in comparison. This information can now be used to partition the trained map into three clusters, as presented in Figure 8.58, where the original SOM panel for the class (last panel in Figure 8.55) is shown along with the map partitioned into three clusters with the Ward method. This level of clustering is indicated by the thicker horizontal line in the dendrogram in Figure 8.56.

Figure 8.58 reveals that the new portioned map closely resembles the original class map. Recall that the class attribute was not used in training.

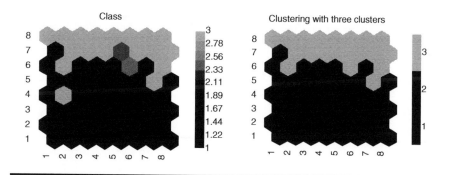

Figure 8.58   Original class map using class indicator (left) and the map partitioned into three unknown clusters using Ward method (right).

Because for this problem class attribute is available, it is advantageously used for comparing with the partitioned map. In many practical situations, the class attribute is not available and the partitioned map must be solely relied upon. Therefore, testing the method for a case where the class information is available sheds light on the appropriateness and robustness of the clustering process.

With the clustering of codebook vectors of neurons, three clusters of vectors are obtained. Correspondingly, there are three cluster centers. These are superimposed on the original data in Figure 8.59, where the large gray dots denote cluster centers and smaller solid circles denote two attributes of input data colored according to species class. As can be seen, clustering has resulted in final cluster centers that are located at the center of original data clusters.

It can now be determined whether the Ward clustering has actually improved the classification accuracy. The clustering accuracy for the map is presented in the form of a confusion matrix and three-dimensional plot in Figure 8.60. There is perfect (100 percent) classification accuracy for class 2 (*Iris versicolor*) as before, but class 1 and 3 rates are now 92 and 80 percent (for *Iris setosa* and *Iris virginica*, respectively). These values from the original map were 86 and 88 percent, respectively, for neighborhood size 0, and 90 and 94 percent, respectively, for neighborhood size 2 (see Figure 8.47) when the class labels were used, as in panel 'Class' of Figure 8.55. The total classification accuracy of the Ward/SOM is 90.7 percent, which is slightly

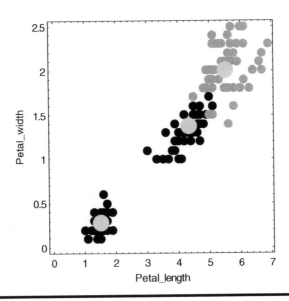

Figure 8.59 Centers of clusters formed by Ward clustering of trained SOM codebook vectors.

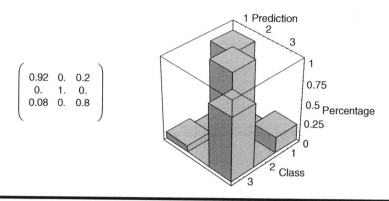

$$\begin{pmatrix} 0.92 & 0. & 0.2 \\ 0. & 1. & 0. \\ 0.08 & 0. & 0.8 \end{pmatrix}$$

Figure 8.60 Percentage classification accuracy of map after Ward clustering presented as a confusion matrix and three-dimensional bar chart.

lower than that achieved with the original SOM (91.7 percent) for neighborhood size 0 and 94.7 percent for neighborhood size 2 (Figure 8.47). However, because class labels are not known a priori in many problems, an approach such as this is required for performing classification, and the classification accuracy reached by the map with Ward clustering is remarkable.

8.5.5.4 Finding Optimum Clusters by K-Means Clustering

K-means is a set of nonhierarchical statistical clustering methods that find exactly k different clusters of greatest distinction by optimizing an appropriate clustering criterion [28]. The method assumes that the number of clusters has already been fixed by the investigator. The method associates an error criterion (as in Equation 8.31) with each partitioning of n data points into the required number of k groups, and this criterion is used to compare the partitions. The error criterion uses concepts of homogeneity and separation in that data points within a cluster should have a cohesive structure and individual clusters should be well separated from each other. The best partition into k clusters is the one that best satisfies these. There are several variants of K-means clustering. In the implementation of a clustering process for finding k clusters simultaneously, cluster centers are first initialized and objects within a predefined threshold distance are assigned to the nearest centers. As the process evolves, threshold distance can be adjusted to include fewer or more objects in the clusters.

K-means clustering has been used here to cluster the original data into three groups; the resulting cluster centroids are shown in Figure 8.61, superimposed on data whose color signifies the species class. The classification accuracy of the K-means method on the training input data is

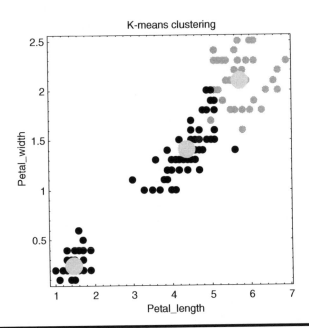

Figure 8.61 Cluster centers obtained from K-means clustering superimposed on original data (large gray dots denote cluster centers and smaller solid dots denote input data for two attributes with color indicating species class).

presented in the form of a confusion matrix and three-dimensional bar chart depicting class and percentage classification accuracy in Figure 8.62.

Figure 8.62 indicates that K-means clustering has accurately classified class 2, which is the bottom cluster in the data plot in Figure 8.43 (*Iris versicolor*). The SOM and SOM/Ward also classified this class with 100 percent accuracy. Classification accuracy for class 1 (*Iris setosa*) from the K-means

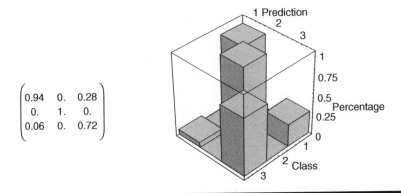

Figure 8.62 Confusion matrix and three-dimensional bar chart depicting percentage classification accuracy from K-means clustering of original input data.

method is 94 percent and for class 3 (*Iris virginica*) is only 72 percent. The total accuracy is 88.7 percent. This rate for the SOM map before Ward clustering was 94.7 percent and with Ward was 90.7 percent. The results indicate that the SOM map before Ward clustering provides higher accuracy than either SOM/Ward or K-means clustering and more importantly, accuracy from SOM/Ward clustering is higher than that from K-means clustering. Note that in this example, the class label was available and was therefore utilized. However, in many clustering problems, the number of clusters in not known a priori and must be determined by clustering the trained map neurons using an appropriate clustering method. Here, the Ward method was used and it provided an accuracy comparable to the case where class labels were known, indicating the effectiveness of the method.

To compare the SOM/Ward clustering with K-means clustering, we plot cluster centers from both methods, superimposed on the original data in Figure 8.63. In this figure, larger dots are cluster centers from the two methods with black representing SOM/Ward clustering and gray representing K-means clustering. The smaller solid dots are original data, colorized according to which cluster they belong from each of the clustering methods.

In terms of cluster centers, the two panels in Figure 8.63 show that SOM/Ward and K-mean cluster centers are not identical but very close to each other. According to the classification results, SOM/Ward has resulted in clusters that are better representations of the top and middle clusters than K-means clustering.

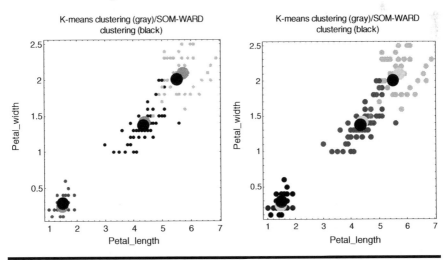

Figure 8.63 Cluster centers obtained from SOM/Ward clustering and K-means clustering superimposed on original data: (a) original data colorized according to SOM/Ward classification, (b) original data colorized according to K-means classification (cluster centers from both methods are shown in both panels for comparison).

8.5.6 Validation of a Trained Map

Ideally, the map should be robust and reliable. In practical applications, the map should be validated. One way of accomplishing this is by testing it on unseen data. To do so, a portion of the data for map training and another portion for validation could be used. The iris dataset could be randomly divided into two portions, with 70 percent of data in the training set and 30 percent in the validation set. The map trained with training data is shown in Figure 8.64. It shows that the new map, is very similar to the previous map, where all data were used for training (see Figure 8.44). Therefore, removal of a portion of data does not affect the overall information content in the map.

The map performance will now be assessed using test data that has been set aside. The classification accuracy on the test data is shown in the following confusion matrix that shows that cluster 2 has been perfectly classified, but clusters 1 and 3 have been classified with 93.7 and 92.3 percent accuracy, respectively. Total accuracy achieved is 95.4 percent. Thus the map is robust against unseen data

$$\begin{bmatrix} 0.937 & 0 & 0 \\ 0.667 & 1 & 0.667 \\ 0 & 0 & 0.923 \end{bmatrix}$$

Next the codebook vectors were grouped into three clusters on the trained map using the Ward method. The following confusion matrix shows the results on test data. One hundred percent classification accuracy has now been achieved on class 2, 93.75 percent on class 1 (as before), but only 84.6 percent on class 3. Total accuracy is 93.2 percent, slightly lower than that obtained from the original map. There is more error in classifying class 3 (*Iris virginica*). These results are for input patterns not seen by the network before and therefore reflect the generalization ability of the network to unseen data:

$$\begin{bmatrix} 0.937 & 0 & 0.125 \\ 0 & 1 & 0 \\ 0.077 & 0 & 0.846 \end{bmatrix}$$

8.5.6.1 n-Fold Cross Validation

For a rigorous validation of a map, an *n*-fold cross validation with the leave-one-out method could be used to test the robustness of the map. The average classification accuracy on validation data of several maps obtained from *n*-fold cross validation is a useful approach for testing the generalization ability or the goodness of a map. A ten-fold cross validation was conducted where the original dataset was randomly divided into

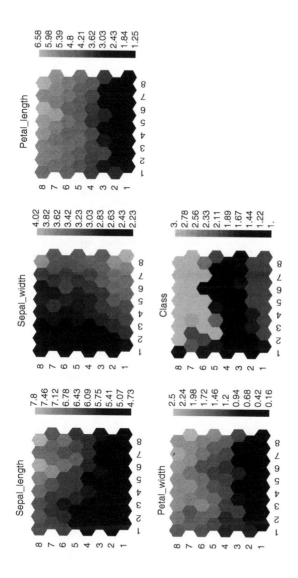

Figure 8.64 Map performance on training data.

ten portions and nine randomly chosen portions (90 percent of data) were used each time for training a map, and one group (10 percent of data) was retained for map validation. The experiment was repeated 10 times with different training and validation sets, each time training on 90 percent of the data and testing on 10 percent. The classification accuracy on the validation sets for the ten trials ranged from 0.75 to 1.0 as shown in Figure 8.65a. The combined classification accuracy for the ten maps over all classes was 93 percent. Combined accuracy for individual classes is presented in the form a confusion matrix and bar chart in Figure 8.65b.

The codebook vectors of each of the ten maps were further clustered using Ward clustering and total classification accuracy for each map over all classes based on all data is presented in Figure 8.66. All data was used for validating this SOM/Ward classification because it was not possible to extract the exact validation sets used by the program for each map. Therefore, a direct comparison of classification accuracy of SOM alone presented in the previous figure, and SOM/Ward presented here, is not totally appropriate; however, the results are still interesting and the idea of map validation still applies. The accuracy of individual maps in this case varies from 90.2 to 95 percent as shown in Figure 8.66, and the combined total accuracy of the 10 maps over all classes is 92.5 percent, which is almost identical to the accuracy of the original map (93 percent) in ten-fold cross validation. It is notable that the combined class accuracy for individual maps is more consistent than that for the original map shown in Figure 8.65a.

Four examples of the ten maps after Ward clustering are shown in Figure 8.67 to illustrate their broad similarity and small variations in cluster boundaries.

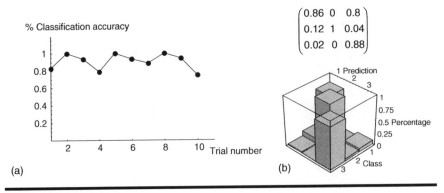

(a)

(b)

Figure 8.65 **Classification accuracy of ten maps developed from ten-fold cross validation: (a) performance of each map on corresponding validation dataset and (b) combined total classification accuracy of the ten maps for each class.**

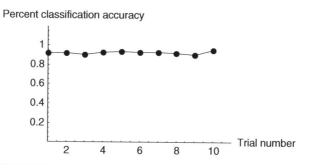

Figure 8.66 Overall classification accuracy for ten maps clustered by the Ward method in ten-fold cross validation.

The ten cluster centers obtained from Ward clustering of the ten maps trained using ten-fold cross validation are superimposed on the original data in Figure 8.68 for the dimensions of petal length and petal width. As can be seen, the map is robust against random sampling and generalizes well to unseen data.

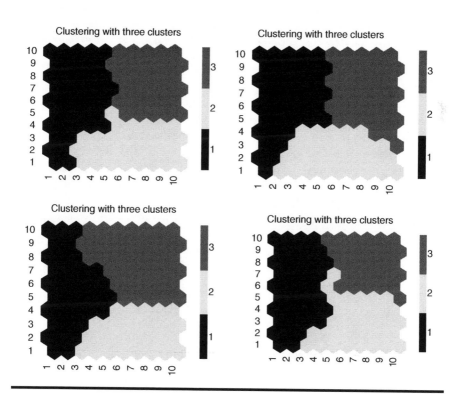

Figure 8.67 Four examples of ten-fold cross validated maps subsequently portioned by Ward clustering into three clusters.

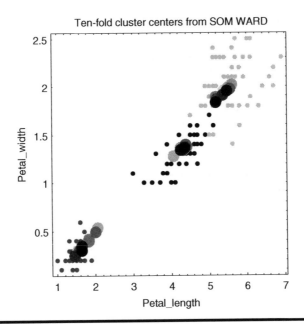

Figure 8.68 Centers for the clusters obtained from Ward clustering of maps in ten-fold cross validation superimposed on original data.

This section presented in detail SOM formation and clustering results for an example iris problem. It illustrates that by creating a map with many neurons and subsequent partitioning, it is possible to achieve high classification accuracy. This is because many neurons can traverse the input space, thus approximating the internal relations in data and thereby providing a better framework for subsequent clustering into a limited number of clusters. Thus, there are two stages of data reduction: the first is when data is mapped to codebook vectors and the second is when codebook vectors are grouped into clusters based on the distance between them. These classes can be further represented by other methods such as fuzzy C-means clustering to introduce fuzzy boundaries between classes and provide a fuzzy classification rather than a crisp classification [15]. From these, some if-then-else rules may be extracted [15], if necessary, to differentiate the classes. For some problems, especially simple ones, these rules may be meaningful and easily interpreted; for problems with many variables, however, interpretation can be extremely difficult.

Although not discussed here to keep the presentation simple, an important first step in map formation is to choose important variables that have discriminating power in clustering. Using irrelevant inputs can reduce the ability of the map to cluster and generalize. The same approaches used for selecting inputs to a multilayer network, as presented in Chapter 5, can be used for selecting inputs to an SOM. The more independent the inputs and the more discriminating power they possess, the more efficient the SOM clustering [28].

8.6   Evolving Self-Organizing Maps

As illustrated in the previous section, SOMs are very useful in clustering and visualizing high dimensional data. The clusters are formed such that not only the inputs that are similar are brought closer together, but also the clusters that are closer in the original data space are brought closer together in the feature map. This is possible due to the neighborhood feature used in training. However, there are several limitations of the method. First, the map structure (i.e., the number of neurons and height/width ratio of the map) must be predetermined. It is likely that a predetermined map size is either too small or too large. In the iris example demonstrated in the previous section, smaller map sizes were tried before deciding on the 8×8 map because this size demonstrated cluster structures better on the trained map. It can be difficult to predetermine the map size for some problems and some trial-and-error process is necessary.

A trained SOM can reveal possible cluster regions but clusters are often not well separated. If the objective is to determine distinct clusters in the data, further clustering of the neurons on the map is needed to transform the map into unique clusters so that a cluster to which an input belongs can be determined. Ward and K-means clustering presented in the previous section are two such clustering methods. For example, for the iris problem discussed in the previous section, visualization of clusters was only possible after the Ward clustering of codebook vectors. This aspect of the map can be improved so that the cluster structure is more apparent on the map itself. Furthermore, in the classical SOM discussed thus far, when a trained map is used to project input vectors onto the map, the inputs are only projected to the best matching neuron, i.e., to a single point on the map. Section 8.5.4.6, which discusses the accuracy of retrieval of input data, illustrates that this can be enhanced with a neighborhood topology where the average of the winning vector and those of its neighbors is used in mapping an input onto the trained map. With this method, an input can be mapped to a location between neurons, as shown in Figure 8.53b. However, the accuracy of the retrieval can be further improved.

Some of the above limitations of the SOM have been addressed by growing SOMs [16]. They are especially useful in reducing the map size and can eliminate the trial and error required in determining map structure. Several methods have been proposed as growth mechanisms to grow SOMs during training as required by the data. Some of these are growing neural gas [18], incremental grid growing (IGG) [19], tree structure SOM (TS-SOM) [22], growing hierarchical SOM, growing cell structure (GCS) [16,17], growing

SOMs (GSOMs) [20], and evolving tree [21]. Most of these growing maps attempt to dynamically evolve the map structure rather than train a predetermined grid. The map grows when some predetermined criteria, such as the number of hits indicating the frequency that a neuron becomes a winner, or the accumulated error for a neuron that indicates the total distance between a neuron and all the inputs that fall closest to it, are exceeded. Neural Gas and TS-SOM have fixed structures, but have efficient approaches to creating connections between neurons during training as required by the data. In what follows, GCS, GSOMs, and evolving tree are explored. However, a brief introduction of the other methods is given here before the three aforementioned methods are discussed.

The neural gas approach proposed by Martinez and Schulten in 1991 [18] produces a self-generating network in that it starts with no connections and a fixed number of neurons floating in data space. As inputs are presented during training, neurons are adapted and connections are created between the winning neuron and its closest competitor. Neurons can be removed using a removal mechanism. The deficiency of this method is that it has a fixed number of neurons that must be predetermined. Therefore, the method has similar limitations to that of the regular SOM. Furthermore, the neural gas structure is not necessarily two-dimensional and its dimensionality depends on the locality of input data. This means that the network can have different dimensionality in different regions in the data space, making it difficult to visualize [20].

The IGG is a method [19] that builds the network incrementally by dynamically changing its structure and connectivity according to the nature of the input data. The network starts with a smaller number of initial neurons and generates new neurons from the boundary using a growth mechanism (some growth mechanisms will be discussed in detail later). Connections are added when the codebook vector difference (distance) between two neurons falls below a threshold value (i.e., they are close to each other). Analogously, connections are removed when the distance between neurons exceeds a threshold. Because only boundary neurons can grow, the map contains a two-dimensional structure; therefore, the internal structure of data can be visualized on the map.

The TS-SOM [22] has several different-sized SOMs arranged in a pyramidal shape. The topmost layer is first trained in the usual way using the methods of regular SOM. Every neuron in the first layer has children in the second layer. When the second layer is trained, the winning neuron in the first layer is found first. Then only the children of that neuron are searched to find the winning neuron in the second-layer SOM. Other than this difference, the rest of the training is similar to that of regular SOM. If necessary, more layers can be added to the second layer, third layer, etc. This hierarchical organization greatly reduces the number of operations needed to find a winner; as such, this makes possible the training of very large maps required by large databases with high dimensionality [21]. This map also has a fixed structure that is predetermined.

In the growing hierarchical SOM developed by Dittenbach et al. [23], the system starts with a simple SOM and neurons are added as necessary. Specifically, a neuron is expanded when a certain criterion is reached and a new SOM is placed below the neuron. The new SOM is trained using only the input vectors that map to the parent neuron. In this section, the three selected approaches to growing SOMs are discussed in detail. They contain some of the features already discussed in this brief introduction.

8.6.1   Growing Cell Structure of Map

The GCS presented here is proposed by Wu and Yen [16]. This method is based on two-dimensional SOMs, but training starts with a small grid size, usually three units forming a triangle. There are two basic approaches to inserting nodes during training: (1) Add neurons in areas that receive a high number of inputs, or (2) add neurons in areas with the highest accumulated error of a neuron (i.e., the largest sum of the distance between a neuron's codebook vector and all inputs belonging to that neuron).

After inserting new neurons, connections are adjusted between them so that the triangular connectivity is maintained. The basic process of map growing is schematically illustrated in Figure 8.69 [16]. In the figure, the initial map is the three-neuron map on the left. The cross indicates the position where a neuron is to be added based on the criteria stated above. The second image in Figure 8.69 shows the new map with the new neuron added and the map reorganized to maintain a triangular structure. The third image shows the map extended after adding a neuron at the position of the cross in the second map and reorganized to maintain the triangular structure. The last image in Figure 8.69 shows a map that has grown further. This process continues until it no longer is necessary to grow the map further.

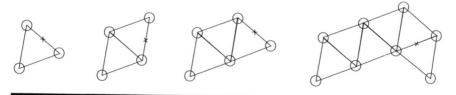

Figure 8.69   Schematic illustration of map growing in growing cell structure (GCS) method.

As already stated, two general approaches have been used for adding a neuron to a growing map. One is based on the number of inputs received by a neuron and the other is based on the average error measured as the average distance from a neuron to the inputs it represents. The basic idea of the two methods is similar except for the criterion used to add neurons. In GCS, focus is on the accumulated error as the criterion. Starting with the initial triangular three-neuron map, training proceeds for a certain number of steps and a neuron is added in the region where the largest accumulated error is found to better represent the input data. The idea is that the density of codebook vectors should represent the probability density of the input space. By inserting new neurons in the region where the error is large, an attempt is made to even out the error on all neurons so that codebook vectors evenly represent the input space. It is expected that over time, such uniform network structure will emerge with the GCS method. The process continues with more neurons being added as necessary during training. Existing nodes are also eliminated if they receive few or no inputs. These neurons are removed, together with all connections associated with that neuron, while still maintaining triangular connectivity of the whole map.

The whole learning procedure that involves the error measure to insert neurons can be described as follows [16].

Map initialization. The initial triangular map is created with three neurons. Initialize the codebook vectors randomly from the input probability distribution. For every neuron i, define error E_i. The error is the sum of distance from a neuron to all inputs that are closest to it.

Map training. Train the map with randomly selected inputs as follows: for every input vector find the winning neuron and add to its E_i the squared distance between the input and that winning neuron. Weights are updated for the winner and the direct topological neighbors of the winner. Constant learning parameters β_c, β_n are used in updating weights for the winner and neighbor neurons, respectively. In contrast, in the original SOM, a neighborhood can be very large initially and decrease with iterations as training continues while strength of adaptation also decreases over time in a specified manner.

After some training, identify the neuron with the maximum accumulated error. Suppose that this neuron is denoted k. Then, insert a new neuron m by splitting the connection between neuron k and its most dissimilar (i.e., farthest) neighboring unit n. This is depicted in Figure 8.69. Remove the old connection between k and n and make new connections $k–m$ and $m–n$. The new neuron now has to be initialized; this is accomplished by allocating the average of the codebook vectors of k and n. This is performed so that there is a smooth transition between the weights of neighboring neurons.

Basically, the addition of a new neuron creates a new Voronoi region for that neuron where the inputs closest to this new neuron fall. This redistribution of the inputs is reflected in the GCS method by redistributing the errors of the two existing neurons with the newly created one. Specifically, the error E_m of the new neuron m is set to the mean error of the neighboring neurons k and n and it is assumed that this represents the error of the new neuron if it had existed from the beginning. The error on the neighbors of the new neuron m is decreased proportionately to indicate their error if the new neuron had existed from the beginning. Then, the error on all neurons on the map is decreased by a fraction to indicate their error if the new neuron had existed from the beginning, and to make the recently altered region have a greater influence. This training process continues with more training patterns introduced to the network and growing the map until some stopping criteria are met. Stopping criteria may be the map size or some performance measure such as minimum average error.

Neurons may be deleted if no or only few inputs fall near a codebook vector. In this case, its error can be redistributed to the neighbors or ignored if too small. This is useful when the input probability density comprises several disjointed regions with very low probability in some regions [17]. The neurons in these regions receive only few inputs and their error may not decrease over time. The removal of neurons can enable the network to model evenly structured distributions to high accuracy. The resulting map is a two-dimensional network of neurons connected to form triangles. The position (x, y coordinates) of the neurons and their corresponding final codebook vectors are stored after training.

If the number of inputs closest to a codebook vector is used as the criterion for map growing, a new neuron is added at the point of highest frequency of inputs. After inserting a new neuron, its input frequency is made equal to the average frequency of the neighbors; this is assumed to be the frequency of that neuron if it had existed from the beginning. The input frequency of the neighbors is reduced proportionately to indicate their frequency if the new neuron had existed from the beginning. Neurons are removed if they have zero or very low input frequency. This has been used in a prior version of GCS [16].

A map trained by the GCS method follows the spread of the input data and therefore can be expected to reveal cluster structures better than a fixed grid. When an input is presented to a trained map, the corresponding winning neuron will respond; its position is the most appropriate position for that input vector in the data space that is well represented by the map. However, mapping is still performed on the best matching neuron only and therefore is discontinuous. However, the mapping can be made more precise; in Section 8.6.1.1, a centroid method for projection of inputs onto positions between neurons on a map trained by the GCS method is discussed.

8.6.1.1 Centroid Method for Mapping Input Data onto Positions between Neurons on the Map

The trained map should represent the input distribution well with its codebook vectors that span the input space. However, mapping of the inputs will still be performed based on the best-matching winner neuron. Section 8.5.4.6 illustrated a method where codebook vectors of the winner and its neighbors are averaged to achieve better mapping of the inputs onto the trained map. Wu and Yen [16] have proposed a centroid method based on projecting an input vector on all neurons on the map and finding the centroid of the projections as described next.

In the centroid method, when an input is mapped onto the trained map, its projection on all codebook vectors is found first. This indicates the degree of closeness of the input to all the codebook vectors. The projection is the highest for the neuron whose codebook vector is the most similar to the input. The projection of an input onto a codebook vector can be found by multiplying the input vector by the codebook vector. This is the dot product, which is equal to the weighted sum of inputs. If the inputs have q attributes, an input vector $\mathbf{x}$ takes the form $\{\mathbf{x}_1, \mathbf{x}_2, ..., \mathbf{x}_q\}$. The codebook vectors will have the same number of q components and a codebook vector $\mathbf{w}$ can be represented as $\{\mathbf{w}_1, \mathbf{w}_2, ..., \mathbf{w}_q\}$. The projection of input vector $\mathbf{x}$ on $\mathbf{w}$ is denoted by $\mathbf{P}_x$. Then, $\mathbf{P}_x$ is simply $\mathbf{w}_1\mathbf{x}_1 + \mathbf{w}_2\mathbf{x}_2 + ... + \mathbf{w}_q\mathbf{x}_q$, which is the scalar magnitude of the projection. By repeating this computation for all K codebook vectors, K projections of the input vector $\mathbf{x}$ on all codebook vectors are obtained. The total projections of one input vector can be visualized as a spatial response across all the neurons on the map. The centroid method is then used to find the center of gravity of all these projections, whose location defines the best location for the input on the map. The negative elements of the projection, however, are first zeroed to stabilize centroid calculation. For simplicity of illustration, this idea is presented schematically in Figure 8.70 for a two-dimensional square grid.

The centroid of the spatial response of projections is computed as the weighted average of the coordinates (x, y) of neurons and their corresponding projections averaged over the total sum of the projections. Suppose that the x and y coordinates of K neurons are organized as a $2 \times K$ matrix denoted by $\mathbf{C}_{xy}$ as

$$\mathbf{C}_{xy} = \begin{bmatrix} x_1 & x_2 & ... & x_K \\ y_1 & y_2 & ... & y_K \end{bmatrix}, \tag{8.35}$$

where x_i and y_i are the x and y coordinates of neuron i. The projections of an input vector $\mathbf{x}$ on all codebook vectors are denoted by $\mathbf{P}_x$, defined as

$$\mathbf{P}_x = \begin{bmatrix} \mathbf{p}_1 & \mathbf{p}_2 & ... & \mathbf{p}_K \end{bmatrix}^{\mathrm{T}} \tag{8.36}$$

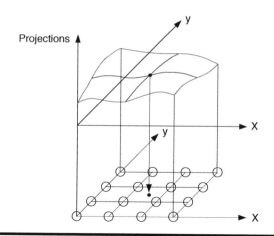

Figure 8.70 **Mapping of an input vector to the centroid of the spatial response of projections of an input vector on all codebook vectors on the map. (Adapted from Wu, Z. and Yen, G., *International Journal of Neural Systems,* World Scientific Publishing 13 (5), 353, 2003. With permission.)**

where $\mathbf{p}_i$ is the projection of $\mathbf{x}$ onto a single vector $\mathbf{w}_i$ of neuron i. The T indicates transpose, meaning that $\mathbf{P}_x$ is a column vector of size $K \times 1$, with K rows and one column. The centroid CG_x of the projections of input $\mathbf{x}$ is given by

$$CG_x = \frac{\mathbf{C}_{xy}\mathbf{P}_x}{\sum_i \mathbf{P}_i}. \tag{8.37}$$

Multiplication of $\mathbf{C}_{xy}$ of size $2 \times K$ with $\mathbf{P}_x$ of size $K \times 1$, results in a 2×1 vector containing the x and y coordinates of the center of gravity of the projections. Figure 8.70 shows the projections of an input on all codebook vectors of the map and the central point where an input is projected onto the map. As can be seen, a new input is mapped to points between neurons, leading to the possibility of high resolution in the retrieval of inputs from the map.

The process of finding the center of gravity of projections for many inputs can be performed simultaneously and efficiently using matrix manipulations. First, N input vectors are organized with q attributes as columns in a matrix $\mathbf{I}$ of size $q \times N$ (i.e., q rows and N columns). K codebook vectors are also organized in a matrix $\mathbf{W}$ of size $K \times q$, where q components of each codebook vector are organized in a row. The two matrices can simply be multiplied ($\mathbf{W} \times \mathbf{I}$) to obtain a projection matrix of size $K \times N$, where each column corresponds to projections of one input vector on all codebook vectors, such as $\mathbf{P}_x$ above with K components. The projection matrix has N such columns, one for each input vector. The coordinate matrix

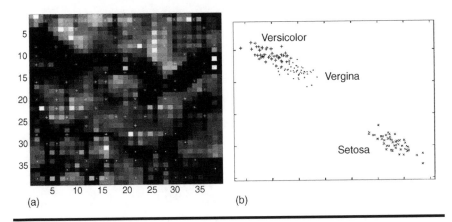

(a)   (b)

Figure 8.71 Comparison of projections of Iris data on trained maps: (a) simple projection involving average of neighbor vectors on a regular SOM grid (shown on the U-matrix) and (b) centroid based mapping on a map obtained from growing cell structure method. (From Wu, Z. and Yen, G., *International Journal of Neural Systems*, World Scientific Publishing, 13 (5), 353, 2003. With permission.)

C_{xy} $(2 \times K)$ can then be multiplied by the projection matrix to perform the operation depicted in Equation 8.37 to simultaneously obtain coordinates of all centers of gravity corresponding to the input vectors. The result is a $2 \times N$ matrix, where each column contains the x and y coordinates of the centroid of spatial response of the projections of one input vector on all codebook vectors. The whole matrix thus contains centroid coordinates for all N input vectors.

A comparison of simple input projections involving the average of neighbor vectors drawn on the U-matrix obtained for iris data from a regular SOM, such as the one trained previously and shown in Figure 8.44 and Figure 8.49, with the centroid-based projections obtained from the GCS method, is presented in Figure 8.71. The SOM map size is 40×40, as shown in Figure 8.71a. The projections of inputs on the map trained with the GCS method are shown in Figure 8.71b. In the GCS method, an additional five neurons have been incrementally added to the initial three-neuron map.

In Figure 8.71b, the projections of inputs belonging to different categories are labeled with different symbols. The figure shows that centroid mapping after training with the GCS method has separated the three different species and revealed internal structure. In contrast, for the projections on the regular SOM grid shown in Figure 8.71a, also labeled with the same symbols used in Figure 8.71b, the separation of the top two species (*Iris versicolor* and *Iris virginica*) is not as distinct. These two, however, are clearly separated from *Iris setosa* in both figures.

8.6.2 Dynamic Self-Organizing Maps with Controlled Growth (GSOM)

Dynamic SOMs with controlled growth is another recent variant of evolving SOMs proposed by Alahakoon et al. [20,24]. This map grows dynamically; its growth can be controlled using a spread factor such that a smaller map showing broader (macro) cluster regions can initially be obtained. This is very useful for large datasets where the computing burden can be very high and a broader picture is initially essential for taking the analysis further. If more accuracy is needed, the whole map can be grown further until desired accuracy is achieved in terms of the visibility of interesting clusters. This map can then be used to further grow only selected areas of interest (zoom in) so that attention can be focused on important regions of the map such that more and more clusters can be brought into light for understanding or reviewing possible subclusters within larger cluster structures. One interesting feature of this map is that it takes the shape of the clusters as it grows. In other words, as clusters are more and more represented by an expanding map, the map grows in specific directions dictated by the clusters. Consequently, different clusters make the map grow in different directions, thereby making the identification of cluster regions easier.

The initial structure of the map is four neurons organized in a two-dimensional grid, as shown in Figure 8.72. The map grows to represent the input data by adding new neurons to the boundary neurons with the highest accumulated error. Thus, the error is the mechanism of growth. Because all four neurons are boundary neurons, all have the freedom to grow in the outward direction as shown in Figure 8.72a. By growing only from boundary nodes, the two-dimensional structure of the map is maintained. A spread factor can be defined to control the growth selectively so that growth can take place in stages or in different regions. A low factor is given initially and then gradually increasing spread factors can be used for further growth of selected regions of interest.

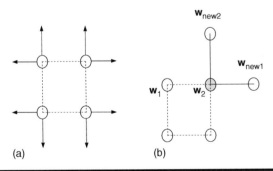

Figure 8.72 Original map and map growth from boundary neurons in GSOM.

The map training proceeds as follows.

Initialization phase. Initialize the four codebook vectors with random values. These may be four input vectors from the input space. For input variables with different scales, it is recommended that data be normalized in the range 0–1. Initial codebook vectors can also be four vectors with components (attributes) in this range. This allows the map to grow freely without restriction, solely based on input data.

Growth phase. The inputs are processed by the network and the accumulated error for each neuron is calculated. The four codebook vectors can be thought of as dividing the input space into four regions (Voronoi cells). The neuron with the largest accumulated error indicates a region that is underrepresented and is an area for generating a new neuron. If the accumulated error of a neuron exceeds a growth threshold (*GT*), then new neurons are grown from this neuron. New neurons are added in all free directions of a boundary neuron (i.e., in the vertical or horizontal outward directions) for simplicity of implementation. This is illustrated in Figure 8.72b, where the neuron with the largest accumulated error is the top right neuron from which two neurons with weights $\mathbf{w}_{new1}$ and $\mathbf{w}_{new2}$ have grown. The codebook vectors of the new neurons are assigned such that they smoothly flow with the neighboring neurons. For example, if two neighbor neurons, as shown in Figure 8.72b, have weights such that $\mathbf{w}_1 > \mathbf{w}_2$, then to maintain the smooth flow of weights, the new weight $\mathbf{w}_{new}$ will be initialized to $\mathbf{w}_1 + (\mathbf{w}_1 - \mathbf{w}_2)$ so that $\mathbf{w}_1 > \mathbf{w}_2 > \mathbf{w}_{new}$. Similarly, if $\mathbf{w}_1 < \mathbf{w}_2$, then the $\mathbf{w}_{new}$ is initialized to $\mathbf{w}_1 + (\mathbf{w}_2 - \mathbf{w}_1)$ so that $\mathbf{w}_1 < \mathbf{w}_2 < \mathbf{w}_{new}$.

The initial learning rate during growth phase requires consideration. Because there are only four neurons initially, the same neuron can be the winner for very different input vectors. This will cause the weights of the same neuron to fluctuate in completely different directions. To improve the situation, the learning rate is modified as follows:

$$\eta(t + 1) = \alpha\, \varphi(n)\, \eta(t) \tag{8.38}$$

where $\eta(t+1)$ is the learning rate at iteration $t+1$, $\eta(t)$ is the learning rate for iteration t, α is constant between 0 and 1, and $\varphi(n)$ is a function that gives a low value when the network size (i.e., total number of neurons n) is smaller and high value when it is bigger so that the initial learning rate is proportional to the network size. This function can take the form of $(1 - R/n(t))$ where $n(t)$ is the network size at iteration t and R is a constant. Alahakoon et al. [20] have arbitrarily used a value of 3.8 for R because the initial network size is four. Thus, the learning rate is initially reduced by the function; as the network grows and there is reasonable representation of the data, the weights can adapt faster with a higher learning rate.

The neighborhood size of a regular SOM is initially large and shrinks with iterations. The reason for a larger initial neighborhood is to order the codebook vectors in the initial iterations. Because weights of new neurons are ordered at the time of initialization in the GSOM such that weights in a neighborhood flow smoothly, an ordering phase is not necessary. Therefore, repeated passes of data over a very small neighborhood (2–4) that reduces to unity are used. Specifically, the GSOM initializes the learning rate and neighborhood size to a starting value for each new input. Weights are then adapted with decreasing neighborhood and learning rate until neighborhood is unity. After this update, learning rate and neighborhood size are initialized again for the next input.

In a regular SOM, the map grows and spreads around the data if it has enough neurons. In GSOM, because new neurons are added as necessary, the growth is initiated by a mechanism that allows the growth of boundary nodes. As presented earlier, this is when the accumulated error of a neuron exceeds a growth threshold. A limitation of the approach is that the map does not grow if the error on a nonboundary (interior) neuron exceeds GT due to high input density in this region. In this case, the map cannot grow and results in a congested map that does not give a proportional representation of the input data. It is argued that such proportional representation of the frequency of occurrence by area is very useful in identifying high-frequency regions in the map. Therefore, it will be useful to have a mechanism that allows nonboundary neurons to initiate growth of boundary neurons by spreading their error outwards. This is accomplished by halving the error of the nonboundary winner and increasing the error of its immediate neighbors by a fraction as

$$E_{t+1}^{w} = GT/2$$

$$E_{t+1}^{i} = E_{t}^{i} + \gamma E_{t}^{i}, \tag{8.39}$$

where w indicates the winner and i indicates an immediate neighbor that can take a value from 1 to 4.

The parameter γ, the factor of distribution, controls the increase in error accumulation of neighbors. Over time, the error from interior neurons will spread outwards towards boundary neurons and growth will be initiated due to this spread.

The GT specifies map growth. When the accumulated error in a neuron exceeds this threshold, new neurons are grown. To obtain a vale for this, a spread factor SF, with a value between 0 and 1, that specifies the amount of map growth is defined. Using this factor, the GT is defined as

$$GT = -D \ln(SF), \tag{8.40}$$

where D is the number of dimensions (attributes) of the input data. For a given spread factor SF, the GSOM approach calculates GT for a dataset with dimension D. The dimension D becomes relevant because the error measured as distance between input and weight vectors involves taking the square difference of the components of the two vectors. From Equation 8.40, GT is a logarithmic function of SF, which means that GT varies exponentially with SF and linearly with dimension D. Thus, by specifying a smaller spread factor, GT for error is larger and will result in a map with a lower number of neurons. Such a map is useful in initial data exploration. For further refinement, a larger spread factor can be used, resulting in smaller GT that allows greater growth of neurons resulting in finer resolution of the whole map or selected regions of the map.

Smoothing phase. When the new neuron growth phase is over, identified by low frequency of growth, weight adaptation is continued with a low learning rate. This is to smooth out any quantization error that may still undergo change in the neurons that were grown in the latter stage of growth. The training data is presented to the network and weights are updated with a smaller learning rate that decays until convergence, when the error values of the neurons become appreciably small. The neighborhood in this phase is constrained to only the immediate neighbors. Thus, in this phase, neighborhood size as well as the initial learning rate are smaller than those in the growth phase.

8.6.2.1 Example: Application of Dynamic Self-Organizing Maps

The main features of the GSOM described in the previous section are controlled growth of the map using a spread factor in a hierarchical fashion and the facilitation of growth of isolated map regions of interest to further investigate their cluster characteristics. These aspects are schematically illustrated in Figure 8.73 as reported by Alahakoon et al. [20].

Alahakoon et al. [20] demonstrate the application of GSOM on a widely used zoo dataset containing 16 attributes of animals and their type— mammal, fish, bird, etc. A record from this dataset is presented in Table 8.2.

Clustering of the zoo dataset with a small spread factor of 0.1 is illustrated in Figure 8.74, where animals are clustered into groups. Additionally, similar subgroups have been mapped near each other. With this small spread factor, it is still possible to see that there are three to four main groups in this dataset. The figure also illustrates one of the main features of GSOM: it indicates the groupings in the data by its shape, which is evident even at this low spread factor. For example, the map in Figure 8.74 has branched out in three directions, indicating the three main subgroups in data: mammals, birds, and fish. Insects have been grouped together with some other animals, but this group is not well separated due to a low SF value.

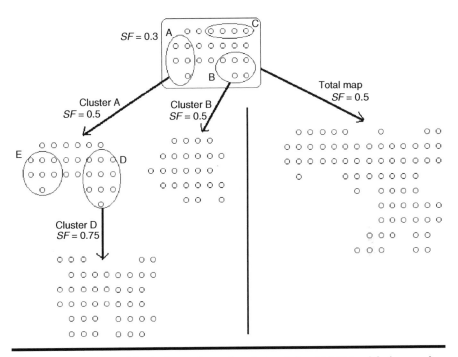

Figure 8.73 Hierarchical clustering of a dataset by GSOM with increasing spread factor, *SF*. (From Alahakoon, D., Halgamuge, S.K., and Srinivasan, B., *IEEE Transaction on Neural Networks*, IEEE Press, 601–614, 2000. With permission.)

When the spread factor is increased substantially to 0.85, clusters spread out further and become well separated, as shown in Figure 8.75. For example, the clusters of birds, mammals, fish, and insects have been clearly separated in the map. At this high value of spread factor, even the subgroups have appeared in the map. For example, predatory birds have been separated into a subcluster, separate from other birds. The mammals have

Table 8.2 Record from Zoo Dataset

Name	Has Hair	Lay Eggs	Feeds Milk	Airborne	Aquatic	Predator	Toothed
Wolf	1	0	1	0	0	1	1

Breathes	Veno-mous	Has Fins	No. of Legs	Has Tail	Dom-estic	Is Cat Size	Type
1	0	0	4	1	0	1	1

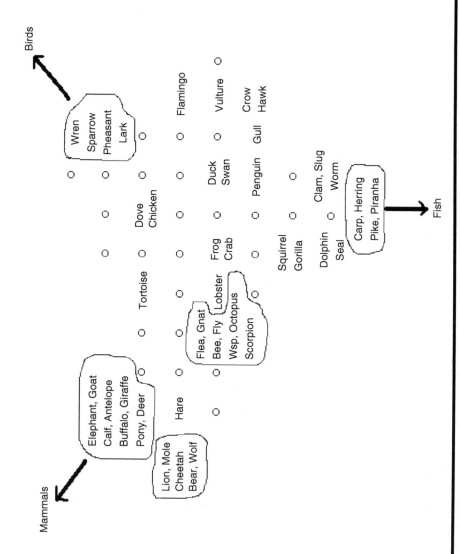

Figure 8.74 GSOM clustering of zoo data with *SF* = 0.1. (From Alahakoon, D., Halgamuge, S.K., and Srinivasan R *IEEE*

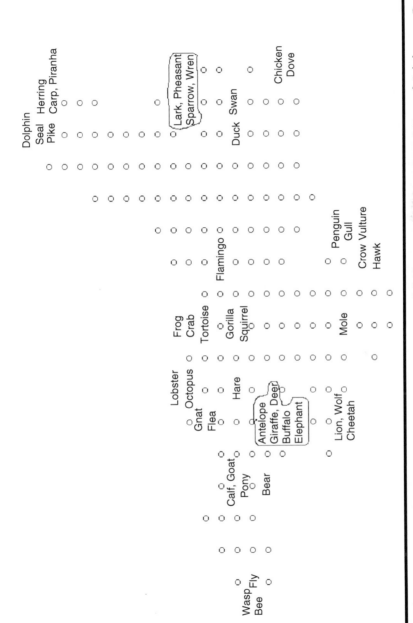

Figure 8.75 GSOM clustering of zoo data with *SF* = 0.85. (From Alahakoon, D., Halgamuge, S.K., and Srinivasan, B., *IEEE Transaction on Neural Networks*, IEEE Press, 601–614, 2000. With permission.)

been separated into predators and nonpredators, and nonpredators have been further separated into wild and domestic subgroups. Furthermore, other subgroups, such as airborne and nonairborne (chicken and dove), as well as aquatics, have also appeared in the map. The flamingo, as the only large bird, has found itself a separate place in the map.

Further growth of the map is possible with even higher *SF* values, but it would sometimes be more advantageous and necessary to expand only some selected areas of interest. This hierarchical clustering process can be continued on selected areas until the desired clarity of cluster structure is attained. Alahakoon et al. [20] illustrate this hierarchical clustering process as depicted in Figure 8.76, where the top right part shows a region selected from the map shown in Figure 8.74 for further expansion. The two clusters shown in circles in this part are expanded with an *SF* value of 0.6 to obtain a more detailed view of clusters. The subgroups inside these clusters are now clearly visible. For example, nondomestic and nonpredatory animals have all been

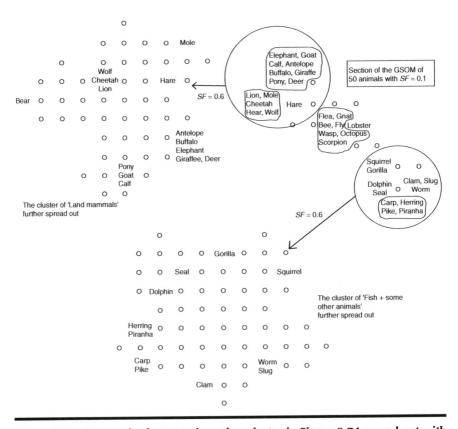

Figure 8.76 **Mammals cluster and another cluster in Figure 8.74 spread out with SF value of 0.6. (From Alahakoon, D., Halgamuge, S.K., and Srinivasan, B., *IEEE Transaction on Neural Networks*, IEEE Press, 601–614, 2000. With permission.)**

mapped together due to their similarity of attributes. The predators and domestic animals have also been separated. Even such details as why a bear, a predatory animal, has been separated from the other predators can be investigated; in this case, the bear is the only predator without a tail. This distinction was possible due to the shape that GSOM attains while branching out in the direction of data.

8.6.3 Evolving Tree

The evolving tree proposed by Pakkanen et al. [21] is another variant of evolving SOMs designed to grow a map flexibly and reduce the effort required to find a winning neuron, especially when the map size is large. In this approach, neurons are arranged in a tree topology that grows when any given branch receives a high number of input vectors. The method is efficient because the search for the winning neuron and its neighbors is conducted along the tree.

The basic mechanism of the evolving tree is that the map starts with one neuron that is sensibly placed at the center of the mass of the data distribution. This neuron is then split into a predetermined number of neurons that become its children. Their weights are initialized to that of the parent neuron. In this manner, a tree is created with a trunk neuron and several leaf neurons. All subsequent operations are performed on leaf neurons. For the next input pattern, the winning neuron is chosen randomly from among the children neurons so that latter neurons' weights differentiate from each other during training. When a leaf node reaches a splitting threshold indicated by the number of times it has become the winner, it splits again and the tree thereby grows recursively during training. Trunk neurons are static and their task is to maintain the connection between leaf neurons and other trunk neurons in the tree.

The process of finding the winner in a tree is illustrated in Figure 8.77, where every trunk node has two children. In the figure, leaf neurons are shown as filled circles and parent neurons as empty circles. Fundamentally, it is a top-down process where search starts with the root neuron and then follows on to its children to find the codebook vector that is closest to the input vector. If it is a leaf neuron, it is the winner; if not, its children are examined in turn until the best matching neuron is found, as shown in Figure 8.77. Thus, the tree works as a hierarchical structure from trunk to leaves located at several levels along its height.

After the winning neuron is found, its codebook vector is updated along with the neighbors using the same method as that used in the regular SOM. The neighbor strength function used is Gaussian with the width of the Gaussian shrinking with iterations. The usual weight update formula is

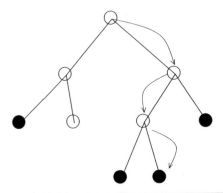

Figure 8.77 The structure of an evolving tree. Arrows indicate the path followed in finding the winner.

repeated in Equation 8.41:

$$\mathbf{w}_i(t+1) = \mathbf{w}_i(t) + \beta(t)\,\mathrm{NS}(d,t)\,[\mathbf{x}(t) - \mathbf{w}_i(t)], \tag{8.41}$$

where $\beta(t)$ is the learning rate at iteration t, and $\mathrm{NS}(d,t)$ is the neighbor strength function for iteration t and distance d from the winner to a neighbor. These are identical to those in regular SOM.

In addition, the frequency of hits, or the accumulated number of times a neuron has become the winner, is updated at each iteration. The frequency after iteration t is denoted by $f(t+1)$, expressed as

$$f(t+1) = f(t) + 1 \tag{8.42}$$

and a leaf growth or splitting of a neuron takes place if

$$f(t+1) > \theta, \tag{8.43}$$

where θ is the splitting threshold. The parameter θ controls the speed of growth of the tree.

The neighbor strength function is the commonly used Gaussian function, expressed as

$$\mathrm{NS}(t) = \mathrm{Exp}\left[\frac{\|\mathbf{r}_c - \mathbf{r}_i\|^2}{2\sigma^2(t)}\right], \tag{8.44}$$

where vectors $\mathbf{r}_c$ and $\mathbf{r}_i$ indicate the location of the winner c and a neighbor i on the map grid. As before, $\sigma(t)$ is the width of the Gaussian distribution. The learning rate and the width of the Gaussian are adjusted in the same manner as in the regular SOM. However, the distance between the winning neuron c and a neighbor i, given by $\|\mathbf{r}_c - \mathbf{r}_i\|^2$, is not as straightforward as in

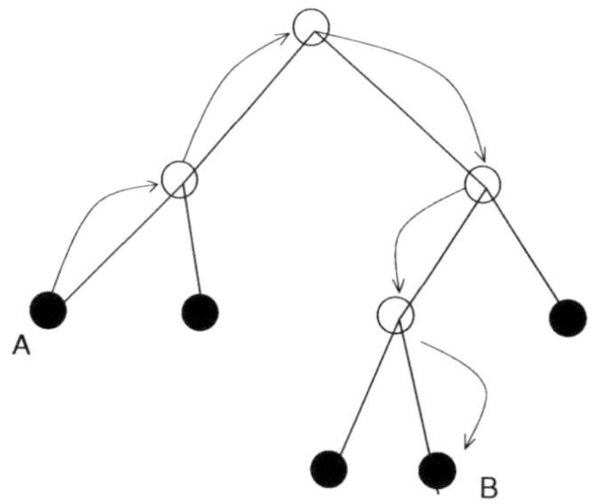

Figure 8.78 Finding the distance between the winner and a neighbor in an evolving tree.

the regular SOM where the structure is symmetric, regular, and static. In the tree structure, the distance cannot be predefined due to the evolving nature of the tree. Therefore, the distance in the shortest path from the winner and a neighbor is used.

The tree is evolved such that there is a unique path between any two neurons; therefore, it must also be the shortest distance between them. This distance can be calculated as the number of steps that must be taken to go from one to the other. This is illustrated in Figure 8.78, where an arrow represents a single step which is the distance between two adjacent neurons. The actual distance is the number of steps in the shortest path minus one. One is subtracted because only the distance between leaf nodes is of interest. However, the closest possible leaf nodes have a common parent, so two steps are required to go from one leaf to another. These closest neighbors should have a distance of 1. Therefore, one is subtracted from the total number of steps in the shortest path to obtain the actual distance between the winner and a neighbor. In Figure 8.78, it takes five steps to go from neuron A to B; the tree distance between them is therefore four.

Because the tree structure defines a unique path, it is intuitive to assume that different branches of the tree grow to represent different areas in the data space. Therefore, as a tree grows, the distances within a branch become smaller than those between branches. The tree grows, freely splitting one neuron at a time where and when necessary. This method is more flexible than the previously discussed methods that are used to evolve maps where some structure, either triangular, rectangular, etc., is imposed on the evolving map.

Pakkanen and Oja [21] have applied the method for two problems with very successful results. The first involves a picture consisting of the letters of

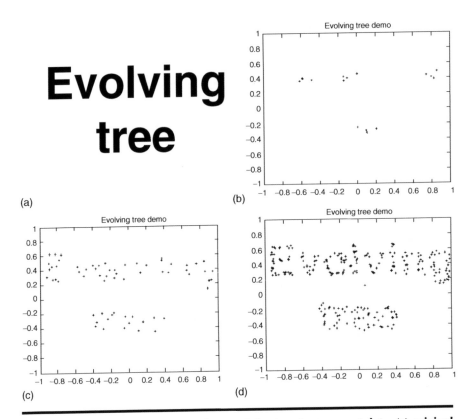

(a)

(b)

(c)

(d)

Figure 8.79   Progress in training of an evolving tree to represent data: (a) original data distribution, (b–d) the position of leaf neurons at several stages during training; the last figure is that obtained after ten epochs. (From Pakkanen, J., Iivarinen, J., and Oja, E., *Neural Processing Letters*, pp. 199–211, 2004. With permission from Springer Science and Business Media.)

evolving tree in the middle of it, as shown in Figure 8.79a. Using 1000 two-dimensional coordinate vectors in the region of the letters, a tree was evolved to represent the expression in the image. Any leaf neuron whose frequency counter exceeded the threshold $\theta = 60$ was split to create four new neurons from it. The training was continued for 10 epochs, but after the presentation of only a few hundred training patterns in the first epoch, the tree appears to have been split several times and neurons have found the general shape of the data cloud as shown in Figure 8.79b. However, the representation is still crude.

Further training involves more splitting of the neurons; slowly, individual letters start appearing, as shown in Figure 8.79c, although it has not captured the overall appearance of the letters yet. After ten epochs, the evolving tree has adapted so closely to the data that the original text can be read from the position of the leaf neurons of the tree (Figure 8.79d).

The final structure is very interesting because the leaf neurons have assumed positions along the contours of the letters and have gone to the extent that even the dot above the letter *i* has been represented separately.

Pakkanen and Oja [21] have further subjected the evolving tree to a rather complex classification task of identifying, from digital images, defects in sheets of paper obtained from an online visual quality control system on a paper machine. There were 14 defects, and even a professional paper maker would find it hard to classify some defects due to subtle variations in them. The defects were various-sized imperfections, such as slits, holes, openings, etc. There were a total of 1308 defective gray images. Input patterns have been extracted from images in three ways, based on color structure, edge histogram, and texture. All three methods have been used separately to evolve trees to classify the defects in images.

In the use of the evolving tree, two parameters must be set: the splitting threshold θ and a splitting factor indicating how many children to grow from a parent. This is problem dependent and for this experiment with defects in paper, θ has been set to 50 and a neuron is split into three new neurons if the splitting threshold has been reached. Tests were conducted with ten-fold cross validation. For comparison, a regular SOM and a TS-SOM have also been trained to classify the same images. The SOM map size was 19×9 with a total of 171 neurons. The evolving trees ended up having approximately 240 leaf nodes, but due to the fact that the tree maintains a very efficient search tree from trunk to leaf neurons, training of the evolving tree was roughly ten times faster. Recall from Section 8.6 that the TS-SOM is a layered map, where regular SOM maps are organized hierarchically with neurons in the first layer having connections to specific neurons in the grid at the lower level so that search for the winner follows these links. The TS-SOM had 16×16 (i.e., 256) neurons in the lowest layer, which makes it more complex than the evolving tree. Results have revealed that, of all the methods, evolving tree achieved the highest average classification accuracy over all classes for the three input extraction methods. The evolving tree performs better than the regular SOM. The TS-SOM has variable results and its overall performance is approximately similar to the regular SOM. It is noteworthy that the evolving tree has had much superior classification accuracy for the two most difficult to classify defects, which is desirable where good representation of difficult regions is required.

8.7 Summary

In this chapter, a comprehensive introduction to SOMs is presented. Starting with competitive learning, the operation of competitive networks is introduced with examples of one-dimensional networks. This leads to one- and two-dimensional SOMs, where a neighbor feature is incorporated for preservation of spatial organization of the input data in the map.

The formation of two-dimensional SOMs is illustrated extensively using a well-known example. The topics include map training, quality of representation of input data by a map, projecting input data onto a trained map for retrieval of input data, distribution of codebook vectors with respect to the original data distribution, classification with maps, as well as grouping the neurons in a trained map into the most likely number of clusters and the resulting classification accuracy. Evolving SOMs are introduced as an extension to regular SOMs with the intention of evolving a map to the desired size and complexity rather than using a fixed map grid. Several approaches to evolving maps are presented with examples.

Problems

1. The two-dimensional vectors $\mathbf{x}$, $\mathbf{w}_1$ and $\mathbf{w}_2$ are given by

$$\mathbf{x} = [0.2 \quad -1.4]$$
$$\mathbf{w}_1 = [0.6 \quad -4.0]$$
$$\mathbf{w}_2 = [0.1 \quad -1.0]$$

   (a) Normalize the input and codebook vectors so that they have a unit length.
   (b) Find the codebook vector the $\mathbf{x}$ is closest to in terms of Euclidean distance.
   (c) Update the weight of the winner using self-organizing learning algorithm with a learning rate of 0.5.
   (d) Update the codebook vector of the neighbor with a neighbor strength that is half of the winner.

2. Repeat Problem 1 with original vectors but this time using the activation of neurons, which is the dot product giving weighted sum of inputs, to find the winner.

3. Two extra inputs and a codebook vector are added to the vectors in Problem 1 to get

$$\mathbf{x}_1 = [0.2 \quad -1.4]$$
$$\mathbf{x}_2 = [0.4 \quad 0.9]$$
$$\mathbf{x}_3 = [0.5 \quad -3.2]$$
$$\mathbf{w}_1 = [0.6 \quad -4.0]$$
$$\mathbf{w}_2 = [0.1 \quad -1.0]$$
$$\mathbf{w}_3 = [0.5 \quad 0.8]$$

   (a) Find the winning codebook vector using Euclidean distance for all input vectors and update weights with a learning rate of 0.1.

For each input vector, update the weights of the neighbors using a neighborhood size of 1 and neighbor strength of 0.5 of the winner.

(b) Repeat the training for another epoch and plot the input and weight vectors to check if the network indicates topology preservation.

4. Group the following input vectors into two clusters using self-organizing learning method.

$$[1, 1, 0, 0]; \quad [0, 0, 0, 1]; \quad [1, 0, 0, 0]; \quad [0, 0, 1, 1]$$

Initial learning rate $\beta(0) = 0.6$:

$$\beta(t + 1) = 0.5 \, \beta(t),$$

Neighborhood size is 0. Assume random initial codebook vectors.

5. The learning rate decay formula for an SOM during the first 1000 iterations is given by

$$\beta(t) = 0.2(t - 1/1000).$$

where t is the iteration. How many iterations will it take for the learning rate to decay to 0.01?

6. The learning rate decay formula for an SOM during the first 1000 iterations is given by an exponential formula

$$\beta = 0.5 \, \text{Exp}[-t/300].$$

where t is the iteration. How many iterations are required for the learning rate to decay to 0.01?

7. If a square neighborhood geometry is used in training an SOM, how many neighbor neurons will be updated if the radius is 1? How many will be updated if the radius is 2?

8. Answer Problem 7 if the neighborhood geometry is hexagonal.

9. Explain the purpose of the initial ordering phase in SOM training.

10. Explain the purpose of shrinking neighborhood size and learning rate decay with iterations.

11. In Section 8.5.2.2 of the chapter, it was demonstrated that the map orientates in the major direction of data. Explain what makes this happen.

12. Explain three methods for initialization of the map. Discuss the advantages and disadvantages of each.

13. How can input data be mapped onto a trained map? What is the outcome of such mapping?

14. How can the quality of input mapping onto a trained SOM be measured?

15. What is U-matrix and what does it tell about cluster structures?

16. Explain distortion measure of a map and how it can be useful in assessing the adequacy of a trained map.

17. What further operations must to be performed after training a map if the objective is classification of inputs? What can be said about the quality of such classification based on trained maps?
18. Discuss how a trained SOM can be validated. What is the objective of map validation?
19. What are the advantages of evolving SOMs over regular fixed grid maps? In what way are evolving maps useful?
20. What fundamental concepts drive evolving SOMs?
21. On a dataset of choice, train an SOM and perform as many operations illustrated in the chapter as possible. Draw as many conclusions and gain as much insight as possible about the cluster structures on the map.

References

1. Kohonen, T. *Self-Organizing Maps*, 3rd Ed., Springer-Verlag, Berlin, Heidelberg, New York, 2001.
2. Rosenblatt, F. The perceptron: A probabilistic model for information storage and organization in the brain, *Psychological Review*, 65, 386, 1958.
3. *Mathematica-Neural Networks*, Wolfram Research, Inc., Champaign, IL, 2002.
4. Kohonen, T. The self-organizing map, *Proceedings of IEEE*, 78, 1464, 1990.
5. Mind and brain, *Readings from "Scientific American Magazine"*, W.H. Freeman, New York, 1993.
6. Haykin, S. *Neural Networks: A Comprehensive Foundation*, 2nd Ed., Prentice Hall, Upper Saddle River, NJ, 1999.
7. *Minitab 14*, Minitab Inc., State College, PA, www.minitab.com, 2004.
8. Kraaijveld, M.A., Mao, J., and Jain, A.K. *Proceedings of the Eleventh ICPR, International Conference on Pattern Recognition*, IEEE Computer Society Press, Los Alamitos, CA, 1992.
9. UCI Machine learning repository, http://www.ics.uci.edu/~mlearn/MLRepository.html.
10. *Machine Learning Framework for Mathematica*, uni software plus, www.unisoftwareplus.com, 2002.
11. Vesanto, J. and Alhoniemi, E. Clustering the self-organizing map, *IEEE Transactions on Neural Networks*, 11, 586, 2000.
12. Buhman, J. and Kuhnel, H. Complexity optimized data clustering by competitive neural networks, *Neural Computation*, 5, 75, 1993.
13. *Cluster Analysis*, StatSoft Inc. 2004. http://www.statsoft.com/textbook/stcluan.html
14. Ward, J.H. Jr., Hierarchical grouping to optimize an objective function, *Journal of the American Statistical Association*, 58, 236, 1963.
15. Drobics, M. and Winiwarter, W. Mining clusters and corresponding interpretable descriptions—a three-stage approach, *Expert Systems*, 19, 224, 2002.
16. Wu, Z. and Yen, G. A SOM projection technique with the growing structure for visualizing high dimensional data, *International Journal of Neural Systems*, 13, 353, 2003.

17. Fritzke, B. Growing cell structures—a self organizing network for unsupervised and supervised learning, *Neural Networks*, 7, 1441, 1994.
18. Martinetz, M. and Schulten, K.J. A neural-gas network learns topologies, *Proceedings of International Conference on Artificial Neural Networks*, T. Kohonen, K. Makisara, O. Simula, and J. Kangas, eds., North Holland, Amsterdam, p. 397, 1991.
19. Blackmore, J. and Miikklulainen, R. Incremental grid growing: encoding high-dimensional structure into a two-dimensional feature map, *Proceedings of the International Conference on Neural Networks*, p. 450, 1993.
20. Alahakoon, D., Halgamuge, S.K., and Srinivasan, B. Dynamic self-organizing maps with controlled growth for knowledge discovery, *IEEE Transactions on Neural Networks*, 11, 601, 2000.
21. Pakkanen, J., Iivarinen, J., and Oja, E. The evolving tree—a novel self-organizing network for data analysis, *Neural Processing Letters*, 20, 199, 2004. (Figure reprinted with kind permission of Springer Science and Business Media.)
22. Koikkalainen, P. and Oja, E. Self-organizing hierarchical feature maps, *Proceedings of International Joint Conference on Neural Networks*, Vol. 2, p. 279, 1990.
23. Dittenbach, M., Rauber, A., and Merkl, D. Recent Advances with the growing hierarchical self-organizing map, *Third Proceedings of the Third Workshop on Self-Organizing Maps*, Vol. 140, 2001.
24. Alahakoon, L.D. Controlling the spread of dynamic self-organizing maps, *Neural Computation and Application*, 13, 168, 2004.
25. Kohonen, T. *Self-Organizing Maps*, Springer, Berlin, 1995.
26. Lopez, M., Samarasinghe, S., and Hickford, J. Use of neural networks for diagnosis of mastitis in dairy cattle, *Proceedings of the International Joint Conference on Neural Networks (IJCNN)*, Vol. 5, Portland, OR, p. 100, 2003.
27. Wang, E. and Samarasinghe, S. On-line detection of mastitis in dairy herd using neural networks, *Proceedings of International Congress on Modelling and Simulation (MODSIM'05)*, A. Zerger and R.M. Argent, eds., Modelling and Simulation Society of Australia and New Zealand, Melbourne, Australia, p. 273, 2005.
28. Hair, J.F., Anderson, R.E., Tatham, R.L., and Black, W.C. *Multivariate data analysis*, 5th Ed., Prentice Hall, Upper Saddle River, NJ, 1998.
29. *SPSS 10.0 for Windows*, SPSS Inc., Chicago, IL, USA.
30. *NeuroShell® 2*, Ward Systems Group, Frederick, MD, 1997.

Chapter 9

Neural Networks for Time-Series Forecasting

9.1 Introduction and Overview

Forecasting future events is an important task in many practical situations. These situations may include the forecasting of river inflow; temperature, rainfall or weather; electricity load or energy; economic, stock market or currency exchange rate; and numerous other scenarios in which future events need to be predicted based on available current and past information. In general, this involves a series of event outcomes that are related or correlated, and correlation dynamics are used to predict the next one or several outcomes in the series. Many cases involve a time-series consisting of the observations of an event over a period of time, and require a prediction of the event for a particular time (hour, day, month), based on the outcome of the same event for some past instances of time. Time-series forecasting is an important area of forecasting in which past observations are analyzed to develop a model describing the underlying relationship in a sequence of event outcomes. The model is then used to forecast future events that have not yet been observed (Figure 9.1).

A time-series can be either linear or nonlinear as shown in Figure 9.1. In a linear time-series, the next outcome is linearly related to the current outcome (Figure 9.1a and Figure 9.1b). In contrast, in a nonlinear time-series, the next outcome has a nonlinear relationship to the current outcome (Figure 9.1c and Figure 9.1d). A main feature of any time-series is

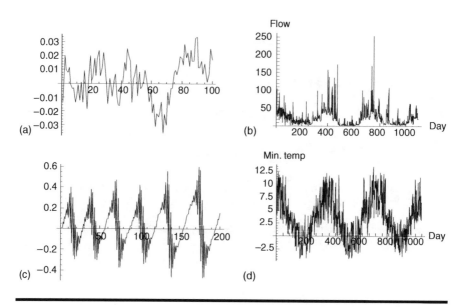

Figure 9.1   Some example time-series: (a) synthetic linear time-series, (b) real linear time-series of flow into a river, (c) synthetic nonlinear time-series, and (d) real nonlinear time-series of minimum temperature variation in a river catchment.

autocorrelation, which is simply the correlation of the value of the observation at a particular time with observations at previous instances in time. In general, the autocorrelation is higher for events in the immediate past and decreases for observations in the distant past. The number of steps that are significantly related are known as lags, and for a time-series, these steps are called time lags. If only the immediate past event has a significant relation to the next observation, time lag is one, and so on. Depending on the problem, the related time lags can all be past observations from the current time to a specific time in the past, or they can be some intermittent past observations between a current and past instance of time. For example, for the forecasting of the daily inflow into a river in a catchment, the inflow of the past ten days may be significant, depending on the characteristics of the catchment. For accurate forecasting, it is essential to capture this dynamic autocorrelation. This is called temporal forecasting, in which only the past observations of a variable are used to forecast its future state; such models are called temporal models.

It can be intuitively expected that the effects of the past outcomes are captured in the current observation of a variable, and that therefore, the current observation should be a good predictor of future observations.

However, in practice, it is known that time lags improve forecasting accuracy.

A spatio-temporal model incorporates not only time lags of the variable whose future observations are of interest but also other influencing variables, and possibly their lags. For example, temperature can have a direct influence on the energy requirement of a building and therefore incorporating time lags of temperature, along with the current energy requirement and possibly its lags, can improve the accuracy of energy-demand forecasting. This is because the model can extract information from the extra variables that is not embedded in the current observation and the lags of the variable of interest. In other words, the total influence of the significant variables may not be completely represented in the current observation or lags of a variable and therefore, additional variables must be used in some cases to capture the underlying dynamics of the series. Spatio-temporal models are required for these situations.

This chapter presents an extensive treatment of the use of neural networks for time-series forecasting, with examples and case studies. It covers linear and nonlinear models for time-series forecasting, long-term forecasting with neural networks, in-depth analysis of model error, and input selection for time-series forecasting. The chapter begins with an introduction to forecasting with linear neuron models and presents an example to demonstrate its short-term and long-term forecasting ability in Section 9.2.1.

Section 9.3 introduces nonlinear networks for time-series forecasting. These networks are modified backpropagation using short-term memory filters represented by input lags (focused time-lagged feedforward networks) and recurrent networks with feedback loops that capture long-term memory dynamics of a time-series and embed it in the network structure itself. Three practical examples (regulating the temperature in a furnace, forecasting the inflow into a river, and forecasting the levels of air pollution) are solved to illustrate modified backpropagation and its ability to perform short-term and long-term forecasting in Section 9.3.2 and Section 9.3.3. As an alternative, generalized neuron models are presented in detail with an application case study involving electricity demand forecasting in Section 9.6.

Recurrent networks are formally treated in Section 9.7. This includes a detailed presentation of three types of recurrent networks: Elman, Jordan, and fully recurrent networks (Section 9.7.1 through Section 9.7.3). In Elman networks, the hidden layer activation is fed back as input at the next time step; in Jordan networks, the output is fed back as the input. In fully recurrent networks, both hidden neuron activation as well as the output are fed back to themselves in the next time step. These sections show how the networks encapsulate long-term memory of a time-series and how they are trained. For each recurrent network type, examples and case studies are presented;

these range from stream flow to rainfall runoff to energy and temperature forecasting in various applied situations. These examples highlight the advantages of these networks in terms of short-term and long-term forecasting accuracy, and compare them with feedforward networks.

The bias and variance issues still apply to time-series models, and Section 9.8 treats the decomposition of error into bias and variance components, using Monte Carlo simulations to assess the contribution of each to model error with respect to number of inputs and hidden neurons in focused time-lagged feedforward networks. This decomposition helps identify which component of error the researcher must pay attention to in improving models for the best generalization. Accuracy of long-term forecasting nevertheless degrades with the forecast horizon. Section 9.9 presents an approach to improve long-term forecasting with neural networks, using multiple neural networks that form a combined set of sub-models with shorter prediction horizons. A practical case study involving hourly gas flow rate prediction is presented in Section 9.9.1 to highlight the advantage of such models.

In the last section, the topic of input selection for neural networks for time-series forecasting is treated, with an example and a practical application case study (Section 9.10). Two methods are used to illustrate the selection of inputs from nonlinearly correlated data. The first method is Partial Mutual Information (PMI) in conjunction with Generalized Regression Neural Networks (GRNNs) (Section 9.10.1.1 and Section 9.10.1.2); the second method uses a self-organization map in conjunction with genetic algorithms and GRNNs (Section 9.10.1.3 and Section 9.10.1.4). These methods are applied to synthetic data and a river salinity problem, and their performance is compared with respect to the adequacy of the inputs selected and forecasting accuracy.

9.2 Linear Forecasting of Time-Series with Statistical and Neural Network Models

There is a large amount of literature dealing with linear statistical models that have been widely used for linear time-series forecasting [1]. In a linear time-series, the next observation is linearly related to the past observation(s). For example, a simple linear time-series can be expressed as

$$y_t = a_0 + a_1 y_{t-1} + \varepsilon_t \tag{9.1}$$

where y_t is the observation at time t and y_{t-1} is that at time $t-1$. The coefficients a_0 and a_1 are constant model parameters, with a_0 indicating the intercept and a_1 indicating the strength of the relationship between the current and next observation. The last component, ε_t, is the error component at time t and is assumed to be from an independent and identically normal distribution (i.i.d.) with zero mean and a variance of σ^2. The coefficients are determined from the linear Least Square Error (LSE) estimation used in regression analysis. By incorporating lags for the variable and the error, Autoregressive Integrated Moving Average (ARIMA) models and their variants can be constructed. A general form of an ARIMA model assumes that the underlying process that generates a time-series has the form:

$$y_t = a_0 + a_1 y_{t-1} + a_2 y_{t-2} + \cdots + a_p y_{t-p} + \varepsilon_t$$

$$- b_1 \varepsilon_{t-1} - b_2 \varepsilon_{t-2} - \cdots - b_q \varepsilon_{t-q} \tag{9.2}$$

where $y_{t-1},\ldots,y_{t-p}$, are past observations of the variables up to p lags and $\varepsilon_{t-1},\ldots,\varepsilon_{t-q}$ are q lags of the error. The coefficients $a_0, a_1,\ldots,a_p$ and $b_1,\ldots,b_q$ are model parameters that are determined using linear LSE estimation. The (p, q) is the order of the ARIMA model. The central task of the ARIMA model building process is to first determine the appropriate model order (p, q). Then the parameters are estimated using LSE. ARIMA is a family of models, and Equation 9.2 encompasses several special cases of ARIMA models [1]. For example, when $q = 0$, the equation becomes an Auto-regressive (AR) model of order p. When $p = 0$, the model reduces to a Moving Average (MA) model of order q.

The neural equivalent of a linear model for time-series forecasting is the linear neuron presented in Chapter 2. For example, Figure 9.2 illustrates a linear neuron with p lags of the variable z, and this neuron, trained using delta rule, is an AR model of order p. The weights $a_0,\ldots,a_p$ are the model parameters estimated from delta rule. When q error lags are incorporated as inputs, the linear neuron represents Equation 9.2. As with ARIMA, the model order (p, q) must be estimated prior to estimating the weights of a linear neuron using delta rule. The trained linear neuron can predict the value of the next observation z_t, and its predictions must be identical to those from ARIMA because the LSE is identical to delta rule as presented in Chapter 2.

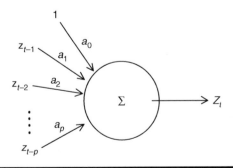

Figure 9.2 Linear neuron (AR) model for linear time-series forecasting.

The central theme of the above ARIMA, AR, and MA models is that they are linear models. They have been used successfully in many situations where there is a strong linear relationship between the current and next outcome [1,2]. However, when there is a strong nonlinear relationship, the forecasting accuracy can be severely affected because the linear models cannot capture nonlinear trends in the data. The traditional linear models are not adequate in such cases, and neural networks can be very useful in extending the capabilities of the linear models.

9.2.1 Example Case Study: Regulating Temperature of a Furnace

A small example will be used to understand the concepts of linear time-series processing. In this example, the temperature of an electric furnace must be accurately monitored. The temperature is maintained at the desired level by adjusting the electric current to the furnace. The data for this exercise was sourced from Neural Connection™, a neural networks software program [3]. The database consists of 503 current and temperature measurements made at regular time intervals of one minute. The current and temperature variation with time is presented in Figure 9.3, which shows that the current has been increased in steps and that the temperature has responded accordingly. The last part of the data, from 250 minutes onwards, has been specially created to test the forecasting model on unseen data.

The task of the model is to predict the temperature at the next time step. It can be seen that temperature is strongly related to the current. When the current increases or decreases, the temperature increases or decreases accordingly; when the current oscillates, the temperature oscillates as well. It can also be seen that there is a strong linear relationship between the temperature and its previous value, as shown in Figure 9.4. Points that are far from the line correspond to jumps in the

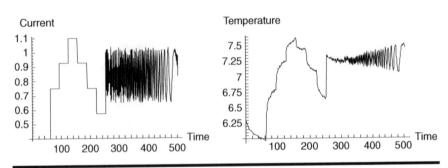

Figure 9.3 Current and temperature variation of the furnace with time.

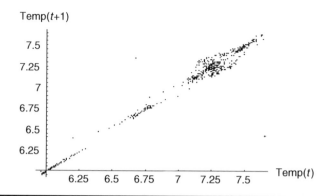

Figure 9.4 Relationship of the temperature to its value at the previous step.

current. The linear relationship indicates that a linear forecasting model should produce accurate forecasts of temperature.

Although it is possible to perform various preprocessing with the input data, this example will demonstrate the model behavior using raw data. The example will use temperature and current as inputs (two time lags each), and develop a linear neural network model with these four inputs. This model contains no hidden neurons* and the four inputs are fed into a linear output neuron, as shown in Figure 9.5. The T and C denote temperature and current, a_0 is the bias weight, a_1 and a_2 are the weights associated with lagged temperature, and a_3 and a_4 are the weights associated with the lagged current. This model is a neural equivalent of an autoregressive with external variables (ARx) model. To use the four inputs, the data must be organized such that the previous temperature and current are in the same row as the present temperature and current. The output is the next temperature.

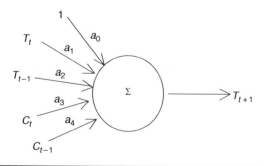

Figure 9.5 Autoregressive linear neuron model with external input (ARx) for forecasting temperature.

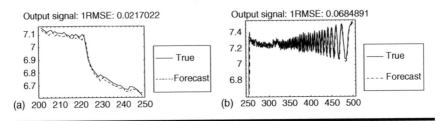

Figure 9.6 Linear neural ARx model performance on test data for single-step ahead forecasting: (a) the first test set, which is similar to the training data, and (b) the second test set with rapid oscillations of current and temperature.

This linear model is expressed by

$$T_{t+1} = a_0 + a_1 T_t + a_2 T_{t-1} + a_3 C_t + a_4 C_{t-1} \qquad (9.3)$$

and the objective is to find the unknown coefficients. The model can be easily trained with least-squares error estimation using delta rule. The first 200 data points were used to train the network and the rest of the dataset was divided into two sets, one ranging from 200 to 250 min and the other from 251 to 500 min. The first set is similar in character to the training data and the second set is dissimilar in that the current oscillates rapidly. The model performance on the first test set is shown in Figure 9.6a, which shows that the model output closely follows the true data with a root means square error (RMSE) of 0.0217. For the second set, the model forecasts are still very accurate, but have a higher RMSE of 0.0685, which is three times as large as that for the first set (Figure 9.6b).

9.2.1.1 Multistep-Ahead Linear Forecasting

In the previous analysis, the temperature was predicted one time step in advance. Now it is possible to see whether the model is able to do more than one-time-step-ahead forecasting, or multistep-ahead forecasting. In this case, the actual temperature is not available as an input at future time steps, so in forecasting mode, the predicted temperature has to be used as input. In practice, the true future values of other external inputs are also not available and therefore they must also be estimated before using a model for multistep forecasting. This example, however, will use the true electric current data in the database for the estimates. Figure 9.7 shows the model performance for two-step, three-step, ten-step, and complete multistep forecasting for the first test set. In two-step forecasting, for example, the predicted output in the last step is used as input for forecasting the next step. In ten-step-ahead forecasting, the previously predicted outputs are recursively used nine times. In complete forecasting, predicted output is used throughout.

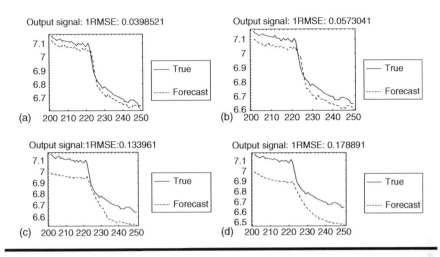

Figure 9.7 Multistep-ahead forecasting performance with linear neural ARx model on the first test set; top two figures from left to right are for two-step-ahead and three-step-ahead forecasting and the bottom two from left to right are for ten-step and complete (50-step) forecasting.

Figure 9.7 shows that the forecasting ability deteriorates as the prediction horizon increases, as indicated by the increasing RMSE with the length of the prediction horizon. This is partly due to the compounding effects of the errors in the predicted output that are then used as input.

The long-term forecasting accuracy of the model on the second, more variable dataset is shown in Figure 9.8 for 10- and 50-step-ahead forecasting. Comparison of Figure 9.7 and Figure 9.8 reveals that the accuracy on the second test set is lower than that on the first test. This is due to the fact that the second test set is more variable than the training data, and also the fact that it is far from the training data on the time scale.

In order to see if it is better to use the whole dataset in forecasting for the last 250 hours, the single- and multistep-ahead forecasting for the whole

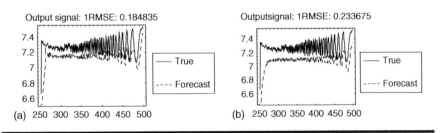

Figure 9.8 Multistep-ahead forecasting for the second, more variable, set of test data. (a) ten-step and (b) 50-step-ahead forecasting.

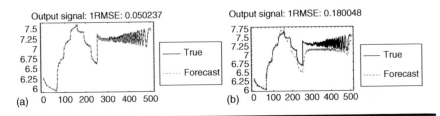

Figure 9.9 Single- and multistep-ahead forecasting using the entire dataset: (a) single-step and (b) complete (250-step) forecasting.

dataset is given in Figure 9.9. It shows that when the whole time-series is used in forecasting mode, the long-term forecasts for the last 250 hours improve considerably, as indicated by the lower RMSE in this figure compared to the long-term forecasting error for the second test set shown in Figure 9.8. Accuracy now is similar to that from the first test set shown in Figure 9.7.

9.3 Neural Networks for Nonlinear Time-Series Forecasting

A recurrent theme throughout the book has been that the power of neural networks lies in their ability to capture nonlinear relationships inherent in data. Whereas linear models depict a linear relationship between the current and next observations, neural networks portray a nonlinear relationship between the two. This can be described as

$$y_t = f(y_{t-1}, y_{t-2}, \ldots y_{t-p}) + \varepsilon_t \tag{9.4}$$

where $f(y_{t-1}, \ldots, y_{t-p})$ is the nonlinear function that maps a series of past observations nonlinearly to the next outcome. This function is the neural network model. The last component in Equation 9.4 is the error, which is expected to be a random variable with a mean of zero and variance σ^2.

9.3.1 Focused Time-Lagged and Dynamically Driven Recurrent Networks

Several network configurations that capture temporal aspects of inputs have been proposed [4]. Two types of neural network models that have been successfully used for various time-series forecasting tasks are focused time-lagged feedforward networks and dynamically-driven recurrent (feedback) networks. Both are based on the feedforward, multilayer networks presented in Chapter 3 and Chapter 4. Figure 9.10 presents a schematic diagram of the focused time-lagged feedforward network, in which the time is embedded externally using a short-term memory filter

Figure 9.10   Focused time-lagged (short-term memory) feedforward network.

that acts as input to a static feedforward network such as Multi-Layer Perceptron (MLP) [4]. The short-term memory filter basically incorporates time lags. Thus a short-term history of the time-series is presented as inputs to a static network.

The concept of a recurrent network is entirely different from that of a focused time-lagged network. It attempts to incrementally build the auto-correlation structure of a series into the model internally, using feedback connections relying solely on the current values of the input(s) provided externally [4–6]. The idea behind such networks is that a network should learn the dynamics of the series over time from the present state of the series, which is continuously fed into it, and that the network should then use this memory when forecasting.

Figure 9.11 presents three generic, dynamically-driven recurrent networks. The first network, in which the network output is fed back through the feedback connections into the input layer with one unit time delay, is called an input–output recurrent model, and represents a nonlinear autoregressive (NAR) model. In this model, the forecast output is fed back into the network as input at the next time step. When the network is based solely on the current state of the variable for which a forecast is made, the external input is only the current state of the variable. During use of this model, the network recursively uses its own output as input for the length of time for which a forecast is necessary. When other relevant external inputs

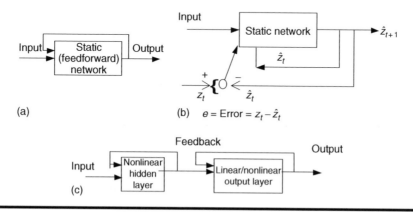

Figure 9.11   Recurrent networks for time-series forecasting: (a) input–output recurrent network NARx model, (b) input–output error recurrent network (NARIMAx model), and (c) fully recurrent network.

are presented to the network in addition to those fed back from the output layer, this is called a nonlinear autoregressive with exogenous inputs (NARx) model. When the error at each time step (target–predicted) is also fed back to be used as an input at the next time step in addition to the output, this yields a nonlinear autoregressive moving average with exogenous inputs (NARIMAx) model, as shown in Figure 9.11b.

The third model (Figure 9.11c) shows another generic recurrent network in which the outputs from both the hidden layer and the output layer are fed back through the feedback connections into the input and hidden layers, respectively, with one unit time delay [4–6]. These hidden and output layer outputs, which are fed back in time step t, become input for the respective layers at time step $t+1$. This creates a network of feedforward and feedback connections, which hold the long-term memory of the network. In this way, a network is able to capture and embed long-term memory dynamics into its internal structure. For a given current observation of a variable, a recurrent network uses its built-in, long-term memory to predict the next outcome, which is expected to be accurate and reliable. Several variants of this recurrent network structure have been used successfully in practical applications, as will be examined in detail later.

In contrast with static linear or nonlinear ARIMA, AR, and MA models that use a specific number of lags as input, a recurrent network builds into its structure all past time lags in forecasting; however, the more recent lags have a larger influence than do the distant lags. Thus the network itself finds the lag structure for the problem during learning from the autocorrelated data. These aspects will be treated in depth throughout this chapter. Now the properties of the two broad types of network for nonlinear time-series forecasting—focused time-lagged networks and recurrent networks—will be explored in more detail.

9.3.1.1 Focused Time-Lagged Feedforward Networks

Conceptually, this is an easy model to understand, because its structure is that of the feedforward network presented in detail in Chapter 3 and Chapter 4. The difference is that the inputs to the network are the lags of the variable to be forecast. In focused time-lagged (or time-lagged) networks, input variables presented to a feedforward network, such as an MLP, at time t are fed again with unit time delay at the next time step. This is repeated recursively up to the required lag length. In this manner, a short-term memory of autocorrelation structure is built into the network by externally feeding inputs as lags. Thus, in time-lagged networks, a model of the form depicted in Equation 9.4 is sought. The network is called focused time-lagged because the entire memory structure is located at the input end of a neuron, as shown in Figure 9.12. The neuron receives lagged inputs and embeds that memory structure into its own weights. Thus the neuron acts as a filter of short-term

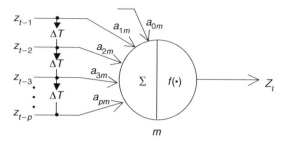

Figure 9.12 **Focused time-lagged neuron (short-term memory filter).**

response of the variable. The $f(\cdot)$ represents the nonlinear function that filters the short-term memory to produce an accurate forecast z_t.

Figure 9.13 shows a focused time-lagged feedforward neural network consisting of neurons that act as short-term memory filters. Each input feeds the network with one time unit delay and the network sees the same inputs a number of times equal to their lag length. Thus, an input vector consists of the p chosen lags of the variable $(z_{t-1},\ldots,z_{t-p})$, and the output is the outcome of the same variable at time $t(z_t)$. Because this model has short-term memory extending to the length of the lags, it is capable of temporal processing. Basically, this is forcing a short-term memory (represented by lags) into a network that is inherently static Recall that the feedforward networks are used for static mapping of inputs to an output(s).

Because time is integrated externally in time-lagged networks, these networks are essentially static and therefore can be modeled using the methods discussed in Chapter 4; similar criteria and issues that are relevant to feedforward networks apply to these models. However, inputs must be applied sequentially, as opposed to the random presentation that is commonly used with feedforward networks, because of the unit time delay input representation, which maintains the correlation of the lagged inputs.

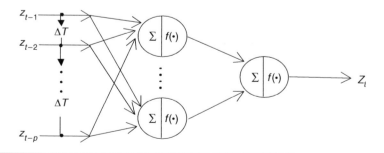

Figure 9.13 **Focused time-lagged (short-term memory) network.**

Usually lags are selected using the correlation of an observation for a variable to its previous values and selecting the lags that show a significant correlation with the current observation.

9.3.1.2 Spatio-Temporal Time-Lagged Networks

In some situations, the response of a time-series is affected by one or more time-series. Accurate forecasting of such interaction requires the incorporation of memory dynamics of each of the time-series. For example, the energy requirement of a building on a particular day is correlated with that of previous time steps, as well as the temperature over the past several days. Inflows into a river may depend on past rainfall and temperature, which affects the forecasting accuracy. To incorporate such interacting time-series effects, a network that distributes the memory across itself is required. Spatio-temporal time-lagged networks implement this task by filtering short-term memory response for each spatial dimension. Such a neuron is depicted in Figure 9.14.

The neuron in Figure 9.14 receives time delay inputs of various orders (leg length) from several inputs. In time-series forecasting, one of these input variables is the variable to be forecast. The other variables are those that influence forecast accuracy. The neuron m receives filtered short-term memory of several variables via the corresponding weights a_{mi}.

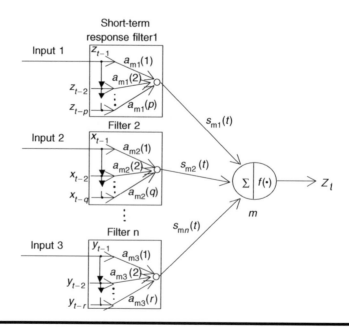

Figure 9.14 Multi-input time-lagged neuron (or spatio-temporal memory filter).

The short-term memory of one variable is embedded in the weighted sum of the time lags $s_{mi}(t)$. This weighted sum for input one, for example, is

$$s_{m1}(t) = \sum_{l=1}^{p} a_{m1}(l) z_{t-l} \tag{9.5}$$

where $a_{m1}(l)$ is the weight from input one with lag l to neuron m and z_{t-l} is the input corresponding to lag l with l extending from 1 to p, the order of the short-term memory. The total weight vector a_{m1} and the input vector of input one are

$$\begin{aligned} a_{m1} &= [a_{m1}(1), a_{m1}(2), ..., a_{m1}(p)] \\ z &= [z_{t-1}, z_{t-2}, ..., z_{t-p}] \end{aligned} \tag{9.6}$$

The other inputs can have a different order or lag length and therefore can have a different number of inputs and weights. When all the sums are calculated for n input variables, there are n weighted sums, $s_{m1}(t)$, $s_{m2}(t), ..., s_{mn}(t)$. The neuron m simply sums these to get the total weighted sum of short-term responses for time t, $S_m(t)$, for that neuron as

$$S_m(t) = \sum_{i=1}^{n} s_{mi}(t) \tag{9.7}$$

The total weighted sum is processed nonlinearly, using a selected nonlinear transfer function. Once trained, the neuron embeds the temporal dynamics of a several nonlinearly interacting time-series in its weights.

Now this whole multi-input time-lagged neuron structure can be nicely embedded in a feedforward network, aptly called a spatio-temporal time-lagged network, as shown in Figure 9.15. This model has spatio-temporal modeling capabilities, owing to the incorporation of spatial dimension (several variables) and temporal dimension (time lags). It captures the nonlinear spatio-temporal dynamics of the time-series. The nonlinear response comes from the nonlinear processing in the hidden neurons. Because the structure of the network is similar to a static feedforward network, the methods described in Chapter 3 and Chapter 4 can be used to train multi-input time-lagged networks; consequently, no further discussion of training is included here.

The spatio-temporal time-lagged network can be thought of as a nonlinear extension of an AR model with exogenous input variables, called a NARx model. The input vector for a time step now has the required lags for the desired variable, as well as the lags for the extra inputs. For the next time step, the first lag of each input variable is replaced by the actual values of the input variables, the other lag values are pushed one step down, and the last lag is eliminated. This process is repeated at each time step. Therefore, the database must be organized in

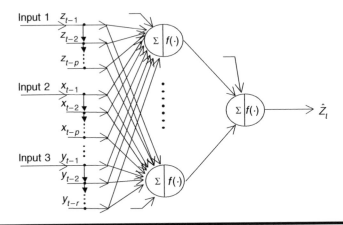

Figure 9.15 Multi-input time-lagged neural network with distributed short-term memory for spatio-temporal processing (NARx model).

such a way that all the required lags for the input variables for each time step are available in a single row, representing one input vector in the database. The corresponding output for the next time step must be associated with each input vector. Once this is done, the input vectors are presented sequentially, as opposed to randomly, and the weights are updated in batch or example-by-example mode. Usually testing for generalization behavior and adequacy of training is performed on a part of the series adjacent to the training series to eliminate overfitting. Another part of the series is used to validate the model by testing its predictions on unseen data. Many aspects of validation will be discussed throughout this chapter.

For ARIMA and related models, the selection of input variables and relevant lags has been an active area of research [1,29]. These issues apply to time-delay neural networks as well because a neural network model can be more accurate if only the essential inputs are fed to the network, as described in Chapter 5. This results in a network with simpler configurations and a smaller number of inputs and neurons and therefore a smaller number of connection weights. The issue of input selection will be revisited in a separate section.

9.3.2 Example: Spatio-Temporal Time-Lagged Network—Regulating Temperature in a Furnace

Recall the problem involving the regulation of the temperature in an electric furnace originally presented in Section 9.2.1. The required temperature profile is obtained by controlling the electrical current; these profiles are shown in Figure 9.3. In Section 9.2.1, a linear neuron model with external inputs (ARx) was used; here, a NARx will be utilized. Although the problem is fairly linear with respect to temperature, the step changes in current introduce oscillations in temperature that makes it deviate from the straight line, as shown in Figure 9.4. It will be shown how a nonlinear model addresses the same problem. Here again, the network uses four inputs: the lags of temperature and current. The output is the temperature at the next time step. There are now hidden neurons; it will be assumed there are three neurons with sigmoid activation functions and one linear output neuron. The Levenberg–Marquardt (LM) method, an efficient second order learning method discussed in Chapter 4, will be used.

The network configuration is shown in Figure 9.16. The same training, test, and validation datasets previously used in the linear analysis are used here as well. Specifically, the first 200 data points were used to train the network and the rest of the dataset was divided into two sets, one ranging from 200 to 250 minutes and the other from 251 to 500 minutes. The first set is similar in character to the training data and the second set is dissimilar in that the current oscillates rapidly. The first test set is used to halt training, using a stopped search method (early stopping) when the error on the test increases. This approach is discussed in detail in Chapter 6. The second set is used for validating the model by testing it on independent data not previously seen by the network.

The training performance is shown in Figure 9.17, which indicates that the best model for good generalization is obtained at epoch four. Test set one, which is adjacent to the training data, is used to obtain the model with the best generalization, as indicated by the dashed line in the figure. Here, iteration and epoch have the same meaning, namely one pass of the whole training dataset.

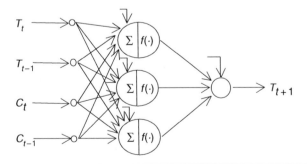

Figure 9.16 **Nonlinear autoregressive neural network model with external inputs (NARx) for regulating furnace temperature.**

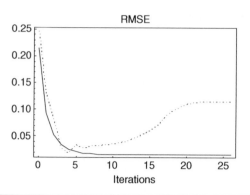

Figure 9.17 **Training performance of the Neural NARx model. Solid line indicates error on training data and dashed line indicates error on test data set 1 (validation data).**

9.3.2.1 Single-Step Forecasting with Neural NARx Model

The one-step-ahead forecasting ability of the nonlinear neural network is shown in Figure 9.18 for the two test sets. Recall that test set one is the closest to the training data and is used to prevent overfitting during training (dashed line in Figure 9.17); test set two is an independent test dataset for assessing generalization ability of the model. The RMSE on test set one is 0.01247 and prediction and target data for this set are plotted in Figure 9.18a. The RMSE on the same data for the linear model was 0.0217. This shows that the nonlinear model is superior to the linear model with a 42 percent reduction in error compared to the linear model. This can also be ascertained from the comparison of Figure 9.18a with Figure 9.6a. Note that in the linear model, test set one is used as an independent test set because it is not used at all during training because overfitting is not an issue with the linear model. However, in the nonlinear network, it is used intermittently to monitor the onset of overfitting, so it is not an entirely independent test set. For the second test set, the linear model that had RMSE of 0.068 outperforms the nonlinear model with RMSE of 0.077 by about 13 percent. This test is an independent validation set for both cases.

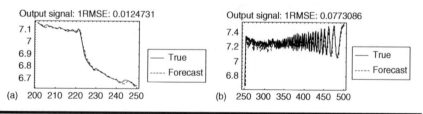

Figure 9.18 **Nonlinear neural model (NARx) performance in single-step-ahead forecasting: (a) test set 1 (validation data) adjacent to training data and (b) test set 2 with rapid oscillations of current and temperature.**

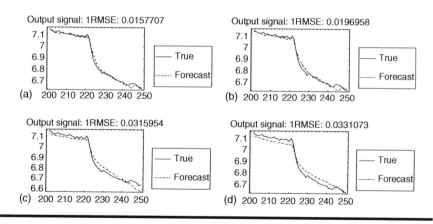

Figure 9.19 Multistep forecasting with neural NARx model on test set 1: (a) two-step, (b) three-step, (c) ten-step and (d) complete (50-step) forecasting.

9.3.2.2 Multistep Forecasting with Neural NARx Model

The multistep-ahead forecasting ability of the nonlinear network for two-step, three-step, ten-step, and complete forecasting for test set one is shown in Figure 9.19. The figure shows that the ability of the nonlinear model in multistep forecasting is far superior to that of the linear model. (Refer to Figure 9.7 for the same results for the linear model.) The deterioration of the forecast with an increasing prediction horizon is greatly reduced in the nonlinear model, indicating that individual predictions are more accurate and lead to a more robust model.

The performance of the nonlinear network for multistep forecasting with test dataset two (validation set) is shown in Figure 9.20 for 10-step and 50-step (complete) ahead forecasting. This reveals interesting features of the nonlinear model. Although the single-step forecast is slightly inferior to that of the linear model, as discussed previously, both the 10-step and the 50-step-ahead forecasts are much better than those for the linear model (see the equivalent results for the linear model in Figure 9.8). A comparison of the

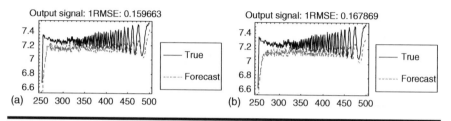

Figure 9.20 Multistep-ahead forecasting of neural NARx with the test dataset 2: (a) ten-step-ahead and (b) complete 50-step-ahead forecasting.

errors for the two models indicates that the nonlinear model error is 13 percent and 28 percent lower for the 10-step- and 50-step-ahead forecasts, respectively, than those for the linear model. What is evident is the increase in the resilience of the nonlinear network, especially when the forecast horizon increases, even when the variation of the inputs is more rapid and different to training data as is the case with the second test set. This superior performance is compatible with that observed for test set one shown previously in Figure 9.19.

Is it better to use the whole dataset when forecasts are made with the network? Figure 9.21 presents the single-step and 250-steps ahead forecasting performance of the nonlinear network when the whole dataset is used from the beginning but prediction errors are calculated for the actual forecasting period of the last 250 minutes. The RMSE for the single-step and 250-steps ahead forecasts are 0.055 and 0.121, respectively, as indicated on the top of the figures. These values for the second test dataset alone are 0.077 and 0.167, respectively (see Figure 9.18b and Figure 9.20b). The comparison indicates a 28 percent reduction in RMSE for each of the forecasts. For the linear model, these improvements are 27 percent and 23 percent, respectively, for the one-step and multistep-ahead forecasts. Thus, as for the linear model presented in Figure 9.9, both the short-term and the long-term forecasting ability of the nonlinear network improve when the whole dataset is used when making forecasts, as opposed to starting afresh with an unseen set of test data.

How do the linear and nonlinear models compare when the whole dataset is used for forecasting? As shown earlier in Figure 9.9, the RMSE for the linear model for 1-step- and 250-step-ahead forecasting with the whole dataset are 0.0502 and 0.18, respectively. These values for the nonlinear network are 0.055 and 0.121, respectively, as shown in Figure 9.21. Although the single-step forecast error of the nonlinear model is 9 percent higher, the 250-step-ahead forecast error is 18 percent lower than that for the linear model. This improvement is significant because accurate extended forecasts are very important for many practical forecasting tasks.

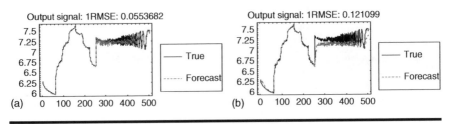

(a) Output signal: 1RMSE: 0.0553682 — True / Forecast; 0 100 200 300 400 500

(b) Output signal: 1RMSE: 0.121099 — True / Forecast; 0 100 200 300 400 500

Figure 9.21 Single- and multistep forecasting of neural NARx model with the whole dataset: (a) single step-ahead forecasting and (b) complete (250-step) ahead forecasting.

In summary, linear and nonlinear autoregressive neural networks were used for forecasting temperature variation in a furnace. It was found that the forecasting ability of the nonlinear network is superior to that of the linear model, especially in long-term forecasting. This idea was tested on two datasets: One with similar characteristics to those of the training data and another with very different characteristics. On the first set of test data, the nonlinear model outperforms the linear model in all aspects. The single-step forecast using data with different characteristics (test set two) from that of the training data was slightly inferior for the nonlinear model, but the multistep-ahead forecasting was superior to that performed by the linear model. For both models, the use of the entire dataset provided a considerable improvement of short-term forecasting accuracy (about 28 percent) as well as long-term accuracy. The improvement of the latter is greater for the nonlinear model (28 percent vs. 23 percent for the linear model).

9.3.3 Case Study: River Flow Forecasting

River flow forecasting is essential for planning and management of water resources for electricity generation, irrigation, recreation, and so on. This example is going to forecast flows into a river. The daily data has been collected for ten years for a river in New Zealand; the data consists of year, month, day, precipitation (mm), minimum temperature (°C), maximum temperature (°C) and river flow (cubic meters/second). The flow into the river is a dynamic time-series, as shown in Figure 9.22.

The relationship of the current flow to the previous day's flow is presented in Figure 9.23, which shows some linear association but indicates a highly variable and noisy relationship, especially for high flows.

A precipitation scenario in the river catchment is presented in Figure 9.24. There is high daily variation in precipitation, as shown in Figure 9.24a. Most precipitation in the river catchment during the winter months (especially in June and July) falls as snow and therefore does not contribute

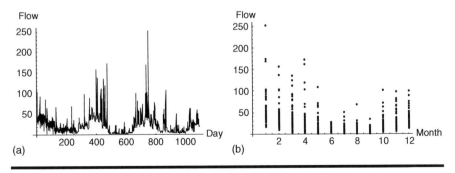

Figure 9.22   River flow from 1960 to 1963: (a) daily flow and (b) monthly flow.

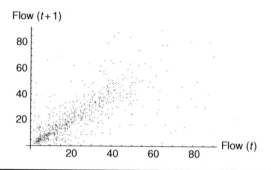

Figure 9.23   Relationship of current flow to previous day's flow.

significantly to the river flow, as indicated in Figure 9.24b. Flow during the hotter months is higher, which is partially attributed to the snowmelt during these months.

The temperature variation in the river region has definite cyclic behavior. This is illustrated for the minimum temperature in Figure 9.25. The daily temperature variation is a nonlinear time-series with yearly cycles

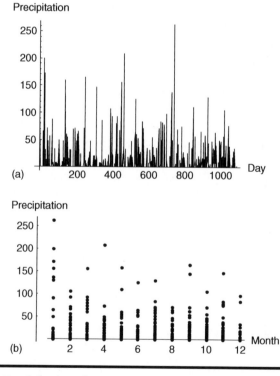

Figure 9.24   Precipitation from 1960 to 1963: (a) daily precipitation and (b) monthly precipitation.

(Figure 9.25a) and within a year, seasonal variations can be seen (Figure 9.25b). A similar plot with higher values was observed for the maximum temperature.

Intuitively, precipitation should be highly correlated with flow. The correlation of flow to the previous day's precipitation was 0.36. Moving averages of precipitation in the recent past, however, failed to significantly improve the relations. Flow was more highly correlated with the minimum temperature, with a coefficient of 0.54. Because both minimum and maximum temperature are highly correlated ($r = 0.8$), the minimum temperature can be used as one of the predictor variables. Various combinations of variables were tested, including moving averages of minimum temperature and precipitation. Of all the models, the one with the highest accuracy incorporated the last three days of flow, the previous day's precipitation, and the minimum temperature as inputs. The correlation of these data to the flow at time t is shown in Table 9.1, where due to symmetry, only the bottom left triangle is shown. In terms of strength of relationship, the table shows that the previous flows are the most important, followed by minimum temperature and precipitation. Specifically, the current flow is most strongly correlated to the previous day's flow, with a correlation coefficient of 0.7962, and to the next two flow lags with coefficients of 0.613 and 0.53. Correlation coefficients

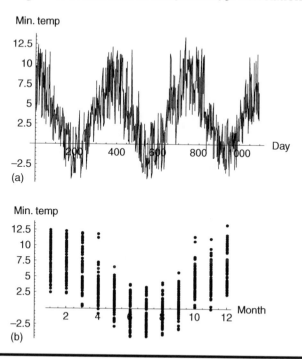

(a)

(b)

Figure 9.25 Minimum temperature from 1960 to 1963: (a) daily minimum temperature and (b) monthly minimum temperature.

Table 9.1 Input Variables and Correlations for River Flow Forecasting

	Precip-1	Min. Temp-1	Flow-3	Flow-2	Flow-1	Flow
Precip-1	1.0					
Min. Temp-1	0.1387	1.0				
Flow-3	0.0159	0.4071	1.0			
Flow-2	0.0486	0.4265	0.7960	1.0		
Flow-1	0.183	0.499	0.6128	0.796	1.0	
Flow	0.361	0.54	0.53	0.613	0.7962	1.0

of flow with the previous day's minimum temperature and precipitation are 0.54 and 0.361, respectively. Minimum temperature is also moderately correlated with flow lags, but weakly correlated with precipitation.

9.3.3.1 Linear Model for River Flow Forecasting

Linear and nonlinear neural network models will now be developed using the inputs. The linear model is equivalent to an ARx model of order three with three lags and two external variables. This is a single linear neuron model. The data was divided into three sets. The first set, with 730 records, is equivalent to two years of data. The rest of the data for the third year was divided into two equal (half-year) portions for validation and testing. The data was standardized by subtracting the corresponding mean from each observation and dividing by the standard deviation so that the range of input variables was similar. In fact, testing was carried out using raw data and scaled data, and the scaled data gave a 2 percent increase in R^2, which can be useful in improving the forecasting accuracy. The models were developed using *Mathematica— Neural Networks* [7].

The model performance on the validation and test sets is shown in Figure 9.26a and Figure 9.26b. Only the training data were used in the model development. Figure 9.26 shows a reasonably good forecast that is superior for the second half of the data. This could be due to the fact that the last half-year training data is similar to the test data, although shifted by one year. The higher error for the first half is due to the discrepancy at high flows.

The quality of the model can be assessed in various ways using visual and statistical means. For example, a comparison between the forecasts and the actual flow is shown in Figure 9.27, which indicates that there is a reasonably strong linear agreement between the two. This means that the model predictions are good.

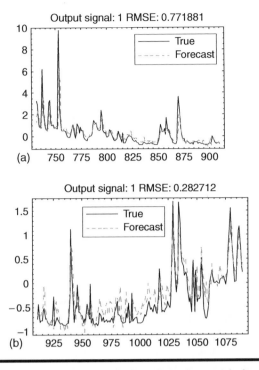

Figure 9.26 Observed and forecasted river flow for two independent datasets:
(a) validation set (first half of the year) and (b) test set (second half of the year).

The performance measures of RMSE, mean absolute error (MAE), relative
MAE (RMAE) and persistence index (PI) are presented here, expressed as

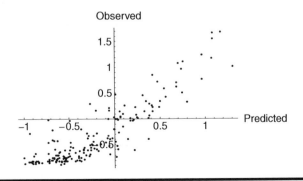

**Figure 9.27 Comparison between the predicted flow from the linear model and
the observed river flow for the independent test data.**

$$\text{RMSE} = \sqrt{\frac{1}{n}\sum_{t=1}^{n}(z_t - T_t)^2}$$

$$\text{MAE} = \frac{1}{n}\sum_{t=1}^{n}|z_t - T_t|$$

$$\text{RMAE} = \frac{\text{MAE}}{\bar{T}}$$

$$\text{PI} = 1 - \frac{\displaystyle\sum_{t=1}^{n}(z_t - T_t)^2}{\displaystyle\sum_{t=1}^{n}(T_t - T_{t-1})^2}.$$

In the formulae, z is the forecast value and T is the corresponding target or observed value. The RMSE has been previously described. The MAE is the average of the absolute error of the prediction over all input patterns. The RMAE is the MAE with respect to the average of the observed (target) data. The PI is particularly suited to time-series because it compares the model errors to a naive model, which is the difference between the current and previous observations $(T_t - T_{t-1})$. For the river problem, these performance measures for the single-step forecasting are

$$\text{RMSE} = 0.283$$

$$\text{MAE} = 0.227$$

$$\text{RMAE} = 0.688$$

$$\text{PI} = 0.369$$

$$R^2 = 0.750$$

The lower the errors and the higher the PI and R^2, the better the model.

How good is the model for multistep forecasting? The performance in three-step-ahead forecasting is shown in Figure 9.28.

With three-step-ahead forecasting, all performance measures deteriorate; these values are

$$\text{RMSE} = 0.393$$

$$\text{MAE} = 0.325$$

$$\text{RMAE} = 0.986$$

$$\text{PI} = -0.219$$

$$R^2 = 0.52$$

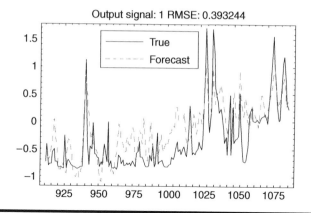

Figure 9.28   Three-step-ahead forecasting using a linear model for independent test data.

9.3.3.2   Nonlinear Neural (NARx) Model for River Flow Forecasting

Next, this section develops a nonlinear neural network model with the same five inputs used in the linear model. Because this model uses external variables, the objective is to see whether there is any advantage to treating the problem as a nonlinear one, and also to demonstrate nonlinear forecasting with neural networks. Four hidden neurons performed best with the LM training method. The validation set was used to prevent overfitting using the stopped-search method, and the test set was used for testing the developed model. The network performance on the validation and test sets is shown in Figure 9.29. The RMSE for the validation set (Figure 9.29a) is 0.91, which is higher than that for the linear model (0.77). However, for this model, the true test of the accuracy is its performance on the test set, because the validation set has already been used to prevent overfitting. Figure 9.29 shows that the RMSE of the predictions from the nonlinear model has improved over the linear model for the independent test set (0.259 compared to 0.283 for the linear model).

The linear correlation between the observed and predicted flows for the independent data is depicted in Figure 9.30, which indicates a strong linear relationship between the two.

Statistical measures of the nonlinear network are

$$\text{RMSE} = 0.259$$

$$\text{MAE} = 0.18$$

$$\text{RMAE} = 0.564$$

$$\text{PI} = 0.471$$

$$R^2 = 0.79$$

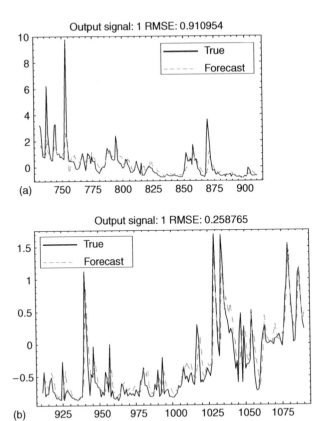

Figure 9.29 Observed and forecasted river flow: (a) validation data and (b) independent test data.

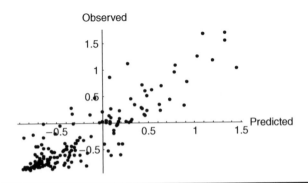

Figure 9.30 Observed river flow and flow predicted by the nonlinear autoregressive network.

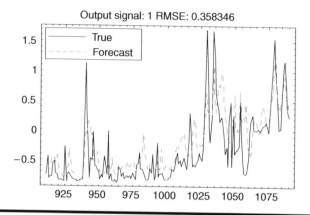

Figure 9.31 Three-step-ahead river flow forecasts from the nonlinear neural network.

All the performance measures have increased for the nonlinear model, especially the PI and R^2. Now the robustness of the model in multistep-ahead forecasting will be examined. To compare with the linear model, three-step-ahead forecasts obtained from the nonlinear model and its RMSE on the independent test set are shown in Figure 9.31. It indicates that the three-step-ahead forecast error has decreased by 9 percent for the nonlinear model (0.358 compared to 0.393 from the linear model).

The linear correlation between the three-step-ahead forecasts and the observed river flow is presented in Figure 9.32. There is still evidence for a linear relationship between the two, but it is not as strong as in the case of single-step forecasts.

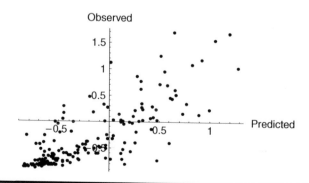

Figure 9.32 Three-step-ahead observed and predicted flow from the nonlinear model.

Now the ways in which the performance measures are affected in three-step-ahead forecasting will be examined. These are as follows:

$$RMSE = 0.3583$$

$$MAE = 0.2737$$

$$RMAE = 0.8301$$

$$PI = -0.0139$$

$$R^2 = 0.60$$

As can be expected, the forecasting accuracy, along with the other performance measures, drop as the forecast horizon increases. However, these results indicate that the nonlinear model is more robust in multistep-ahead forecasting. For example, the R^2 drop for the nonlinear model is 24 percent, whereas that for the linear model is 31 percent. The increase in RMSE for the linear model is 39 percent, compared with 5.4 percent for the nonlinear model. This is illustrated clearly in the error histograms for the two models, as shown in Figure 9.33. The figures indicate that the error

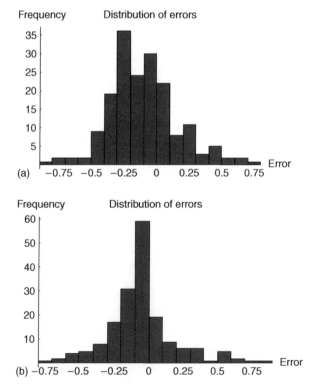

Figure 9.33 Forecasted error distributions on the independent test dataset: (a) linear model and (b) nonlinear neural network model.

distribution for the nonlinear neural network model more closely resembles a symmetric Gaussian distribution and is more compact, in the sense that most of the error is close to zero.

9.3.3.3 Input Sensitivity

After training, the neural network is an analytical expression of the model, which in this case is a nonlinear function of the three recent flows and the previous day's precipitation and minimum temperature. The model for the current river flow $z(t)$ developed by the network is

$$z(t) = 16.5 + \frac{0.63}{1 + e^{2.73 - 1.6a + 0.046b + 0.057c - 0.44d - 1.17e}}$$

$$- \frac{3.44}{1 + e^{1.8 + 0.44a + 0.91b - 0.03c - 0.41d - 0.19e}}$$

$$+ \frac{1.67}{1 + e^{-12.38 + 0.61a - 0.13b + 1.38c + 1.94d + 0.03e}}$$

$$- \frac{21.46}{1 + e^{-1.5 + 0.09a - 0.06b + 0.017c + 0.12d + 0.06e}} \tag{9.8}$$

where $a = z(t-1)$, $b = z(t-2)$, $c = z(t-3)$, $d = precip(t-1)$, and $e = minTemp(t-1)$.

By differentiating $z(t)$ with respect to the inputs, it is possible to study the sensitivity of the one-day ahead forecast to various inputs. Because a nonlinear neural network model is being used, the derivative with respect to an input is not constant as for the linear model, but varies with the input i.e., is situation dependent. In this case, one way to study the sensitivity of the output to the inputs is to find the sensitivity of the output at the average values of the inputs. Although not shown here, it can be seen from the expression for $z(t)$ that the derivative with respect to one input depends on the values of the other variables of the input vector. Substituting the mean values for the variables in the derivative with respect to an input, yields the sensitivity at the mean value of that input, as shown in the first row of Table 9.2 for the whole dataset (i.e., for three years).

Table 9.2 indicates that the output is most sensitive to the previous day's flow, followed by the precipitation, the flow two days back, the previous day's temperature and the flow three days back. This highlights the effect of precipitation. In the previous correlation analysis results, shown in Table 9.1, the correlation of the precipitation to the next day's flow was not prominent. However, the sensitivity of the model to precipitation is higher than its sensitivity with respect to the minimum temperature and flow two days back. This could be attributed to the fact

Table 9.2 Sensitivity of Output to Inputs and Their Relative Contribution from Nonlinear Network

	$z(t-1)$	$z(t-2)$	$z(t-3)$	Precip $(t-1)$	Min-temp $(t-1)$
Sensitivity at the mean value	0.528	0.177	0.04	0.217	0.152
Mean sensitivity $(\Sigma \partial z/\partial x)/n$	0.541	0.214	0.04	0.214	0.167
Input contribution (%)	53	12	11	15	9

that linear correlation analysis cannot capture nonlinear correlations in the data but that the nonlinear model can capture these trends.

It is also possible to find the contribution of an input variable to the output by observing the change in sensitivity of $z(t)$ when the input is varied from its minimum to maximum value while the values of the other input variables are held at their mean values. The mean of these sensitivities, which is the sensitivity averaged over all n observations of an input x, is presented in the second row of Table 9.2. By repeating this for each input variable and presenting its effect relative to the total effect of all the variables, the relative contribution of each variable to the output can be calculated. These contributions are also shown in Table 9.2. The trend followed by the mean sensitivity is similar for the most part to that of the sensitivity at the mean value, but the values are higher, indicating that the nonlinear effects are relevant for this model. Results show that the greatest contribution comes from previous day's flow, z_{t-1} (53 percent), followed by precipitation (15 percent), the flow during the previous two and three days (12 percent and 11 percent, respectively), and the minimum temperature (9 percent).

To highlight the variability of the input sensitivity, it is illustrated here for the two most important variables, $z(t-1)$ and $Precip(t-1)$, while keeping the other four input variables at their mean values. These are presented in Figure 9.34 and Figure 9.35. The sensitivity is variable and depends on the magnitude of the input with respect to which the sensitivity is being analyzed. The two long spike regions in the figures correspond to peak flows during the two summer periods.

9.4 Hybrid Linear (ARIMA) and Nonlinear Neural Network Models

The performance of time-series forecasting can be improved by combining linear and nonlinear models. The advantage of such an approach is that

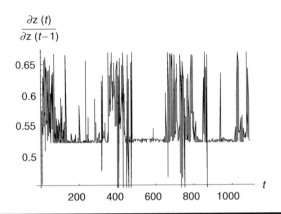

Figure 9.34 Variation of the sensitivity of the output $z(t)$ to the previous day's flow $z(t-1)$.

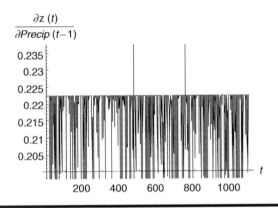

Figure 9.35 Variation of the sensitivity of the output $z(t)$ to the previous day's precipitation, *Precip* $(t-1)$.

the linear component in the data can be modeled by the linear model and the nonlinear component by the nonlinear model. This can lead to simpler models and higher forecasting accuracy. Considering the nature of a time-series, it is reasonable to decompose it into a linear autocorrelation structure and a nonlinear component [1] as

$$y_t = L_t + N_t \tag{9.9}$$

where L_t denotes the linear component and N_t is the nonlinear component. The two components need to be estimated from the data. The linear model is developed first; the residuals from the linear model contain the inherent nonlinear relationship modeled by the neural network. This approach is depicted in Figure 9.36.

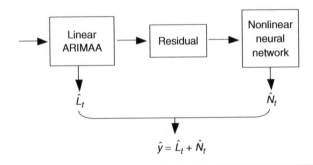

Figure 9.36 Hybrid linear (ARIMA) and nonlinear neural network models.

Zhang [1] proposed the use of ARIMA to model the linear component. Let the residual from the linear model be e_t. Then

$$e_t = y_t - \hat{L}_t \qquad (9.10)$$

where $\hat{L}_t$ is the forecast from the linear model at time t. A linear model is not sufficient if linear correlations are still present in the residuals. Residual analysis, however, is not able to detect any nonlinear patterns in the data. Currently, there is no diagnostic statistic for detecting nonlinear auto-correlations in data [1]. Any significant nonlinear pattern in the residuals indicates the limitations of ARIMA. The residuals can then be modeled by a neural network to discover nonlinear relationships in them. The ANN model for the residuals with n past values is

$$e_t = \hat{N} + \varepsilon_t = f(e_{t-1}, e_{t-2}, ..., e_{t-n}) + \varepsilon_t \qquad (9.11)$$

where $\hat{N}$ is the nonlinear network model prediction of the residuals, represented by f, which is the function to be approximated by the network, and ε_t is the random error. A good ANN model will produce random errors. The combined forecast from linear ($\hat{z}$) and nonlinear ($\hat{N}$) models is

$$\hat{z} = \hat{L} + \hat{N} \qquad (9.12)$$

which is an approximation of the actual data.

9.4.1 Case Study: Forecasting the Annual Number of Sunspots

Zhang [1] shows that for three time-series with nonlinear characteristics, the hybrid model outperforms both ARIMA and ANNs. One time-series is the Wolf's sunspot data series, regarded as nonlinear and non-Gaussian, and widely used for evaluating the effectiveness of nonlinear models. This series

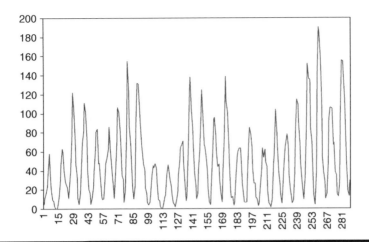

Figure 9.37 Sunspot series (1700–1987). (From Zhang, G.P., *Neurocomputing*, 50, 159, 2003. With permission from Elsevier.)

contains the annual number of sunspots from 1700 to 1987 with a total of 288 observations, as presented in Figure 9.37.

The best linear model for the data was an autoregressive model of order nine. The neural model contained four input lags of forecast error or residuals and four hidden neurons, as has been used by several previous researchers [1]. In addition, a separate ANN was trained to evaluate the effectiveness of ANN alone to forecast the number of sunspots. The models were trained for one-step-ahead forecasting and evaluated for 35- and 67-steps-ahead forecasting.

The models were trained with the first 221 observations and tested with the last 67 data, obtaining 35- and 67-step-ahead forecasts. The 67-step-ahead forecast is shown in Figure 9.38. The results show that the hybrid model outperformed the linear model by 16.13 percent and the ANN model by 9.89 percent in terms of RMSE for 35 steps ahead forecasting. In 67-steps-ahead forecasting, the hybrid model again outperformed linear model by 8.4 percent and the ANN model by 20 percent. ANN had a higher error at some data points.

9.5 Automatic Generation of Network Structure Using Simplest Structure Concept

The most serious draw back of feedforward networks such as MLP is the trial and error involved in the selection of model parameters, particularly the number of hidden neurons. This section includes a case study in which the number of hidden neurons is automatically selected using the

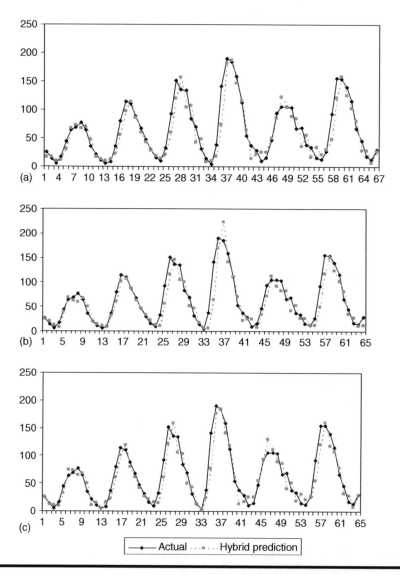

Figure 9.38 Sixty-seven-step ahead sunspot data forecasting: (a) ARIMA prediction, (b) neural network prediction, and (c) hybrid ARIMA-neural network prediction. (From Zhang, G.P., *Neurocomputing*, 50, 159, 2003. With permission from Elsevier.)

concept of the simplest structure. The simplest network structure is essentially equivalent to the linear regression model; the most complicated MLP structure will have small training errors, but high generalization errors. In between the simplest and the most complicated structures, there exists the optimum structure for a network such as an MLP. For many problems, the optimum structure can be simple because although the variable interaction may be complex, most phenomena may not be highly nonlinear, which is accounted for by the large number of neurons in a network.

The search for the optimum structure of a network is based on the generalization error. The data is divided into training, validation, and testing sets. Training is started with a simple structure with one neuron. The trained network performance is intermittently tested against validation data. If the validation error does not change at some point, this is considered to be a local optimum. Training is stopped and a new neuron is added. Training continues with this structure until the validation error stops changing. Another local minimum has been reached, and another neuron is added to the structure. This process continues until the optimum structure is found when the validation error does not decrease further; this point is considered to be the global minimum. This final model is the optimum model. Because the whole process is automated, the structure is obtained without the intervention of the model developer. This approach is applied to solve a practical air pollution forecasting problem below.

9.5.1 Case Study: Forecasting Air Pollution with Automatic Neural Network Model Generation

Air pollution has become a serious health concern due to excessive amounts of pollutants being released to the environment because of various human activities. Some of these pollutants are respirable particulate matter (PM_{10}), sulfur dioxide (SO_2), and nitrogen dioxide (NO_2). Countries have adopted various methods for the daily forecasting of pollution. The study presented here was reported by Jian et al. [8] for forecasting the Air Polution Index (API) in China. For reporting to the public, the daily average pollutant data are converted to an API, representing the levels of PM_{10}, SO_2, and NO_2 in the air at a given time. For each pollutant, the value of the API varies from zero to 500. This range has been divided into six air quality classes: 0–50 is class I, 50–100 is class II, 100–200 is class III, 200–300 is class IV, 300–400 is class V, and 400–500 is class VI. For example, the API for PM_{10} is class I if the mass concentration of PM_{10} is between 0.000–0.050 mg m^{-3} and class VI for the range between 0.500–0.600 mg m^{-3}.

The air pollution levels are closely related to atmospheric conditions. In this model, inputs include the date, the maximum and average surface temperatures (measured at 10 m height), the surface pressure, the surface humidity, the wind speed, and the daily precipitation. The date accounts for periodic changes in meteorological systems, as well as differences in emissions between weekdays and weekends. The date has been converted to four input nodes: Sine and cosine for the time of the week and the year. Wind is the dominant factor in the transport of pollutants. The surface wind data are collected four times a day. Surface pressure, temperatures and humidity account for the effects of the weather; these inputs are daily averages. The precipitation is an effective mechanism to remove pollutants from the air. The daily precipitation was classified into five levels: Zero to four, corresponding to no rain, light, medium, heavy rain, and storms. In total, there are 16 external variables and with the previous day's API as an input, the model has 17 variables.

The structure of the network for forecasting air pollution is shown in Figure 9.39. A separate network has been trained for each of the pollutants PM_{10}, SO_2, and NO_2.

The model was developed based on 457 days of observations collected over about 14 months from June 2000. The original dataset with 457

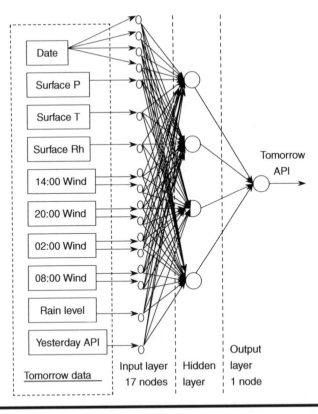

Figure 9.39 The network structure, in which the number of hidden neurons is determined automatically during training. (From Jiang, D., Zhang, Y., Hu, X., Zeng, Y., Tan, J., and Shao, D., *Atmospheric Environment*, 38, 7055, 2004. With permission from Elsevier.)

observations was divided into training, validation, and testing sets with 397, 30, and 30 observations, respectively. These were obtained by selecting one week's data for training, the next day for validation and the day after for testing, and repeating this process for the entire dataset. The hidden neurons used sigmoid transfer functions and the output neuron is linear. The backpropagation with a smaller learning rate has produced a smooth convergence. Incremental (example-by-example) training has produced better results than batch training.

Using the procedure described above for adding neurons incrementally based on the generalization error of the existing network, the optimum network was found to have four hidden neurons. To compare it using the usual trial and error process, an extensive trial was conducted using network configurations including one or two hidden layers and 1–16 neurons in each layer. Of the 200 possible configurations, more than half were trained and the best performance was obtained for networks with three to six neurons, supporting the simpler structure principle.

The model has been in use since 17 September 2002 and the correlation coefficients for the predictions of the three pollutants (PM_{10}, SO_2, and NO_2) were found to be 0.61, 0.70, and 0.63, respectively, in relation to the application data for this date and the following eight months. These correlation coefficients were considered to be satisfactory. Although the model was trained using observed meteorological data, the model application is based on meteorological data provided by a model, which obviously has some effect on the accuracy of the model forecasts. The results from the application of the model for PM_{10} to the test data are shown in Figure 9.40. With the exception of the high peak on 12 November 2002, which resulted from a sand storm, the information for which was not in the training data, the model performs well. This has also affected the correlation coefficients presented earlier.

Currently, the operators retrain the model on a monthly basis using new data. The model has been tested against a linear autoregressive time-series model based on the previous API values. Although the error was slightly lower for the autoregressive model for the training data, the slope of the observed and predicted output based on the application data was 0.48 for the MLP model and 0.28 for the autoregressive model, as illustrated in Figure 9.41. The intercepts were correspondingly 36.5 and 50 for the two models. These suggest that the MLP is more accurate for individual predictions.

9.6 Generalized Neuron Network

The principle behind the generalized neuron network (GNN) is that a model should consist of simple as well as higher order neuron

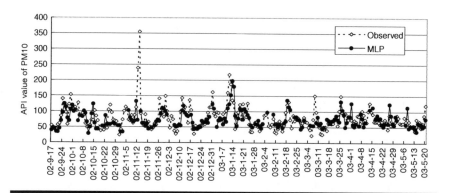

Figure 9.40 Results from the model obtained from simple structure principle for PM$_{10}$ on application data. (From Jiang, D., Zhang, Y., Hu, X., Zeng, Y., Tan, J., and Shao, D., *Atmospheric Environment*, 38, 7055, 2004. With permission from Elsevier.)

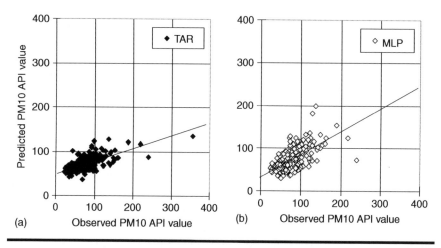

Figure 9.41 Plots of observed and predicted data for the application period of eight months from September 2002 to May 2003 for the autoregressive time-series model (TAR) and the neural network (MLP) model. (From Jiang, D., Zhang, Y., Hu, X., Zeng, Y., Tan, J., and Shao, D., *Atmospheric Environment*, 38, 7055, 2004. With permission from Elsevier.)

characteristics and that there should be no need to select the hidden layers and neurons, meaning that the model complexity must be reduced. This model attempts to overcome the drawbacks of conventional ANN in terms of the extensive trial and error during model building. This section will study the time-series forecasting ability of a GNN model. However, before that, the section will present a detailed look at the principles behind GNN and how the network processes data.

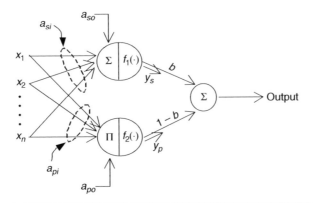

Figure 9.42 Configuration of a generalized neuron network (GNN) model. (From Chaturvedi, D.K., Mohan, M., Singh, R.K., and Karla, P.K., *Soft Computing*, 8, 370, 2004. With permission from Springer Science and Business Media.)

The generalized neuron model has a simple structure with two hidden neurons, as shown in Figure 9.42. One does the summation operation of the weighted inputs as with conventional networks; the other neuron computes the product of the weighted inputs. The former incorporates simpler linear characteristics and the latter, more nonlinear characteristics. The idea is based on the well-known fact that any function can be approximated with the use of sum of products and products of sums.

In the GNN model in Figure 9.42, the first neuron sums the weighted inputs (Σ) and processes it by a sigmoid activation function. The product neuron computes the product of the weighted input and processes it by a Gaussian function. The final output is the weighted sum of the output of the two neurons. These weights are as indicated in the figure.

For the general model in Figure 9.42, the output y_s of the summation neuron is

$$y_s = f_1\left(\sum_{i=1}^{n} a_{si}x_i + a_{s0}\right) = f_1(a_{s1}x_1 + a_{s2}x_2 + \cdots + a_{s0}) \qquad (9.13)$$

where n is the number of inputs, a_{si} is the weight vector associated with input x_i, and a_{s0} is the bias weight. The activation function is denoted by f_1. This is the usual operation of a neuron. The product neuron output y_p is

$$y_p = f_2\left(\left(\prod_{i=1}^{n} a_{pi}x_i\right)a_{p0}\right)$$

$$= f_2(a_{p1}x_1 \times a_{p2}x_2 \times \cdots \times a_{p2}x_2 \times a_{p0}) \qquad (9.14)$$

where Π denotes product, a_{pi} is the input-hidden weight associated with input x_i feeding the product neuron. The activation function of this neuron is denoted by f_2. The output is basically the product of the weighted inputs, including the bias weight a_{p0}, transformed by f_2. The final output z is the weighted sum of the hidden neuron outputs expressed as

$$z = y_s b + y_p (1 - b) \tag{9.15}$$

where b is the hidden-output weight of the summation neuron and the hidden-output weight of the product neuron is taken as $(1-b)$. This process constitutes the forward pass. With this structure, any learning algorithm discussed in Chapter 4 can be used for weight update and the process is transparent.

The basic steps of GNN are illustrated using an example two-input network, shown in Figure 9.43.

The operation of GNN can be summarized as follows: The square error for one input pattern E is

$$E = \frac{1}{2}(t - z)^2 \tag{9.16}$$

where t is the target for the input pattern and z is the model output. The sum of the square error is minimized during training. The forward processing of the network produces the output of the summation and product neurons. The summation neuron uses a sigmoid activation function, and that used by the product neuron is Gaussian. The sum u_s and output y_s of the summation neuron are

$$u_s = x_1 a_{s1} + x_2 a_{s2} + a_{s0}$$

$$y_s = \frac{1}{1 + e^{-u_s}}. \tag{9.17}$$

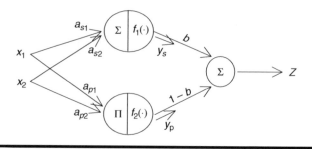

Figure 9.43 Example of a two-input GNN.

The product u_p and output y_p of the product neuron are

$$u_p = x_1 a_{p1} \times x_2 a_{p2} \times a_{p0}$$

$$y_p = \frac{1}{e^{u_p^2}}$$

(9.18)

where y_p is the output from the Gaussian function. The output of the network is

$$z = b y_s + (1 - b) y_p.$$

(9.19)

Using the error for one input pattern for clarity, it is possible to obtain the weight adjustments using the gradient descent concepts introduced in Chapter 4. First it is necessary to determine the gradients of error with respect to the weights. The gradient with respect to weight b is

$$\frac{\partial E}{\partial b} = \frac{\partial E}{\partial z} \frac{\partial z}{\partial b} = -(t - z)(y_s - y_p).$$

(9.20)

For the whole training set with n patterns, the batch weight adjustment Δb_{new} is

$$\Delta b_{new} = \eta \sum_{i=1}^{n} (t_i - z_i)(y_s - y_p) + \alpha \Delta b_{old}$$

$$b_{new} = b + \Delta b_{new}$$

(9.21)

where η is the learning rate and α is the momentum, which is a constant expressing the proportion of the previous weight increment passed to the new increment. As discussed in Chapter 4, this amounts to the exponential average of the past increments and is done to stabilize the search process.

The gradient with respect to weight a_{s1} can be computed as follows

$$\frac{\partial E}{\partial a_{s1}} = \frac{\partial E}{\partial y_s} \frac{\partial y_s}{\partial u_s} \frac{\partial u_s}{\partial a_{s1}},$$

$$\frac{\partial E}{\partial y_s} = \frac{\partial E}{\partial z} \frac{\partial z}{\partial y_s} = -(t - z)b,$$

$$\frac{\partial y_s}{\partial u_s} = y_s(1 - y_s), \tag{9.22}$$

$$\frac{\partial u_s}{\partial a_{s1}} = x_1,$$

$$\frac{\partial E}{\partial a_{s1}} = -(t - z)by_s(1 - y_s)x_1.$$

Similarly, the derivative with respect to a_{s2} and the bias a_{s0} weights are

$$\frac{\partial E}{\partial a_{s2}} = -(t - z)by_s(1 - y_s)x_2$$

$$\tag{9.23}$$

$$\frac{\partial E}{\partial a_{s0}} - (t - z)by_s(1 - y_s).$$

The weight increments for the three weights based on gradient descent based error backpropagation are

$$\Delta a_{s1-new} = \eta \sum_{i=1}^{n}(t_i - z_i)by_{s_i}(1 - y_{s_i})x_{1_i} + \alpha \Delta a_{s1-old}$$

$$\Delta a_{s2-new} = \eta \sum_{i=1}^{n}(t_i - z_i)by_{s_i}(1 - y_{s_i})x_{2_i} + \alpha \Delta a_{s2-old}$$

$$\tag{9.24}$$

$$\Delta a_{s0-new} = \eta \sum_{i=1}^{n}(t_i - z_i)by_{s_i}(1 - y_{s_i}) + \alpha \Delta a_{s0-old}$$

$$a_{si,new} = a_{si} + \Delta a_{si,new}.$$

The only new addition in this model is the modification of the weights of the product neuron; the processing is presented in detail here. Suppose that the product of the weighted inputs is denoted by u_p. Then:

$$u_p = \prod_{i=1}^{n} a_{pi}x_i \times a_{p0} = a_{p1}x_1 \times a_{p2}x_2 \times a_{p0}. \tag{9.25}$$

The error derivative with respect to weight a_{pi}, which links input x_i with the product neuron, using the usual process described in Chapter 4 is

$$\frac{\partial E}{\partial a_{pi}} = \frac{\partial E}{\partial y_p}\frac{\partial y_p}{\partial u_p}\frac{\partial u_p}{\partial a_{pi}}. \tag{9.26}$$

The first component in Equation 9.26 is

$$\frac{\partial E}{\partial y_p} = \frac{\partial E}{\partial z}\frac{\partial z}{\partial y_p} = -(t-z)(1-b). \tag{9.27}$$

The Gaussian activation function, described in more detail in Chapter 3, takes the form of the first expression in Equation 9.28 and from it the second component in Equation 9.26 can be obtained as

$$y_p = e^{-u_p^2}$$

$$\frac{\partial y_p}{\partial u_p} = e^{-u_p^2}(-2u_p) = -2y_p u_p. \tag{9.28}$$

The last component of the error derivative in Equation 9.26 is

$$\frac{\partial u_p}{\partial a_{p1}} = \frac{\partial(a_{p1}x_1 \times a_{p2}x_2 \times a_{p0})}{\partial a_{p1}} = \frac{u_p}{a_{p1}}. \tag{9.29}$$

For the second weight, $\partial u_p/\partial a_{p2} = u_p/a_{p2}$, and for the bias weight, $\partial u_p/\partial a_{p0} = u_p/a_{p0}$.

Therefore,

$$\frac{\partial E}{\partial a_{p1}} = -(t-z)(1-b)(-2y_p u_p)\frac{u_p}{a_{p1}} = 2(t-z)(1-b)y_p\frac{u_p^2}{a_{p1}}. \tag{9.30}$$

Similarly, for the second and bias weights the error derivatives are

$$\frac{\partial E}{\partial a_{p2}} = 2(t-z)(1-b)y_p\frac{u_p^2}{a_{p2}}$$

$$\frac{\partial E}{\partial a_{p0}} = 2(t-z)(1-b)y_p\frac{u_p^2}{a_{p0}} \tag{9.31}$$

and the weight increments based on gradient descent are:

$$\Delta a_{\text{p1-,new}} = -2\eta \sum_{i=1}^{n} (t_i - z_i)(1 - b)y_{\text{p}_i} \frac{u_{\text{p}i}^2}{a_{\text{p1}}} + \alpha \Delta a_{\text{p1,old}}$$

$$\Delta a_{\text{p2-,new}} = -2\eta \sum_{i=1}^{n} (t_i - z_i)(1 - b)y_{\text{p}_i} \frac{u_{\text{p}i}^2}{a_{\text{p2}}} + \alpha \Delta a_{\text{p2,old}} \qquad (9.32)$$

$$\Delta a_{\text{p0-,new}} = -2\eta \sum_{i=1}^{n} (t_i - z_i)(1 - b)y_{\text{p}_i} \frac{u_{\text{p}i}^2}{a_{\text{p0}}} + \alpha \Delta a_{\text{p0,old}}$$

$$a_{pi,\text{new}} = a_{pi} + \Delta a_{pi,\text{new}}.$$

9.6.1 Case Study: Short-Term Load Forecasting with a Generalized Neuron Network

Short-term forecasting of electricity (load) demand is important for maintaining a reliable power supply at a reasonable cost. The gap between generation and demand leads to voltage fluctuations/drops and a shortage of electrical power. To bridge this gap, short-term load forecasting is vital for predicting the load demand pattern in advance. In this study, reported by Chaturvedi et al. [9], a GNN is used for short-term load forecasting.

Chaturvedi et al. [9] tested the generalized neuron model on seven variations of weighted inputs and transfer functions. The usual procedure is to use the inputs weighted by weight i.e., $w \times x$, where w is the weight associated with input x. The authors tried a variety of different weighted inputs. These were power functions of the form $(x + w)^n$. Recall from Chapter 3 that the usual weighted sum of inputs produces a plane which linearly separates the inputs (linear neuron). The power functions produce closed separating surfaces. Using $n = 2$, for example, the net input to the summation neuron is $(x + w)^2$. For $u = 0$, which describes the classification boundary, $(x + w)^2 = 0$ produces an ellipse. Thus, it is possible to train nonlinear and non-separable problems with ease. In this case study, the authors attempted to use $(x + w)$ and $(x + w)^2$ as the specific forms of the weighted inputs for one or both of the hidden neurons. In addition, linear transfer functions in both hidden neurons were also tested as an alternative to the above combination of sigmoid and Gaussian functions. All together, seven models were tested for the suitability of short-term load forecasting and these are presented in Table 9.3. The generalized neuron models in Table 9.3 were trained on actual electricity load demand data with four lags as input to predict the present demand. The data were normalized in the range from 0.1–0.9.

The models were trained using the gradient descent method described in the previous section (Equation 9.24 and Equation 9.32), incorporating both a

Table 9.3 Weighted Input and Activation Function for Summation and Product Neurons of GNN Models and Corresponding Test Error for Electricity Load Forecasting

Model	Weighted Input for Neurons		Activation fn		Training Error	Test Error
	Sum	Product	Sum	Product		
1	$w \times x$	$w \times x$	Sigmoid	Gaussian	0.0420	0.0648
2	$w \times x$	$w \times x$	Linear	Linear	0.0268	**0.0504**
3	$(x+w)^2$	$w \times x$	Sigmoid	Gaussian	0.0322	**0.0566**
4	$(x+w)^2$	$w \times x$	Linear	Linear	0.0495	0.0677
5	$(x+w)^2$	$(x+w)$	Sigmoid	Gaussian	0.0329	**0.0572**
6	$(x+w)$	$(x+w)$	Sigmoid	Gaussian	0.0371	0.0601
7	$(x+w)^2$	$(x+w)^2$	Sigmoid	Gaussian	0.0328	**0.0571**

learning rate (0.05) and momentum (0.65) for 200 epochs. The training performance and comparison of the predicted and target forecasts for independent data for the best model are given in Figure 9.44 (model 2 in Table 9.3). The root mean square error for the models on independent test

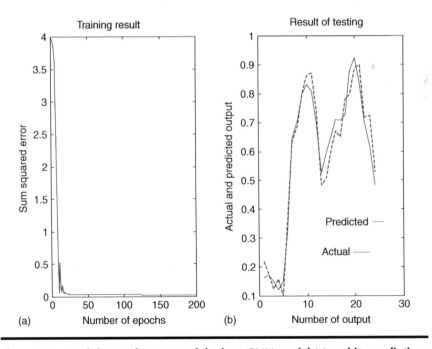

Figure 9.44 Training performance of the best GNN model (a) and its predictions against independent test data (b). (From Chaturvedi, D.K., Mohan, M., Singh, R.K., and Karla, P.K., *Soft Computing*, 8, 370, 2004. With permission from Springer Science and Business Media.)

data ranged from 0.0504 to 0.0677, as summarized in Table 9.3; it can be seen that the error range for the models is small, indicating that they all perform adequately. The best model (model 2) used regular weighted inputs and linear activation functions in both neurons. In the worst model (model 4), the weighted input to the summation neuron was $(x + w)^2$, whereas that to the product neuron was the regular weighted input, i.e., $w \times x$.

Table 9.3 indicates that the performance of models 3, 5, and 7 is almost as good as that of the best model. These all use $(x + w)^2$ for summation and $x \times w$, $(x + w)$, and $(x + w)^2$ for the product. The ability of the GNN with sum and product neurons to cope with nonlinearities in data can be investigated by evaluating the output of the neurons. Figure 9.45 shows the output of the summation and product hidden neurons as well as the output neuron for the best model (model 2 in Table 9.3) and two second-best models (models 3 and 5). Figure 9.45 further illustrates the ability of models 3 and 5 to easily handle nonlinear and non-separable problems by generating closed, nonlinear output surfaces without requiring additional model complexity involving a substantial number of neurons to generate such closed surfaces, as in MLP.

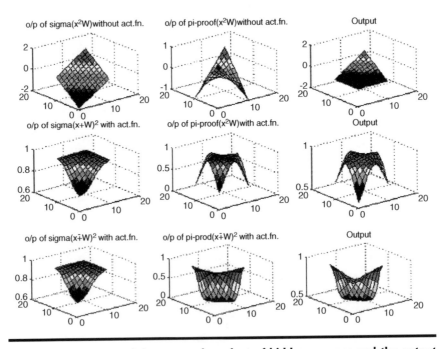

Figure 9.45 Output of the sum and product of hidden neurons and the output neuron of the GNN for three variations of the sum and products of weighted inputs (top: model 2; middle: model 3; bottom: model 5 in Table 9.3). (From Chaturvedi, D.K., Mohan, M., Singh, R.K., and Karla, P.K., *Soft Computing*, 8, 370, 2004. With permission from Springer Science and Business Media.)

9.7 Dynamically Driven Recurrent Networks

In focused time-lagged networks, a short-term memory is incorporated through externally fed input lags. For processes with shorter memory lengths (correlated lag lengths), these networks are appropriate. However, it would be useful if the network, through its internal dynamics, could discover by itself the long-term history of a series so that it does not have to depend on externally provided memory filters. Dynamically-driven recurrent networks with feedback loops attempt to capture such long-term history in data. The structure of three generic recurrent networks is presented in Figure 9.11. The major difference between these and the time-lagged networks is that there are feedback connections designed to remember past states of a network and feed them back into the network recursively so that at each time step, in addition to the current inputs, a memory trace of all previous inputs is presented to the network. Over time, the network stores long-term memory structure in its feedback (recurrent) and regular connections, whose weights are adjusted during training. The features of several recurrent networks will first be explored, then a small recurrent network will be trained by hand.

In recurrent networks, the lag structure and corresponding memory dynamics are determined by the network itself. Therefore, only the input(s) at the current time step are used by the network. The memory is built in through recurrent or feedback loops, which carry the output of the hidden or output neurons to feed them back to the network at the next time step. Various recurrent network structures can be designed using this idea. Elman [5] and Jordan [6] networks are two such networks that have gained popularity in the applied sciences; the following sections will treat these networks in detail. In an Elman network, hidden neuron outputs are fed back as input to the hidden neurons at the next time step, whereas in the Jordan network shown in Figure 9.11a, output neuron output(s) is fed back as input to hidden neuron at the next time step. Among other possibilities is a network in which both the output and the prediction error are fed back as input at the next time step, resembling a nonlinear autoregressive moving average (NARIMA) recurrent network, shown in Figure 9.11b, and a fully recurrent network in which both the hidden layer and the output layer feed their outputs back to themselves with time delay, as shown in Figure 9.11c.

9.7.1 Recurrent Networks with Hidden Neuron Feedback

9.7.1.1 Encapsulating Long-Term Memory

The way memory is built in to a network can be illustrated using a simple linear neuron presented in Figure 9.46. At time t, the neuron receives two

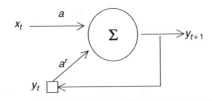

Figure 9.46 A simple illustration of the encapsulation of long-term memory by a recurrent linear neuron.

inputs: One external input (x_t) and the delayed output of the neuron at the previous time step. The neuron predicts the state of the system y_{t+1} at the next time step $t+1$. The output then becomes a delayed input to the neuron at the next time step through the feedback loop and the process is repeatedly recursively. The delayed input is stored in a context unit, depicted by the square in Figure 9.46, until it is presented at the next time step. The weight associated with this delayed input is called a recurrent weight and is denoted by a^r. This neuron represents a linear autoregressive model.

Suppose that at time t the neuron receives the concatenated inputs y_t and x_t. The network output y_{t+1} can be expressed as

$$y_{t+1} = a^r y_t + a x_t. \tag{9.33}$$

At the next time step $t+1$, y_{t+1} becomes an input with the external input x_{t+1}. The network output for this step becomes:

$$y_{t+2} = a^r y_{t+1} + a x_{t+1}. \tag{9.34}$$

Substituting Equation 9.33 into Equation 9.34,

$$y_{t+2} = a^r(a^r y_t + a x_t) + a x_{t+1} = (a^r)^2 y_t + a^r a x_t + a x_{t+1}. \tag{9.35}$$

To interpret the neuron response, suppose that $a^r = 0.2$ and $a = 0.3$. The output then is

$$y_{t+2} = (0.2)^2 y_t + (0.2)(0.3) x_t + 0.3 x_{t+1}$$

$$= 0.04 y_t + 0.06 x_t + 0.3 x_{t+1}. \tag{9.36}$$

It can be seen that the output has built both of the last external inputs x_t and x_{t+1} into the structure, as well as the last predicted output y_t. In time-series forecasting, variable x is the current observed value of the series. This yields a built-in structure for encapsulating the memory of a time-series with the feedback loop mechanism of a recurrent neuron. In this particular linear neuron case, the output of the neuron is an AR model in which lags are captured by the model itself. The recurrent weight determines the weight put onto the lags. It can also be seen that the more recent inputs lags are more heavily weighted than the distant lags. The output at time $t+n$, for example, is

$$y_{t+n} = (a^r)^n y_t + (a^r)^{n-1} ax_t + (a^r)^{n-2} ax_{t+1} + \cdots + ax_{t+n-1} \qquad (9.37)$$

which in its expression has inputs at all time steps fed externally to the neuron from the first (x_t) to the last input (x_{t+n-1}). This allows a recurrent network to use only the current input and build the memory recursively over time. During training, both the recurrent and regular weights are adjusted, as will be demonstrated shortly, and the role of the recurrent weight is to determine the appropriate influence of the recent and distant inputs.

Incorporation of nonlinear processing into the neuron results in NAR models. For example, if the hidden neuron in the example network has a nonlinear activation function f, the network output at time step $t+2$ becomes:

$$y_{t+2} = f[a^r f(a^r y_t + ax_t) + ax_{t+1}] \qquad (9.38)$$

in which past time lags are nonlinearly processed at each step and recursively embedded in the long-term memory structure.

If no exogenous inputs (other external variables) are used, then the network depends entirely on its own memory of the time-series, along with the current value. However, in some situations, exogenous variables (and possibly their lags) may improve forecasting accuracy and therefore it could be beneficial to use them in modeling. To represent these situations, only the present value of the external variables is needed because, when concatenated with the delayed inputs, the network also creates a lag structure for external inputs within the network in exactly the same way as described here for the input x_t. This results in a NARx.

It is also possible to use the network error as an additional input at each time step. This provides a NARIMA model, which also can use exogenous inputs (NARIMAx). At each time step, the predicted output and error are fed back to the input layer for use as inputs at the next step. The prediction error is obtained by subtracting the model output from the actual value. This concept is presented in the network shown in Figure 9.11b.

To build a recurrent network, one or more of the recurrent neurons presented here is incorporated as a hidden neuron(s) in a multi-layer network with feedback loops originating from the hidden neurons. When there are more neurons in the hidden layer, they create several variants of the same autoregressive process. This can help differentiate the significance of lags during training and can aid in finding the most appropriate lag structure for the time-series.

9.7.1.2 Structure and Operation of the Elman Network

A conceptual diagram of the Elman network [5] with only the hidden layer feeding its output into itself with a unit time delay is shown in Figure 9.47a. The manner in which the delayed feedback is applied is demonstrated in Figure 9.47b for a small example network with one externally fed input, two hidden neurons and one output. The externally-fed input is the value of the time-series at time $t-1$, z_{t-1}, and the output is the prediction $\hat{z}_t$ for the time-series at the next time step t. In this model, the outputs of the hidden neurons at a time step t not only feed the output neuron but also are stored in context units, which bear this name because they represent the memory of the network; these units are depicted by rectangular boxes in Figure 9.47b. At the next time step, context units feed the delayed hidden neuron outputs to the hidden neurons as inputs through the recurrent connections represented by dashed lines in Figure 9.47b, in which a superscript r is used to denote the recurrent weights. The recurrent connections facilitate encapsulation of the long-term memory structure of the data. The portion of the output that is fed back into the context units can be controlled by using a constant weighting factor k between zero and one. When the weighting factor is equal to one, the whole output is fed back.

The operation of the network is time-linked because of the feedback connections. To demonstrate how operations at successive time steps t,

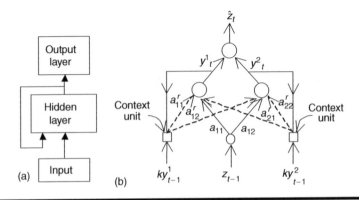

Figure 9.47 A simple recurrent network: (a) conceptual diagram of an Elman network and (b) a small example network illustrating the operation of the network.

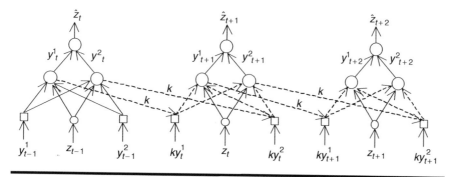

Figure 9.48 Illustration of the time-linked operation of the Elman network in Figure 9.47(b) for three time steps.

$t+1$ and $t+2$ are linked, the processing in the small network shown in Figure 9.47b is demonstrated for three time steps in Figure 9.48.

As shown in Figure 9.48, the hidden neuron output is fed back to the context units for use as input in the next time step. The parameter k is a weighting factor that determines the portion of the hidden neuron output to be stored in context units until the next time step. These feedback loops are depicted by dashed lines between the network states. The dashed lines within a network denote the recurrent connection through which the stored portion of the hidden neuron outputs is fed into each of the hidden neurons at the next time step. Thus, at each time step, the network sees a copy (or a portion) of the previous state of the network, which is fed simultaneously with the actual input z for the current step. This process is repeated recursively, until at the end of the training. The initial input of the context units is set to a random value because the hidden neuron output is not yet available during the first time step. The hidden neuron outputs are expected to reach appropriate values within a few time steps, making the influence of the initial random values negligible.

In the operation of this small network, the weighted input to the two neurons in the hidden layer can be expressed as

$$u_t^1 = a_{11}z_{t-1} + a_{11}^r ky_1(t-1) + a_{21}^r ky_2(t-1)$$
$$u_t^2 = a_{12}z_{t-1} + a_{12}^r ky_1(t-1) + a_{22}^r ky_2(t-1)$$
$$y_1(t) = f(u_t^1)$$
$$y_2(t) = f(u_t^2)$$

(9.39)

where u_t^1 and u_t^2 are the weighted input to hidden neurons one and two, respectively. Thus each hidden neuron sees the total previous state of the system depicted by $y_1(t)$ and $y_2(t)$, which are combined with the current state, enabling the network to recall the past state. The output of the neurons, $y_1(t)$ and $y_2(t)$ is the nonlinear transformation of the weighted inputs, as depicted by the last two expressions in Equation 9.39. The output of the network is generated in the usual way.

9.7.1.3 Training Recurrent Networks

The training of recurrent networks requires the update of recurrent weights [5,13]. This requires attention in the training method. Except for this portion, training is similar to MLP networks. This section refers back to and borrows basic concepts from Chapter 4, in which feedforward network training was presented in detail. Using the available information, the network can be trained with any method discussed in Chapter 4. For simplicity, the steps for basic gradient descent learning are illustrated in this section. Appropriate learning parameters apply for other gradient descent methods, and appropriate second derivatives and relevant parameters apply to second-order learning methods. Early stopping and regularization methods, discussed in Chapter 4 for removing bias and variance and assuring the best possible generalization, are applicable to recurrent networks as well. The difference is that the network contains more weights (recurrent) and associated inputs.

To illustrate the workings of these networks, the steps in the training of a small recurrent network will be studied. The network consists of one hidden neuron and one output neuron, as shown in Figure 9.49. It has two external inputs (x_1 and x_2) and a delayed input y, which is the output of the hidden neuron at the previous step. Generally, one of these external inputs is the value of the time-series to be modeled. At time step t, the inputs are $x_1(t)$, $x_2(t)$ and $y(t)$. Note that the time step upwards from the hidden neuron is taken to be $t+1$. There is one context unit to store the delayed input. The total output of the hidden neuron is fed back during the next step.

The hidden neuron weights, corresponding to the external inputs, are denoted by $a_1(t)$ and $a_2(t)$, and $a^r(t)$ denotes the weight of the recurrent connection. Processing from the hidden neuron and up is considered to belong to time step $t+1$. Thus output neuron weight is $b(t+1)$. For clarity, bias weights are ignored. Now the input vectors at time t will be fed and the activation will be propagated through the layers, one by one.

The weighted sum $u(t+1)$ and output $y(t+1)$ of the hidden neuron are:

$$u(t + 1) = a_1(t)x_1(t) + a_2(t)x_2(t) + a^r(t)y(t)$$
$$y(t + 1) = f(u(t + 1))$$

$$(9.40)$$

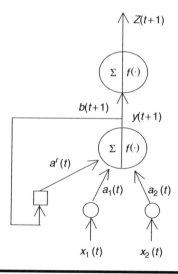

Figure 9.49 Training of a simple Elman recurrent network.

where $f(\cdot)$ is a nonlinear transfer function (sigmoid, hyperbolic tangent, etc.), which processes the weighted sum. For the output neuron, the weighted sum $v(t+1)$ and output $z(t+1)$ are

$$v(t + 1) = b(t + 1)y(t + 1)$$
$$z(t + 1) = f(v(t + 1))$$

(9.41)

where $z(t+1)$ is the predicted value of the target $T(t+1)$ and $z(t)$ is the predicted value of $T(t)$. For one input pattern, the error for a time step is

$$E(t + 1) = \frac{1}{2}[z(t + 1) - T(t + 1)]^2.$$

(9.42)

For the sigmoid transfer function, as derived in Chapter 4, the weight change Δw for any weight w, which in this case is either a_1, a_2, a^r, or b, is

$$\Delta w = -\varepsilon d$$

(9.43)

where ε is the learning rate and d is the error derivative with respect to a weight w. Removing time from the equations for clarity yields:

$$d = \frac{\partial E}{\partial w}.$$

(9.44)

A detail treatment of derivation of error derivatives is given in Chapter 4. Therefore, only an outline is given here and the derivative for recurrent weights will be the only new addition to the training method. From Chapter 4, the derivative of error with respect to any output neuron weight was given, using the chain rule of differentiation:

$$\frac{\partial E}{\partial b} = \frac{\partial E}{\partial z}\frac{\partial z}{\partial v}\frac{\partial v}{\partial b}. \tag{9.45a}$$

For actual target value T, the components of the error derivative can be expressed as

$$\frac{\partial E}{\partial z} = z - T$$

$$\frac{\partial z}{\partial v} = z(1 - z)$$

$$\frac{\partial E}{\partial z}\frac{\partial z}{\partial v} = (z - T)z(1 - z) = p \tag{9.45b}$$

$$\frac{\partial v}{\partial b} = \begin{cases} y & \text{regular weight} \\ 1 & \text{bias} \end{cases}.$$

Therefore, the error derivative is

$$\frac{\partial E}{\partial b} = \begin{cases} py & \text{regular weight} \\ p & \text{bias} \end{cases}. \tag{9.46}$$

Incorporating time,

$$\frac{\partial E(t + 1)}{\partial b(t + 1)} = p(t + 1)y(t + 1). \tag{9.47}$$

The error gradient with respect to hidden neuron weights can be expressed as

$$\frac{\partial E}{\partial a} = \frac{\partial E}{\partial y}\frac{\partial y}{\partial a}$$

$$\frac{\partial E}{\partial y} = \frac{\partial E}{\partial z}\frac{\partial z}{\partial v}\frac{\partial v}{\partial y} = pb. \tag{9.48}$$

The second component of the error gradient can be expressed as

$$\frac{\partial y}{\partial a} = \frac{\partial y}{\partial u}\frac{\partial u}{\partial a}$$

$$\frac{\partial y}{\partial u} = y(1-y). \tag{9.49}$$

Therefore,

$$\frac{\partial y}{\partial a} = y(1-y)\frac{\partial u}{\partial a} \tag{9.50}$$

and

$$\frac{\partial E}{\partial a} = pby(1-y)\frac{\partial u}{\partial a}. \tag{9.51}$$

The difference between a static MLP and a recurrent network is that in the recurrent network, an input vector consists of regular inputs and delayed feedback from the hidden units. Therefore, the weighted sum u for the hidden neuron, with time incorporated into the expression, is

$$u(t+1) = a_1(t)x_1(t) + a_2(t)x_2(t) + a^r(t)y(t). \tag{9.52}$$

The error derivative with respect to inputs depends on whether the input is regular or recurrent. For regular inputs, the derivation is the same as for a static MLP, as presented in Chapter 4. For a recurrent weight, the differentiation of products must be used, because both $u(t+1)$ and $y(t)$ are functions of a^r. Therefore, the derivative of net input u with respect to weights are

$$\frac{\partial u(t+1)}{\partial a_1(t)} = x_1(t),$$

$$\frac{\partial u(t+1)}{\partial a_2(t)} = x_2(t), \tag{9.53}$$

$$\frac{\partial u(t+1)}{\partial a^r(t)} = y(t) + a^r(t)\frac{\partial y(t)}{\partial a^r(t)}.$$

The $y(t)$ which appears in the last derivative defines the dynamic state of the network. By substituting $f'[u(t+1)]$ into Equation 9.50 for $y(1-y)$, which is the derivative of the hidden output with respect to net input for time step $t+1$,

$$\frac{\partial y(t+1)}{\partial a_1(t)} = f'(u(t+1))\frac{\partial u(t+1)}{\partial a_1(t)} = f'(u(t+1))x_1(t)$$

$$\frac{\partial y(t+1)}{\partial a_2(t)} = f'(u(t+1))\frac{\partial u(t+1)}{\partial a_2(t)} = f'(u(t+1))x_2(t)$$

$$\frac{\partial y(t+1)}{\partial a^r(t)} = f'(u(t+1))\frac{\partial u(t+1)}{\partial a^r(t)} = f'(u(t+1))\left[y(t) + a^r(t)\frac{\partial y(t)}{\partial a^r(t)}\right].$$

$$(9.54)$$

Let's denote $\pi(t) = \frac{\partial y(t)}{\partial a^r(t)}$. Then $\pi(t+1) = \frac{\partial y(t+1)}{\partial a^r(t+1)}$. By making $\pi(t+1) = \frac{\partial y(t+1)}{\partial a^r(t+1)} \approx \frac{\partial y(t+1)}{\partial a^r(t)}$, it is possible to write $\pi(t+1)$ using $\pi(t)$ and the last expression in Equation 9.54 as

$$\pi(t+1) = f'(u(t+1)[y(t) + a^r(t)\pi(t)]. \qquad (9.55)$$

Therefore, the error derivative with respect to the recurrent weights from Equation 9.51, Equation 9.54, and Equation 9.55 is

$$\frac{\partial E(t+1)}{\partial a^r(t)} = p(t+1)b(t+1)\pi(t+1). \qquad (9.56)$$

For a given weight update, $\pi(t)$ (see Equation 9.55) must be known. Because initially $y(0)$ is not available, $\pi(0)$ is assumed to be zero, making the initial gradient zero. In the next time step, $\pi(t+1)$ is needed, but it has already been calculated by Equation 9.55 in the previous step and can therefore be used recursively as a delayed gradient during each step to update the recurrent weight.

Now that all of the gradients are known, the weight update can be performed in the usual way. For the output neuron,

$$\Delta b(t+1) = -\varepsilon_1 p(t+1)y(t+1) \qquad (9.57)$$

where ε_1 is the learning rate for output weights. For the regular hidden neuron weights and the recurrent weight, the weight updates are

$$\Delta a_1(t) = -\varepsilon_2 p(t+1)b(t+1)f'(u(t+1))x_1(t)$$

$$\Delta a_2(t) = -\varepsilon_2 p(t+1)b(t+1)f'(u(t+1))x_2(t) \qquad (9.58)$$

$$\Delta a^r(t) = -\varepsilon_2 p(t+1)b(t+1)\pi(t+1)$$

where ε_2 is the learning rate for hidden neuron weights.

The weight update can be done in either incremental or batch mode. In incremental mode, the weights are updated at each time step using the procedure described. In batch update, the total error over all time steps E_{total} is minimized, and the cost function is obtained as

$$E_{total} = \sum_{t=0}^{T} E(t + 1) \tag{9.59}$$

and the weight change would be based on the total accumulated error gradient for each weight. The weights are updated after the whole input series has been processed by the network. In batch mode, the error derivative d is replaced by $d_m = \sum_i \left(\frac{\partial E}{\partial w}\right)_i^m$ where i denotes the input pattern number and m is the epoch number. The weight change for any given weight would be the average of the required weight changes for all input patterns n in the series. Thus the weight change can be written as

$$\Delta \bar{b} = \sum_{t=0}^{T} \Delta b(t + 1)$$

$$\Delta \bar{a} = \sum_{t=0}^{T} \Delta a(t) \tag{9.60}$$

$$\Delta \bar{a}^r = \sum_{t=0}^{T} \Delta a^r(t)$$

where the bar above the letter indicates the summed original weight changes. The new weights are obtained by adding the weight changes to the original weights.

9.7.1.4 Network Training Example: Hand Calculation

Now a hand calculation will be performed to illustrate the training of the above network (Figure 9.49). A three-step time-series from the furnace problem (Section 9.2.1 and Section 9.3.2) will be used, in which the next temperature of an electric furnace must be accurately forecasted from the current temperature and electric current. The output from the hidden neuron is fed back to a context unit that sends it as a delayed input back to the hidden neuron. The profiles of current and temperature for the furnace are given in Figure 9.3; as the figure shows, the current is increased in steps to regulate the temperature.

The first three input patterns and the corresponding outputs are given below; T denotes temperature and C denotes current, and these are the inputs $x_1(t)$ and $x_2(t)$, respectively. The desired output at the current step $T(t + 1)$ becomes the input in the next step.

$$x_1 = T(t) \quad x_2 = C(t) \quad z(t+1) = T(t+1)$$

	$x_1 = T(t)$	$x_2 = C(t)$	$z(t+1) = T(t+1)$
Step 1	7.3047	0.7019	7.2852
Step 2	7.2852	1.0193	7.2632
Step 3	7.2632	0.6555	7.2876

Assume the following initial random weights: $a_1(1) = 0.7$; $a_2(1) = -0.8$; $a^r(1) = 0.6$; $b(2) = 0.7$; $y_1 = 0.5$ and the constant learning rates $\varepsilon_1 = \varepsilon_2 = 0.5$.

Step 1. Using Equation 9.40 and Equation 9.41, previously derived in Section 9.7.1.3, $u(t+1)$, $y(t+1)$, and $z(t+1)$ are obtained as follows:

$$u(t+1) = a_1(t)x_1(t) + a_2(t)x_2(t) + a^r(t)y(t)$$

$$u(2) = a_1(1)x_1(1) + a_2(1)x_2(1) + a^r(1)y(1)$$

$$= 0.7 \times 7.3047 - 0.8 \times 0.7019 + 0.6 \times 0.5 = 4.8517;$$

$$y(t+1) = f(u(t+1)) = \frac{1}{1 + \text{Exp}(-u(t+1))}$$

$$y(2) = \frac{1}{1 + \text{Exp}(-u(2))} = \frac{1}{1 + \text{Exp}(-4.8517)} = 0.9922;$$

$$v(t+1) = b(t+1)y(t+1) = 0.7 \times 0.9922 = 0.6946;$$

$$z(2) = v(2) = 0.6946.$$

Denoting the MSE by $E(t+1)$:

$$E(t+1) = \{(\text{Target}(t+1) - z(t+1)^2\}/2$$

$$E(2) = \{(\text{Target}(2) - z(2)^2\}/2$$

$$= \{(7.2852 - 0.6946)^2\}/2 = 21.72.$$

Weight adjustment. The output neuron weight (Equation 9.45a and Equation 9.45b, and Equation 9.57) is

$$p(t+1) = [z(t+1) - \text{Target}(t+1)]z(t+1)(1 - z(t+1));$$

$$p(2) = [(z(2) - \text{Target}(2)] \times z(2) \times (1 - z(2))$$

$$= (0.6946 - 7.2852) \times (0.6946) \times (1 - 0.6946)$$

$$= -1.398;$$

$\Delta b(t + 1) = -\varepsilon_1 p(t + 1) y(t + 1);$

$\Delta b(2) = -\varepsilon_1 \times p(2) \times y(2) = -0.5 \times (-1.398) \times 0.9922 = 0.6936;$

$b(3) = b(2) + \Delta b(2) = 0.7 + 0.6936 = 1.3936.$

The hidden-neuron weights (Equation 9.58) are:

$\Delta a_1(t) = -\varepsilon_2 p(t + 1) b(t + 1) f'(u(t + 1)) x_1(t)$

$\Delta a_2(t) = -\varepsilon_2 p(t + 1) b(t + 1) f'(u(t + 1)) x_2(t)$

$\Delta a^r(t) = -\varepsilon_2 p(t + 1) b(t + 1) \pi(t + 1);$

$\Delta a_1(1) = -\varepsilon_2 \times p(2) \times b(2) \times y(2) \times (1 - y(2)) \times x_1(1)$

$\qquad = (-0.5) \times (-1.398) \times (0.7) \times (0.9922) \times (1 - 0.9922) \times 7.3047$

$\qquad = 0.0275;$

$a_1(2) = a_1(1) + \Delta a_1(1) = 0.7 + 0.0275 = 0.7275;$

$\Delta a_2(1) = -\varepsilon_2 \times p(2) \times b(2) \times y(2) \times (1 - y(2)) \times x_2(1)$

$\qquad = (-0.5) \times (-1.398) \times (0.7) \times (0.9922) \times (1 - 0.9922) \times 0.7019$

$\qquad = 0.00264;$

$a_2(2) = a_2(1) + \Delta a_2(1) = -0.8 + 0.00264 = -0.7973;$

$\Delta a^r(1) = -\varepsilon_2 \times p(2) \times b(2) \times \pi(2).$

From Equation 9.55, repeated below:

$$\pi(t + 1) = f'(u(t + 1)[y(t) + a^r(t)\pi(t)]$$
$$\pi(2) = y(2) \times (1 - y(2)) \times [y(1) + a^r(1)\pi(1)].$$

The values for $\pi(1)$ and $y(1)$ must be initialized to start the training process. Assuming that $\pi(1) = 0$ and using $y(1) = 0.5$ from Step 1:

$$\pi(2) = 0.9922 \times (1 - 0.9922) \times [0.5 + 0.6 \times 0] = 0.00385.$$

Therefore,

$$a^r(1) = (-0.5) \times (-1.398) \times 0.7 \times 0.00385 = 0.00188$$
$$a^r(2) = a^r(1) + \Delta a^r(1) = 0.6 + 0.00188 = 0.60188.$$

The weights and parameters carried forward to Step 2 are

$$a_1(2) = 0.7275; \quad a_2(2) = -0.7973; \quad a^r(2) = 0.60188;$$
$$\pi(2) = 0.00385; \quad y(2) = 0.9922;$$
$$b(3) = 1.3936.$$

Step 2. Using the results from Step 1, the inputs, and desired output yields

$$u(3) = a_1(2)x_1(2) + a_2(2)x_2(2) + a^r(2)y(2)$$

$$= 0.7275 \times 7.2852 - 0.7973 \times 1.0193 + 0.6019 \times 0.9922$$

$$= 5.0863,$$
$$y(3) = f(u(3))$$

$$= \frac{1}{1 + \text{Exp}(-u(3))} = \frac{1}{1 + \text{Exp}(-5.0863)} = 0.9938,$$

$$v(3) = b(3)y(3) = 1.3936 \times 0.9938 = 1.3851,$$
$$z(3) = v(3) = 1.3851,$$

$$E(3) = \{\text{Target}(3) - z(3)^2\}/2$$

$$= \{(7.2632 - 1.3851)^2\}/2 = 17.276.$$

Weight adjustment. The output neuron weight is

$$p(3) = [z(3) - \text{Target}(3)] \times z(3) \times (1 - z(3))$$

$$= (1.3851 - 7.2632) \times (1.3851) \times (1 - 1.3851)$$

$$= 3.1351,$$

$$\Delta b(3) = -\varepsilon_1 \times p(3) \times y(3)$$

$$= -0.5 \times (3.1351) \times 0.9938 = -1.5579,$$

$$b(4) = b(3) + \Delta b(3)$$

$$= 1.3936 - 1.5579 = -0.1643.$$

The hidden neuron weights are

$$\Delta a_1(t+1) = -\varepsilon_2 p(t+1)b(t+1)f'(u(t+1))x_1(t)$$
$$\Delta a_2(t+1) = -\varepsilon_2 p(t+1)b(t+1)f'(u(t+1))x_2(t)$$
$$\Delta a^r(t+1) = -\varepsilon_2 p(t+1)b(t+1)\pi(t+1)];$$
$$\Delta a_1(2) = -\varepsilon_2 \times p(3) \times b(3) \times y(3) \times (1-y(3)) \times x_1(3)$$
$$= (-0.5) \times (3.1351) \times (1.3936) \times (0.9938) \times (1-0.9938) \times 7.2852$$
$$= -0.0973,$$
$$a_1(3) = a_1(2) + \Delta a_1(2)$$
$$= 0.7275 - 0.0973 = 0.6301,$$

$$\Delta a_2(2) = -\varepsilon_2 \times p(3) \times b(3) \times y(3) \times (1-y(3)) \times x_2(2)$$
$$= (-0.5) \times (3.1351) \times (1.3936) \times (0.9938) \times (1-0.9938) \times 1.019$$
$$= -0.0136,$$

$$a_2(3) = a_2(2) + \Delta a_2(2)$$
$$= -0.7973 - 0.0136 = -0.8109,$$

$$\Delta a^r(2) = -\varepsilon_2 \times p(3) \times b(3) \times \pi(3);$$
$$\pi(t+1) = f'(u(t+1)[y(t) + a^r(t)\pi(t)],$$
$$\pi(3) = y(3) \times (1-y(3)) \times [y(2) + a^r(2)\pi(2)],$$
$$\pi(3) = 0.9938 \times (1-0.9938) \times [0.9922 + 0.6019 \times 0.00385] = 0.0061.$$

Therefore,

$$\Delta a^r(2) = (-0.5) \times (3.1351) \times 1.3936 \times 0.0061 = -0.0133,$$
$$a^r(3) = a^r(2) + \Delta a^r(2) = 0.60188 - 0.0133 = 0.58859.$$

The weights and parameters carried forward to Step 3 are

$$a_1(3) = 0.6301; \quad a_2(3) = -0.8109; \quad a^r(3) = 0.5886;$$
$$\pi(3) = 0.0061; \quad y(3) = 0.9938;$$
$$b(4) = -0.1643.$$

Repeating the process for input three yields the following:

$u(4) = 4.6303;$ $y(4) = 0.9903;$ $v(4) = z(4) = -0.1627;$ $E(4) = 27.75;$

$\Delta b(4) = -0.6978;$

$b(5) = -0.8621;$ $\pi(4) = 1.409;$ $\Delta a_1(3) = 0.00804;$ $a_1(4) = 0.6382;$

$\Delta a_2(3) = 0.00072;$

$a_2(4) = -0.8102;$ $\Delta a^r(3) = 0.0011;$ $a(4) = 0.5897.$

The MSEs for the three iterations are $[(21.72+17.276+27.75)/3] = 22.24.$

9.7.1.5   Recurrent Learning Network Application Case Study: Rainfall Runoff Modeling

Estimating and forecasting rainfall runoff is important for stream flow forecasting, flood management, reservoir operation, and watershed management. In watersheds with high mountains and steep slopes, heavy rainfalls can flood downstream cities within a few hours. These watershed rainfall runoff processes can be quite complex, highly nonlinear, and dynamic in nature, posing serious challenges to flood forecasting. This complexity is due mainly to the interaction of a number of complex rainfall runoff processes in a watershed.

Chiang et al. [14] compare static feedforward and recurrent networks for rainfall runoff modeling in Taiwan where thunderstorms and typhoons, which occur around four times per year, can bring heavy rainfalls, creating flashfloods in a matter of hours. The data has been collected from four rainfall gauges and a stream-flow station along a major river in Taiwan. The hourly data obtained from 1981 to 2000 consist of 23 storm events and 1632 observations. The dataset was divided into four cases with differing data length, storm events, and so on. Case 1 and case 2 have 15 storm events in training and three in the testing set. Cases 3 and 4 include only six storm events in training but 11 in the testing set. Each case has a different length of data records; cases 1 and 2 include more training data than do cases 3 and 4. Thus cases 3 and 4, which contain less data but more events in the testing set than in the training set, require a network to learn using less data and then make forecasts for unknown events in the testing set. Each case also included a validation dataset. Validation data is used to reduce overfitting; test data is used to assess the performance of a model.

Figure 9.50 shows the architecture of the recurrent network used in this case study, called a real time recurrent learning (RTRL) network. The network has only five inputs, four current rainfall measurements and one stream flow reading. The current values were used because of the short

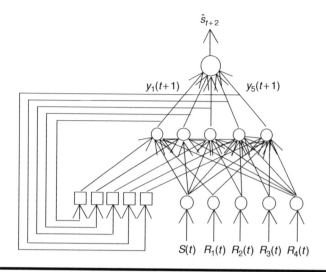

Figure 9.50 The architecture of the real-time recurrent learning (RTRL) network for rain-fall runoff forecasting (From Chiang Y.M., Chang L.C., and Chang F.J., *Journal of Hydrology*, 290, 2004, 297.)

timeframe involved in flash flooding. The hidden layer contains five neurons using sigmoid activation functions.

To assess and compare models, three criteria were used: MAE, RMSE, and RMAE. These measures are described in Section 9.3.3.1.

Two training algorithms were compared for the focused, time-lagged feedforward networks: A standard backpropagation and conjugate gradient, a second-order error minimization method [4]. The number of hidden neurons was determined by trial and error. For static networks, the inputs must be selected prior to modeling, because the next outcome usually depends on the current outcome and its lags, as well as the lags of other important input variables.

Four input schemes were tested:

(1) The current value of the four rainfall measurements and the stream flow (5 inputs)
(2) The current value and one lag of the four rainfall measurements and the stream flow (10 inputs)
(3) The current value and two lags of the four rainfall measurements and the stream flow (15 inputs)
(4) The current value and three lags of the four rainfall measurements and the stream flow (20 inputs)

The results for case 1 have shown that both the standard back-propagation and conjugate gradient (CG) methods provided the best results using 15 inputs consisting of two lags. This provides evidence that the average lag time is no more than 3 hours. However, CG is superior to backpropagation in terms of error criteria. For example, the MAE and RMSE based on test data for CG are 48 and 105, whereas for standard back propagation, these measures are 90 and 171. This indicates that the error from CG is about half of that from BP. BP provides best performance with 11 hidden neurons, whereas CG uses only 5 hidden neurons, indicating the effect of the learning method.

To compare a CG-based method with recurrent network performance, five neurons were used in the hidden layer of the recurrent network. The rest of the analysis was conducted using feedforward networks with 15 inputs and a CG learning method only. The results for the recurrent network and a feedforward network based on CG are superimposed on actual observations in Figure 9.51 and Figure 9.52; RTRL refers to recurrent network, CG refers to conjugate gradient.

The results in Figure 9.51 and Figure 9.52 indicate that the RTRL outperforms the feedforward network in forecasting peak performance in all cases, which is an important outcome because peak flow estimates for a typhoon are critical for flood management. For cases 1 and 2, in which the training includes more events than the testing set, error measures are slightly smaller for CG. However, for case 3 and case 4, which have less data in the training set and more unknown events in the testing set, RTRL has smaller

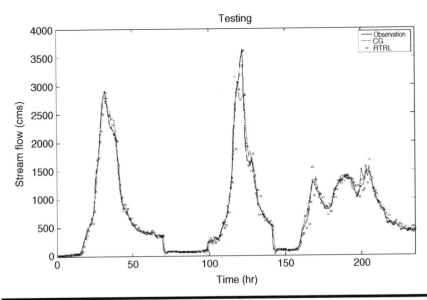

Figure 9.51 Forecasted and observed stream flow for the test period for case 1. (From Chiang, Y.M., Chang, L.C., and Chang, F.J., *Journal of Hydrology*, 290, 297, 2004. With permission from Elsevier.)

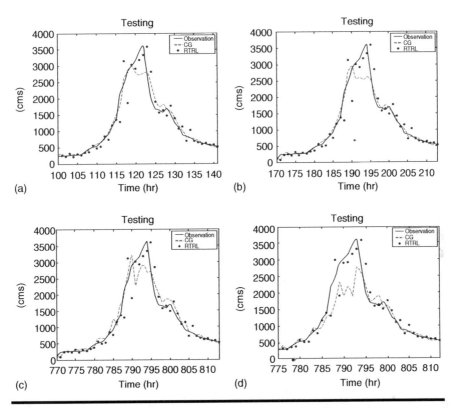

Figure 9.52 Observed and forecasted stream flow in the vicinity of peak flow: (a) case 1, (b) case 2, (c) case 3 and (d) case 4. (From Chiang, Y.M., Chang, L.C., and Chang, F.J., *Journal of Hydrology*, 290, 297, 2004. With permission from Elsevier.)

error measures, indicating that RTRL is robust in capturing general underlying temporal dynamics in the data so that even unseen events can be forecast with more reliability. The error measures for the hourly forecasts for case 4 are shown in Table 9.4, along with the observed peak flow Q_p and the forecast peak flow $\hat{Q}_p$. The table indicates that RMAE for RTRL on test data is only 65 percent of that from the feedforward network based on CG.

9.7.1.6 *Two-Step-Ahead Forecasting with Recurrent Networks*

In many practical engineering problems, such as flood warning systems and reservoir operation, it is crucial that the models provide multistep-ahead forecasts. The problem with multistep-ahead prediction is that, due to the absence of future data, a network has to rely on its own predictions at time t to forecast the outcomes at time $t+1$, and so on. The error in successive steps accumulates, thereby making multistep-ahead predictions more challenging than single-step predictions. Single-step-ahead recurrent

Table 9.4 Error Measures for a Feedforward Network with Conjugate Gradient (CG) Learning and RTRL for Case 4

	Observed Peak Flow	Predicted Peak Flow	MAE	RMAE	RMSE
Training:					
CG	2916	2725	40	0.082	70
RTRL		2818	47	0.097	84
Validation:					
CG	3020	2739	54	0.107	112
RTRL		2755	37	0.074	78
Testing:					
CG	3640	2790	113	0.168	249
RTRL		3315	74	0.110	158

Source: From Chiang, Y.M., Chang, L.C., and Chang, F.J., *Journal of Hydrology*, 290, 297, 2004. With permission from Elsevier.

networks can be modified to perform multistep prediction. For example, by choosing the output at time $t+2$ for the output and choosing inputs at time t, an RTRL can be easily modified to get the desired two-step-ahead prediction. The algorithm is the same as that for single-step-ahead forecasting, except that the output for time $t+2$, instead of $t+1$, as was used for single-step prediction.

The architecture of this recurrent network is shown in Figure 9.53. The weights are adjusted using the procedure for one-step-ahead forecasting.

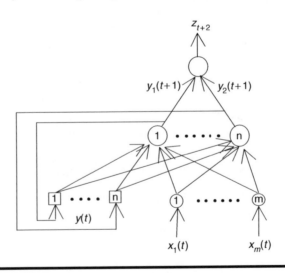

Figure 9.53 Real-time recurrent learning (RTRL) architecture for two-step-ahead forecasting.

The difference is that, in one-step-ahead RTRL, the hidden to output layer performs a simultaneous mapping process i.e., from $y(t+1)$ to $z(t+1)$, but in the two-step-ahead RTRL, the hidden-output layer performs one-step-ahead mapping from $y(t+1)$ to $z(t+2)$.

9.7.1.7 Real-Time Recurrent Learning Case Study: Two-Step-Ahead Stream Flow Forecasting

Chang et al. [15] used a recurrent network for two-step-ahead stream flow forecasting into a major river in Taiwan. Because a power plant operates on the river, accurate forecasting is extremely important. As discussed earlier in Section 9.7.1.5, the high peak flows due to typhoons and heavy rainfall can create flash floods in cities downstream in a matter of just a few hours; however, typhoons are a major source of water.

The recurrent network takes five inputs consisting of the current four readings of rainfall gauges [$R_1(t)$, $R_2(t)$, $R_3(t)$, and $R_4(t)$], and one reading of stream flow $S(t)$, and forecasts the stream flow two steps ahead $\hat{S}(t+2)$, which is an estimate of the actual flow $S(t+2)$. The network configuration is shown in Figure 9.54. The network has five hidden neurons and is trained using learning rates of 0.4 and 0.8 for the output layer weights and hidden layer weights, respectively.

For comparison, an additional model has been trained: a linear autoregressive moving average with exogenous variables (ARMAX) model using the current value and two lags of rainfall, current stream flow, and error e at time $t+1$, as given in Equation 9.61.

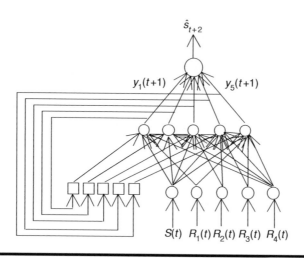

Figure 9.54   Two-step-ahead RTRL configuration for stream-flow forecasting.

$$\hat{S}(t+2) = a_1\hat{S}(t+1) + a_2 S(t) + \sum_{i=0}^{2} b_{1i}R_1(t-i) + \sum_{i=0}^{2} b_{2i}R_2(t-i)$$

$$+ \sum_{i=0}^{2} b_{3i}R_3(t-i) + \sum_{i=0}^{2} b_{4i}R_4(t-i) + c_1 e(t+1) \qquad (9.61)$$

where $\hat{S}(t+2)$ and $\hat{S}(t+1)$ are the estimated stream flow for time steps $t+2$ and $t+1$, respectively, $S(t)$ is the observed stream flow at time t, $R_1(t-i)$, $R_2(t-i)$, $R_3(t-i)$, $R_4(t-i)$ are the rainfall from the four gauges at time $t-i$, respectively, and $e(t+1)$ is the model error at time $t+1$. The model parameters a_i, b_{ij}, and c_1 are to be obtained from linear least-squares estimations.

The error measures used to compare models were MSE and MAE, as previously described in Section 9.3.3.1, and an additional measure of the quality of the fit with respect to a benchmark G_{bench}, expressed as

$$G_{\text{bench}} = 1 - \frac{\sum_{i=1}^{n}(Q_i - \hat{Q}_i)^2}{\sum_{i=1}^{n}(Q_i - \tilde{Q}_i)^2} \qquad (9.62)$$

where Q_i is the observed flow, $\hat{Q}_i$ is the forecast flow in step i, and $\tilde{Q}_i$ is the previous observed flow used in step i, which, in this case, is Q_{i-2}. The G_{bench} measure compares the current error with respect to the difference between the current and previously observed values. In other words, it compares the model with a benchmark model which has the previous observed value as output. A higher G_{bench} value indicates that the model is better than the benchmark.

It was found that RTRL can approximately predict stream flow 2 h in advance after 30 training steps. Both RTRL and ARMAX models were then tested on data from six different years. Table 9.5 shows the summary of the results, which indicates that RTRL has outperformed ARMAX for all years in terms of error measures and peak flow predictions. The predictions for the year 2000 are shown in Figure 9.55, in which the period affected by the typhoon is highlighted. The observed peak flow is 1150 (m³/s or cm s), the RTRL predicted flow is 923 cms and the ARMAX prediction is 859 cm s. The G_{bench} for RTRL is 0.221, which is much higher than that for ARMAX (0.017), indicating that the RTRL predictions are better overall.

Table 9.5   Annual Mean Observed Stream Flow Q_0, Observed Annual Peak Flow Q_p, Estimated Peak Flow Q_p and Error Measures for Six Years

Year	Q_0	Q_p	ARMAX				Two-step RTRL			
			MAE	RMAE	$\hat{Q}_p$	G_{bench}	MAE	RMAE	$\hat{Q}_p$	G_{bench}
1995	19.01	105	0.440	0.023	101	0.209	0.388	0.020	96	0.266
1996	24.04	1090	0.886	0.037	987	0.036	0.676	0.028	1089	0.376
1997	24.04	798	0.925	0.038	667	0.009	0.757	0.031	762	0.192
1998	32.24	713	1.090	0.034	677	0.037	0.918	0.028	691	0.338
1999	21.55	203	0.738	0.034	194	0.138	0.686	0.032	194	0.217
2000	33.92	1150	1.658	0.049	859	0.017	1.548	0.046	923	0.221

Source: From Chang, F.J., Chang, L.C., and Huang, H., *Hydrological Processes*, 16, 257, 2002. With permission. Copyright John Wiley & Sons Ltd.

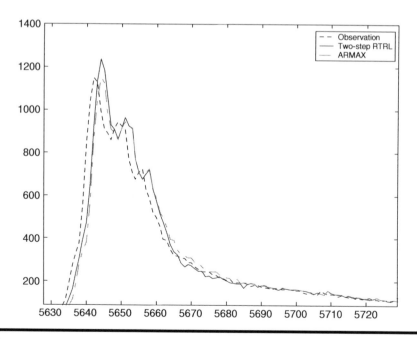

Figure 9.55   Forecast for year 2000 highlighting the typhoon affected period. (From Chang, F.J., Chang, L.C., and Huang, H., *Hydrological Processes*, 16, 257, 2002. With permission. Copyright John Wiley & Sons Ltd.)

9.7.2 Recurrent Networks with Output Feedback

When the output is fed back as input after one unit time delay, this yields another variant of the partially recurrent network, known as a Jordan network [5]. A schematic diagram of this network is shown in Figure 9.11a. In this network, output at time t is fed back as input at time $t+1$. This network also builds an autoregressive model with input lags, but owing to the processing in the output layer, it can create more complex autoregressive structures. The portion of the output that is fed back can be controlled using a constant between zero and one for the feedback loop. If the weight is one, the total output is fed back; otherwise, only a portion of the output is propagated. This delayed output is presented as input to the hidden layer via connections represented by recurrent weights. During training, these are updated in the same way as the regular weights.

9.7.2.1 Encapsulating Long-Term Memory in Recurrent Networks with Output Feedback

This section will study a small network with one linear hidden and one output neuron, shown in Figure 9.56, to illustrate how memory is built into recurrent networks with output feedback. (Linear processing is used only for simplicity of illustration; nonlinear processing endows the network with the power to model any complex nonlinear time-series). At time step t, the external input x_t and the previous network output z_t are concatenated and presented as input. In response, the network produces the hidden neuron output y_{t+1} for the next time step $t+1$ and the output neuron output z_{t+1}, also for time step $t+1$. The recurrent weight is a^r, the regular weight is a, and the hidden-output weight is b.

The network output and hidden-neuron processing for input x_t is given by

$$z_{t+1} = by_{t+1}$$
$$y_{t+1} = a^r z_t + ax_t$$

(9.63)

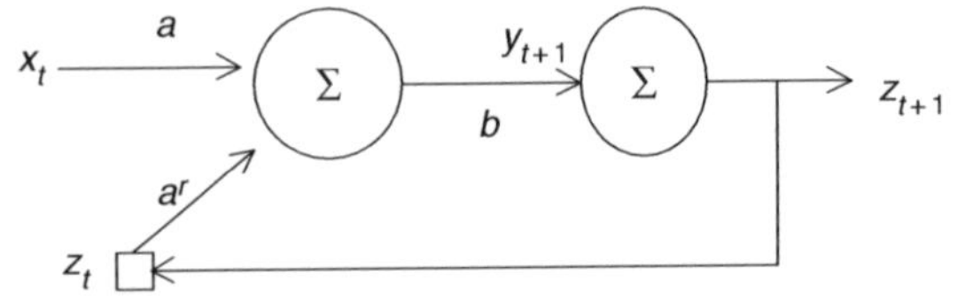

Figure 9.56 An example recurrent network with output feedback to illustrate the encapsulation of long-term memory.

Substituting y_{t+1} into the first expression in Equation 9.63 yields:

$$z_{t+1} = b(a^r z_t + ax_t) \tag{9.64}$$

This output is fed back through the feedback loop mechanism with unit time delay to be used as input at the next time step, along with the corresponding external input, x_{t+1}. The output at the next time step is

$$z_{t+2} = by_{t+2}$$
$$y_{t+2} = a^r z_{t+1} + ax_{t+1}. \tag{9.65}$$

By substituting y_{t+2} into z_{t+2}:

$$z_{t+2} = b(a^r z_{t+1} + ax_{t+1}) \tag{9.66}$$

Because Equation 9.64 yields an expression for z_{t+1}, the network output can be expressed as

$$z_{t+2} = b[a^r \{b(a^r z_t + ax_t)\} + ax_{t+1}] = b[b(a^r)^2 z_t + ba^r ax_t + ax_{t+1}]. \tag{9.67}$$

To clarify the meaning of the output, suppose that $b = 0.4$, $a^r = 0.2$, and $a = 0.3$. Substituting these values into the expression for z_{t+2} yields

$$z_{t+2} = 0.0064z_t + 0.0096x_t + 0.12x_{t+1}. \tag{9.68}$$

The output is an AR model, in which input lags are recursively incorporated by the network, with only the current input presented to the network at any given time. The past history has the form of an exponential average. For example, the immediate lags have a higher weighting and the distant lags have a lower weighting, as can be seen from the coefficients for these lags. The recurrent and output weights determine the weighting in the exponential averaging process. Thus, the network recursively builds an autoregressive model for predicting the next output.

When there are more neurons in the hidden layer, they create several variants of the same autoregressive process. This can help differentiate the significance of lags during training and can help find the most appropriate lag structure for the time-series.

Incorporation of nonlinear processing into the network results in NAR models. For example, if the hidden neuron in the example network has a nonlinear activation function f, the network output becomes

$$z_{t+2} = b[f[a^r bf(a^r z_t + ax_t) + ax_{t+1}]]. \tag{9.69a}$$

where, at each step, the past time lags are nonlinearly processed and recursively embedded in the long-term memory structure. When the output layer also uses nonlinear processing, another layer of processing is added to the network, creating more complex NAR models.

If no exogenous inputs (other external variables) are used, then the network depends entirely on its memory of the time-series, along with its current value. However, in some situations, exogenous variables and possibly their lags may improve forecasting accuracy and it could therefore be beneficial to use them in modeling. To represent these external variables, only their present value is needed because, when concatenated with the delayed inputs, the network also creates an internal lag structure for them, in exactly the same way as described for the input x_t. This results in a NARx model. It is also possible to use the network error as an additional input at each time step. This provides a NARIMA model, which can also use exogenous inputs (NARIMAx). At each time step, the predicted output and error are fed back to the input layer for use as inputs at the next step. The prediction error is obtained by subtracting the model output from the actual value. The next section includes a case study describing a recurrent network with output and error feedback, representing a NARIMAx model. The case study is followed in Section 9.7.2.3 by a detailed description of learning in a Jordan network.

9.7.2.2 Application of a Recurrent Net with Output and Error Feedback and Exogenous Inputs: (NARIMAx) Case Study: Short-Term Temperature Forecasting

Accurate temperature forecasting can be useful in many practical situations. It can be used in models that require estimates of temperature for forecasting weather, energy demand, and so on. The outside temperature is affected by the season, hour, wind, solar intensity, rain, and other factors. In this case study, reported by Lanza et al. [10] temperature forecasts, useful for maintaining the internal environment of a building, are made with a NARIMAx recurrent network. The model configuration is shown in Figure 9.57. The model has three inputs: The hour (external variable), the

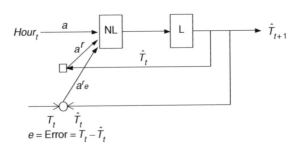

Figure 9.57 NARIMAx recurrent network for temperature forecasting. (From Lanza, P.A.G. and Cosme, J.M.Z., *Engineering Applications of Artificial Intelligence*, 15, 459, 2002. With permission from Elsevier.)

previously predicted temperature, and the prediction error. The output is the temperature at the next step.

The model includes regular weights associated with the hour, and recurrent weights associated with the predicted output and error from the previous step. These weights must be simultaneously updated; the approach for adjusting recurrent weights is described in detail in the next section. The same approaches used to train a feedforward network can be used once the recurrent weights are built into the equations.

The network was trained using two datasets, shown in Figure 9.58. One temperature profile (PROBEN1) has a very irregular pattern and a descending trend during the winter. The two datasets are known to have been used as benchmark test data for building energy predictions.

The data for training were formed using a moving window of 28 hours. Inputs were normalized between -0.9 and 0.9. Five hidden neurons with logistic transfer functions and a linear output neuron were used in the network. The training criterion is the MAE, expressed as

$$\text{MAE} = \frac{1}{n} \sum_{t=1}^{n} |\hat{T}_t - T_t| \qquad (9.69b)$$

where n is the total number of hours. The training was conducted using a random optimization method [11,12]. The basic approach is that an initial weight increment is randomly selected; if a reduced error is obtained by performing an update with plus or minus the selected increment, the weights are updated in the direction of the reduced error. The random weight increment is extracted from a Gaussian distribution whose center is also updated with each weight increment in the direction of the weight update. This process is repeated until the termination criterion is reached.

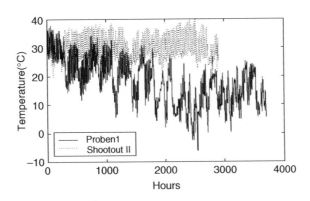

Figure 9.58 Temperature profile of two datasets used in forecasting. (From Lanza, P.A.G. and Cosme, J.M.Z., *Engineering Applications of Artificial Intelligence*, 15, 459, 2002. With permission from Elsevier.)

In addition, the weights of the trained network are updated daily during use; therefore, using the moving window, the inherent nonstationary behavior of temperature is reflected in the daily updating of weights. Initially, the network is trained with 1000 epochs to obtain good accuracy in its first prediction. Then, daily the network is retrained with a smaller number of epochs. As a validation measure, MAE is checked during each daily retraining so that the model improves with time. An MAE of about 0.5°C or lower is considered adequate.

Alongside the recurrent neural network, a linear ARMAX model was also developed with the same inputs. This had the form:

$$\hat{T}_t = b_1 H_{t-1} + a_1 T_{t-1} + c_1 e_{t-1} \tag{9.70}$$

where H is hour, $\hat{T}$ is the predicted temperature, T is the observed temperature, and e is the error. The predicted temperature for next time step t is predicted from the inputs for current step $t-1$. The coefficients b_1, a_1, and c_1 are the constants to be determined from the linear least squares. The MAE for the linear model on the two datasets was 0.5579 and 0.5187, respectively. These values for the recurrent network were 0.4438 and 0.4242, respectively.

Figure 9.59 shows the results from the recurrent network superimposed on data for 14 continuous days from the Proben1 dataset. The recurrent network is depicted by a state space neural network (ssNN), another name used for networks that use the feedback from the state of the system. The predictions are very good, irrespective of the irregular temperature profile, indicating that the forecaster is able to follow the changes very closely.

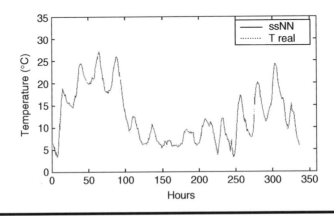

Figure 9.59 Forecasting performance of the recurrent network on dataset 1 (Proben1). (From Lanza, P.A.G. and Cosme, J.M.Z., *Engineering Applications of Artificial Intelligence*, 15, 459, 2002. With permission from Elsevier.)

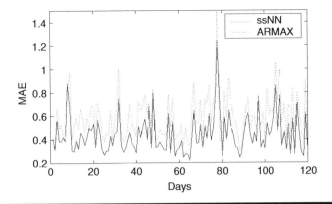

Figure 9.60 Comparison of MAE for linear ARMAX and the nonlinear ARMAX recurrent network for dataset 1 (Proben1). (From Lanza, P.A.G. and Cosme, J.M.Z., *Engineering Applications of Artificial Intelligence,* **15, 459, 2002. With permission from Elsevier.)**

The MAE for daily updates from the two models (ARIMAX and recurrent net) for the same dataset is shown in Figure 9.60. Clearly, the recurrent network error is significantly lower than that for the linear model.

Now that the advantages of a recurrent network with output and error feedback have been demonstrated, the training of these networks will be studied.

9.7.2.3 Training of Recurrent Nets with Output Feedback

This section includes a detailed treatment of the learning aspects of recurrent networks with output feedback; this network structure is referred to as a Jordan network. [6] Recall that in this network, there are feedback loops from the output layer to the input layer, as shown in Figure 9.11a and Figure 9.56, and that the network output at time t is used as an input at time $t+1$. Therefore, the network output is recursively embedded in the temporal structure of the network, as shown in Section 9.7.2.1. The structure of a Jordan network and a simple network chosen to illustrate training are shown in Figure 9.61.

The training follows as for the previously presented Elman network, in which the hidden neuron output is fed back as input (Section 9.7.1). However, because the delayed input comes from the output layer, the recurrent weight update must be modified. Other weights are updated following the usual procedure. The recurrent error derivative can be written in the form of

$$\frac{\partial E}{\partial a^r} = \frac{\partial E}{\partial z}\,\frac{\partial z}{\partial a^r} = (z - d)\frac{\partial z}{\partial a^r} \tag{9.71}$$

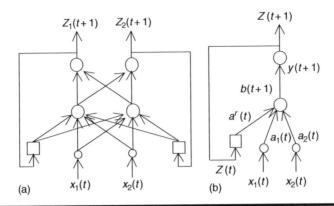

Figure 9.61 Jordan network configuration: (a) generic network structure and (b) a simple network selected to illustrate training.

where z is the network output and d is the desired output. The second component of Equation 9.71 consists of:

$$\frac{\partial z}{\partial a^r} = \frac{\partial z}{\partial v}\frac{\partial v}{\partial y}\frac{\partial y}{\partial u}\frac{\partial u}{\partial a^r} = z(1-z)by(1-y)\frac{\partial u}{\partial a^r} = Q\frac{\partial u}{\partial a^r} \qquad (9.72)$$

where $Q = z(1-z)by(1-y)$. Because

$$u = a_1 x_1 + a_2 x_2 + a^r z \qquad (9.73)$$

the derivative in Equation 9.72 for time $t+1$ is (note that z is a function of a^r)

$$\frac{\partial z(t+1)}{\partial a^r(t)} = Q(t+1)\frac{\partial u(t+1)}{\partial a^r(t)} = Q(t+1)\frac{\partial [a^r z(t)]}{\partial a^r(t)}$$

$$= Q(t+1)\left[z(t) + a^r(t)\frac{\partial z(t)}{\partial a^r(t)}\right]. \qquad (9.74)$$

Denoting $\frac{\partial z(t)}{\partial a^r(t)} = \Phi(t)$ and assuming that:

$$\frac{\partial z(t+1)}{\partial a^r(t)} \approx \frac{\partial z(t+1)}{\partial a^r(t+1)} = \Phi(t+1). \qquad (9.75)$$

Equation 9.74 can be written as

$$\Phi(t+1) = Q(t+1)[z(t) + a^r(t)\Phi(t)]. \qquad (9.76)$$

Therefore, the error derivative with respect to recurrent weight a^r from Equation 9.71 is

$$\frac{\partial E(t+1)}{\partial a^r(t)} = (z(t+1) - d(t+1))\frac{\partial z(t+1)}{\partial a^r(t)}$$

$$= (z(t+1) - d(t+1))\Phi(t+1). \qquad (9.77)$$

Therefore, the recurrent weight update is

$$\Delta a^r(t) = -\varepsilon_2 \frac{\partial E(t+1)}{\partial a^r(t)} = -\varepsilon_2(z(t+1) - d(t+1))\Phi(t+1).$$ (9.78)

As for the Elman network, the value of $\Phi(t)$ at time Step 1 is assumed to be zero. At the next step, it is replaced by $\Phi(t+1)$, which is calculated from Equation 9.76.

9.7.3 Fully Recurrent Network

When both the output layer and the hidden layer feed their delayed outputs back to themselves, a more complex network is obtained. This structure is shown in Figure 9.11c, in which two levels of feedback are illustrated. A conceptual diagram and a small example network are demonstrated in Figure 9.62 to highlight the features of this network.

The recurrent network shown in Figure 9.62b includes a context unit at the hidden layer level to store the delayed output of the network until it is presented as input at the next time step. It also includes a context unit at the level of the input layer for temporary storage of the hidden neuron output. As before, the recurrent weights are denoted by a subscript r. At a particular instance of time, both the actual inputs and the delayed hidden neuron outputs are fed as input into the hidden layer. Similarly, the actual hidden-neuron output and the delayed output of the output neuron are fed as inputs to the output neuron. This process is repeated in a time-linked manner in which both the hidden neuron and the output neuron are linked to themselves (feed themselves) at each iteration.

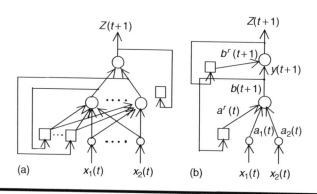

Figure 9.62 Fully recurrent network with feedback loops from both the output and the hidden layer feeding delayed outputs back to themselves: (a) conceptual diagram and (b) a small example network illustrating the operation of the network with two levels of feedback.

The output of the hidden neurons is produced in a way similar to that previously presented for the Elman network. The output now includes an extra input: Its own delayed output. The weighted sum v and output z of the output neuron for time $t+1$ are:

$$v(t + 1) = b(t + 1)y_1(t + 1) + b^r(t + 1)z(t)$$
$$z(t + 1) = f(v(t + 1)) \tag{9.79}$$

In this network, a long-term history is captured at both the hidden and output layer levels. At the hidden layer level, the current state is combined with the previous state, which is the delayed output of the hidden neurons, fed as input through the recurrent connections. This way, at every time step, the network sees the whole history embedded in the previous state of the network. At the output layer level, the output neuron(s) sees it own output recursively fed as delayed input; therefore, at each time step, it also sees the whole history of its output embedded in its delayed input. Thus, this network creates a complex, dynamic structure for embedding temporal characteristics in the data.

Because there is a recurrent connection to the output neuron, it has to be adjusted. The method used to do this is similar to that used for adjusting the recurrent weight to the hidden neuron. Because this is the only new addition, only the adjustment of the hidden-output recurrent weight is shown here.

The net input $v(t+1)$ to the output neuron is

$$v(t + 1) = b(t + 1)y(t + 1) + b^r(t + 1)z(t) \tag{9.80}$$

where $z(t)$ is the previous output of the network. The network output is $z(t + 1) = f(v(t + 1))$.

The error derivative with respect to the output layer weights, removing time for clarity, is

$$\frac{\partial E}{\partial b} = \frac{\partial E}{\partial z} \frac{\partial z}{\partial v} \frac{\partial v}{\partial b} \tag{9.81}$$

where the first term is equal to $(d - z)$. The second two terms can be expressed as

$$\frac{\partial z}{\partial b} = \frac{\partial z}{\partial v} \frac{\partial v}{\partial b} \tag{9.82}$$

which, for the recurrent connection with time incorporated, is

$$\frac{\partial z(t + 1)}{\partial b^r(t + 1)} = \frac{\partial z(t + 1)}{\partial v(t + 1)} \frac{\partial v(t + 1)}{\partial b^r(t + 1)}$$

$$= z(t + 1)(1 - z(t + 1))\left[z(t) + b^r(t + 1)\frac{\partial z(t)}{\partial b^r(t + 1)}\right]. \tag{9.83}$$

Denoting the last derivative on the right-hand side of the expression in Equation 9.83 by $\Pi(t+1)$,

$$\Pi(t + 1) = \frac{\partial z(t)}{\partial b^r(t + 1)}.$$

Assuming that $\frac{\partial z(t+1)}{\partial b^r(t+1)} \approx \frac{\partial z(t+1)}{\partial b^r(t+2)} = \Pi(t + 2)$, we have:

$$\Pi(t + 2) = z(t + 1)(1 - z(t + 1))[z(t) + b^r(t + 1)\Pi(t + 1)]. \qquad (9.84)$$

Therefore, the recurrent weight update incorporating Equations 9.84 and 9.81 is

$$\Delta b^r = -\varepsilon_1(d(t + 1) - z(t + 1)\Pi(t + 2). \qquad (9.85)$$

Using an initial guess for $\Pi(t+1)$, the required $\Pi(t+2)$ can be recursively calculated from Equation 9.84 for the rest of the steps, as was done for the hidden recurrent weight. The update of the regular output neuron weights follows the usual procedure.

9.7.3.1 Fully Recurrent Network Practical Application Case Study: Short-Term Electricity Load Forecasting

In electricity generation, short-term load (power) forecasts (STLF), ranging from 1 hour to one week are desired so that generation can be adapted to meet the demand of the customers. The STLF has significant effects on power system operations, because forecast errors result in increased operation costs. Therefore, it is extremely important to forecast with low error. Although there is a daily pattern to load demand, the visual observation of the demand over time shows different characteristics. Therefore, it is difficult to design a single model to cover all such characteristics. Data clustering can be a useful approach in such situations. Topalli and Erkmen [16] propose a hybrid clustering and fully recurrent network approach for short-term load forecasting; a summary of their work is presented here to illustrate the performance of fully recurrent networks (FRNN). The hourly data for each day is clustered such that each cluster contains the load demand for a particular hour. Thus, there are 24 clusters, corresponding to the 24 hours in a day. A fully recurrent network is developed for each cluster. To forecast one day in advance for a specific hour, the present and previous two days' load for that hour are used as input. The day of the week is also given as input to the model, to account for variations in daily characteristics. This approach to model development is shown in Figure 9.63.

For the 24 hours, 24 fully recurrent networks are trained. The learning rate is 0.5, and the hidden neuron uses a sigmoid activation function. Training is done with data from the year 2000 and the model is tested on data from the year 2001. The average training error over all clusters is 2.31 percent. The model is tested on three cases.

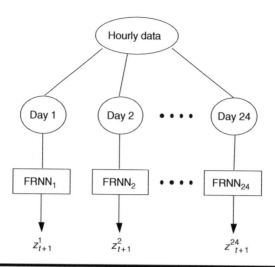

Figure 9.63   Data clustering and subsequent development of multiple FRNNs.

(1) *Forecasting with the static model.* In this model, the fixed static weights obtained from training are used in forecasting. The real-time application using data from the year 2001 produces an average forecast error of 2.717 percent.

(2) *Off-line learning followed by continuous real-time weight adjustment during forecasting.* In this case, the trained model is used in forecasting, but the weights are continually adjusted using the 2001 data. The average error is 2.453 percent, which is about 10 percent lower than that obtained using fixed weights. This indicates the

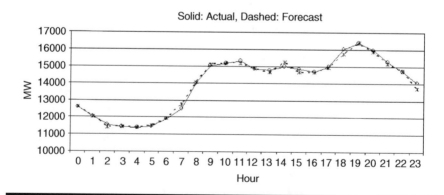

Figure 9.64   Actual and forecasted hourly power demand from an FRNN using on-line learning subsequent to off-line learning for the day with the least error in 2001. (From Topalli, A. and Erkmen, I., *Neurocomputing,* 51, 495, 2003. With permission from Elsevier.)

advantage of continual weight adjustment. The day with the best performance had 0.698 percent error, as shown in Figure 9.64. Thus, the hybrid off-line learning, followed by real-time learning, yields better results because it takes advantage of the new data. Because the near-optimal weights are available prior to forecasting, the model can adapt to real situations quickly with fewer oscillations due to abrupt environmental changes.

(3) *Real-time learning and forecasting.* In this case, the model begins with random weights and uses data from the year 2001 for real-time learning, and does not use prior (off-line) learning at all. This model predicts and learns at the same time. The average error for the 24 models is 9.534 percent, much higher than the hybrid model with off-line and subsequent on-line learning. This shows the impossibility of starting a real-time application with non-trained networks. The model takes longer to respond to real situations, with more oscillations due to abrupt environmental changes.

(4) *Effect of clustering.* To test the effect of clustering, all data without clustering is used to model a recurrent network. The training error on data from the year 2000 is 10.645 percent; the real-time application using data from 2001 yields an error of 12.96 percent. These are much higher than the errors for all the above cases in which a single model was used for each hour. The error is even higher than that for the simultaneous real-time learning and forecasting shown in Case 3, in which no trained model was used. Thus, clustering simplifies the data and models, and makes it possible to achieve much higher forecasting accuracy than does use of the full dataset.

9.8 Bias and Variance in Time-Series Forecasting

When fitting models to data, it is desirable to find an optimum model that perfectly captures the trends in the data and generalizes well to unseen data. This is generally called the bias–variance dilemma. Too little learning makes a model incapable of representing the data, thereby limiting its power, which introduces error due to bias. Too much learning makes the model too rigid, thus making it unable to generalize well, and thereby introducing error due to variance or overfitting. As a data-driven approach to pattern recognition, neural networks can often suffer from overfitting, as discussed in detail in Chapter 6. Overfitting results from a network's ability to fit limited training data with high accuracy and generalize poorly on unseen data. This can make a forecasting model unstable and therefore unreliable in real-world applications [17].

Overfitting is reduced by using cross-validation, also called early stopping, which is addressed in detail in Chapter 6. In cross-validation, the dataset is divided into training and validation sets. The training set is used to develop the model and the validation set is used to monitor the effect of overfitting. The validation sample also serves as the basis for final model selection in cases where several candidate models exist. In many situations, a third testing sample is also used to assess the capability of the model for accurate out-of-sample forecasting.

Cross-validation or early stopping has been extensively used in neural networks and statistical modeling. The limitation of this approach is that it monitors only overall performance measures such as the Sum of Squared Error (SSE) or the MSE. Although these are useful in improving overall generalization performance, they do not shed light on how and why overfitting problems arise in a particular forecasting model and therefore do not provide a complete picture of a model's ability to learn and generalize. Moreover, the overall measures do not provide much insight into how the forecasting performance of a model could be improved. For example, they do not reveal which error component, bias or variance, has a more degrading effect on the model's forecasting accuracy (bias is a lack of fit and variance is overfitting of a model). Therefore, a useful approach would be to systematically study the time-series forecasting ability of neural networks from a bias–variance perspective [17]. This can be done by studying the bias and variance components of MSE.

The total error depicted by MSE can be decomposed into bias and variance. In a real sense, bias is a measure of the systematic error of a forecasting model in learning the underlying trends in a time-series. Variance refers to the stability of models developed from different samples of data generated by the same physical process; variance therefore provides a measure of the ability of a model to generalize. Decomposing the total error into bias and variance components is a useful approach when investigating the learning and generalization abilities of neural networks.

Bias and variance are important statistical properties of an empirical model [17]. This is particularly important for time-series problems in which time-indexed data cannot be replicated and each data point represents one observation at a particular instance of time. In contrast, in static prediction, observations can be replicated and more data can be collected from the same generating process in a static timeframe. Moreover, time-series data is highly correlated. With the known effects of bias and variance, efforts to improve model accuracy can be focused on the specified source of error. These efforts may include improving the model algorithms and the data handling methods.

AR models are widely used for time-series forecasting. What is the difference between AR models and neural networks in term of bias and variance? AR models are parametric, in that the form of the model is specified before fitting the data, and consequenty they do not overfit, but can have a large bias due to inadequate modeling capacity. However, they have a good ability to generalize because they do not have too much flexibility. Neural networks, on the other hand, are nonparametric in that the form of the model is defined by the data, and usually involves some trial and error. With their flexibility, they can learn too much by relying on a particular training sample, thereby reducing the bias but rendering the model sensitive to different samples and thereby increasing the variance. Using a fixed dataset, it is not possible to improve both bias and variance, because reducing one leads to an increase in the other. Cross-validation attempts to find a balance or a trade-off between the two components, without regard to the relative importance of each to forecasting accuracy.

9.8.1 Decomposition of Total Error into Bias and Variance Components

Bias-plus-variance decomposition is well studied in relation to classification with neural networks [18–20]. Consider a time-series denoted by:

$$y = f(\mathbf{x}) + \varepsilon \qquad (9.86)$$

where $\mathbf{x}$ is the input vector, consisting of lags of y (i.e., y_1, y_2, etc.), $f(\mathbf{x})$ is the underlying generating process of the time-series, and ε is a random error component. The $f(\mathbf{x})$ is thus the target output. For example, if two lags are used as input to the network, the first input pattern $x_1 = (y_1, y_2)$ and the target $f(\mathbf{x})$ is y_3. The task of the model is to approximate $f(\mathbf{x})$ in such a way that it follows $f(\mathbf{x})$ closely while leaving out the error component. Bias indicates how closely the model follows the actual pattern, so higher the bias, the lower the ability of the model to capture the underlying pattern. Variance is a measure of how closely the model follows the noise. Therefore, high variance means that the model shows high errors on different samples generated by the same process. A good model should perform well on all samples obtained from the same process. To study the bias and variance, therefore, models should be built and tested on more than one sample.

Given a set of time-series data D consisting of input $\mathbf{x}$ and a model $f'(\mathbf{x})$ to approximate $f(\mathbf{x})$, the overall MSE for the model on a separate test set containing N observations is

$$\text{MSE} = \frac{1}{N} \sum_{i=1}^{N} (f'(\mathbf{x}_i) - f(\mathbf{x}_i))^2 \qquad (9.87)$$

where $\mathbf{x}_i$ is the input vector, which may contain one or more lags, $f'(\mathbf{x}_i)$ is the model output for input vector $\mathbf{x}_i$, and $f(\mathbf{x}_i)$ is the actual or target output for input vector $\mathbf{x}_i$. Suppose that more datasets are generated from the same time-series using a different random error generator and that models are developed to approximate $f(\mathbf{x})$ such that the total number of datasets, and thus the total number of models, is M. The overall MSE for all the models calculated on the same test set containing N observations can be expressed as

$$\text{MSE} = \frac{1}{MN} \sum_{j=1}^{M} \sum_{i=1}^{N} (f'(\mathbf{x}_i)_{D_j} - f(\mathbf{x}_i))^2 \qquad (9.88)$$

where, D_i indicates the dataset number or model number. The bias indicates how close a model is to the underlying process that generated the data, and can be expressed as

$$\text{Bias} = \frac{1}{N} \sum_{i=1}^{N} \left[\bar{f}'(\mathbf{x}_i) - f(\mathbf{x}_i) \right]^2$$

$$(9.89)$$

$$\text{Variance} = \frac{1}{MN} \sum_{j=1}^{M} \sum_{i=1}^{N} \left[f'(\mathbf{x}_i)_{D_j} - \bar{f}'(\mathbf{x}_i) \right]^2$$

where $\bar{f}'(\mathbf{x}_i)$ is the average of the predictions from all M models for input vector $\mathbf{x}_i$. The bias measures the average model prediction with respect to the expected value (mean) for $\mathbf{x}_i$. The variance is a measure of the overall variability of the performance of the models with respect to the average model performance for input vector $\mathbf{x}_i$. Thus with M time-series generated from the same underlying process, M models are developed; their mean performance is compared with the target performance in the bias component, and the overall deviation of each model from the mean of all models is captured in the variance component. With these two components, bias and variance can be separated from overall MSE, and the two components can be studied separately. This approach to generating multiple datasets and developing models to assess the overall behavior is called Monte Carlo Simulations, first introduced in this book in Chapter 6 where it is used for generating weight probability distributions and sensitivity distributions.

9.8.2 Example Illustrating Bias–Variance Decomposition

It is of practical benefit to know what the bias and variance components of a model are and how they are affected by model parameters such as inputs and the number of hidden neurons. This section reports an example Monte Carlo simulation study conducted by Berardi and Zhang [17] to highlight these aspects of neural networks in time-series forecasting. They generated data from three different time-series:

(1) *Simple autoregressive (AR1) process with one lag*:

$$y_t = 0.7y_{t-1} + \varepsilon_t. \qquad (9.90)$$

(2) *Autoregressive Moving Average with two input lags and one error lag, ARMA (2, 1) process:*

$$y_t = 0.6y_{t-1} + 0.3y_{t-2} + \varepsilon_t + 0.5\varepsilon_{t-1}. \tag{9.91}$$

(3) *Smooth Transition Autoregressive (STAR) process:*

$$y_t = 0.3y_{t-1} + 0.6y_{t-2} + (0.1 - 0.9y_{t-1} + 0.8y_{t-2}) \\ \times x[1 + \text{Exp}(-10y_{t-1})]^{-1} + \varepsilon_t. \tag{9.92}$$

In the above process ε_t is an independent and i.i.d. normal distribution with a mean of zero and a standard deviation of 0.01 [i.e., $N(0, 0.01^2)$]. Figure 9.65 and Figure 9.66 present two realizations each of the first and third time-series, given by Equation 9.90 and Equation 9.92.

Berardi and Zhang [17] generated 100 independent time-series ($M = 100$) from each process; each series had 200 observations. Furthermore, a smaller series consisting of 25 observations was extracted from the end of each series to investigate the effect of smaller sample size. Feedforward (MLP) neural network models were generated for input lag numbers varying from one to five and a number of hidden neurons ranging from one to five.

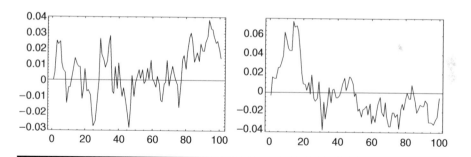

Figure 9.65   Two realizations of an AR process.

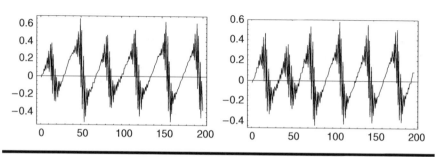

Figure 9.66   Two realizations of a STAR process.

A total of 30 random weight initializations and optimizations were performed for each input lags-hidden neuron parameter combination based on each of the 100 datasets to obtain the best model. The conjugate gradient method, a second order training method, was used for training the networks. From the results, the total error (MSE), bias, and variance were calculated. An Analysis of Variance (ANOVA) test was first conducted to test the significance of the input lags for a fixed number of hidden neurons, and the significance of the hidden neurons for a fixed number of input lags. The results depended on the type of series, but generally indicated that both input lags and number of hidden neurons have a significant effect on the error. As sample size increases, inputs had an even higher significance, as indicated by the F statistic.

Figure 9.67 shows the overall error, bias, and variance components for the AR(1) process using varying input lags. The first figure shows results for an AR1 process modeled with no hidden neurons, using the larger sample (200 data). Note that AR1 is a linear process; the figure therefore depicts a linear model used with various lags. It shows that the total error increases with the number of lags, and that the minimum error is produced by the first lag. This indicates that the model has identified the correct autocorrelation structure for this process. The bias is insensitive to lags, but the variance increases drastically as the number of lags is increased; it is completely responsible for the total error. Over-specification of lags, therefore, affects the bias (i.e., learning underlying relationship) very little, but affects the variance (generalization) greatly. For a smaller sample size, all three error terms are much higher, as shown in Figure 9.67b. The fact that the bias is higher indicates the greater difficulty of estimating the correct model using a smaller sample size.

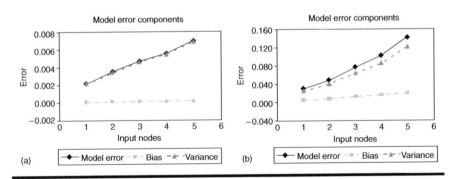

(a)   (b)

Figure 9.67 Absolute model error components for zero hidden neurons for an AR(1) time-series: (a) sample size 200 and (b) sample size 25. (From Berardi, V.L. and Zhang, P.G., *IEEE Transactions on Neural Networks*, 14, 668, 2003. With permission. Copyright from IEEE Press.)

Results for the ARMA process (Equation 9.91), which originally had two input lags and was modeled by a network with one hidden neuron, are shown in Figure 9.68. The left figure, based on the larger sample, shows that the bias is high for all lags and that the variance is negligible compared to the bias. Overall, the error consists almost entirely of bias. This indicates a limited model that has difficulty capturing the underlying relationship, which suggests that, for a complex long-term memory ARMA series, input lags alone may not be sufficient to accurately approximate the underlying generation process. The bias is smallest for the two-input lags, which correctly reflects the autocorrelation structure of inputs. However, the original series included one error lag as an input; its absence in the neural network has led to a high bias. For the smaller sample (Figure 9.68b), it is even harder for the network to model the time-series (note the difference in scales in the two figures), and both the bias and the variance are higher than in the model generated using a larger sample size.

For the third time-series (STAR) (Equation 9.92), results for a network with two hidden neurons are shown in Figure 9.69. The left figure for the larger sample shows that one lag is not able to capture the underlying relationship, producing high bias error. This process requires at least two lags, but over-specification of more lags has little effect on the network's performance indicating lack of model complexity. For a smaller sample size (Figure 9.69b), the bias and variance contribute equally to the error when the model input is under-specified.

The effect of the hidden neurons, based on the best input lags found in the previous analysis, is shown in Figure. 9.70 for the AR1 process.

Figure 9.70 shows that the model has learned the underlying linear relationship without the use of hidden neurons, and that over-specification of hidden neurons does not affect the error components. For the larger sample size, the total error consists entirely of variance. Thus, although the

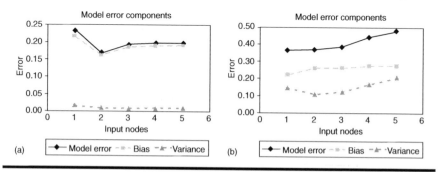

Figure 9.68 Absolute model error components for one hidden neuron for an ARMA(2,1) time-series: (a) sample size 200 and (b) sample size 25. (From Berardi, V.L. and Zhang, P.G., *IEEE Transactions on Neural Networks*, 14, 668, 2003. With permission. Copyright from IEEE Press.)

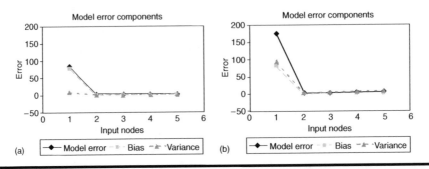

Figure 9.69 Absolute model error components for two hidden neurons for a STAR time-series: (a) sample size 200 and (b) sample size 25. (From Berardi, V.L. and Zhang, P.G., *IEEE Transactions on Neural Networks*, 14, 668, 2003. With permission. Copyright from IEEE Press.)

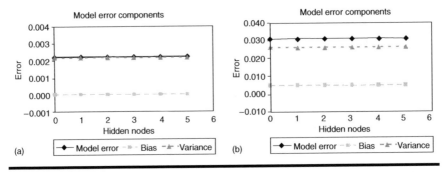

Figure 9.70 Effect of hidden neurons on the error terms for an AR(1) time-series with one input lag: (a) sample size 200, and (b) sample size 25. (From Berardi, V.L. and Zhang, P.G., *IEEE Transactions on Neural Networks*, 14, 668, 2003. With permission. Copyright from IEEE Press.)

correct model has been identified, the generalization error (variance across replications) is high, indicating that, for this model, effort should be focused on reducing the variance. For the smaller sample size, the bias is higher than that produced by the larger sample and it is also insensitive to over-specification of hidden neurons. This indicates that the best model cannot be found from the smaller sample size. For this sample, variance accounts for about 80 percent of the total error, the rest being that due to bias.

Figure 9.71 illustrates how a network with two input lags, to represent the ARMA (2,1) process, is affected by the number of neurons used. This original series was generated with two input lags and one error lag, and is a linear model. Earlier, it was shown that with one hidden neuron, the bias is the smallest when two input lags are used, but that it is still very high.

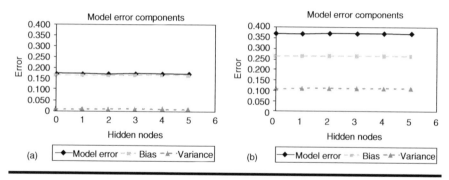

Figure 9.71 Effect of hidden neurons on the error terms for an ARMA(2,1) time-series with two input lags: (a) sample size 200, and (b) sample size 25. (From Berardi, V.L. and Zhang, P.G., *IEEE Transactions on Neural Networks*, 14, 668, 2003. With permission. Copyright from IEEE Press.)

The errors are insensitive to the number of neurons, regardless of the sample size. For the larger sample (Figure 9.71a), the variance is almost zero and the total error consists entirely of bias, whereas for the smaller sample, depicted in Figure 9.71b, the variance is larger and the bias comprises about 70 percent of the total error. This shows that the larger sample size has helped to reduce the variance. The bias dominates in both samples. Figure 9.71a shows that the model has identified the linear characteristics of the process because the result for zero neurons is the same as that for other neurons. However, without all the relevant inputs, the bias remains high, indicating again that input lags alone are not sufficient to model this process.

Figure 9.72 shows the effect of hidden neurons on the error for the STAR process, for both larger and smaller sample sizes. Both samples show that a

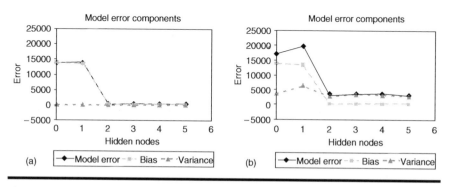

Figure 9.72 Effect of hidden neurons on the error terms for a STAR time-series with two input nlags: (a) sample size 200, and (b) sample size 25. (From Berardi, V.L. and Zhang, P.G., *IEEE Transactions on Neural Networks*, 14, 668, 2003. With permission. Copyright from IEEE Press.)

minimum of two neurons is needed, which indicates that the model has identified the nonlinear process that was used to generate the data. Over-specification of neurons does not the effect the errors, but the errors vary for between two sample sizes. For the larger sample, variance is zero throughout, meaning that the results across the various samples do not vary, regardless of the number of neurons. However, owing to a lack of flexibility in models with zero and one hidden neurons, the bias is higher for these models but drops to zero for models including two hidden neurons. Beyond this number of neurons, both the bias and the variance are zero, indicating perfect model predictions. For the smaller sample, both the bias and variance are initially higher. The optimum number of neurons (two) is found when zero bias error is reached, but the variance remains high throughout. For this model, attention must be focused on reducing the variance, for example, by increasing the sample size.

This case study shows that both inputs and hidden neurons can significantly affect the learning of a neural network when forecasting a time-series. However, the ways in which these two parameters affect model bias and variance is quite different. Over-specifying input lags does not affect the bias, but it can impact variance. Under-specification of either the input lags or the hidden neurons can have a severe negative effect on model bias, leading to poor forecasting performance. The larger samples have the advantage of reducing the total error and the effects of overfitting.

The overall error, as well as its bias and variance components, depends on the interaction between the sample size and the complexity of the time-series. For simple models, such as the AR(1) process, variance dominates the total error, which is greatly influenced by the sample size, with smaller samples producing a much higher error. For complex processes such as STAR time-series, the overall error is less affected by the sample size. In general, for smaller samples, variance becomes the dominant concern. For larger samples, however, variance is less of a concern, and model bias should be the major focus.

9.9 Long-Term Forecasting

There are many situations that require a longer-term forecast than simply one or two steps. Economic and weather forecasts, as well as forecasts for the demand on resources such as water, energy, and so on, usually have a long-term horizon. A common problem with time-series forecasting models is that when they are used recursively for making multistep-ahead predictions, their long-term forecasts have a low level of accuracy. Estimates may be reliable for the short term, but are not accurate enough for the long term. There may be several reasons for this low accuracy. The environment in which the model was developed may have changed over time, rendering current inputs invalid. Another possible explanation is that the model itself may be inadequate, due to insufficient structure or training of the model, or to a lack of appropriate data. Low accuracy may also be due to the propagation of errors that occurs during recursive model predictions. If a network designed for one-step-ahead forecasting is used recursively n times for n-steps ahead forecasting, the error in each step can accumulate beyond acceptable levels. This situation can be improved by training networks specifically for the purpose of multistep-ahead or long-term forecasting.

In multistep forecasting, a model must recursively make forecasts for the n required time steps of $t+1$, $t+2$, ..., $t+n$, using only the inputs at time t. The network relies on its own prediction in one step to recursively predict the outcome for the next step. In essence, the network learns to look many steps ahead into the future based solely on current inputs and its own predictions at each step. During training, the actual output for each step is available from a historical dataset. However, the real problem with such forecasting is that, during use, predictions must be made for many steps without being able to measure the accuracy of the prediction at each step by comparison with the actual output. This makes accurate, long-term forecasting a difficult problem.

As previously discussed in this chapter, the static feedforward and recurrent networks can be used for multistep-ahead forecasting (please refer to Section 9.3.2 and Section 9.7). In feedforward networks in which several lags of the forecast variable are used as inputs, the network output is used recursively as a delayed input to obtain multistep forecasts. After a while, the network effectively sees only its own outputs as inputs. If there are other external variables, they are fixed for all time steps, unless there are other ways to obtain their forecasts. The network is trained in the usual manner, using the actual outputs to calculate the error at each step.

In recurrent networks for multistep forecasting, only the current input is used. Jordan, Elman, and fully recurrent networks (Section 9.7.1 through Section 9.7.3) can all be used for this purpose. The network state (hidden neuron output), the output neuron output, or both are repeatedly fed back as delayed input to make the next prediction. These networks also rely completely on their own output to make long-term forecasts. For both static and recurrent networks, the prediction error at each step contributes to the error in the next step, and as the forecasting horizon becomes longer, this compounding effect can severely degrade the forecast.

Nguyen and Chan [21] propose a Multiple Neural Network (MNN) model to address the problem associated with compounding errors in long-term forecasting. This model combines neural networks for short-term and long-term forecasting to accommodate a wide range prediction time periods and to reduce the error by decreasing the number of recursions necessary to make a long-term forecast. The approach basically involves combining several neural networks built to make short- to long term predictions into a single model. As reported by Nguyen and Chan [21], this model is demonstrated using the prediction of hourly flow rates at a gas station.

In MNN, each neural network has a different forecast horizon, meaning that it is developed to make predictions for a different time period ahead. The prediction time is of order two. For example, a zero-order network is used for one-step-ahead (2^0) prediction. A first-order network predicts two steps (2^1) ahead, and a second-order network is used for four steps (2^2) ahead prediction, and so on. A network that predicts 2^n times ahead is referred to as an n-ordered network. When designing an MNN, the range of the prediction horizon is first determined. Then, an adequate number of networks of different orders are assembled so that together they provide the required prediction horizon. For example, to make a three-step-ahead prediction, two sub-networks of orders zero and one can be used, as shown in Figure 9.73. Here the first zero-order network predicts one step ahead using the actual inputs. Its prediction is fed into a first-order network, along with the other required inputs. This second network makes a two-time-step-ahead prediction, but the first network has already made the one-step-ahead prediction, and therefore, the final output is for three steps ahead.

The two sub-models in Figure 9.73 are connected to and dependent on each other, and their weights must therefore be estimated at the same time. For networks with a higher lead time, the combinations of networks can be much more complex, with several layers and connections among sub-ANNs. This structure can require a long training time. Therefore, MNN breaks the training down to sub-ANNs, which are trained separately. To make a

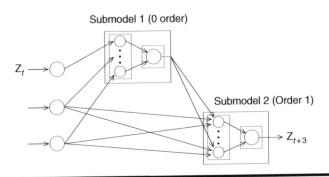

Submodel 1 (0 order)

Z_t

Submodel 2 (Order 1)

Z_{t+3}

Figure 9.73 A simple MNN model consisting of two sub-models for three-step-ahead forecasting. (From Nguyen, H.H. and Chan, C.W., *Neural Computing and Applications*, 13, 90, 2004. With permission from Springer Science and Business Media.)

prediction, the highest-order network is used first. For example, to predict seven steps ahead, the network of order two (four-step-ahead) is used first. Then lower order networks are used appropriately. For example, assuming that for the current time t, x_t, and x_{t-1} are known, and every network has two inputs. Then the seven-step-ahead prediction can be made with a network of order two using inputs x_{t+3} and x_{t+2} as

$$x_{t+7} = f_2(x_{t+3}, x_{t+2}) \qquad (9.93)$$

where f_2 indicates a model of order two. Because a two-order model predicts four steps ahead, it will advance the prediction for x_{t+3} to x_{t+7}. However, because the two required inputs, x_{t+3} and x_{t+2} are not yet known for this model, it is necessary to create sub-networks to predict them. To predict x_{t+3}, the first order (f_1) model (two-steps ahead), which uses x_{t+1} and x_t, can be used. However, because x_{t+1} is still unknown, another network of order zero (f_0) must be created using x_t and x_{t-1} as inputs. Similarly, a first order model is needed to predict x_{t+2} from x_t and x_{t-1}. This process of building an MNN with sub-nets can be presented as follows:

$$x_{t+7} = f_2(x_{t+3}, x_{t+2}) = f_2(f_1(x_{t+1}, x_t), f_1(x_t, x_{t-1}))$$

$$= f_2(f_1(f_0(x_t, x_{t-1}), x_t), f_1(x_t, x_{t-1})). \qquad (9.94)$$

There are four sub-networks in the above MNN: One of order two, two of order one, and one of order zero. The higher-order networks use the outputs of lower-order networks. The sub-networks can be trained separately or together with other networks, depending on the complexity of the entire set. Multistep validation can be used in the model development. For example, a simple measure for a network of order n is the RMSE of the 2^n to $(2^{n+1} - 1)$ steps-ahead predictions. To calculate these predictions, higher-order networks have to use predictions from lower ordered networks.

Nguyen and Chan [21] describe a system development environment for building, training, validation and testing of MNN. Here, feedforward, MLP networks trained with backpropagation are used as sub-networks. For each network, parameters can be specified and the networks can be trained separately. The higher-order networks receive outputs from lower-ordered networks as required for training and validation.

9.9.1 Case Study: Long-Term Forecasting with Multiple Neural Networks (MNNs)

This section will present a case study to illustrate the MNN and to compare its long-term forecasts with those of a single network. This case study involves forecasting the hourly flow rate of gas through a compressor station, which is part of a gas pipeline distributions system. The objective is to obtain accurate predictions for customer demands to help dispatchers satisfy customer demands with minimal operating costs. Forecasts should help operators make decisions to turn compressors on or off appropriately and to maintain the necessary pressure while minimizing waste. The study was reported by Nguyen and Chan [21]. The hourly flow rates were collected from December 2001 and August 2002. The variation of the hourly flow rate on a typical day is given in Figure 9.74. The dataset was divided into training, validation, and testing sets in the ratio of 5:1:1. The training set contained 2500 observations and the validation and test sets comprised 500 observations each.

The input vector consists of 6 hours of data (i.e., for a quarter day). Predictions are made for 24 hours ahead. For a 24-step-ahead prediction, it is possible to design various combinations of sub-networks ranging from two to four. These possible networks were trained and the best combination, based on test data, was used as the final MNN. During training, the weights of the first network were set randomly and the weights of the subsequent nets were initialized with the previous network's weights to reduce training time. Five combinations of networks, 1, 2, 3, 4, and 5, were trained and validated using the validation data. Network one is the single network for 24-step-ahead forecasting. Out of the last four MNNs, the one with the lowest validation error was selected as the final MNN; it happened to be the one with five networks, producing 8.76 percent mean absolute percentage error (MAPE). The single network MAPE was 11.7 percent, much higher than that of the best MNN model. To compare the single network with the MNN models, the 24-hour forecasts from the two models and the corresponding actual outputs for the independent test set are presented in Figure 9.75; this figure clearly illustrates the superiority of the MNN model, showing that the MNN consistently outperforms the single network. The MNN forecasts

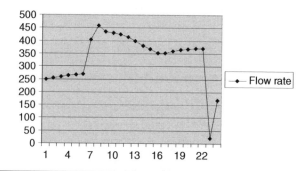

Figure 9.74 Hourly gas flow rate variation in a day. (From Nguyen, H.H. and Chan, C.W., *Neural Computing and Applications*, 13, 90, 2004. With permission from Springer Science and Business Media.)

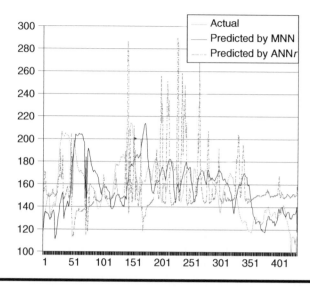

Figure 9.75 Twenty-four-hours ahead forecasts from single and multiple networks in comparison to actual data. (From Nguyen, H.H. and Chan, C.W., *Neural Computing and Applications*, 13, 90, 2004. With permission from Springer Science and Business Media.)

seem like a delayed version of the actual outputs, but the range of values are close to the actual data. The single network forecasts are rather large random. This indicates that the MNN is superior for long-term forecasting.

As calculated by summing the 24 predictions and averaging, the average MAPE for the 24 predictions of the single network and the MNN, were 12.38 percent and 8.73 percent, respectively, which indicates that the MNN performs better than the single network. Neither model, however, is completely satisfactory, which could be due to an insufficient amount of data to allow for the incorporation of seasonal variations and other effects into the prediction. Because the available data was limited, predictions for 6 h ahead were also obtained for the two models, and the average MAPE for the single network and the MNN on the test data was reduced to 5.75 percent and 4.97 percent, respectively. Although both model predictions are close, the comparative plots of the outputs revealed that, overall, MNN results are still better.

9.10   Input Selection for Time-Series Forecasting

Selecting the most appropriate inputs to a model is an important first step in model building; it is especially important for neural networks that are powerful, nonlinear processors. Input selection for static networks is dealt with in detail in Chapter 5. For time-series, inputs also include lags. As the input dimensionality increases, model complexity increases and learning becomes more difficult, leading to poor convergence. With a smaller number of relevant inputs, a network can focus on developing the required relationships with more efficiency. The challenge is to select from all the potential inputs a subset of inputs that will lead to a superior model. For d potential inputs, there are $(2^d - 1)$ input subsets. For a larger number of inputs, it is impossible to test all of the possible combinations. Therefore, when the potential inputs are large, efficient methods are required to select the inputs. As stated earlier, the problem is exacerbated when time lags are used in time-series applications. If there are several input time-series, then it is necessary to find the appropriate lags that are significant to the output for each time-series.

Cross-correlation is the most popular analytical technique for selecting inputs and the number of lags (memory length). The major disadvantage of cross-correlation is that it is only able to detect linear dependence between two variables. Because cross-correlation is unable to capture nonlinear dependence between variables, it can lead to omission of important inputs that are nonlinearly related to the output.

Another approach is the stepwise selection of inputs, designed to avoid the necessity of working with all inputs at the same time. Two standard stepwise approaches are forward selection and backward elimination. In forward selection, the best single input is found and selected for the model. In each of the subsequent steps, the input that most improves the model performance is added to the model. In backward elimination, the model starts with all the inputs and sequentially deletes inputs that reduce the model performance the least [31].

Another method that is used is based on the sensitivity analysis of trained neural networks. Note that Chapter 6 and Chapter 7 treat the topic of sensitivity analysis of networks in detail. Plots of sensitivity for each input are inspected, and significant inputs are chosen using judgment based on the high sensitivity of an input to the output. The practical difficulty here is choosing a reasonable value by which to perturb the inputs, and selecting the appropriate cut-off point for input significance [22].

Yet another method for finding the significance of inputs from the trained network is the saliency analysis. In this method, variables are removed from the network (i.e., made equal to zero) one at a time, and the forecast error is computed. By repeating the process for each input variable, it is possible to determine the relative importance of each variable. The disadvantage of this approach is that the network is not retrained after the removal of each input. This can lead to erroneous results if zero is not a reasonable value for the input. The results can be particularly questionable if the inputs are statistically dependent, because in general, the effects of different inputs cannot be separated [23].

In time-series analysis, Box-Cox transformation is another widely used method for selecting the number of lagged dependent and independent variables. In this method, a transfer function, as described by Box and Jenkins [32], is fitted to the data to determine whether a Box-Cox transformation of the data is necessary and to determine the important inputs and lags [22]. However, as with cross-correlation, this method has a major limitation in solving nonlinear problems because it captures only the inputs that are linearly related to the data.

In general, input determination can be divided into two stages. The objective of the first stage is to reduce the dimensionality of the original inputs to obtain an independent set of inputs. These inputs may not necessarily be related to the output. At this stage of input preprocessing, model-free approaches i.e., methods that do not depend on pre-existing models, are commonly used. During the second stage, this subset is used to determine which inputs are related to the output. For this purpose, model-free or model-based approaches can be used. In model-based approaches, the significance of various subsets of inputs are tested on an appropriate model to find the set that produces the least error. For a large number of inputs, the process can be time consuming. In model-free approaches, some statistical measure of independence, such as correlation, is used. If there is significant dependence between the output and an input, the input is retained in the final subset; otherwise, it is discarded. It is very useful to have a method that accounts for both linear and nonlinear dependencies.

9.10.1 Input Selection from Nonlinearly Dependent Variables

9.10.1.1 Partial Mutual Information Method

The stepwise partial mutual information (PMI) method, which is an extension of the concept of mutual information, can account for both the linear and nonlinear dependence between variables. This approach does not require that the inputs be preprocessed to ensure independence. The mutual information (MI) is a measure of the dependence between two random variables [24]. For a bivariate sample, the MI score can be expressed in discrete form as

$$MI = \frac{1}{N} \sum_{i=1}^{N} \ln \left[\frac{P_{X,Y}(x_i, y_i)}{P_X(x_i)P_Y(y_i)} \right] \tag{9.95}$$

where x_i and y_i are the ith bivariate sample pair, N is the sample size, $P_{X,Y}(x_i, y_i)$ is the joint probability density at the sample point, and $P_X(x_i)$ and $P_Y(y_i)$ are the univariate marginal (simple) probability densities at the sample point. The idea is that if two variables are independent, their joint probability is the product of the marginal probabilities, making the expression within brackets equal to one. Because ln(1) is zero, for independent variables, MI is equal to zero. However, to use this method, the probability densities must be estimated from the data. A simple measure would be to approximate the densities using histograms. However, a Gaussian kernal density estimate can be more stable and efficient [22].

The MI criterion can be used to identify significant input variables; however, it cannot identify redundant variables. As Bowden et al. [22] point out, if x is an important variable and has a high MI score, $z = 2x$ will also have a high score. Therefore, z will also be selected, although it is a redundant variable, because it is already described by the existing variable x. The partial mutual information (PMI) was introduced to avoid this problem [25]. The PMI is a measure of the partial or additional significance added to an existing model by a new variable. The PMI between a dependent variable y and an independent variable x when a set of inputs z has already been selected is given in discrete form as

$$\text{PMI} = \frac{1}{N} \sum_{i=1}^{N} \ln \left[\frac{P_{X',Y'}(x_i', y_i')}{P_{X'}(x_i') P_Y(y_i')} \right] \tag{9.96}$$

and

$$x' = x - E[x|z]; \quad y' = y - E[y|z]$$

where x_i' and y_i' are the ith residuals in a sample of size N after z inputs have already been used in regression; $P_{X'}(x_i')$, $P_{Y'}(y_i')$ and $P_{X',Y'}(x_i', y_i')$ are the respective marginal and joint probability densities. The $E(y|z)$ denotes the expected value or mean, which is a conditional estimate of y for the pre-existing input set z (i.e., regression of y on z); this estimate is used to obtain the residual y' that remains to be modeled. A method to obtain the conditional estimate will be discussed shortly. Similarly, $E(x|z)$ denotes the expectation or mean of x for the pre-existing input set z. This is the regression of x on, z and accounts for the dependence between x and z. This estimate indicates the overlap or correlation between x and z, and can be used to find the residual x', which is tested for its capacity to significantly affect the residual y'. In this way, correlated variables can be used and their correlation systematically removed. Thus x' and y' contain only residual information after the effect of the pre-existing variable set z has been taken into consideration.

To estimate PMI, it is necessary to estimate the marginal and joint probabilities and the conditional values. Bowden [22] shows that for probability estimates, the city block distance kernel shown in Equation 9.97 is a good alternative to the Gaussian probability density function due to its similarity to the Gaussian function, but with the added advantage of computational efficiency. The function in Equation 9.97 is basically the multivariate density function, which is the product (denoted by Π) of the univariate functions:

$$\hat{f}_X(x) = \frac{1}{N(2\lambda)^d} \sum_{i=1}^{N} \prod_{j=1}^{d} \exp^{-|x_j - x_{ij}|/\lambda}$$

$$= \frac{1}{N(2\lambda)^d} \sum_{i=1}^{N} \exp\left[-\frac{1}{\lambda}\sum_{j=1}^{d}|x_j - x_{ij}|\right] \tag{9.97}$$

where $\hat{f}_X(x)$ is the multivariate kernel density estimate of a d-dimensional variable set X at coordinate location x. The x_{ij} is the jth component of the ith multivariate input pattern in a sample of size N. The expression in the square brackets in Equation 9.97 calculates the distance of an arbitrary vector x to the ith input vector in the database in terms of city block distance (see Chapter 8, Section 8.4.2.1). City block distance computes the sum of the absolute value of the difference between vector components. The larger the absolute sum, the larger the distance. The probability estimate given by Equation 9.97 is based on this distance of a vector x to all input vectors in the sample space. For an input vector located at the center of gravity of the sample, the distance will be approximately zero and will have the highest probability. In the univariate case, $d = 1$ and the density estimate becomes

$$\hat{f}_X(x) = \frac{1}{2N\lambda} \sum_{i=1}^{N} \exp^{-|x - x_i|/\lambda} \tag{9.98}$$

which closely resembles the univariate Gaussian density distribution. The parameter λ is a smoothing parameter, known as the bandwidth of the kernel density estimate. An accurate density estimate requires careful attention be given to λ. Small values of λ tend to emphasize details of individual data points whereas large values of λ have the tendency to gloss over the details resulting in over-smoothing of the distribution. The Gaussian reference bandwidth, for example, is

$$\lambda = \left(\frac{4}{d+2}\right)^{1/(d+4)} N^{(-1/(d+4))} \tag{9.99}$$

which can also be used for the city block estimate, due to its closeness to the Gaussian estimate. The parameter d is the dimensionality of the variable set. The above procedure can be used to obtain the required probabilities. The last required quantity is the conditional values $E(y|z)$ and $E(x|z)$. The simplest method for obtaining these values is to estimate them using a linear regression; however, there is a problem, as this model is not dealing with nonlinearities. Bowden et al. [22] proposes the use of the generalized regression neural network (GRNN) to estimate this quantity, as explained in the next section.

9.10.1.2 Generalized Regression Neural Network

The GRNN is a supervised feedforward neural network with the capability to universally approximate smooth functions. Developed by Specht [26], it uses a nonparametric estimate of the probability density function of the observed data. The advantage of the model is that it models nonlinear relationships, and more importantly, its structure is fixed and training is therefore more rapid.

The GRNN network can be used to obtain the conditional mean of y given $\mathbf{x}$ $E(y|\mathbf{x})$ (nonlinear regression of dependent variable y on the set of independent variables $\mathbf{x}$). Suppose that an input vector $\mathbf{x}$ of n independent random variables is used to predict y, the dependent scalar variable. If $\mathbf{x}$ is a particular value of the input vector, and if the joint density $f(\mathbf{x},y)$ is known, then the conditional mean of y given $\mathbf{x}$ is

$$E\left[y|\mathbf{x}\right] = \frac{\int\limits_{-\infty}^{\infty} y.f(\mathbf{x},y)\mathrm{d}y}{\int\limits_{-\infty}^{\infty} f(\mathbf{x},y)\mathrm{d}y} \tag{9.100}$$

If the joint density is not known, then the conditional mean can be estimated based on sample observations. The GRNN uses a class of nonparametric estimators known as Parzen window estimators, which Specht [26] used to obtain a sample estimate $\hat{y}(\mathbf{x})$ of the population conditional mean $E(y|x)$ for a particular measured values of $\mathbf{x}$ as follows:

$$\hat{y}(\mathbf{x}) = \frac{\sum\limits_{i=1}^{N} y_i . \mathrm{Exp}\left(\frac{-D_i^2}{2\sigma^2}\right)}{\sum\limits_{i=1}^{N} \mathrm{Exp}\left(\frac{-D_i^2}{2\sigma^2}\right)} \tag{9.101}$$

where $D_i^2 = (\mathbf{x}-\mathbf{x}^i)^T(\mathbf{x}-\mathbf{x}^i)$, and T is transpose. The D_i^2 is a measure of the Euclidean distance between $\mathbf{x}$ and $\mathbf{x}_i$, the ith input pattern. In Equation 9.101, the sum of an exponential form of such distance from $\mathbf{x}$ to all other inputs $\mathbf{x}_i$ is used in the denominator. The numerator multiplies the exponential distance by the corresponding observed or target values y_i, indicating a moment. The resulting fraction is similar to the center of gravity in that the predicted $\hat{y}$ balances the sum total distance between $\mathbf{x}$ and $\mathbf{x}_i$ and is therefore the conditional mean of y for $\mathbf{x}$ [i.e., $E(y|\mathbf{x})$]. The parameter σ is the smoothing parameter and N is the sample size.

Equation 9.101 can be easily implemented by a GRNN that has four layers, as shown in Figure 9.76. In this figure, the first layer is the input layer and the second (pattern) layer stores the input vectors. When an input is presented, each neuron in this layer calculates the distance between this input pattern and the pattern it stores. Third is the summation layer, in which the first neuron computes the numerator of Equation 9.101, and the second neuron

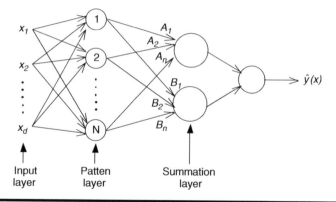

Figure 9.76 Generalized regression neural network (GRNN).

computes the denominator. Thus the weights of the first neuron of the summation layer are set equal to the outputs i.e., $\mathbf{A}_i = y_i$, $i = 1,\dots,N$ and those of the second neuron are set to one i.e., $\mathbf{B}_i = 1$, $i = 1,\dots,N$. The output neuron computes the $\hat{y}(\mathbf{x})$ by finally implementing Equation 9.101. The optimum value of the parameter σ must be found through trial and error, but as stated by Bowden et al. [22], the curve of the RMSE versus σ is typically very flat near the minimum and it is therefore not difficult to find a suitable value for σ.

Once the conditional expectations are thus estimated from the GRNN, the residual values y' and x' can be calculated. The PMI score is then computed from Equation 9.96, using the probabilities computed from the assumed city block distance kernal representing the marginal and joint probability densities. The variable with the highest PMI score is selected and a 95th percentile randomized sample PMI score for the identified input is estimated. This measure of significance is used to test whether the selected input is a significant predictor of the output being modeled. This is accomplished by using, for example, 100 randomly selected sequences of the input variable. These can be created by rearranging or bootstrapping the input variable. In this way, the output is made independent of the input due to randomization, and a randomized sample of PMI scores can be obtained. The input is selected as a significant predictor if the sample PMI score is greater than the 95th percentile of the randomized score. Basically, this measure tests the sample PMI score with a score that is obtained if the input is not related to the output.

9.10.1.3 Self-Organizing Maps for Input Selection

The self-organizing feature map (SOM) is a powerful multivariate data clustering method, and is treated in detail in Chapter 8. In SOMs, multidimensional data are projected onto a one- or two-dimensional map in such a way that not only are the similar input vectors clustered, but the inputs that are closer in the input space are closer in the map space as well. The latter quality is called topology preservation. The SOM usually consists of an input layer connected to an output layer of neurons that make up the map, as shown for the case of a two-dimensional map in Figure 9.77. The set of connection weights from inputs to each of the neurons are represented by a weight or codebook vector with the dimension equal to the input dimension. During training, inputs are presented to the network and the neuron with the weight vector closest to the input vector is declared the winner. A common distance measure is the Euclidean distance. Once the winner is found, the weights of the winner and usually those of a neighborhood of neurons around the winner are moved closer to the input vector. The process is repeated with all the inputs until the map converges and the weights no longer change appreciably.

In the trained map, each neuron represents a cluster of similar inputs. By selecting one input variable from each cluster, it is possible to remove highly correlated inputs from the original inputs [22]. The selected input dimension is the one that is closest to the corresponding component of the weight representing the cluster center. To ensure that an appropriate number of inputs are selected, the map size should be kept large enough to form a suitable number of clusters. The selected set of inputs is still not optimized, but it is a subset of the larger original input set. From this preliminary set, the best set can be found using a method such as the GRNN, which is a nonlinear function approximator whose structure is fixed and training is therefore fast, as has been previously discussed. For a smaller set of inputs, GRNN can be directly applied. However, when the selected set of inputs is large, it is necessary to use more efficient methods to arrive at an optimum set of inputs faster. The use of genetic algorithms (GAs) is one such approach to speed up the selection of inputs from the GRNN.

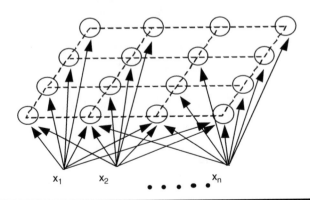

x_1 x_2 • • • • • x_n

Figure 9.77   Two-dimensional, self-organizing map.

9.10.1.4   Genetic Algorithms for Input Selection

The concept of GAs is simple. It is a powerful optimization technique based on the principles of natural evolution and selection [28]. In the specific case of selecting the optimum set of inputs from a larger set, GAs can be used to search through a large number of input combinations with interdependent variables. The power of GAs lies in the fact that they simultaneously test a population of solutions i.e., a large number of possible sets of inputs, and by using genetic operators they eventually find an optimum set of inputs that best satisfies a relevant fitness or selection criterion. This can be done by applying each candidate solution as inputs to a GRNN and computing the RMSE between the model output and the target output. The RMSE is the selection criterion. The set that produces the minimum RMSE is selected as the optimum. The question is how to select the candidate sets of inputs appropriately while avoiding an exhaustive search of all possible combinations. The GAs are designed for this purpose. They use three genetic operators: Selection, crossover, and mutation; the generic steps of the operation of GAs are illustrated in Figure 9.78, along with the equivalent specific steps taken when applied to input selection, as shown within brackets in each block.

The following discussion walks through the steps in Figure 9.78, highlighting the operations done in each step to find a solution that is the optimum set of inputs. Before starting, possible solutions are coded into binary strings, which are equivalent to biological chromosomes. Basically, a string is a set of bits and the length of the strings in the input selection problem is equal to the total number of inputs from which a subset is sought. Each bit in the string represents one input variable, and is biologically equivalent to a gene. In general, a bit can have a value of either 0 or 1, where 0 indicates the absence and one indicates the presence of an input. This way, each possible input combination can be represented by a unique string. For example, if there are three inputs, the string {1, 0, 0} represents the case in which only input 1 is in the selected set, and the string {0, 1, 1} represents the case in which both inputs two and three are in the set. In the case of a large number of inputs, the number of strings or chromosomes i.e., possible solutions, is large. In GAs, search is initialized with a random population of chromosomes (i.e., a random set of input combinations). Then the inputs represented by each of these solutions are tested on the GRNN and the RMSE (fitness) of each chromosome is found. From these, the best chromosomes are selected for mating in a process called crossover.

While various selection methods exist, the most commonly used is tournament selection [27,28]. In this method, chromosomes are randomly paired, and the one with the higher fitness (lower RMSE) in the pair moves to the next generation. Because only half of the chromosomes are selected this way, another tournament is held using all the original chromosomes, but this

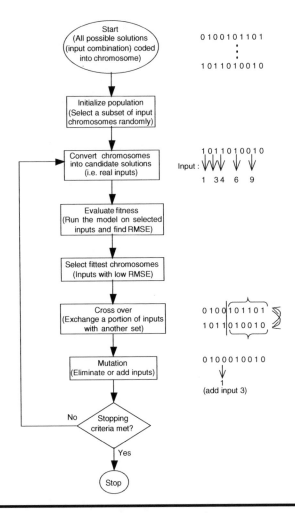

Figure 9.78 Steps in a genetic algorithm highlighting the equivalent tasks involved in the selection of an optimum set of inputs from a large number of inputs.

time the tournament is between a different set of random chromosome pairs. The winners from this round make up the other half of the crossover pool. In this process, two copies of the best chromosomes and no copies of the worst ones are replicated in the pool. These are called parent chromosomes.

After chromosomes are selected, a crossover or mating process takes place, in which two randomly paired chromosomes in the pool partially exchange genetic information. If the crossover probability is less than the predetermined crossover probability parameter, an exchange does not take place, and the two unchanged parent chromosomes enter the new population of candidate solutions. If the crossover probability is greater than a predetermined crossover probability parameter, an exchange will take place between the two chromosomes. If an exchange is to take place, the crossover site where the chromosome will be split is selected at random; genes after the crossover site are then exchanged between the two chromosomes. Specifically, after the split, the first portion of one chromosome is combined with second portion of the second chromosome and vice versa to create two new child chromosomes of the same length as the parent chromosomes; these recombined chromosomes then enter the new population of candidate solutions. There are several variants of crossover that can be used.

The final genetic operator is mutation, in which randomly selected bits change their state from 0 to 1 or vice versa. This basically changes the structure of the candidate solutions. The mutation is designed to allow solutions that may have been discarded prematurely to have an opportunity to enter the process, as well as to keep the population diverse to prevent premature convergence to a local minimum solution. The common approach to mutation is to use a mutation probability parameter to determine the probability that each bit in the chromosome will be mutated. The selected bits are mutated by flipping their value. This ends one genetic cycle in the generation of a population, and the candidate chromosomes are again tested for fitness. In the input selection case, these are used as inputs to the GRNN, and the RMSE is obtained. The chromosomes with a low RMSE have high fitness. This process is repeated over numerous cycles until some termination criterion such as the number of generations is reached or the process converges to a final solution. The final solution in the input selection problem is the best set of inputs that are significant for predicting the output.

9.10.2 Practical Application of Input Selection Methods for Time-Series Forecasting

This section illustrates the methods discussed in the previous section for input selection using an example study conducted by Bowden et al. [22]. The partial mutual information (PMI) and self organization map (SOM) followed by genetic algorithm (GA) based generalized regression neural network (GAGRNN) methods were tested on four datasets with known dependence structure. The data was generated from models; and the first three sets are from time-series models used by past researchers [25]. and the fourth was generated by a nonlinear model. These models are

1. *AR1*:

$$x_t = 0.9x_{t-1} + 0.866\,e_t \tag{9.102}$$

where e_t is a Gaussian random deviate with zero mean and unit standard deviation.

2. *AR9*:

$$x_t = 0.3x_{t-1} - 0.6x_{t-4} - 0.5x_{t-9} + e_t. \qquad (9.103)$$

3. *TAR2—Threshold autoregressive of order two*:

$$x_t = \begin{cases} -0.5x_{t-6} - 0.5x_{t-10} + 0.1e_t, & \text{if } x_{t-6} \leq 0 \\ 0.8x_{t-10} + 0.1e_t, & \text{if } x_{t-6} > 0 \end{cases}. \qquad (9.104)$$

4. *Nonlinear system*:

$$y = (x_2)^3 + \cos(x_6) + 0.3\sin(x_9). \qquad (9.105)$$

For the first three models, 500 data points were generated from each time-series. The first 15 lags were arbitrarily chosen as potential model inputs. From these, the final subset of inputs was selected using the two methods: PMI and SOM-GAGRNN. If the input selection methods are robust, they should select the appropriate lags, as shown in the models. For the nonlinear system, which is static, 15 standard Gaussian random variables were generated with 500 data points for each. These 15 variables were used as the potential model inputs and the final subset was determined from the two input selection methods. In this case, only the random variables two, six, and nine are used to generate the actual output. The selection methods should find these three inputs.

The SOM was trained using standardized data, and after training, the input dimension closest to the cluster center was selected from each neuron on the map. This helps to exclude highly correlated inputs. The GAGRNN was implemented on a commercially available neuro-genetic optimizer. The population size was 50, and 100 generations were run with a fitness criterion of RMSE.

The PMI algorithm has been specially developed to implement Equation 9.96 for this study. The PMI results for the four models are presented in Table 9.6, along with the variables selected based on their high PMI scores. The variables at the cut-off points are denoted by the un-highlighted PMI scores that are less than the 95th percentile randomized sample PMI. Variables above this have PMI scores greater than the corresponding 95th percentile value, and are selected as significant. For all models, PMI has selected the appropriate inputs. For example, the first model has x_{t-1} only and the second has x_{t-1}, x_{t-4}, and x_{t-9}. These have been selected by the PMI method. Similarly, for the third and fourth models, the correct inputs have been selected.

Table 9.6 also reveals that the PMI method can indicate the importance of the variables. For example, the coefficient associated with an input to a model indicates its significance for predicting the output. For instance, in the second model, the coefficient for the second input in Equation 9.103 (x_{t-4})

Table 9.6 Input Subsets Selected from SOM-GAGRNN Method for the Four Models

Model	Inputs	PMI	95th Percentile Randomized Sample PMI
AR1	x_{t-1}	0.604	0.080
	x_{t-9}	**0.029**	**0.033**
AR9	x_{t-4}	0.354	0.080
	x_{t-9}	0.204	0.059
	x_{t-1}	0.104	0.044
	x_{t-3}	**0.026**	**0.034**
TAR2	x_{t-10}	0.363	0.098
	x_{t-6}	0.077	0.064
	x_{t-9}	**0.053**	**0.060**
Nonlinear	x_2	0.558	0.097
	x_9	0.082	0.031
	x_6	0.072	0.026
	x_{10}	**0.009**	**0.013**

is greater than that for the third input (x_{t-9}), which in turn is greater than that for the first input (x_{t-1}). The values of these coefficients indicate the order of significance of the inputs. This has been identified by the method by allocating correspondingly decreasing scores of 0.354, 0.204, and 0.104 for these inputs. The PMI results for the other two models also indicate the proper order of significance of inputs. Thus, PMI scores are a powerful input selection method when both linear and nonlinear dependencies are present in the data.

The results for SOM-GAGRNN are presented in Table 9.7. The table shows the preliminary sets of inputs selected from the SOM model for the first three models and the subsequent final set chosen using the GAGRNN model. For the nonlinear model, SOM was not trained to exclude correlated variables because inputs are independent by construction. The table shows that the method selects the correct input for model 1. The SOM selects two out of three correct inputs for the second model (x_{t-1} and x_{t-9}), but

Table 9.7 Results from PMI Input Selection for the Four Models

Model	After SOM (unsupervised)	After GAGRNN (supervised)
AR1	$x_{t-1}, x_{t-4}, x_{t-6}, x_{t-9}, x_{t-12}, x_{t-14}$	x_{t-1}
AR9	$x_{t-1}, x_{t-3}, x_{t-5}, x_{t-6}, x_{t-7}, x_{t-8},$ x_{t-9}, x_{t-11}	$x_{t-1}, x_{t-3}, x_{t-5}, x_{t-7}, x_{t-8}, x_{t-9}$
TAR2	$x_{t-1}, x_{t-3}, x_{t-4}, x_{t-5}, x_{t-6}, x_{t-9}$	$x_{t-3}, x_{t-4}, x_{t-5}, x_{t-6}$
Nonlinear	not applicable	x_2, x_{14}

includes additional variables that are not used. The other required input of lag four is replaced by lag three, because it is closer to the weight of the cluster center from which the input was selected. The GAGRNN keeps the two correct inputs, but still retains some of the additional variables. A similar observation is made for the other two models. For the TAR2 model, only one (x_{t-6}) out of two correct inputs are selected, and for the nonlinear model, only one (x_2) out of three correct inputs are selected. Thus this method, while it deals with both linear and nonlinear dependencies by nonlinearly clustering the data, can select some, but not all, of the relevant inputs. This is because it can select inputs that are closer to the actual input but are selected because they happen to be closer to the weights of the cluster. Improvements to this selection scheme can lead to better results.

9.10.3 Input Selection Case Study: Selecting Inputs for Forecasting River Salinity

Salinity in rivers can be a major problem for all water usage activities. Salinity can adversely affect health and crop yield and cause corrosion in pipes. Salinity is particularly a problem in South Australia, where it is predicted that salinity levels in drinking water may go beyond acceptable levels in the next 50–100 years [29]. Salinity is measured in electrical conductivity (EC) units and the desirable EC threshold for drinking water is 800 EC units.

Salinity is a time-series with seasonal irregular characteristics; salinity is typically high during the first half of the year and low during the second half. It is considered important to forecast salinity about 14 days in advance so that pumping policies can be adjusted to pump more water during periods of low salinity. This case study, reported by Bowden et al [29], presents input selection for forecasting salinity at a pumping location in a river in South Australia.

Salinity is mainly affected by upstream salinity, flow rate, and river level. Daily readings at 16 locations upstream of the river were used as variables. These readings included salinity measurements at five upstream locations, flow measurements in three locations and river level measurements at eight locations. In addition, 60 lags of each variable were selected as potential inputs, resulting in a total of 960 inputs. The river flow and the river level are inversely related to salinity. To select the best inputs for use in a forecasting model based on feedforward networks, Partial Mutual Information (PIM) and SOM methods were used, followed by a genetic algorithm-based GAGRNN. The data selected was from December 1986 to June 1992; 80 percent of the data was used for model calibration and 20 percent for model testing. Of the calibration data, 20 percent was used for model validation, that is, to prevent overfitting.

Due to the large number of inputs, bivariate PIM was run separately for each of the 60 lags of the 16 variables. It was shown that the first lag was important for all 16 variables. The second lag selected was a higher-order lag, such as 50; the third important lag intermediate, such as 25, and the fourth lag was again of higher order, such as 60. This is because the information contained in the first input is removed from each of the remaining inputs and the output using nonlinear regression. Because the residuals are used to select the next input, the inputs that still contain additional information are those of higher order. After a higher-order lag is selected, the additional information remaining in the residual will most likely be provided by a lower-order lag, and so on. For most of the 16 inputs, PMI for the first lag was significantly higher than that for the other lags. Based on PMI and a 95 percent randomized sample PMI, the selected number of lags (not the lag number) for each of the 16 variables varied from three to five. In this way, the original set of 960 inputs were reduced to 66 inputs. PMI method was applied again to the reduced set of 66 variables, resulting in a final set of 13 variables. This set included five salinity variables, one flow variable, and seven river level variables. Some were of the first lag, and others were from lags ranging from 11 to 60.

The SOM-GAGRNN produced a final set of 21 variables. There was some overlap between the inputs selected by the two methods, but they were mostly different. There were many lag-1 inputs in the set found by PMI, whereas the set found by SOM-GAGRNN was predominantly composed of longer lags. The forecasting errors based on feedforward networks were similar for the two sets of inputs, but the first set found by PMI was slightly superior. The number of hidden neurons in the two models was 32 and 33, respectively. The validation error for the two sets was 34.0 and 36.2 EC units, respectively. However, PMI had an advantage in that it has produced a more parsimonious model, because it used only 13 inputs. Furthermore, inputs selected from PMI can provide insight into the most important variables. These were found to be better than the results of previous studies performed using inputs selected from two other methods. The first was based on a priori knowledge of salinity travel time, combined with a neural network-based sensitivity study, and the second involved the method of Haugh and Box [30], which relies on cross-correlation, which only can capture linear dependencies. These studies had validation errors of 43.0 and 46.2, respectively.

To compare the PMI scores with inputs identified as important by the neural networks, an input-sensitivity study on the best network (with 32 hidden neurons), which used inputs selected from PMI, was conducted by perturbing each input by 5 percent. The sensitivity was calculated as

$$\text{Sensitivity} = \frac{\text{percent change in output}}{\text{percent change in input}} \times 100. \qquad (9.106)$$

The sensitivities and PMI scores are presented in Figure 9.79, in which the horizontal axis indicates the 13 input variables. The last one or two digits of the input labels denote the lag number (1–60), the last letter denotes the type of input—salinity (S), flow (F) or river level (L)—and the first two letters of the label are an acronym for the name of the upstream location at which the measurement was taken. It can be seen that there is a close agreement between the PMI scores for inputs and their sensitivity to the output of the neural network. The five most important inputs to the neural network are the five inputs with the highest PMI scores. The inputs with the lowest sensitivities have low PMI scores. The PMI provides insights into the significance of the inputs and can therefore provide valuable information about the system.

To compare the forecasting performance of the two models with actual data in real-time forecasting, they were tested on a new dataset for the time period between August 1992 and March 1998 and comprised of 2028 observations. This set had characteristics that were not in the calibration or testing datasets. The 14-day ahead forecasts from the two models superimposed on actual data are presented in Figure 9.80, in which the left figure (a) is for the PIM-based input selection and the right figure (b) is for SOM-GAGRNN-based input selection. In the figures, the regions identified as one and two include the characteristics not found in the data used during model development. The model based on the PMI-selected inputs was better able to forecast unseen events than the

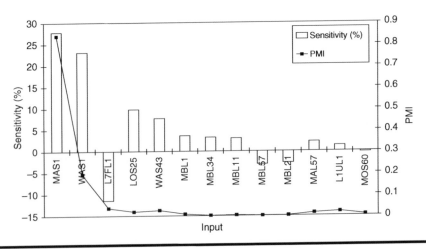

Figure 9.79 Comparison of the relative sensitivity of inputs obtained from the neural network and corresponding PMI scores. (From Bowden, G.J., Maier, H.R., and Dandy, G.C., *Journal of Hydrology*, 301, 93, 2005. With permission from Elsevier.)

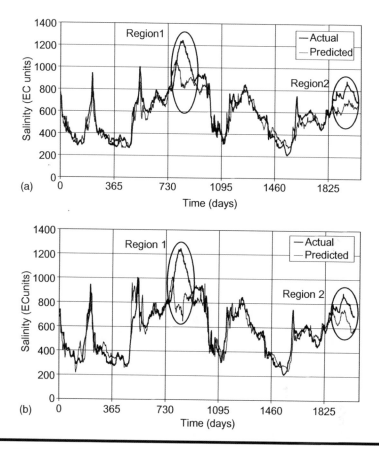

Figure 9.80 Actual and forecasted 14-day ahead salinity values from the networks: (a) with inputs selected from PMI and (b) with inputs selected from SOM-GAGRNN. (From Bowden, G.J., Maier, H.R., and Dandy, G.C., *Journal of Hydrology*, 301, 93, 2005. With permission from Elsevier.)

model based on inputs selected using the SOM-GAGRNN method. The PMI-based model has a considerably lower RMSE than the SOM-GAGRNN-based input model (95 and 112.6, respectively). When the two regions were removed from the data, the first model still outperformed the second model; the RMSE values for the two models were 67.9 and 72.5 EC units, respectively.

9.11 Summary

This chapter presents an extensive discussion of neural networks for time-series forecasting, illustrated by examples and case studies. It covers the topics of linear and nonlinear models for time-series forecasting, long-term forecasting with neural networks, model error decomposition into bias and variance components, and input selection for time-series forecasting. The chapter begins with an introduction to forecasting using linear neuron models and demonstrates that, with input lags, these models are equivalent to auto regressive (AR) models in classical time-series modeling. When error lags are introduced to this model, it represents an autoregressive moving average (ARMA) model. A linear model is illustrated using an example involving the regulation of the teperature in a furnace to demonstrate the short-term and long-term forecasting ability of this model. The results show that the linear neuron model captures the linear temporal dynamics; however, the long-term forecasting ability deteriorates as the prediction horizon increases.

Next the chapter introduced nonlinear networks for time-series forecasting. These networks use modified backpropagation with short-term memory filters, represented by input lags (focused time-lagged feedforward networks) and recurrent networks with feedback loops, which capture the long-term memory dynamics of a time-series and embed it in the network structure itself. The chapter demonstrates that these networks represent variants of nonlinear auto regressive (NAR) models. Three practical examples (regulating the temperature in a furnace, forecasting the inflow into a river, and forecasting the daily air pollution) are solved to illustrate the modified backpropagation method and its ability to perform short-term and long-term forecasting. The results show that use of a nonlinear network improves short-term and long-term forecasting ability compared to linear models.

Three types of recurrent networks are treated in detail: Elman, Jordan, and fully-recurrent networks. In an Elman network, the hidden-layer activation is fed back as an input at the next time step, while in a Jordan network, the output is fed back as an input. In fully recurrent networks, both the hidden-neuron activation and the output are fed back to themselves in the next time step. The chapter demonstrates how these networks encapsulate the long-term memory of a time-series and illustrates that recurrent networks are variants of NAR models, and that when error is incorporated as input, they become nonlinear auto regressive integrated moving average (NARIMA) models and NARIMAx models with exogeneous inputs. The training of these networks is presented in detail.

Examples and case studies are presented for each recurrent network type; these examples range from stream flow and rain fall run-off to energy and temperature forecasting in various applied situations. The results demonstrate that the recurrent networks outperform focused time-lagged feedforward networks and are more robust in long-term forecasting.

Accuracy of long-term forecasting nevertheless degrades with the forecast horizon, and the chapter presents an approach to improve long-term forecasting with neural networks using multiple neural networks (MNN) that amalgamate a set of sub-models with shorter prediction horizons. A practical case study involving hourly gas flow rate forecasting is presented to highlight the advantage of such models in extended long-term forecasting. The bias and variance issues still apply to time-series models, and the chapter addresses the decomposition of error into bias and variance components using Monte Carlo simulations to assess the contribution of each to model error with respect to the number of input lags and hidden neurons in focused time-lagged feedforward networks. This decomposition helps to identify which component of error the researcher must focus on when improving models to produce the best generalization.

The last section treats the topic of input selection for neural networks used in time-series forecasting, including an example and a practical application case study. Two methods used for illustration are partial mutual information (PMI) in conjunction with generalized regression neural networks (GRNN) and a self-organizing maps (SOM) in conjunction with a genetic algorithm (GA) and GRNN for the selection of inputs that are nonlinearly correlated. Applying these methods to synthetic data and a river salinity problem, the chapter demonstrates that the PMI method is superior to the other method and provides insight into the significance of inputs to the prediction of the output. The PMI has the ability to efficiently select the most relevant inputs from a large set of correlated and redundant inputs.

Problems

1. Define linear and nonlinear time-series and the implications of this aspect for modeling.

2. Summarize the main attributes of a focused time lagged feedforward networks (FTLFN).

3. Describe the difference between FTLFN and recurrent networks. What aspect of a time-series do recurrent networks capture and how do they do this?

4. What are the shortcomings of both FTLFN and recurrent networks in relation to long-term forecasting?

5. For a time-series dataset of your choice, plot the time-series and describe its trends and characteristics. Consider what preprocessing may be appropriate for the data.

6. Using available software of your choice or by programming, develop an appropriate one-step-ahead forecasting model for the time-series data in Problem 5 using all possible inputs. Assess the validity of the model and test its performance on unseen data.

7. For the time-series data in Problem 5, use appropriate methods for selecting relevant inputs (and lags) for one-step-ahead forecasting

of the time-series. Incorporate any relevant data pre-processing methods, including those discussed in Chapter 6. If other external variables exist, do they influence the outcome?

8. Several statistical measures are used in this chapter for calibration (i.e., model development and validation) and the testing of time-series. Explain their meaning and the similarities or differences among them. Which ones are particularly relevant to time-series data?

9. Using available software of your choice or by programming, develop an appropriate one-step-ahead forecasting model for the time-series data in Problem 5 using the inputs selected in Problem 7. Assess the validity of the model and test its performance on unseen data. Describe the effect of input selection on forecasting accuracy by comparing results from the network using all possible inputs (in Problem 6) and using only carefully selected inputs.

10. What are the sources of error in long-term forecasting and how can they be minimized?

11. Assess the quality of the model developed in Problem 9 for long-term forecasting. What changes can be made to the model to improve long-term forecasting accuracy? Implement the proposed changes and assess the model's accuracy.

References

1. Zhang, G.P. Time series forecasting using a hybrid ARIMA and neural network model, *Neurocomputing*, 50, 159, 2003.
2. Chang, F.J., Chang, L.C., and Huang, H. Real-time recurrent learning neural network for stream-flow forecasting, *Hydrological Processes*, 16, 257, 2002.
3. *Neural Connection*, SPSS, Chicago, IL, 1991.
4. Haykin, S. *Neural Networks: A Comprehensive Foundation*, 2nd Ed., Prentice Hall, Upper Saddle River, NJ, 1999.
5. Elman, J. Finding structure in time, *Cognitive Science*, 14, 179, 1990.
6. Jordan, M.I. Serial order: A parallel, distributed approach, *Advances in Connectionist Theory: Speech*, J.L. Elman and D.E. Rumelhart, eds., Erlbaum, Hillsdale, NJ, 1989.
7. *Mathematica—Neural Networks*, Wolfram Research, Champaign, IL, 2003.
8. Jiang, D., Zhang, Y., Hu, X., Zeng, Y., Tan, J., and Shao, D. Progress in developing an ANN model for air pollution index forecast, *Atmospheric Environment*, 38, 7055, 2004.
9. Chaturvedi, D.K., Mohan, M., Singh, R.K., and Karla, P.K. Improved generalized neuron model for short-term load forecasting, *Soft Computing*, 8, 370, 2004.

10. Lanza, P.A.G. and Cosme, J.M.Z. A short-term temperature forecaster based on a state space neural network, *Engineering Applications of Artificial Intelligence*, 15, 459, 2002.

11. Matyas, J. Random optimization, *Automation and Remote Control*, 26, 246, 1965.

12. Solis, F. and Wets, R. Minimization by random search techniques, *Mathematics of Operations Research*, 6, 19, 1981.

13. Williams, R. and Zisper, D. A learning algorithm for continually running fully recurrent neural network, *Neural Computation*, 1, 270, 1989.

14. Chiang, Y.M., Chang, L.C., and Chang, F.J. Comparison of static feedforward and dynamic feedback neural networks for rainfall-runoff modeling, *Journal of Hydrology*, 290, 297, 2004.

15. Chang, L.C., Chang, F.J., and Chiang, Y.M. A two-step ahead recurrent neural network for stream-flow forecasting, *Hydrological Processes*, 18, 81, 2004.

16. Topalli, A. and Erkmen, I. A hybrid learning for neural networks applied to short term load forecasting, *Neurocomputing*, 51, 495, 2003.

17. Berardi, V.L. and Zhang, P.G. An empirical investigation of bias and variance in time-series forecasting: Modeling considerations and error evaluation, *IEEE Transactions on Neural Networks*, 14, 668, 2003.

18. Heskes, T. Bias/variance decompositions for likelihood-based estimators, *Neural Computation*, 10, 1425, 1998.

19. Jacobs, R.A. Bias/variance analyzes of mixtures-of-experts architectures, *Neural Computing*, 9, 369, 1977.

20. Kohavi, R. and Wolpert, D.H. Bias plus variance decomposition for zero-one loss functions, *Proceedings of the 13th International Conference Machine Learning*, L. Saitta (ed.), Morgan Kaufmann, San Mateo, CA, p. 275, 1996.

21. Nguyen, H.H. and Chan, C.W. Multiple neural networks for a long term time-series forecast, *Neural Computing and Applications*, 13, 90, 2004.

22. Bowden, G.J., Dandy, G.C., and Maier, H.R. Input determination for neural network models in water resources applications. Part 1— Background and methodology, *Journal of Hydrology*, 301, 75, 2005.

23. Sarle, W.S. *Neural Network FAQ*, 1997. http://www.comp.ai.neural-nets

24. Yang, H.H., Van Vuuren, S., Sharma, S., and Hermansky, H., Relevance of time-frequency features for phonetic and speaker channel classification, *Speech Communication*, 31, 35, 2000.

25. Sharma, A. Seasonal to interannual rainfall probabilistic forecasts for improved water supply management: Part I—A strategy for system predictor identification, *Journal of Hydrology*, 239, 232, 2000.

26. Specht, D.F. A general regression neural network, *IEEE Transactions on Neural Networks*, 2, 568, 1991.

27. Goldberg, D.E. *Genetic Algorithms in Search, Optimization and Machine Learning*, Addison-Wesley, Reading, MA, 1989.

28. Goldberg, D.E. and Deb, K.A. A comparative analysis of selection schemes used in genetic algorithms, *Foundation of Genetic Algorithms*, Rawlins, J.E. (ed.), Morgan Kauffman, San Mateo, CA, 1991.

29. Bowden, G.J., Maier, H.R., and Dandy, G.C. Input determination for neural network models in water resources applications. Part 2—Forecasting salinity in a river, *Journal of Hydrology*, 301, 93, 2005.
30. Haugh, L.D. and Box, G.E.P. Identification of dynamic regression (distributed lag) models connecting two time-series, *Journal of the American Statistical Association*, 72, 121, 1977.
31. Hair, J.F., Anderson, R.E., Tatham, R.L., and Black, W.C., *Multivariate Data Analysis*, 5th Ed., Prentice Hall, Upper Saddle River, NJ, 1998.
32. Box, G.E.P. and Jenkins, G.M., *Time Series Analysis Forecasting and Control*, Holden-Day, San Francisco, CA, 1976.

Appendix

A.1 Linear Algebra, Vectors, and Matrices

Vectors provide an efficient way to compute with multidimensional data. An ordered pair of numbers can be regarded as a point or a vector in two-dimensional space, as shown in Figure A.1.

Similarly, a point in three-dimensional space is represented by a vector with three components. A vector with n dimensions is written using the notation $\mathbf{x} = [x_1, x_2, ..., x_n]$. For example, an input vector with values for n variables is an n-dimensional vector. A vector can be written in row format or column format. For example:

$$\mathbf{x} = [3, \ 2, \ 1]$$

is in row format, while

$$\mathbf{y} = \begin{bmatrix} 3 \\ 2 \\ 1 \end{bmatrix}$$

is in column format. Because the individual values and their order are the same for both $\mathbf{x}$ and $\mathbf{y}$, $\mathbf{y}$ is called the transpose of $\mathbf{x}$, written as $\mathbf{y} = \mathbf{x}^T$, or $\mathbf{x}$ is the transpose of $\mathbf{y}$, written as $\mathbf{x} = \mathbf{y}^T$.

A.1.1 Addition of Vectors

Vectors can be added or subtracted following the usual way addition or subtraction is carried out on an element-by-element basis. Consider two

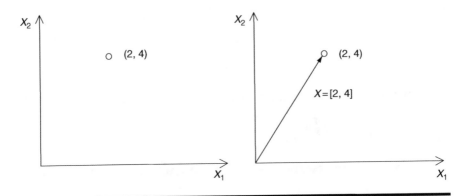

Figure A.1 A point in two-dimensional space (left) and the same point viewed as a vector (right).

vectors, **x** and **y**:

$$\mathbf{x} = [\,2,\ 3,\ 1\,]$$
$$\mathbf{y} = [\,1\ \ 6\ \ 5\,]$$
$$\mathbf{x} + \mathbf{y} = [\,3\ \ 9\ \ 6\,]$$
$$\mathbf{x} - \mathbf{y} = [\,1\ -3\ -4\,].$$

A.1.2 Multiplication of a Vector by a Scalar

A vector can be multiplied by a scalar (a constant value). For example, multiplying $\mathbf{x} = [3\ 2\ 1]$ by 2 produces a vector twice as long, $2\mathbf{x} = [6\ 4\ 2]$.

A.1.3 The Norm of a Vector

The length (also called norm or magnitude) of a vector denotes the distance to the point referred to by its coordinates. The norm or magnitude of a vector **x** with n components is

$$\|\mathbf{x}\| = \sqrt{x_1^2 + x_2^2 + \cdots + x_n^2}.$$

Therefore, for the vector $\mathbf{x} = [3, 2, 1]$, the norm is

$$\|\mathbf{x}\| = \sqrt{3^2 + 2^2 + 1^2} = \sqrt{14}.$$

A.1.4 Vector Multiplication: Dot Products

The dot product provides an efficient method of multiplying two vectors. The dot product of two vectors $\mathbf{x} = [x_1, x_2, \ldots, x_n]$ and $\mathbf{w} = [w_1, w_2, \ldots, w_n]$ is

$$\mathbf{x} \cdot \mathbf{w} = x_1 w_1 + x_2 w_2 + \cdots + x_n w_n,$$

which is the sum of products of vector components. The dot product can also be written as

$$\mathbf{x} \cdot \mathbf{w} = \|\mathbf{x}\| \|\mathbf{w}\| (\cos \theta),$$

where θ represents the angle between the two vectors. This is presented in Figure A.2.

Example. Find the dot product of the vectors $[-1 \ 3 \ 6 \ -2]$ and $[1 \ 2 \ 2 \ -3]$. Find the angle between the two vectors.

Solution. The dot product is

$$[-1 \ 3 \ 6 \ -2] \cdot [1 \ 2 \ 2 \ -3]$$
$$= -1 \times 1 + 3 \times 2 + 6 \times 2 + -2 \times -3 = 23.$$

The angle between the two vectors is (from the above equation)

$$\cos \theta = \frac{23}{\sqrt{50}\sqrt{18}} = 39.94^\circ.$$

The $\sqrt{50}$ and $\sqrt{18}$ in the above expression are norms or the lengths of the two vectors $\mathbf{x}$ and $\mathbf{w}$. When two vectors are perpendicular or orthogonal to each other, their dot product is zero. This is because when $\theta = 90°$, $\cos \theta = 0$.

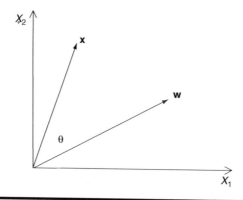

Figure A.2 Two vectors, x and w, and the angle between them.

A.2 Matrices

Operating with many vectors can be greatly simplified by organizing them in matrices. A matrix is a rectangular array of numbers. An $m \times n$ matrix is a matrix with m rows and n columns that represent vectors. Each element of a matrix can be indexed by its row and column position. The order of elements is therefore written as

$$
A = \begin{bmatrix}
a_{11} & a_{12} & a_{13} & \cdots & a_{1n} \\
a_{21} & a_{22} & a_{23} & \cdots & a_{2n} \\
a_{31} & a_{32} & a_{33} & \cdots & a_{3n} \\
\vdots & & & & \\
a_{m1} & a_{m2} & a_{m3} & \cdots & a_{mn}
\end{bmatrix}.
$$

For the 3×2 matrix B, given below, the element $b_{32} = -5$ and $b_{21} = 2$:

$$
B = \begin{bmatrix}
-3 & 3 \\
2 & 9 \\
1 & -5
\end{bmatrix}.
$$

A.2.1 Matrix Addition

Matrices can be added by summing their corresponding elements. For example, if $C = A + B$, then $c_{ij} = a_{ij} + b_{ij}$. For example, if

$$
A = \begin{bmatrix}
3 & 1 \\
2 & 4 \\
1 & 5
\end{bmatrix} \quad \text{and} \quad B = \begin{bmatrix}
-2 & 1 \\
3 & 2 \\
1 & 4
\end{bmatrix},
$$

then

$$
C = A + B = \begin{bmatrix}
3-2 & 1+1 \\
2+3 & 4+2 \\
1+1 & 5+4
\end{bmatrix} = \begin{bmatrix}
1 & 2 \\
5 & 6 \\
2 & 9
\end{bmatrix}.
$$

A.2.2 Matrix Multiplication

To multiply two matrices, A and B (written as AB), the number of columns in A must match the number of rows in B. If A is an $m \times n$ matrix and B is an

$n \times s$ matrix, then the matrix product **AB** is an $m \times s$ matrix. If **C** = **AB**, then an element of this product is defined as

$$\mathbf{C}ij = (i\text{th row vector of } \mathbf{A}) \cdot (j\text{th column vector of } \mathbf{B})$$

$$= \sum_{k=1}^{n} \mathbf{a}_{ik} \mathbf{b}_{kj}.$$

Example. Compute the product

$$\mathbf{C} = \begin{bmatrix} -2 & 3 & 1 \\ 1 & 2 & 4 \end{bmatrix} \begin{bmatrix} 4 & 2 & 1 & 1 \\ -3 & -1 & 4 & 3 \\ 1 & 4 & 5 & 1 \end{bmatrix}$$

Solution. The elements c_{11} and c_{23} are calculated as

$$\mathbf{C} = \begin{bmatrix} -2 \times 4 + 3 \times -3 + 1 \times 1 \ldots & \ldots & \ldots \\ \ldots & \ldots 1 \times 1 + 2 \times 4 + 4 \times 5 & \ldots \end{bmatrix},$$

and the complete solution is

$$\mathbf{C} = \begin{bmatrix} -16 & -3 & 15 & 8 \\ 2 & 16 & 29 & 11 \end{bmatrix}.$$

Using this procedure, any number of matrices can be multiplied together by sequentially carrying out the computation.

A.2.3 *Multiplication of a Matrix by a Vector*

A matrix can be multiplied by a vector in a similar manner to the multiplication of two matrices. For example, for a matrix **w** of size 2×2 and a vector **x** of size 2×1, given as

$$\mathbf{w} = \begin{bmatrix} 1 & 2 \\ 3 & 4 \end{bmatrix}; \quad \mathbf{x} = \begin{bmatrix} 5 \\ 6 \end{bmatrix},$$

multiplication gives a 2×1 vector, **u**, as

$$\mathbf{u} = \begin{bmatrix} 1 & 2 \\ 3 & 4 \end{bmatrix} \begin{bmatrix} 5 \\ 6 \end{bmatrix} = \begin{bmatrix} 5 + 12 \\ 15 + 24 \end{bmatrix} = \begin{bmatrix} 17 \\ 39 \end{bmatrix}.$$

In this manner, many matrices and vectors can be multiplied together.

A.2.4 Matrix Transpose

Similar to vector transpose, matrices are transposed by arranging original rows as columns, or vice versa. If matrix **B** is the transpose of matrix **A**, then the element b_{ij} is the same as element a_{ji}.

Example. Write the matrix **B** if $\mathbf{B} = \mathbf{A}^{\mathrm{T}}$ and

$$\mathbf{A} = \begin{bmatrix} -2 & 3 & 1 \\ 1 & 2 & 4 \end{bmatrix}.$$

Solution.

$$\mathbf{B} = \mathbf{A}^{\mathrm{T}} = \begin{bmatrix} -2 & 1 \\ 3 & 2 \\ 1 & 4 \end{bmatrix}$$

References

1. Callan, R. *The Essence of Neural Networks*, Pearson Education, London, UK, 1999.

Index